S0-BLM-123

Phosphorus Chemistry

Proceedings of the 1981 International Conference

Louis D. Quin, EDITOR
Duke University

John G. Verkade, EDITOR
Iowa State University

Based on the
International Conference on
Phosphorus Chemistry at
Duke University,
Durham, North Carolina,
June 1–5, 1981.

ACS SYMPOSIUM SERIES 171

AMERICAN CHEMICAL SOCIETY
WASHINGTON, D. C. 1981

Library of Congress CIP Data

International Conference on Phosphorus Chemistry (1981: Duke University)
Phosphorus chemistry.
(ACS symposium series, ISSN 0097-6156; 171)

"Based on the International Conference on Phosphorus Chemistry at Duke University, Durham, North Carolina, June 1-5, 1981."

Includes bibliographies and index.

1. Phosphorus—Congresses.
I. Quin, Louis D., 1928– . II. Verkade, John G., 1935– . III. American Chemical Society. IV. Title. V. Series.

QD181.P1I57 546'.712 81-14956
ISBN 0-8412-0663-5 AACR2 ACSMC8 171 1-640 1981

PRINTED IN THE UNITED STATES OF AMERICA

ACS Symposium Series

M. Joan Comstock, *Series Editor*

FOREWORD

The ACS SYMPOSIUM SERIES was founded in 1974 to provide a medium for publishing symposia quickly in book form. The format of the Series parallels that of the continuing ADVANCES IN CHEMISTRY SERIES except that in order to save time the papers are not typeset but are reproduced as they are submitted by the authors in camera-ready form. Papers are reviewed under the supervision of the Editors with the assistance of the Series Advisory Board and are selected to maintain the integrity of the symposia; however, verbatim reproductions of previously published papers are not accepted. Both reviews and reports of research are acceptable since symposia may embrace both types of presentation.

CONTENTS

BIOCHEMISTRY OF PHOSPHORUS COMPOUNDS

Dedicated to Frank H. Westheimer

ORGANIC SYNTHETIC METHODS BASED ON REAGENTS CONTAINING PHOSPHORUS

PHOSPHORUS HETEROCYCLES

PHOSPHAZENES

NEW ORGANOPHOSPHORUS COMPOUNDS OF COMMERCIAL INTEREST

NEW PHOSPHORUS LIGANDS AND COMPLEXES (INCLUDING CATALYTIC PROPERTIES)

SPECTROSCOPY OF PHOSPHORUS COMPOUNDS

PHOTOCHEMISTRY WITH PHOSPHORUS COMPOUNDS

BONDING AND THEORY OF PHOSPHORUS COMPOUNDS

PREFACE

During the first week of June 1981 some 400 chemists actively engaged in research on phosphorus chemistry gathered at Duke University, Durham, North Carolina for the continuation of a series of International Conferences on Phosphorus Chemistry. These chemists came from 29 different nations. During the two decades that these International Conferences have been held, the meetings have taken place in Europe, thus making the Durham Conference the first of its kind in the United States. The growth of the activity in phosphorus-based research in this country, and throughout the world, has been remarkable in recent years. No diminution in this activity is in sight, for new discoveries are constantly being made that open up fresh channels of research. Phosphorus chemistry as a field can only be partially categorized by the traditional lines of organic and inorganic compounds. In modern phosphorus chemistry increasing attention is also being paid to biological involvement of phosphorus compounds. In designing the 1981 Conference, papers describing research in all of these aspects of phosphorus chemistry were included, marking the first time this has been done in any of the international conferences. The Durham meeting therefore attracted a broad variety of participants and papers, from both academic and industrial research laboratories; the opportunities for interaction among persons of diverse backgrounds were abundant.

The program was unique in another sense: with the exception of the opening lecture by Professor Rolf Appel, the oral presentations were of equal length (25 minutes) and all chemists, regardless of age and stature, had an equal opportunity for gaining a place on the oral program. Investigators with research results that were insufficient to fill the alloted time period were encouraged to make use of the poster medium. A further stipulation was that only new research results should be presented; review papers or extensive literature discussions were discouraged.

This volume contains the manuscripts provided by 128 of the participants in the oral part of the program. (A few papers on the original program were not delivered). To publish a volume at a reasonable price with such a large number of papers required a severe restriction in the length of the manuscripts; authors were asked to present their results in "Communication to the Editor" style in a maximum of four pages of text. The authors have complied faithfully with this request, and the Editors thank them all for the care with which they prepared their papers. The Editors also are grateful to the foreign authors for their excellent efforts in preparing their manuscripts in English. With a collection of papers of such diversity in topic and geographic origin, the style is more varied than is normal for such a symposium volume, but these variations do not interfere with the quality of the chemical discussions, and the Editors chose not to enforce strict adherence to detailed uniformity.

The 127 posters on the program presented another fascinating array of research results. It is unfortunate that space limitations do not permit the inclusion of the abstracts of these presentations in this volume. However, the titles and addresses of the authors are provided on pages 623–630 for the convenience of those who may wish to contact the authors for copies of their abstracts.

A feature of the Conference program was the inclusion of sessions recognizing the signal accomplishments of two pioneers in phosphorus chemistry, Nobel Laureate Professor Georg Wittig and Professor Frank H. Westheimer. Papers in these special sessions were solicited by Professor Hans-Jurgen Bestmann and Professor Steven Benkovic, respectively. The work of Professor Wittig in the use of phosphorus compounds in organic synthesis, which among other contributions was responsible for his receiving the Nobel Prize in Chemistry in 1979, was recognized by the session "New Organic Synthetic Methods Based on Reagents Containing Phosphorus." The noteworthy accomplishments of Professor Westheimer on the mechanistic aspects of phosphate ester chemistry stimulated the session "Biochemistry of Phosphorus Compounds."

This Conference was enthusiastically supported by the American scientific community; the chemical industry, the U.S. National Science Foundation, and the Petroleum Research Fund, all made contributions that ensured an adequate financial base, and many chemists worked diligently on the numerous tasks that are associated with such an undertaking. The Editors express their deep appreciation to all.

LOUIS D. QUIN
Duke University
Durham, North Carolina

JOHN G. VERKADE
Iowa State University
Ames, Iowa

June 29, 1981.

1

Phosphorus-Carbon Compounds with $p\pi$-$p\pi$ Bonds

Opening Lecture

R. APPEL

Anorganisch Chemisches Institut der Universität, 5300 Bonn, Gehard-Domagk-Strasse 1, FRG

In this lecture some new routes to phosphorus-carbon compounds with P-C multiple bonds, found in connection with our investigations on reactions of tertiary phosphanes with chlorinated carbon compounds, such as tetrachloromethane, hexachloroethane, phosgene, and isocyanide dichlorides are reported. Furthermore some stereochemical problems concerning this type of compound will be discussed.

Action of tetrachloromethane on trimethylsilyl-substituted methyl-diphenylphosphanes causes quantitative chloroform elimination with formation of trimethylsilylated (TMS) methylenechlorodiphenylphosphoranes. Heating the bistrimethylsilyl substituted

$$Ph_2P\text{-}CH(Tms_3)_2 + CCl_4 \xrightarrow{-HCCl_3} ClPh_2P{=}C(Tms_3)_2$$

$$Tms = SiMe_3 \qquad -Me_3SiCl \downarrow 120°C$$

$$PhP = CPh(Tms^3) \leftarrow [Ph_2P{\equiv}CTms_3]$$

compound causes spontaneous gas evolution of TmsCl at 120°C. The product is identified by elemental analysis, molecular mass determination, and the characteristic ^{31}P nmr shift. It is a greenish-yellow liquid which can be distilled in vacuo without decomposition. The general applicability of this synthesis, based upon migration of an organyl substituent from phosphorus to the ylid-carbon is restricted so far to P-aryl substituted compounds.

Another route to a number of theoretically interesting compounds of two-coordinate phosphorus is the reaction of P-trimethylsilyl-substituted phosphanes with phosgene. The reaction proceeds via several isolable intermediates. The first is the phosphino-substituted methylene-phosphane, which is generated by silyl-group migration from phosphorus to oxygen. Treatment with further phosgene then causes further elimination of CO and chlorotrimethylsilane, yielding a compound with a P-P bond which

0097-6156/81/0171-0001$05.00/0

$$4PhPTms_2 + 2Cl_2CO \xrightarrow{4TmsCl} 2PhP{=}C\text{-}PPhTms$$

$$\downarrow + Cl_2CO \quad - TmsCl_1CO$$

```
          OTms
           |
   PhP —— C —— PPh
    |     |     |
   PhP —— C —— PPh
           |
          OTms
```

no longer shows the characteristic ^{31}P nmr shift of two-coordinate phosphorus. An x-ray structure determination which was carried out by Dr. Halstenberg and Professor Huttner at Konstanz, showed that it is the first 2,3,5,6-tetraphosphabicyclo[2.2.0]-hexane, which contains an asymmetric and distorted bicyclohexane skeleton with two sets of equivalent phosphorus atoms in different environments. Thus far the bicyclohexane skeleton can only be obtained in the reaction of $COCl_2$ with phenyl-bis(trimethylsilyl)phosphane. Initially t-butyl-bis(trimethylsilyl)phosphane indeed reacts analogously to give the phosphino-substituted methylenephosphane, which can be isolated as a pure substance and which can be converted to the P-H phosphine by methanol. What happens in the second step with further phosgene, however, depends very much upon the rate of addition. If the addition occurs very slowly at -80°, the five-membered ring system with a C-C double bond is formed, while rapid mixing of solutions of both components in the molar ratio 1:1 affords a five membered ring containing four phosphorus atoms bridged by a CO moiety.

```
TmsO          OTms
    \        /
     C ==== C                 t-Bu-P —— P-t-Bu
    /        \                     /        \
t-Bu-P        P-t-Bu          t-Bu-P          P-t-Bu
     \      /                        \      /
        P                               C
        |                               ||
       t-Bu                             O
```

The reaction of phenyl(bistrimethylsilyl) phosphine with phenylisocyanide dichloride, the aza-analogue of phosgene yields tetraphosphahexadiene according to elemental analysis, cryoscopic molecular mass determination, and ^{31}P nmr studied. To understand the interesting structural problem of this compound, we must look in detail at the ^{31}P nmr spectrum. At ambient temperature the substructure of an AA'XX' 4-spin system is observed. The downfield signals at 258 ppm are assigned to the two-coordinate phosphorus and the upfield multiplet at -12.3 ppm arises from the trivalent phosphane moiety. Between both groups there is a

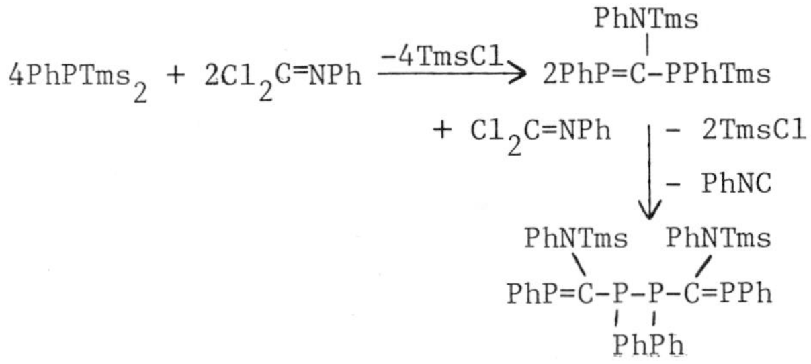

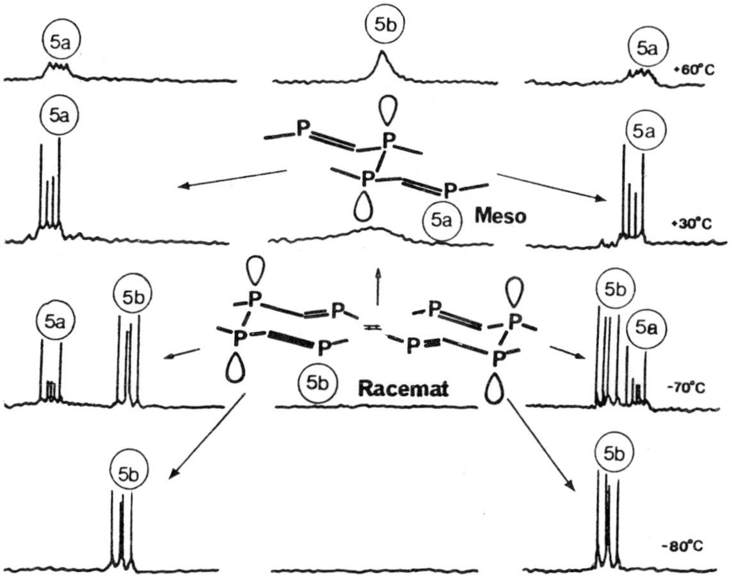

$$4PhPTms_2 + 2Cl_2C{=}NPh \xrightarrow{-4TmsCl} 2PhP{=}C(NPhTms){-}PPhTms$$

$$+\ Cl_2C{=}NPh \quad \downarrow \begin{array}{l} -\ 2TmsCl \\ -\ PhNC \end{array}$$

$$PhP{=}C(NPhTms){-}P(Ph){-}P(Ph){-}C(NPhTms){=}PPh$$

5a 5b 5a +60°C

5a Meso 5a +30°C

5a 5b 5b Racemat 5b 5a -70°C

5b 5b -80°C

diffuse absorption at 115.7 ppm which sharpens on heating to 60°C while the other signals broaden. The significance of the broad signal can be elucidated by cooling to -70°C. In addition to the two sharp multiplets already mentioned, a second 4-spin AA'XX' system appears, the left half indicating two-coordinate phosphorus and the right, at -3.2 ppm., the diphosphane P atoms. Renewed heating to ambient temperature results in coalescence of the two inner multiplets at 115 ppm. Thus the process is reversible. On further cooling to about -80°C, however, the outer multiplets labelled 5a totally disappear. The reason for this is precipitation due to insufficient solubility. Low temperature filtration of the crystals and redissolving them at 30° again gives the complete ^{31}P nmr spectrum as before. Firstly, according to elemental analysis and molecular mass determination, the compound is homogeneous, that is to say, the broad absorption is caused by an isomer. Furthermore the splitting into a second AA'XX' system at -70° indicates a diastereomeric tetraphosphahexadiene, since both the characteristic multiplets of diphosphane- and methylenephosphane-phosphorus can be identified. An explanation, which elucidates the reversible temperature-dependent coalescence of the inner signals and also rationalizes why only one stereoisomer displays this phenomenon is now given.

Attempts to explain the coalescence by migration of the silyl group between the nitrogen and phosphorus atom or by a rotation around the C-N or P=C double-bond are not supported by the experimental data. We came to the conclusion that a pericyclic reaction occurs, which is analogous to the Cope rearrangement for hexadiene-1,5. In a [3.3]-sigmatropic reaction of the tetraphosphahexadiene, the bond between the two phosphorus atoms breaks with simultaneous formation of the P=C double bond and a new P-P bond. The original nmr signals of the P atoms with the coordination number 2 and 3 must collapse in the middle because the rearrangement is a symmetrical one and because it occurs on the nmr time scale. The fact that only one diastereomer shows coalescence can be explained as follows. As is well known, symmetrical, differently substituted diphosphanes which are comparable to our type in substitution have two centers of chirality. Their synthesis usually yields a 1:1 proportion of the _meso_ form and racemic mixture, which can be characterized by their different ^{31}P nmr shifts. In our case the outer set of multiplets must be assigned to one diastereomer and the inner set to the other. From the observation that the spectrum of the crystals filtered at low temperature shows two signal groups again at room temperature, we must conclude that the configurational equilibrium between the _meso_ form and racemic mixture is promptly achieved.

The fast exchange in the ^{31}P nmr spectrum of one isomer at room temperature, which is not observed for isomer _5a_, leads to the conclusion that only one form fulfills the stereochemical demands of the Cope-rearrangement. This is the racemic mixture as shown. If this is correct, the crystals filtered at low temperature, which show no coalescence, should be the _meso_ form. The x-ray analysis of these crystals verifies this assertion since the substituents at the P-P bond are indeed _trans_ to one another.

Of course, we must be careful when transferring conceptions from carbon compounds to other elements. Nevertheless there are interactions between electrons of the 2p and electrons of the 3p level which is considerably higher. Therefore we looked for further proof of this hypothesis. The following experiment seemed to be a linking one between the fields of carbon and phosphorus chemistry. We treated succinic acid dichloride with bis(trimethylsilyl)silylphenyl phosphane. According to our hypothesis a primary halosilane condensation followed by a silyl migration and formation of the P=C double bond, and finally a [3.3]-sigmatropic rearrangement to the diphosphane should occur. In the ^{31}P nmr spectrum. We observe a halosilane condensation to

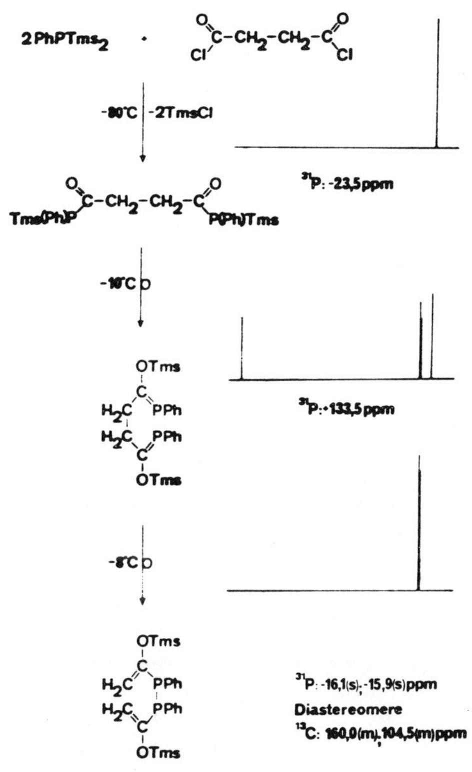

the corresponding diphosphide of the succinic acid, which is stable only below -10°C. The following step is a double 1,3-silylmigraiton resulting in a 1,6-diphosphahexadiene-1,5 which spontaneously passes over by a [3.3]-sigmatropic rearrangement to the substituted 1,2-diphenyl-1,2-divinylphosphane.

The observation that the C-C bond is cleaved in favor of a P-P bond and the formation of 2 olefinic double-bonds surprised us and we decided to obtain further verification of the structure of the divinyl diphosphane by a classical decomposition reaction. Indeed, mild methanolysis yields instead of the C_4 unit of the succinic acid 2, C_2 units in the form of methylacetate. Moreover, both of the other fragments, the 1,2-diphenyldiphosphane and the silylmethylether, could be unambiguously identified.

After the elucidation of the reaction of phenylisocyanide dichloride with phenyl(trimethylsilyl)phosphane, several other differently substituted isocyanide dichlorides react extremely slowly, so that high temperatures are necessary which prevent the

$$\begin{matrix} & OTms & \\ & | & \\ H_2C{=} & C & {-}PPh \\ & & | \\ H_2C{=} & C & {-}PPh \\ & | & \\ & OTms & \end{matrix} + 4MeOH \rightarrow 2H_3CCO_2Me + 2TmsOMe + HPhPPPhH$$

isolation of interesting intermediates and yield only cyclophosphanes. In the reaction of the ring-substituted phenylisocyanide dichlorides absolute agreement with the results of phenylisocyanide dichloride was only observed with the p- and m- monochlorophenyl compounds. The isolable phosphinomethylenephosphanes are formed first, which react with additional isocyanide dichloride with elimination of trimethylchlorosilane and isocyanide to give the tetraphosphahexadiene. The ^{31}P nmr spectra are also very similar. At ambient temperature, a broad absorption is observed in the middle of the two-fold 4-signal spectrum of the AA'XX' type, which splits on cooling to -80° into two signal groups of the same type. The results of the reaction of the o- and o,m-dihalogen substituted phenylisocyanide dichlorides with $PhPTms_2$ are discussed in detail in the paper by Dr. Knoll in this Conference, with special attention given to the question of E and Z isomerism.

The story of the reaction of isocyanide dichlorides with bis(trimethylsilyl)phenylphosphane, however, does not end here. Like the reaction of t-butyl-bis(trimethylsilyl)phosphane with phosgene, the conversions with isocyanide dichlorides depend very much upon the reaction conditions. If the reaction is carried out at ambient temperature in a molar ratio of 1:1 in such a way that the isocyanide dichloride is rapidly dropped into the phosphane solution, the hitherto unknown, 2,4-bis(phenylimino)-1,3-diphenyl-1,3-diphosphetanes are obtained besides the phosphinomethylenephosphanes. Several findings, suggest that their

$$PhPTms_2 + Cl_2C{=}NR \rightarrow (PhP{=}C{=}NR) \rightarrow PhP\begin{matrix} C(=NR) \\ C(=NR) \end{matrix}PPh$$

formation occurs by way of the not yet isolated monophosphacarbodiimide, which readily dimerizes to give the diphosphetane. The ^{31}P nmr spectra of the diphosphetanes exhibit three signals between 84.3 and 74 ppm. All signals are rather broad, although the substances proved to be homogeneous by sharp melting points, crystal structure and mass spectra. The observation of several signals is caused by stereo isomers which are in a dynamic equilibrium at room temperature. This can be confirmed by high and low temperature ^{31}P nmr spectroscopy. If crystals of the dichlorosubstituted compound are dissolved at -80° in deuterated methylene chloride, only one sharp signal is observed. On heating to 30° the broadened 3-line spectrum appears which at 115° shows coalescence to give one broad signal. On cooling to -100° the spectrum shows three relatively sharp signals. The different

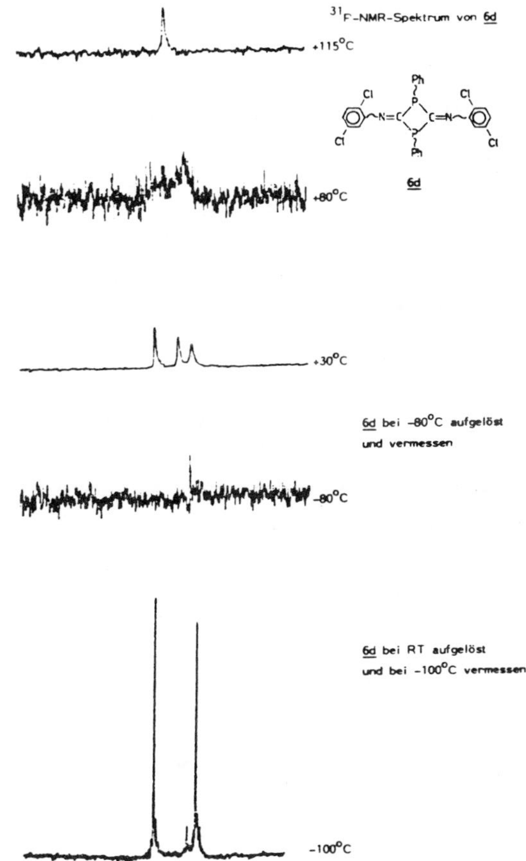

stereoisomers can easily be explained by the possible cis or trans arrangements of the P-phenyl groups or the imino substituents at the C atom relative to the plane of the four-membered ring. The approximate planarity of the 4-membered ring has been proved by an x-ray structure analysis for the 2,4-bis(2,5-dichlorphenyl)imino-1,3-diphenyl-1,3-diphosphetane.

Up to now the unexpected stability of methylidenephosphanes was always thought to be due to the shielding of the P=C bond by bulky substituents at phosphorus or carbon. This assumption is only partly right. Several halogen (Cl,Br,I) methylidenephosphanes could be obtained by deprotonation of α-CH acidic primary dihalogenophosphanes with tertiary amines. The P-chloro-

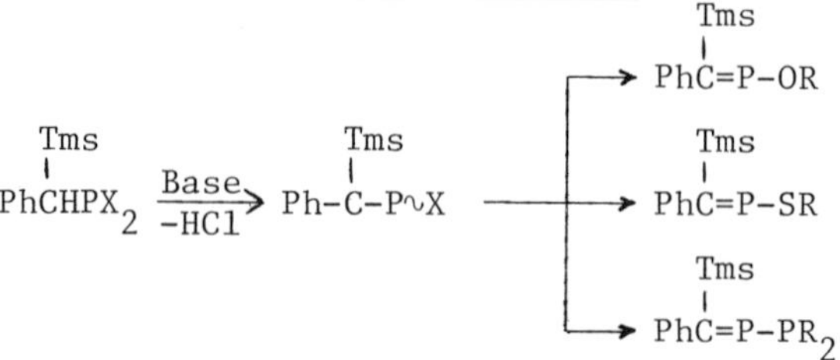

methylenephosphane is an easily accessible key compound for the synthesis of new methylenephosphanes. In addition to compounds with known basic structures, the P-alkoxy and P-alkylthio and P-phosphino methylenephosphanes could be prepared for the first time. Our particular interest in this compound was focused on the trimethylchlorosilane elimination, because this reaction should provide a route to phenylmethinephosphane (phosphaphenylacetylene). Heating a sample to 700°C in the vacuum provided by a Hg-diffusion pump, indeed spontaneously causes this reaction. Having optimized this elimination process by means of a mass

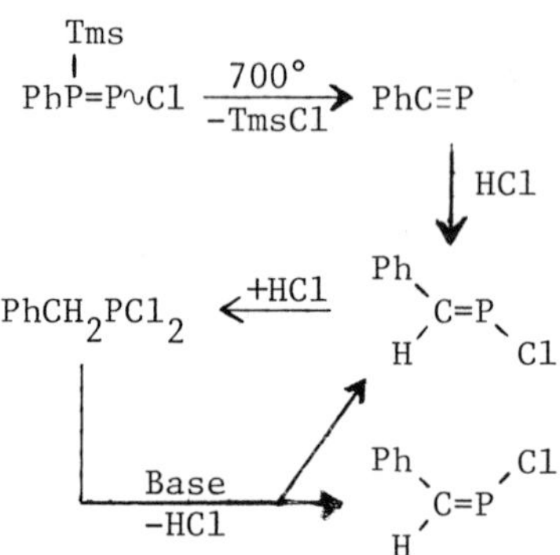

spectrometer directly connected to the pyrolysis apparatus, we readily succeeded in synthesizing the methinephosphane on a preparative scale and characterized it further. The PC triple bond in the phospha-alkyne has been confirmed by its characteristic ^{13}C and ^{31}P nmr data as well as by stepwise HCl addition. The phospha-alkene is obtained first, which is transformed to benzyldichlorophosphane by a second mole of HCl. By ^{31}P nmr spectroscopic studies we could show that benzyldichlorophosphane can be dehydrochlorinated by tertiary amines in reversal of its formation reaction. In addition to the E isomer, the Z isomer is also formed. Yet HCl addition to the phospha-alkyne does not produce any trans-compound. This can be explained in terms of a stereospecific cis-addition to the triple bond.

The direct vacuum pyrolysis of benzyldichlorophosphane also yields some phospha-alkyne in addition to a large amount of unidentified side products. In addition to the phenylphosphaacetylene which has a half-life of 7 minutes at 0°C, we could prepare the considerably more stable trimethylsilylderivative which has a half-life at room temperature of 50 minutes. The t-butyle derivative is entirely stable, as Becker reported recently.

To sum up, we can state that today a surprisingly great number of stable compounds with PC double bonds is known. At the moment we have several methods for the formation of this double bond. In addition to the dehydrohalogenation of halogenophosphines, the thermal elimination of trimethylchlorosilane and the elimination of hexamethyldisiloxane can be utilized. Prior to elimination, the silyl group can migrate from phosphorus to the oxygen, nitrogen, or sulfur of an α-carbonyl or hetero carbonyl function subsequent to initial TmsCl addition to or condensation with silylphosphanes. Most importantly, there is no longer any reason for the preparative chemist to balk at $2p_{\pi}$-$3p_{\pi}$ interactions. Very clearly the ^{13}C nmr spectra support this view, since they beyond any doubt show that the carbon in the methylidenephosphanes is sp^2-hybridized and sp-hybridized in the phosphaalkynes. The existence of E and Z isomers at the PC double bond can also be demonstrated. Moreover, the fluctuating structure of the tetraphosphahexadiene is absolutely analogous to purely olefinic compounds.

Whether the hopes will be fulfilled that these studies of the last five years have turned the first pages of a new chapter of phosphorus-carbon chemistry still remains to be seen. Whether, for instance, it will be possible to construct conjugated PC systems suitable for Diels-Alder reactions, is not yet certain. Exploration of the coordination chemistry of these new species may also prove fruitful.

RECEIVED July 13, 1981.

NEW ORGANIC SYNTHETIC METHODS BASED ON REAGENTS CONTAINING PHOSPHORUS

Dedicated to Georg Wittig

2

Selective Bond Formation of Organophosphorus Acids with Functional Groups of Biological Importance

L. HORNER, R. GEHRING, and H.-W. FLEMMING

Institut für Organische Chemie der Universität Mainz, Johann-Joachim-Becher-Weg 18–20, D-6500 Mainz, FRG

A knowledge of the interaction of nucleophiles with the phosphorylating agent, (modified in its reactivity by the use of different leaving-groups) is a precondition for the selective bonding of biologically important groups (e.g. OH, NH_2, SH) to organophosphorus acids in a controlled manner. The following variations were investigated:

$$R^1\text{-}\overset{\overset{O}{\|}}{\underset{\underset{R^2}{|}}{P}}\text{-}X + RYH + B \longrightarrow R^1\text{-}\overset{\overset{O}{\|}}{\underset{\underset{R^2}{|}}{P}}\text{-}YR + BHX$$

Phosphinic acid derivatives, Phosphonic acid derivatives, Phosphoric acid derivatives

X = Cl, F, CN, N_3, $OC_6H_4NO_2$(p); for RYH: Y = O, NR, S

The relative reactivities were determined by means of competition with different nucleophiles RYH and RY'H.

$$\underset{1}{Ⓟ\text{-}X} + \underset{1}{RYH} + \underset{1}{RY'H} \xrightarrow{NEt_3} \begin{cases} Ⓟ\text{-}YR \\ Ⓟ\text{-}Y'R \end{cases}$$

$$Ⓟ = R^1R^2\text{-}\overset{\overset{O}{\|}}{P}$$

Examples:

X = Cl (a); X = F (b); X = N_3(c); X = CN (d); X = $OC_6H_4NO_2$(p) (e)

RYH = n-$BuNH_2$; RY'H = n-BuOH

(a) Ⓟ-YR = Amide (94 %); Ⓟ-Y'R = Ester (2 %)
(b) Ⓟ-YR = Amide (0 %); Ⓟ- Y'R = Ester (87 %)

0097–6156/81/0171–0013$05.00/0

(c) Ⓟ-YR = Amide (~80 %); Ⓟ-Y'-R = Ester (~20 %)

(d) Ⓟ-YR = Amide (0 %); Ⓟ-Y'-R = Ester (~90 %)

(e) Ⓟ-YR = Amide (0 %); Ⓟ-Y'R = Ester (~90 %)

X = Cl (a); X = N_3 (b); X = CN (c)

RYH = n-BuNH_2; RY'H = n-BuSH

(a) Ⓟ-YR = Amide (~90 %); S-ester (trace)

(b) Ⓟ-YR = Amide (~80 %); S-ester (~20 %)

(c) Ⓟ-YR = Amide (trace); S-ester (~90 %)

With X = F and $OC_6H_4NO_2$: no reaction

Three products are possible when the two competing nucleophiles are situated in the same molecule (e.g. ethanolamine, cysteamine, serine and cysteine).

Ⓟ-X + $HOCH_2$-CH_2-NH_2 → Ⓟ-NH-CH_2-CH_2OH

→ Ⓟ-O-CH_2-CH_2-NH_2

→ Ⓟ-NH-CH_2-CH_2-OⓅ

Ⓟ = $(C_6H_5)_2P(O)$

Several selectivity studies were also carried out for serine (important in the active site of many enzymes) and for cysteamine and cysteine (also in view of their biological importance).

Selectivity in the reaction of serine-N-butylamide with $Ph_2P(O)X$ (X = Cl or F):

$HOCH_2$-CH(NH_2)-C(=O)-NHBu —$Ph_2P(O)F$→ $Ph_2P(O)$-O-CH_2-CH(NH_2)-C(=O)-NHBu

$HOCH_2$-CH(NH_2)-C(=O)-NHBu —$Ph_2P(O)Cl$→ $HOCH_2$-CH(NHP(O)Ph_2)-C(=O)-NHBu

Selectivity in the reaction of cysteamine with diphenylphosphinic acid cyanide:

$$HS\text{-}CH_2\text{-}CH_2\text{-}NH_2 + Ph_2P(O)CN \longrightarrow Ph_2P(O)\text{-}S\text{-}CH_2\text{-}CH_2\text{-}NH_2$$

$$HS\text{-}CH_2\text{-}\underset{\displaystyle CO_2CH_3}{\underset{|}{CH}}\text{-}NH_2 + Ph_2P(O)CN \longrightarrow Ph_2P(O)\text{-}S\text{-}CH_2\text{-}\underset{\displaystyle CO_2CH_3}{\underset{|}{CH}}\text{-}NH_2$$

Diphenylphosphoric acid derivatives $(C_6H_5O)_2P(O)X$ (X = Cl, F, N_3, CN, $OC_6H_4NO_2$(p))are as selective as the corresponding phosphinic acid derivatives $(C_6H_5)_2P(O)X$.

Corresponding competition reactions ($BuNH_2$, BuOH, BuSH) were also carried out with the diphenylthiophosphinic acid derivatives $Ph_2P(S)X$ (X = Cl, F, CN).

Organophosphorus compounds bearing a fluorescent group were specifically introduced into the active sites of the serine-enzymes α-Chymotrypsin, Trypsin and Butyrylcholin-esterase using the agents 2, 3 and 4. This was shown using electrophoresis.

[Structures: 1-substituted 5-dimethylaminonaphthalene (R, NMe_2) and 1-substituted 5-methoxynaphthalene (R, OMe)]

1 R = SO_2Cl
(dansylchloride)

2 = PhP(O)F

3 = $PhP(O)\text{-}OC_6H_4NO_2$(p)

4 R = EtOP(O)F

The compounds 2, 3 and 4 are effective inhibitors of serine enzymes (α -Chymotrypsin, Trypsin, Butyrylcholin-esterase and Acetylcholin-esterase (only 4).

RECEIVED July 13, 1981.

3

Chemical Synthesis and Biological Properties of the 5′-Terminus of Eukaryotic Messenger Ribonucleic Acids (mRNA)

TSUJIAKI HATA, MITSUO SEKINE, IWAO NAKAGAWA, KAZUO YAMAGUCHI, SHINKICHI HONDA, TAKASHI KAMIMURA, and KAZUKO YAMAGUCHI

Department of Life Chemistry, Tokyo Institute of Technology, Nagatsuta, Midoriku, Yokohama 227, Japan

KIN-ICHIRO MIURA

National Institute of Genetics, Mishima 411, Japan

In contrast to prokaryotic mRNAs, mRNAs from various eukaryotic cells and viruses have been found to contain a terminal 7-methylguanosine (m^7G) residue linked from its 5'-position through a triphosphate bridge which was presented commonly as shown in the following Scheme 1 .

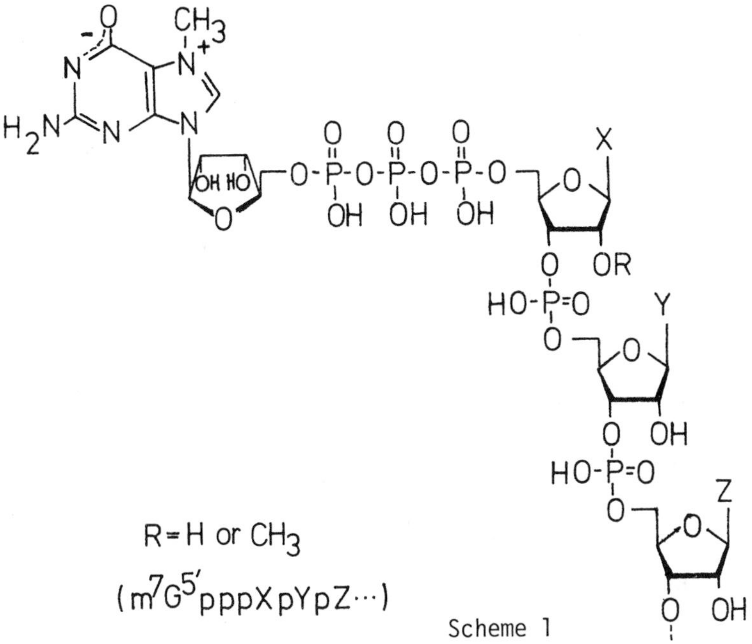

R = H or CH_3

($m^7G^{5'}pppXpYpZ\cdots$)

Scheme 1

0097–6156/81/0171–0017$05.00/0

We first describe the synthesis of unsymmetrical α,γ-dinucleoside triphosphates involving 7-methylguanosine. For the construction of the terminal cap structure, there might be two possible reaction modes where an activatable protecting group (X) was introduced into a nucleotide by direct displacement with phosphate hydroxyl group (Method A) and by pyrophosphorylation between Xp and pN (Method B).

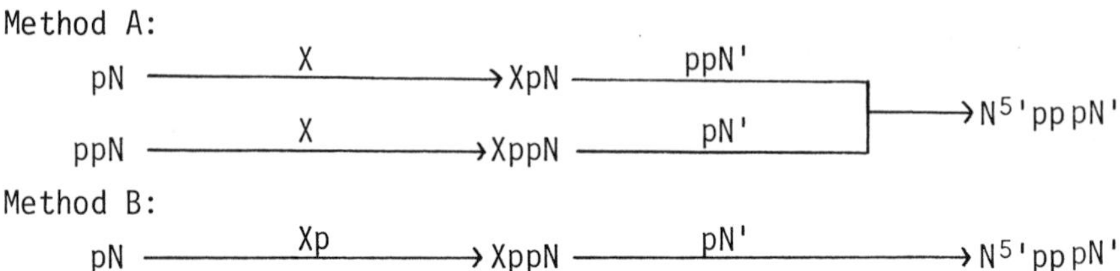

As the former type of reaction, we have developed the triphosphate bond formation by means of phosphinothioyl bromide as shown in the following scheme.

$$\mathrm{HOP(O)(O^-)O\text{-}[ribose(B)(HO)(OH)]} \xrightarrow[\text{pyridine}]{Bu_2P(S)Br} \mathrm{Bu_2P(S)O\text{-}P(O)(O^-)\text{-}O\text{-}[ribose(B)(HO)(OH)]} \xrightarrow[Ag^+]{ppB'} \mathrm{BpppB'}$$

The latter requires a phosphorylating species possessing an activatable protecting group for the preparation of the starting material (XppN). Now, we found a convenient method for the synthesis of this type of pyrophosphorylating reagent by the reaction of methyl phosphorodichloridate with thiophenol in pyridine. A promising capping reagent, P^1-S-phenyl P^2-7-methylguanosine-5'-pyrophosphorothioate, was synthesized as described below.

$$\mathrm{MeOP(O)Cl_2} + \text{pyridine} \xrightarrow{PhSH} \mathrm{PhS\text{-}P(O)(O^-)\text{-}Cl} \xrightarrow{pm^7G} \mathrm{PhSppm^7G}$$

The phenylthio group could serve as the temporary protecting group which was easily activated by treatment with silver acetate, silver nitrate or iodine. By the above methods, we synthesized several kinds of unsymmetrical α,γ-dinucleoside triphosphates involving methylated or nonmethylated cap structures. The above-mentioned methods have been also applied to the synthesis of cap structure containing oligoribonucleotides. Furuich and Miura discovered the terminal structure of cytoplasmic polyhedrosis virus (CPV) mRNA which was represented as $m^7G^{5'}$pppAmpGpU··· In order to confirm the structure by chemical synthesis and investigate the function and the mechanism of its formation *in vitro* $m^7G^{5'}$pppAm, $m^7G^{5'}$pppAmG, and $m^7G^{5'}$pppAmpGpU were synthesized by utilizing the present capping reactions and the phosphotriester approach. However, we have felt during the investigation that 7-methylguanylic acid and its derivatives used for the construction of the cap structure should be appropriately protected because of their extreme instability and poor solubility. In most of mosaic virus mRNAs, there exsists the common terminal structure of $m^7G^{5'}$-pppGpU··· In order to elucidate its biological meaning, we tried the synthesis of this part by using monomethoxytrityl and dimethoxytrityl groups as the highly lipophilic protecting groups of N^2-amino functions of guanosine and 7-methylguanosine. Consequently, m^7GpppGpU was obtained in good yield and some related compounds of $m^7G^{5'}$pppGpUpU and $m^7G^{5'}$ppGpU were also synthesized in a simmilar manner.

We have tested some of biological properties of the cap structure by employing unusual man-made cap analogues which were synthesized by the above methods.

Here, we shall describe the methylation reactions at the 7-position of guanosine residue of the cap structure.

First, the methylation of guanosine moiety of the cap structure with S-adenosylmethionine (SAM) was examined. When only GTP was added as a substrate to the in vitro RNA synthesizing system of CPV in the presence of SAM, GTP was not methylated. However, GTP and ATP were added to the same system, $m^7G^{5'}$pppA and $m^7G^{5'}$pppApG were formed. On the basis of the above facts, chemically synthesized $G^{5'}$pppA was added as a substrate in place of GTP and ATP, $m^7G^{5'}$pppA was obtained expectedly. In this case no methylation took place of 2'-OH of adenosine moiety. In the same system $G^{5'}$pppG was not methylated. Therefore, the cap structure of CPV was formed as follows:

$$\mathrm{GTP} + \mathrm{ATP} \longrightarrow \mathrm{G^{5'}pppA}$$

$$\mathrm{G^{5'}pppA} + \mathrm{SAM} \longrightarrow \mathrm{m^7G^{5'}pppA}$$

$$\mathrm{m^7G^{5'}pppA} + \mathrm{GTP} \longrightarrow \mathrm{m^7G^{5'}pppApG}$$

On the other hand, mRNAs from reovirus were represented as $m^7G^{5'}$pppGmpU··· In the in vitro RNA synthesizing system of reovirus, $G^{5'}$pppG was methylated selectively at the 7-position of one of two guanosine residues, but any methylation is not caused for $G^{5'}$pppA. This shows that the methylation enzymes either in CPV or in reovirus recognized strictly the structure of the confronting nucleoside residues.

Next, we examined the structural requirement for the confronting phosphate bridge in the methylation. The CPV system as described previously was employed.

It was found that $G^{5'}$ppppA and $G^{5'}$ppA were methylated imperfectly about 50% relative to $G^{5'}$pppA in the presence of SAM and no methylation was observed in the case of $G^{5'}$pA.

RECEIVED July 7, 1981.

4

Triphenylphosphane-diethylazodicarboxylate: A Useful System for Directed Structural Variation of Carbohydrates

E. ZBIRAL, H. H. BRANDSTETTER, and E. MARK

Institute for Organic Chemistry, University of Vienna, A-1090 Vienna, Austria

Nucleophilic substitution reactions at alcoholic substrates with HX initiated by activation of the hydroxy-group by means of TPP/DEAD first developed by Mitsunobu (1) are being increasingly utilized (1,2). This principle has been seldom applied until now in the field of carbohydrates. In order to diminish hydrophilicity, the mono- and bis-t-butyldimethylsilylether derivatives of a series of carbohydrates were prepared, which represent very useful substrates with two or even three free OH groups for realizing a series of interesting transformations by means of TPP/DEAD or TPP/DEAD/HX.

The methyl-ß-glucopyranoside 1a can be transformed in this way exclusively to the 3-desoxy-3-X-allose derivatives 2a-2d (X = N_3, Br, J, p-NO_2-C_6H_4COO) (figure 1), whereas the corresponding methyl-α-D-glucopyranoside 1b gives exclusively the analogous 4-desoxy-4-X-galactose derivatives 3a-3d. This useful substitution reaction especially by means of TPP/DEAD/HN_3 can be applied also to the methyl-2-acetamido-6-

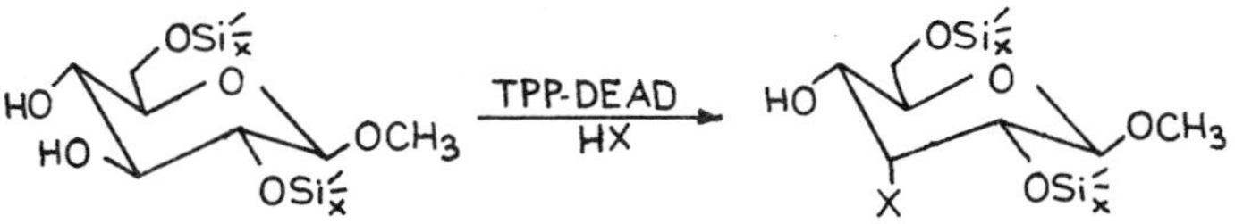

1a 2

2a: X=N_3 2b: X=NO_2-C_6H_4COO 2c: X=Br 2d: X=J

Figure 1.

(t-butyldimethylsilyl)-2-desoxy-α-D-glucopyranoside 4 and methyl-2-acetamido-3,6-bis-O-(t-butyldimethylsilyl)-2-desoxy-α-D-glucopyranoside 5. The former is

0097–6156/81/0171–0021$05.00/0

transformed to the methyl-2-acetamido-3-azido-6-O-(t-butyldimethylsilyl)-2,3-didesoxy-α-D-allopyranoside 6 whereas the latter goes over to the corresponding galactose derivative 7. When 1a, 1b and 4 are treated with TPP/DEAD alone the preparatively useful 3,4-anhydro-galactose derivatives 8 (methyl-3,4-anhydro-2,6-bis-O-(t-butyldimethylsilyl)-β-D-galactopyranosid), 9 (methyl-3,4-anhydro-2,6-bis-O-(t-butyldimethylsilyl)-α-D-galactopyranosid) and 10 (methyl-acetamido-3,4-anhydro-6-O-(t-butyldimethylsilyl)-2-desoxy-α-D-galactopyranoside) arise. Under similar conditions methyl-6-O-(t-butyldimethylsilyl)-α-D-glucopyranoside 11 yields the methyl-2,3-anhydro-6-O-(t-butyldimethylsilyl)-α-allopyranosid 12 which can be subjected additionally an substitution process at C4 by means of TPP/DEAD/HN_3 resp. p-NO_2-C_6H_4COOH without any opening of the oxirane function by HX. Methyl-2,3-anhydro-4-azido-6-O-(t-butyldimethylsilyl)-4-desoxy-α-D-gulopyranoside 13a and methyl-2,3-anhydro-6-O-(t-butyldimethylsilyl)-4-O-(p-nitrobenzoyl)-α-D-gulopyranoside 13b are formed.

The transformations of 2,6-O-bissilyl-methyl-α-D-mannopyranoside are represented in figure 2. The predominating product is the methyl-3,4-anhydro-talopyranoside 15 even when TPP/DEAD/HN_3 is used. Obviously the substitution process leading to the altro-sugar 16 is greatly hindered by the well known 1,3-diaxial interaction. All the epoxy sugars 8, 9, 10 and 15 serve as very useful starting points for further interesting structural modifications by opening the epoxide ring with HX (2, 3).

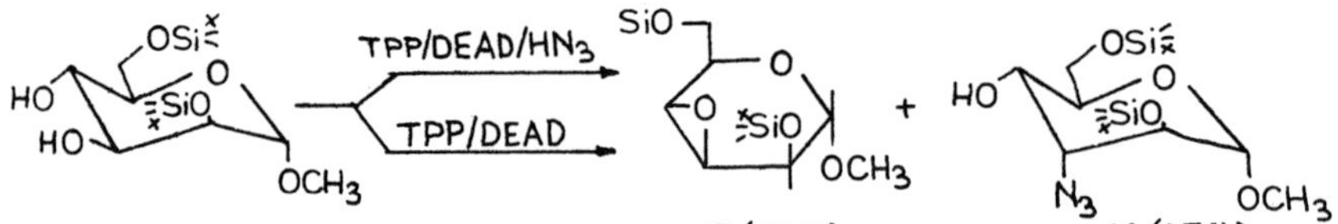

Figure 2.

The behaviour of the D-1,4-gluconolactone derivative 17 towards TPP/DEAD/HX is summarized in figure 3. The exclusive activation of the OH group at C5 using one equivalent TPP/DEAD followed by a substitution by HX opens an interesting approach to the L-1,4-idonolactone derivative 18 whereas the activation of the second OH group at C3 yields by an elimination process the unsaturated sugar lactone 19. By applying such procedures to the analogous 2,6-bis-O-(t-butyldimethyl-silyl)-ether derivative of D-1,4-galactonolactone 20 the corresponding L-altronolactone derivatives 21a ($X = N_3$) and 21b ($X = p\text{-}NO_2\text{-}C_6H_4COO$) on the one and 2,6-bis-O-(t-butyldimethylsilyl)-5-O-p-nitrobenzoyl-L-erythro-hex-2-en-1,4-lactone 22 on the other hand are resulting. A similar pattern can be observed in the case of the 2,6-bis-O-(t-butyldimethylsilyl)-ether

TPP/DEAD HX TPP/DEAD (HX)

17 18 19

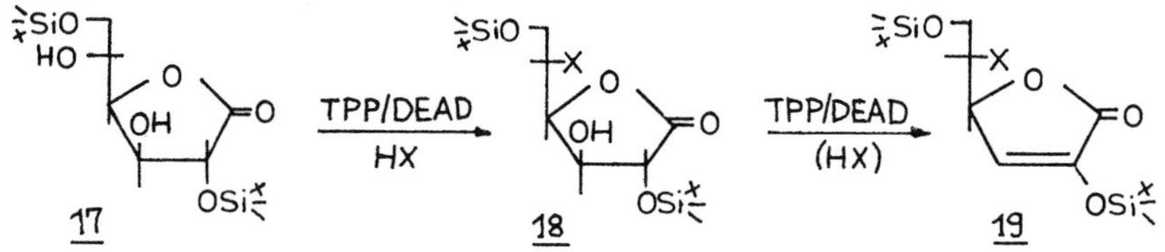

Figure 3.

derivative of L-mannonolactone 23. By omission of HX the olefin sugar lacton 24 (2,6-bis-O-(t-butyldi-methylsilyl)-3-desoxy-L-erythro-hex-2-en-1,4-lactone) is formed without any involvement of the C5-OH. In contrast the D-1,4-gulonolactone derivative 25 is transformed into the 3,6-anhydro-D-gulonolactone 26 (figure 4).

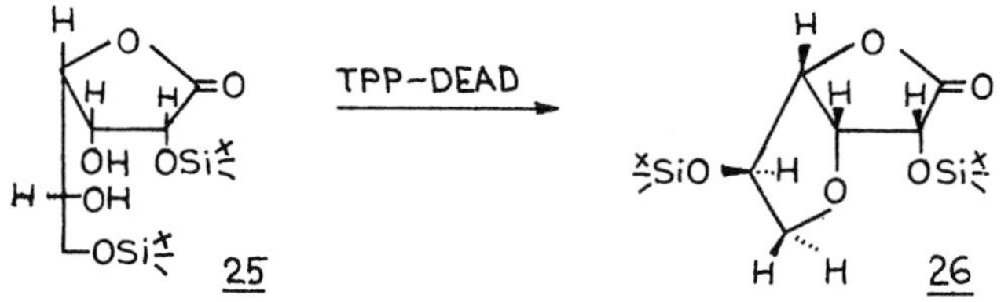

TPP-DEAD

25 26

Figure 4.

The 4,9-bis-O- and 4,8,9-tris-O-t-butyldimethyl-silylether derivatives 27 and 28 of the biologically important neuraminic acid represent useful starting points for the synthesis of the new structurally varied nonulosonic acid derivatives 29 (figure 5) 30 and 31 (figure 6). Opening of the oxirane ring of 29 with HN_3 leads to the 8-desoxy-8-azido-neuraminic acid derivative 30 which corresponds completely 28. By the reaction of the tris-silylether derivative 28 with $TPP/DEAD/HN_3$ an interesting result was observed (figure 6). The primary activated OH group of C7 is attacked obviously as a result of ideal stereoelectronic conditions by the acetamido group of C5. As a consequence of this neighbouring group participation an inversion of the configuration of C7 leading to the D-glycero-L-altro-5-(5'-methyl-1-N-tetrazolo)-nonulosonic acid derivative 31 occured.

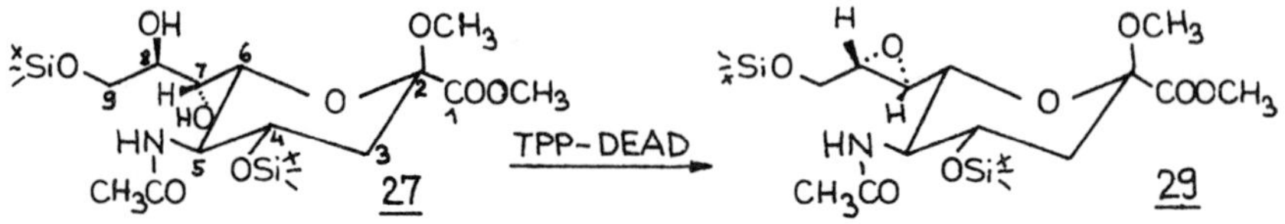

Figure 5.

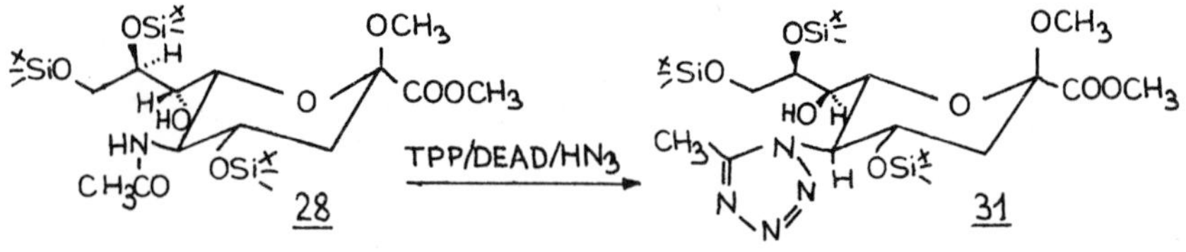

Figure 6.

Literature cited

1. O. Mitsunobu, Synthesis 1981, 1-28
2. H.H. Brandstetter, E. Zbiral, Helv.Chim.Acta 1978, 61, 1832, 1980, 63, 327
3. E. Mark, E. Zbiral, H.H. Brandstetter, Mh.Chem. 1980, 111, 289.

RECEIVED June 30, 1981.

5

Synthetic Application of Element Organic Substituted Phosphorus Ylides

H. J. BESTMANN

Institut für Organische Chemie der Universität Erlangen-Nürnberg, Henkestrasse 42, 8520 Erlangen, FRG

The trimethylsilylated ylides 1 (1), easily generated from trimethylchlorosilane and ylides, react with aldehydes 2 to form vinylsilanes 3 (2,3). The vinylphosphonium silanolates 4 are also formed. Compounds 3 are versatile reagents for further reactions (4). The ylide 1 (with R^1=H) reacts with aldehydes 2 to give the dienes 9. The oxidation of 1 with the adduct 6, from triphenylphosphite and ozone, gives access to a general synthesis of acylsilanes (trimethylsilylketones) 8 (2). The silylated ylides 1 react to form phosphonium salts 7 with halogen compounds. The salts 7 can be desilylated by fluoride ions. The disubstituted ylides 10 formed can be converted in statu nascendi with aldehydes 11 into the tris-substituted olefin 12 (2,3). In the case of R^3=I, vinyl

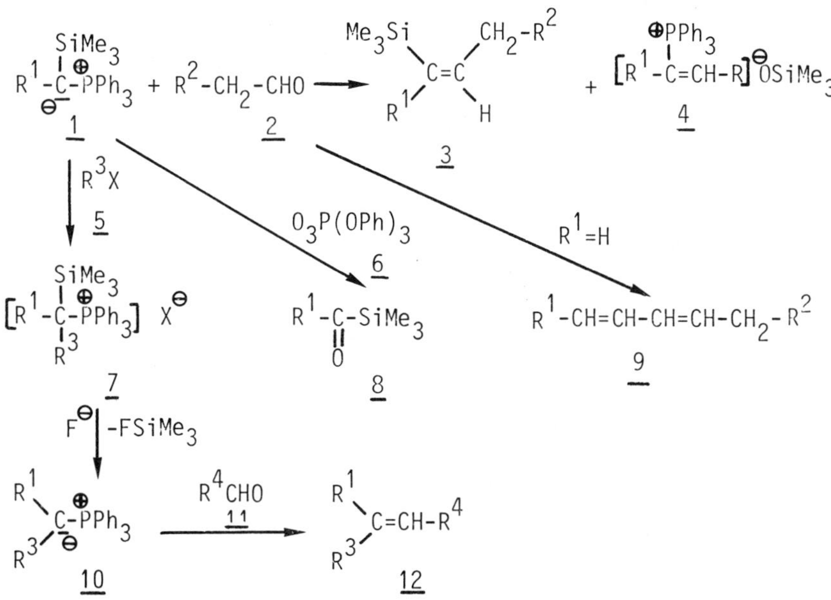

0097–6156/81/0171–0025$05.00/0

iodides 12, R^3=I, are obtained (2). Using deuterated halogen compounds R^3X (and R^1=H) partially selectively deuterated pheromones can be obtained by this method (5). Recent results show, that (tert. butyl-dimethylsilyl)-methylene phosphorane 1 (R^1=H, and tert. butyl-dimethylsilyl instead of Me_3Si) gives a Wittig reaction which affords a vinyl silane with terminal double bond in high yields (2).

The hexaphenylcarbodiphosphorane 13 (6) can be understood as an elementorganically substituted ylide with a particular character. It reacts with S_8 to form CS_2 14, which immediately reacts further with one more molecule of 13 via a betaine intermediate as described previously (7) to make thioketenylidene triphenylphosphorane 15 (8).

$Ph_3\overset{\oplus}{P}-\overset{\ominus}{\underset{\ominus}{C}}-\overset{\oplus}{P}Ph_3$ (13) $\xrightarrow[-2SPPh_3]{S_8}$ CS_2 (14) $\xrightarrow{13}$ $Ph_3\overset{\oplus}{P}-\overset{\ominus}{C}=C=S$ (15)

$H_3COOC-C{\equiv}C-COOCH_3$ (16)

(17) =C=O

(18) $(H_3COOC)C=C=C(COOCH_3)$; $H_3COOC-C^{\ominus}(\overset{\oplus}{P}Ph_3)$; $^{\ominus}C(\overset{\oplus}{P}Ph_3)-COOCH_3$

(19) $=C=\overset{\ominus}{C}-\overset{\oplus}{P}Ph_3$

(20) Ph_3P

Compound 13 reacts with 2 moles of acetylene dicarboxylic ester 16 via a twofold cycloaddition and a subsequent electrocyclic ring opening reaction (9) to form the allene bisylid 18 (8). The radialene 20 is formed (8) from 13 and fluorenylidene ketene 17. Compound 13 probably reacts first with 17 to give the phosphacumulene ylide 19 (10), from which a pentatetraen is formed with 17. This then reacts with 19 to give a cycloaddition yielding 20 (9).

Phosphorus ylides 21 combine with BH_3 22 to yield adducts 23 (11), which rearrange thermally to give the monoalkyl borane-triphenylphosphane adducts 24. On further heating these disproportionate to trialkylboranes 29 and the adduct from BH_3 and triphenylphosphane 30 (12).

$$\underset{\mathbf{21}}{R^1\text{-}\overset{R^2}{\overset{|}{C}}{}^{\ominus}\text{-}\overset{\oplus}{P}Ph_3} + \underset{\mathbf{22}}{BH_3} \longrightarrow \underset{\mathbf{23}}{R^1\text{-}\overset{R^2}{\overset{|}{C}}(\overset{\ominus}{B}H_3)\text{-}\overset{\oplus}{P}Ph_3} \longrightarrow \underset{\mathbf{24}}{R^1\text{-}\overset{R^2}{\overset{|}{C}}H\text{-}BH_2\cdot PPh_3} \xrightarrow[\text{R-CH=CH-R } \mathbf{26}]{PhCH_2I\ \mathbf{25}} \underset{\mathbf{27}}{R^1\text{-}\overset{R^2}{\overset{|}{C}}H\text{-}B(\overset{R}{\overset{|}{C}}H\text{-}CH_2\text{-}R)_2}$$

$$\mathbf{24} \xrightarrow{HCl} \underset{\mathbf{28}}{R^1\text{-}\overset{R^2}{\overset{|}{C}}H\text{-}BHCl\cdot PPh_3} \qquad \mathbf{24} \xrightarrow{\Delta} \underset{\mathbf{29}}{(R^1\text{-}\overset{R^2}{\overset{|}{C}}H)_3B} + \underset{\mathbf{30}}{PPh_3\cdot BH_3}$$

$$\mathbf{27} \xrightarrow{1.\ CO\quad 2.\ H_2O_2} \underset{\mathbf{31}}{R^1\text{-}\overset{R^2}{\overset{|}{C}}H\text{-}C(OH)(\overset{R}{\overset{|}{C}}H\text{-}CH_2\text{-}R)_2}$$

$$\mathbf{28} \xrightarrow{\mathbf{25}\ \mathbf{26}} \underset{\mathbf{32}}{R^1\text{-}\overset{R^2}{\overset{|}{C}}H\text{-}BCl\text{-}\overset{R}{\overset{|}{C}}H\text{-}CH_2\text{-}R} \xrightarrow[2.\ DCME]{1.\ NaOCH_3} \underset{\mathbf{33}}{R^1\text{-}\overset{R^2}{\overset{|}{C}}H\text{-}C(=O)\text{-}\overset{R}{\overset{|}{C}}H\text{-}CH_2\text{-}R}$$

We were able to direct the rearrangement 23→24 so that no disproportion into 29 and 30 occurred (13). The adducts 24 are stable and can now be used for hydroboration reactions whereby a suitable method for the elimination of triphenylphosphane from complex 24 must be used. This can be achieved with benzyl-iodide 25. On addition of the iodo compound 25 and an olefin 26 to a solution of 24 in tetrahydrofuran, the benzyl-triphenylphosphonium iodide precipitates and the free $R\text{-}BH_2$ adds to the olefin 26 forming the trialkylborane 27 (13). The latter can be converted into tertiary alcohols 31 by subsequent treatment with CO and H_2O_2 (14).

The reaction of 24 and HCl yields alkylchloroborane triphenylphosphane complexes 28, which can be converted with olefins in the presence of benzyliodide 25 (15). The dialkylchloroboranes 32 thus formed can be transformed into ketones 33 with sodium methylate and then the DCME-technique of H.C. Brown (16).

These last results combine the Wittig ylide chemistry with Brown's hydroboration reaction. We hope that a preparatively interesting ylide-borane chemistry will arise from this new "alloy".

I thank my coworkers, named in the references, for their enthusiastic engagement in the solution of our common problems.

Literature cited:

1 Seyferth, D.; Singh, G. J. Am. Chem. Soc. 1965, 87, 4156. Schmidbaur, H.; Tronich, W. Chem. Ber. 1967, 100, 1032.
2 Bestmann, H.J.; Bomhard, A. unpublished.
3 Bestmann, H.J. Pure and Appl. Chem. 1980, 52, 771.
4 Colvin, E.W. Chem. Soc. Rev. 1978, 7, 15.
5 Bestmann, H.J.; Hirsch, L. unpublished.
6 Ramirez, F.; Desai, N.B.; Hansen, B.; Mc Kelvie, N. J. Am. Chem. Soc. 1961, 83, 3539.
7 Matthews, C.N.; Birum, G.H. Tetrahedron Lett. 1966, 5707. Matthews, C.N.; Driscoll, J.S.; Birum, G.H. Chem. Comm. 1966, 736.
8 Bestmann, H.J.; Öchsner, H. unpublished.
9 Bestmann, H.J.; Rothe, O. Angew. Chem. 1964, 76, 569; Angew. Chem. Int. Ed. Engl. 1964, 3, 512.
10 Bestmann, H.J. Angew. Chem. 1977, 89, 361; Angew. Chem. Int. Ed. Engl. 1977, 16, 349.
11 Hawthorne, F.M. J. Am. Chem. Soc. 1961, 83, 367.
12 Köster, R.; Rickborn, B. J. Am. Chem. Soc. 1967, 89, 2782.
13 Bestmann, H.J.; Sühs, K.; Röder, Th. unpublished.
14 Brown, H.C. "Organic Synthesis via Boranes" John Wiley Sons, New York, Chichester, Brisbane, Toronto, 1975.
15 Bestmann, H.J.; Röder, Th. unpublished.
16 Brown, H.C.; Ravindran, N.; Kulkarni, S.U. J. Org. Chem. 1979, 44, 2417.

RECEIVED June 30, 1981.

6

Mono-, Di-, and Multi-Ylides in Organometallic Chemistry

HUBERT SCHMIDBAUR

Anorganisch-Chemisches Institut der Technischen Universität München, Lichtenbergstrasse 4, D-8046 Garching, FRG

The ylid function in an organophosphorus compound is described in the various theoretical models as a reactive molecular site with a combination of donor and acceptor capacity - or as a set of adjacent nucleophilic and electrophilic centers at carbon and phosphorus, respectively. The situation is not unlike the characteristics of the carbonyl group, one of the most important reactive functions in classical organic chemistry. Following the pioneering work by G. Wittig (1) and his collaborators, the synthetic potential of phosphorus ylids - and the reaction with carbonyl compounds in particular - has been widely exploited and the literature witnesses an ever increasing range of new preparative uses (2).

In most of these reactions the PC bond of the ylid is cleaved and formally a carbene moiety is exchanged with a corresponding part of the substrate. In contrast, the majority of the reactions of ylids with acceptor sites centered at metals or metalloids occur with conservation of the PC bond and lead to organometallic/organometalloidal products containing M-C-P bridges (3,4,5,). Depending on the nature of M, these carbon bridges between phosphorus and heteroatoms may show CH acidity and may be deprotonated when exposed to strong bases or to an excess of ylid, which acts as a transylidating agent.

$$\left\{\begin{matrix} R_3\overset{\oplus}{P}-\overset{\ominus}{C}H_2 \\ R_3P{=}CH_2 \end{matrix}\right. \xrightarrow[M]{} \left.\begin{matrix} R_3\overset{\oplus}{P}-CH_2-\overset{\ominus}{M} \\ R_3P-CH{=}M \end{matrix}\right\} \text{Base}$$

Bonding in the resulting products again depends strongly on the nature of M, and cases with the character of metallated ylids $R_3P{=}CH{-}M^-$ are also known (3) as are their phosphoniumcarbene metal counterparts R_3P^+ P-CH=M^{2-} (6,7,8).

A second feature, also uncommon in reactions of ylids with organic substrates, is the additional ylidation of the substituents at phosphorus (3,4,5). This alternative provides for an enormous variety of bridged and chelated metal complexes, amply described in a large number of papers in recent years.

0097-6156/81/0171-0029$05.00/0

Recent advances in the author's laboratory emerged from a search for metal complexes with chiral centers at the P-C*-M bridge, with spirocyclic centers generated from cyclopropylides, for multifunctionality of cyclic and non-cyclic ylids and their complexes with weak acceptor metals, such as the alkali and earth alkaline metals, or the d^5 and d^{10} transition metals. Only representative examples can be given, and the reader is referred to related papers quoted for further reference.

Introduction of Chirality (A. Mörtl, B. Zimmer-Gasser). There are examples scattered in the literature, where the ylidic donor center becomes chiral on presenting a dative bond to the metal. Perhaps the simplest case is the substitution product of $(C_6H_{11})_3P{=}CHCH_3$ with $Ni(CO)_4$ (9): $(CO)_3Ni\text{-}^*CH(CH_3)\text{-}P(C_6H_{11})_3$. A metal bridging case is represented by the uranium compound $[(C_5H_5)_2UCH_2(CH)P(C_6H_5)_2]_2$ (10).

A pronounced chirality effect on the chemical reactivity of ylid complexes is to be expected from propeller-type ligands with a C_2 rotation axis. In a first attempt to synthesize pertinent examples, a spirobicyclic ylide 1 was designed (11) and converted into a novel lithio-derivative 2, which on reaction with metal halides yields multi-spiro complexes of unusual geometries:

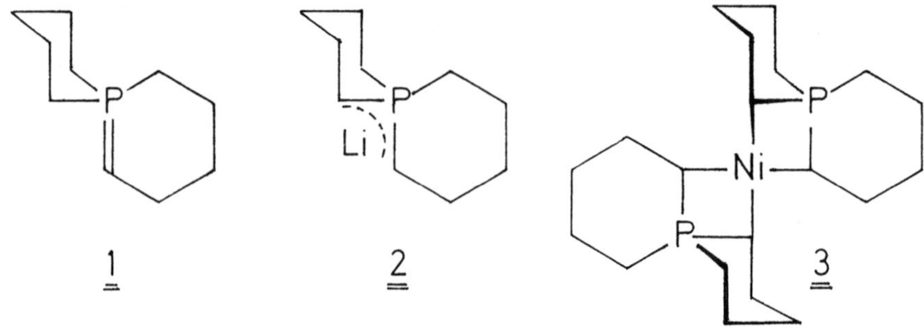

3 is obtained as yellow crystals, which have the ligand donor centers in a DD, LL configuration, yielding a centrosymmetric molecule (12). In Figure 1 the square planar array of CH groups around the Ni atom and the paddle-type position of the phosphorinane rings - all in a chair conformation - are immediately obvious.

The Chemistry of Phosphonium Cyclopropylids (A. Schier). A small number of cyclopropylids are known in the literature (13, 14,) but their coordination chemistry has not been developed. For a detailed study of the effect of the cyclopropyl ring on the donor properties of ylids a series of simple species was prepared in the pure, salt-free state and characterized by multiple-resonance experiments (14, 15).

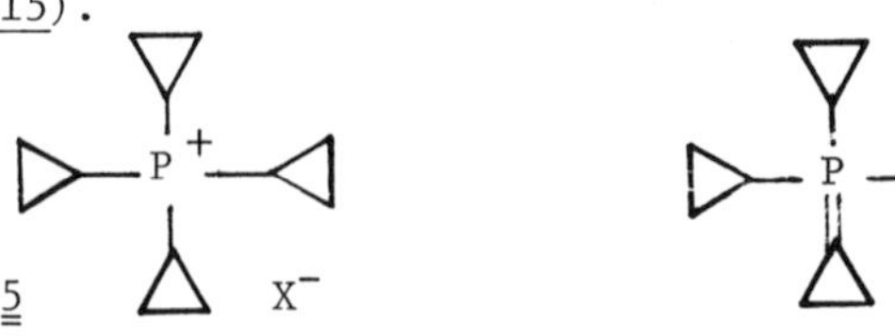

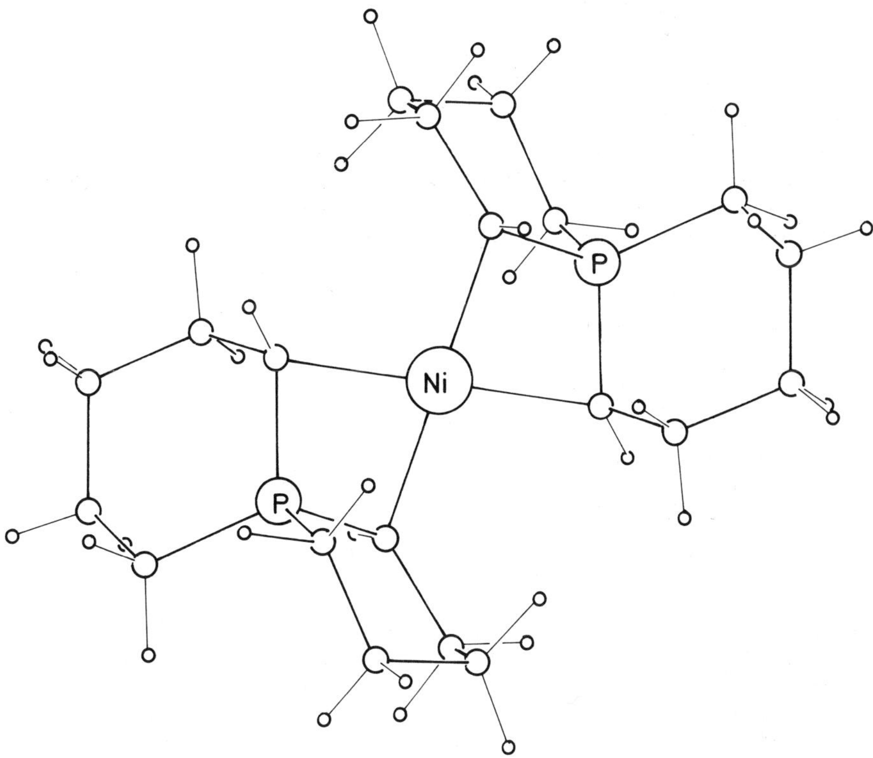

Figure 1. Molecular structure of the nickel complex 3 as determined by x-ray diffraction (12).

The fully cyclopropylated ylid 4 is a fluxional molecule exhibiting rapid α-proton exchange (15).

Moreover, low temperature NMR spectra indicate a pyramidal structure of the carbanion in triphenylphosphonium cyclopropylid 6, and an X-ray diffraction analysis of the crystalline solid confirms this result (14). Contrary to earlier predictions, cyclopropylids are thus found to be the first class of ylids to contain non-planar carbanions with an unusual ylidic bonding (Fig.2). They form stable metal complexes, as illustrated by a gold complex (7):

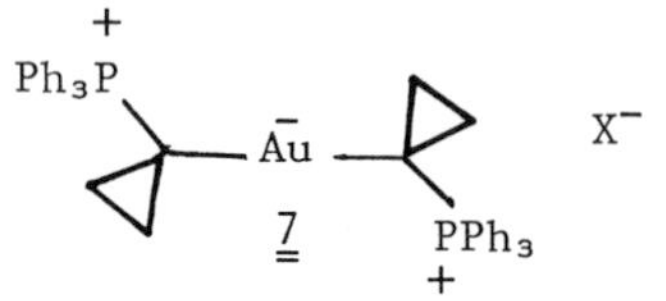

7

Open-chain Double- and Multiple-ylids (U. Deschler and B. Milewski-Mahrla).
In compounds of the type $R_3P{=}CH{-}PR_2$ (8) an additional phosphine donor center is directly attached to the basic ylid carbon atom (16, 17). Novel derivatives of these ligands are, e.g., nickel complexes 9 containing the NiCPCP five-membered ring (18).

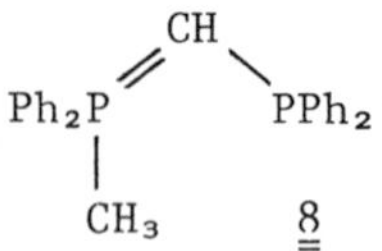

8

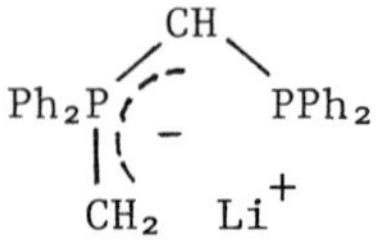

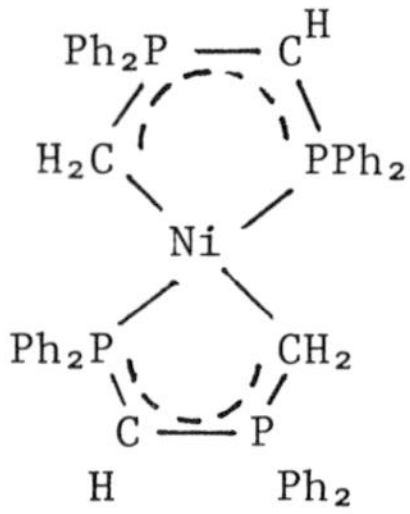

9

The crystal structure of 9 has been determined and a centrosymmetrical rectangular coordination geometry was found around the Ni atom (Fig. 3). All PC bonds of the rings show partial ylidic character. The geometry of the corresponding Ni-complex containing the six-membered rings NiCPCPC, derived from the carbodiphosphorane was established in previous studies (19). Four-membered rings NiCPC (20) and MiPCP (21) were detected in ylid and diphosphinomethanide complexes, respectively. It appears therefore that a whole series of organometallic complexes with alternating C/P ring members is now available, closely related to some of the S/N complexes of metals (22).

For related examples of alkali complexes see (22, 23).

Cyclic Multiple-ylids (Th. Costa and B. Milewski-Mahrla).
Cyclic double ylids with aliphatic ring members occur solely as carbodiphosphoranes (24, 25). The related benzo-heterocycle exists as a 1,3-bis-ylid 10, however, and is easily transformed into a cyclic diphosphonium-triple-ylid 11 (26).

This anion forms a large variety of complexes with uni- and bivalent metals. Among these are alkali and alkaline earth metals, but also Zn, Cd, Mn, Fe, Co and others.

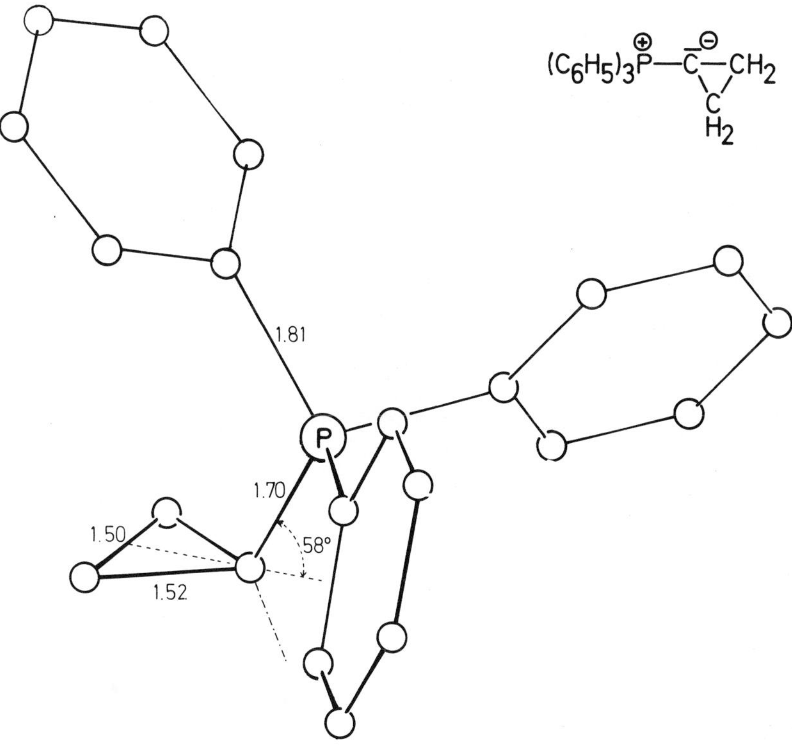

Figure 2. Molecular structure of triphenylphosphonium cyclopropylid 6 as determined by x-ray diffraction (14).

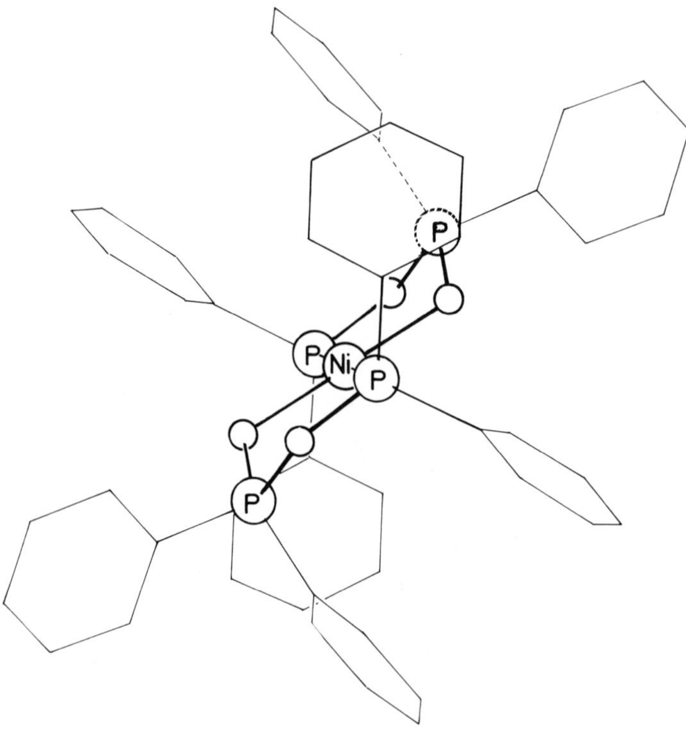

Figure 3. Molecular structure of the nickel complex 9 as determined by x-ray diffraction (18).

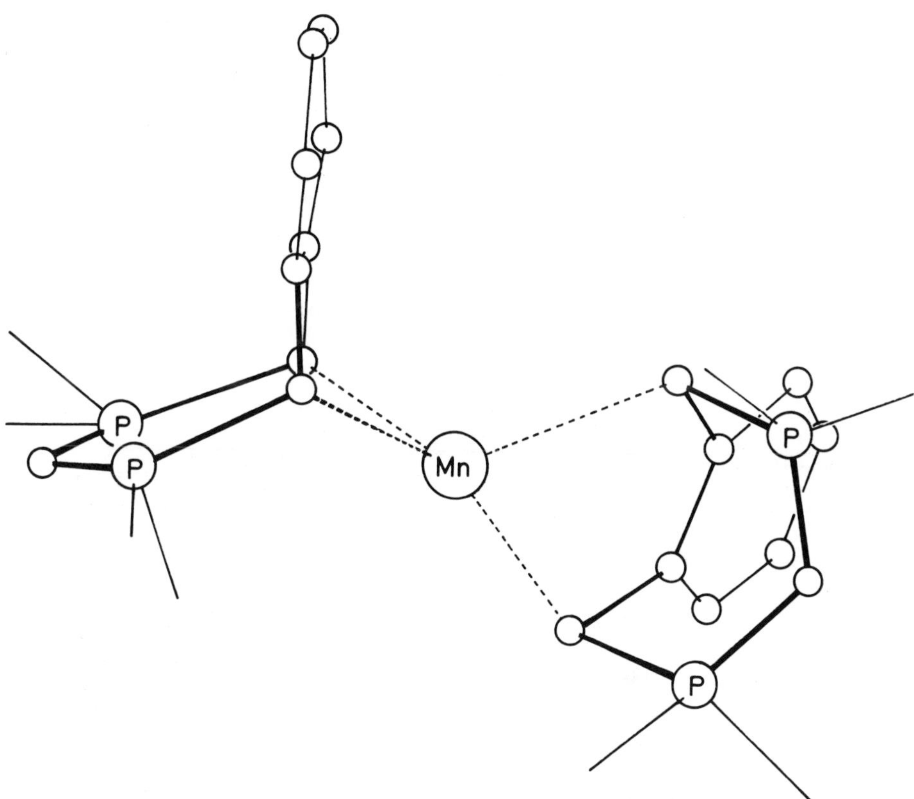

Figure 4. Molecular structure of the manganese complex 12 (M = Mn) as determined by x-ray diffraction (26).

Two crystal structures have been determined so far (25, 26), the most recent result showing the Mn(II) complex to be a quasi-sandwich compound with the metal being attached predominantly to the benzylic carbon atoms (26). As in the Cd(II) complex, the compound has a C_2 axis (Fig. 4) through the metal.

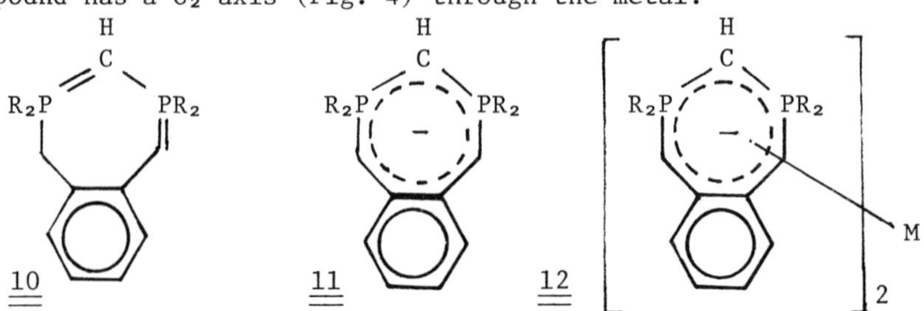

The coordination sphere of the metal is a strongly distorted tetrahedron. After the d^{10} and d^5 cases (Cd(II) and Mn(II), respectively), a number of other structures will have to be elucidated, before the metal-ligand bonding can be properly understood.

The author is indebted to Drs. U. Schubert and C.Krüger for vital assistance with X-ray studies. Support by Deutsche Forschungsgemeinschaft, Fonds der Chemischen Industrie, and Hoechst AG, Knapsack is gratefully acknowledged.

1. Wittig, G., Accounts Chem. Res. 1974, 7, 6.
2. Bestmann, H.J., Zimmermann, R., Fortschr. Chem. Forsch. 1971, 20, 1.
3. Schmidbaur, H., Accounts Chem. Res. 1975, 8, 62.
4. Schmidbaur, H., Pure & Applied Chem. 1978, 50, 19.
5. Schmidbaur, H., Pure & Applied Chem. 1980, 52, 1057.
6. Cramer, R.E., Maynard, R.B., Gilje, J.W., J. Amer. Chem. Soc. 1981, in press.
7. Baldwin, J.C., Keder, N.L., Strouse, C.E., Kaska, W.C., Z. Naturforsch. 1980, 35b, 1289.
8. Schwartz, J., Gell, K.I., J. Organometal. Chem. 1979, 184, C1; Inorgan. Chem. 1981, in press.
9. Krüger, C., Angew. Chem. Int. Ed. Engl. 1972, 12, 387.
10. Cramer, R.E., Maynard, R.B., Gilje, J.W., Inorg. Chem. 1980, 19, 2564.
11. Schmidbaur, H., Mörtl, A., Z. Naturforsch. 1980, 35b, 990.
12. Schmidbaur, H., Mörtl A., Zimmer-Gasser, B., Chem. Ber. 1981, in press.
13. Bestmann, H.J., Hartung, H., Pils, I., Angew. Chem. Int. Ed. Engl. 1965, 4, 957.
14. Schmidbaur, H., Schier, A., Milewski-Mahrla, B. Chem. Ber. 1981, in press.
15. Schmidbaur, H., Schier, A., Chem. Ber. 1981, in press.
16. Schmidbaur, H., Tronich, W., Chem. Ber. 1968, 101, 3545.
17. Issleib, K., Lindner, R., Lieb. Ann. Chem. 1967, 707, (1967), ref. therein.
18. Schmidbaur, H., Deschler, U., Milewski-Mahrla, B. Angew. Chem. 1981, in press.
19. Schmidbaur, H., Gasser, O., Krüger, C., Sekutowski, J.C., Chem. Ber. 1977, 110, 3517.
20. Karsch, H.H., Schmidbaur, H. Chem. Ber. 1974, 107, 3684.
21. Bassett, J.-M., Mandl, J.R., Schmidbaur, H. Chem. Ber. 1980, 113, 1145.
22. Schmidbaur, H., Deschler, U., Zimmer-Gasser, B., Neugebauer, D., Schubert, U., Chem. Ber. 1980 113, 902
23. Schmidbaur, H., Deschler, U., Zimmer-Gasser, B., Milewski-Mahrla, B., Chem. Ber. 1981, 114, 608.
24. Schmidbaur, H., Costa, T., Chem. Ber. 1981, in press.
25. Schmidbaur, H., Costa, T., Milewski-Mahrla, B., Schubert, U., Angew. Chem. Int. Ed. Engl. 1980, 19, 555.
26. Schmidbaur, H., Costa, T., Milewski-Mahrla, B., Chem. Ber. 1981, in press.

RECEIVED July 7, 1981.

7

Synthetic and Spectroscopic Investigations Involving α-Heterosubstituted Phosphonate Carbanions

HANS ZIMMER

Department of Chemistry, University of Cincinnati, Cincinnati, OH 45221

α-Heterosubstituted carbanions are generated by a base-inducted abstraction of a proton from a variety of substituted diethyl (or diphenyl) phosphonic acid esters

$$(RO)_2P(O)CHR'X \xrightarrow{B^{\ominus}} (RO)_2P(O)\bar{C}R'X \quad (I)$$

(R=Et, Ph; R'=alkyl or aryl groups; X=NHAr, $Nalk_2$, Cl, $OSiMe_3$ and $OC(O)Ph\text{-}4\text{-}NO_2$.)

The reaction of I (X-NHAr) with aromatic aldehydes, II, leads to enamines which easily hydrolyze to give deoxybenzoins.[1] Thus, reactions of I with ortho-nitrosubstituted upon hydrolysis and subsequent reduction of the nitro group to an amino group leads to formation of indols.[1] Similarly, reactions of I with ortho-methoxysubstituted II followed by hydrolysis of the enamine and ether cleavage yields benzofurans.[1]

I + R–C₆H₃(CH=O)–NO_2(OCH_3) —steps→ R–indole–Ar or R–benzofuran–Ar

Carbanions of type I derived of ortho-nitrocinnamaldehyde lead in an analogous sequence to 2-benzylquinoline.[2]

R–C₆H₄–CH=C(NHAr)–CH=CH–C₆H₄–NO_2 —1) $H_3O^{\oplus}$ 2) HI→ R–2-benzylquinoline

0097–6156/81/0171–0037$05.00/0

The reactions of I with phenylpropargylaldehyde leads to the expected yne-enamine which upon heating to about 50°C in solution undergoes secondary reactions with the main one being a cyclization to 2-phenyl-4-benzylquinolines.[1]

I + PhC≡C-CH=O ⟶ [O_2N-aryl-NH-C(Ph)=CH-C≡C-Ph] ⟶ [O_2N-substituted 4-benzyl-2-phenylquinoline]

Anions of type I (R'=aromatic, X=Cl) lead to formation of substituted vinyl chlorides, or when two equivalent of base in the H-abstraction step are used, to acetylenes[1]

$(RO)_2P(O)\bar{C}Cl(R)$ + R-C_6H_4-CH=O ⟶ R'ClC = CH-C_6H_4-R

2 equ. $B^{\ominus}$ ⟶ R'C≡C-C_6H_4-R

With I (X-Nalk$_2$; R'=H) the intermediate hydroxyphosphonate(III) is rather stable and undergoes the expected cycloelimination only poorly or not at all. However, III upon acidification undergoes a semi-pinacol type rearrangement to give aldehydes[3]

Ⓟ - CH_2NMe_2 —[1) $B^{\ominus}$; 2) RR'CH=O; 3) H_2O]⟶ H-C(Ⓟ)(NMe_2) - C(R)(OH)-R' —[$H^{\oplus}$]⟶ H - C(Ⓟ)(NMe_2) - C(R)($OH_2^{\oplus}$) - R' ⟶

$Me\overset{\oplus}{N}$=C(H) - C(R)R' —[$+H_2O$; $-H_2\overset{\oplus}{N}Me_2$]⟶ [O=C(H) - C(R)(P) - R'] —[$+H_3O^{\oplus}$; $-\oplus$ -OH]⟶ RR'CH-CH=O

Ⓟ = $(EtO)_2P(O)$-

In reactions with I (R'=aromatic; X=OCOCH$_3$, OCOPh-4-NO$_2$) enol ethers were obtained.[3]

$(RO)_2P(O)\overset{\ominus}{C}(R')$-O-COCH$_3$ + ArCH=O ⟶ ArCH=CR'OCOCH$_3$ —[$H_3O^{\oplus}$]⟶ ArCH$_2$C(O)R'

A totally different course was observed with type I anions (R= aromatic or aliphatic; X=$OSiMe_3$); the expected trimethylsilyl substituted enol ethers were not obtained. Instead, after a 1,4-O.O-trimethylsilyl migration and splitting off of $(RO)_2P\text{-}O^-$ benzoins or acyloines were obtained.[1]

$$(RO)_2P(O)\overset{\ominus}{C}(R)OSiMe_3 + R'R''CH{=}O \longrightarrow \left[(RO)_2P(=O)\text{-}C(R)(O\text{-}SiMe_3)\text{-}C(R')(R'')\text{-}O^{\ominus}Li^{\oplus} \right]$$

$$\longrightarrow R'R''C(OSiMe_3)\text{-}C(=O)\text{-}R \xrightarrow{H_2O} R'R''C(OH)\text{-}C(=O)\text{-}R$$

In all reactions involving I (X - NHAr) and R' an ortho-substituted benzaldehyde no condensation reaction was observed. An X-ray structure of such an α-heterosubstituted carbanion shows that the carbanionic site of such anions is rather effectively shielded against attack by an electrophile.

The intermediacy of enamines in the reaction involving $(EtO)_2P(O)^-CHNalk_2$ is currently questioned. It was thought that a ^{31}P-NMR study of the reaction could solve this question. The following observations were made.

$$(EtO)_2P(=O)\text{-}CH_2\text{-}NR_2 \;(A) \xrightarrow{B^{\ominus}} (EtO)_2P(=O)\text{-}\overset{\ominus}{C}H\text{-}NR_2 \;(B) \xrightarrow{Ph-C(=O)-Ph}$$

$$R_2 = \begin{matrix} -CH_2CH_2 \\ -CH_2CH_2 \end{matrix} > O$$

$$(EtO)_2P(=O)(\text{-}C(H)(NR_2)\text{-}C(Ph)(Ph)\text{-}O^{\ominus}) \;(C) \longrightarrow Ph_2C{=}C(H)NR_2 \xrightarrow{H_3O^{\oplus}} Ph_2C\text{-}H\text{-}C(H){=}O$$

$$+ \; (EtO)_2P(=O)\text{-}O^{\ominus} \quad (D)$$

A had a chemical shift of 22.4 δ;upon addition of base presence of B was indicated by a chemical shift of 49.6 δ. Addition of Ph_2CO to B yielded the expected adduct C with a chemical shift of 32.7δ. Quenching of C with 1 equivalent of H_3O^+ gave the expected hydroxyphosphonate derived of C (29.6δ). Upon standing for 12 hours only traces of D (chemical shift -1.3δ identical with authentic material) was observed. Also the sharp peaks due to B and C had disappeared. The lack of D is taken as an indication that only traces of an enamine are formed under these conditions.

Acknowledgement: Results for this presentation are taken from the Ph.D. theses by D.M. Nene, R.E. Koenigkramer, P.D. Seemuth and the proposed Ph.D. thesis by M.R. Crenshaw.

Literature Cited

1. Koenigkramer, R.E., Zimmer, Hans. J. Org. Chem. 1980, 45, 3994.
2. Nene, D.M. Ph.D. Thesis, University of Cincinnati, 1976.
3. Koenigkramer, R.E. Ph.D. Thesis, University of Cincinnati, 1979.
4. Broekhof, N.L.J.M., Jonkers, F.L. Van der Gen, A. Tetrahedron Letters, 1979, 2433.

RECEIVED September 14, 1981.

A New Approach to Activation of Hydroxy Compounds Using Pentacoordinated Spirophosphoranes

J. I. G. CADOGAN

B. P. Research Centre, Chertsey Road, Sunbury-on-Thames, Middlesex, TW16 7LN, England

I. GOSNEY, D. RANDLES, and S. YASLAK

Department of Chemistry, University of Edinburgh West Mains Road, Edinburgh, EH9 3JJ, Scotland

There are several examples in the literature of activation of hydroxy compounds by various organophosphorus compounds. For example, in the presence of pyridine, n-butyltriphenoxyphosphonium bromide (1) activates carboxylic acids towards reaction with amines or phenols to give, respectively, amides or esters (1). Such condensation reactions are also promoted by certain trivalent phosphorus compounds, e.g. triphenyl phosphite (2) or diphenyl ethylphosphonite (3), or to a lesser extent by phosphonate esters, e.g. diphenyl n-butylphosphonate (3). "Bates' reagent," μ-oxobis[tris(dimethylamino)phosphonium] bis-tetra-fluoroborate (2) may also be used to activate the carboxyl function towards amide bond formation during peptide synthesis (4) and to bring about the Beckmann rearrangement of ketoximes (5).

We wish to report a new approach to condensation reactions of hydroxy compounds related to the Ritter reaction, the Beckmann rearrangement and peptide formation based on easily accessible pentaco-ordinate spirophosphoranes of the type (3) (6 - 9).

Thus, reaction of equimolar amounts of 2-phenyl-2,2′-spirobis(1,3,2-benzodioxaphosphole) (3a) (6) and benzhydrol under anhydrous conditions in boiling acetonitrile gave N-benzhydrylacetamide (40%), bis(benzhydryl)ether (30%) and 2-phenyl-1,3,2-benzodioxaphosphole (4) which was identified by spectroscopic comparison with an authentic sample prepared from phenylphosphonic dichloride and catechol (10).

The formation of N-benzhydrylacetamide from benzhydrol and acetonitrile constitutes a Ritter reaction (11) which, in this case, is accomplished under extraordinarily mild, and above all neutral conditions. Usually, such reactions are carried out in

0097–6156/81/0171–0041$05.00/0

concentrated sulphuric acid. The formation of the by-product, bis(benzhydryl) ether, may be inhibited and the yield of Ritter product greatly increased (to 84%) by slow addition of the alcohol to an excess of the phosphorane (3a) in boiling acetonitrile, other phosphoranes giving similar results. Reaction as in Scheme 1 is assumed, whereby the first step (step A) involves nucleophilic substitution of benzhydrol for one of the catechol oxygens in the phosphorane (3a) to give the phosphorane intermediate (5). This activates the alcohol toward nucleophilic attack by either the solvent, acetonitrile (step B), or another molecule of benzhydrol (step C) to give, respectively, N-benzhydrylacetamide or bis(benzhydryl) ether. Nucleophilic attack on the intermediate (5) has a direct analogy in the reaction shown in Scheme 2, in which the phosphorane (6) acts as a powerful methylating agent towards carboxylic acids and phenols (12).

Equimolar quantities of benzhydrol and the phosphorane (3a) were also reacted in dimethyl sulphoxide solution. Apart from bis(benzhydryl) ether (18%) and catechol monobenzhydryl ether (39%), a small amount of benzophenone (17%) was obtained. It was shown that, in the absence of phosphorane, benzhydrol is not oxidised to benzophenone by dimethyl sulphoxide. Reaction similar to that outlined in Scheme 1 is a likely possibility. Again, the alcohol is activated by reaction with the phosphorane toward nucleophilic attack, in this case by dimethyl sulphoxide. Significantly, oxidation of alcohols by dimethyl sulphoxide is usually carried out using the Pfitzner-Moffatt reagent (dicyclohexylcarbodiimide and anhydrous phosphoric acid in dimethyl sulphoxide) (13) whereas the reaction using the phosphorane (3a) is carried out under neutral conditions. Unfortunately, however, attempts to improve the yield of benzophenone have hitherto failed.

Pentaco-ordinate spirophosphoranes of the type (3) are also capable of activating N-hydroxy functional groups towards Beckmann-type rearrangements. In a typical experiment, reaction of acetophenone oxime in boiling acetonitrile for several days in the presence of phosphorane (3a) afforded acetoacetanilide in 70% yield. Here again, the conditions employed are much milder than the usual conditions required for the Beckmann rearrangement (e.g. phosphorus pentachloride, concentrated sulphuric acid, polyphosphoric acid), and are comparable to the conditions required using Bates' reagent (5).

Preliminary results show that pentaco-ordinate phosphoranes are practical reagents for the formation of the peptide link. Thus, application of the phosphorane (3b) (7) in the stringent Izumiya test (14)

Scheme 1. R = Ph_2CH.

(A)

Ph, O, P, O, O, O → Ph, O, P, O, O⁻, ⁺O, R, H → (5)

ROH (3a)

(B)

Ph, O, P, O, O, HO, O, R, (5), MeC≡N → O, P, O, Ph (4) + OH, O⁻ + $Me\overset{+}{C}=NR$

OH, O⁻ + $Me\overset{+}{C}=NR$ → OH, O–C=NR, Me

acid workup

OH, OH + MeCONHR

(C)

Ph, O, P, O, O, HO, O, R, (5), ROH → O, P, O, Ph + OH, OH + ROR

Scheme 2. R= MeC = O; 2,4,6-trimethylbenzoyl; Ph.

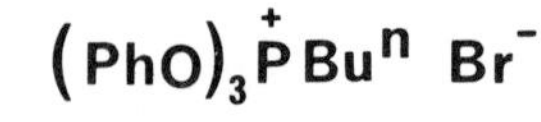

$(PhO)_3\overset{+}{P}Bu^n\ Br^-$

(1)

$(Me_2N)_3\overset{+}{P}-O-\overset{+}{P}(NMe_2)_3$

$2BF_4^-$

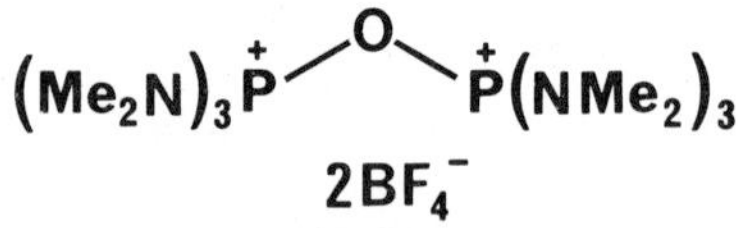

(2)

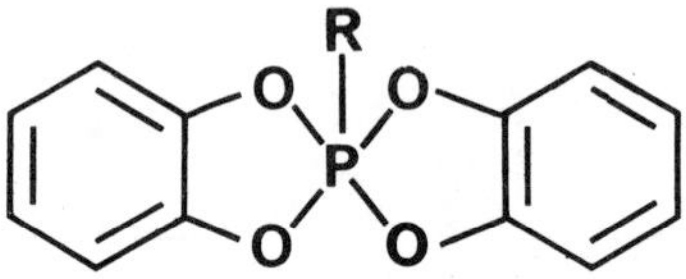

(3) a: R = Ph; b: R = H

c: R = Cl; d: R = PhO

Me

MeO

O

P

MeO

O

Me

RÖH

(6)

→ ROMe + $(MeO)_2\overset{O}{\overset{\|}{P}}CH_2CH_2\overset{O}{\overset{\|}{C}}Me$

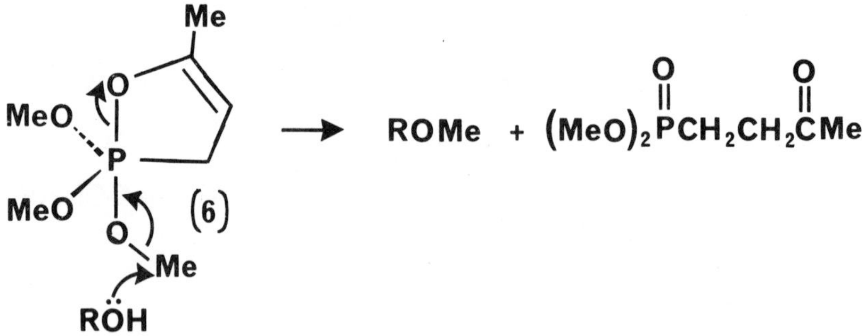

Table 1. Degree of racemisation (15) as measured by the Izumiya test.

Coupling agent	Conditions	Yield%	Racemisation%
Et_3N-THF	20°, 14 days	30	18
	60°, 1 day	50–60	20
	100°, 1 day	50–60	23
N-methylmorpholine (NMM)-THF	60°, 7 days	50–60	21
	40°, 7 days	50–60	12
	40°, 4 days	50–60	11
$(Me_2N)_3\overset{+}{P}\text{-O-}\overset{+}{P}(NMe_2)_3$ $2BF_4^-$ DMF, 20° (4)	Et_3N		43
	NMM		16
	poly Hünig base		9
	+ HOBt		<1

Z.Gly.(L)Ala.OH + H.(L)Leu.OBzl → Z.Gly.(DL)Ala.(L)Leu.OBzl

produced a degree of racemisation [DL/(DL + LL)] x 100, (see Table) comparable to that obtained using Bates' reagent in the absence of the additive, 1-hydroxybenzotriazole (HOBt), which forms activated esters of the carboxyl component not prone to racemisation. A detailed study is now in progress of the effect of bases, solvent, and additives on the viability of pentaco-ordinate phosphoranes as peptide coupling reagents.

Literature Cited

1. Yamazaki, N., Yamaguchi, M., Higashi, F., and Kakinaki, H., Synthesis, 1979, 355.
2. Yamazaki, N., and Higashi, F., Tetrahedron, 1974, 30, 1323.
3. Yamazaki, N., Niwano, N., Kawabata, J., and Higashi, F., Tetrahedron, 1975, 31, 665.
4. Bates, A.J., Galpin, I.J., Hallett, A., Hudson, D., Kenner, G.W., Ramage, R., and Sheppard, R.C., Helv. Chim. Acta, 1975, 58, 688.
5. Galpin, I.J., Gordon, P.F., Ramage, R., and Thorpe, W.D., Tetrahedron, 1976, 32, 2417.
6. Wieber, M., and Hoos, W.R., Monatsh. Chem., 1970, 101, 776.
7. Savignac, P., Breque, A., Bartet, B., and Wolf, R., Comptes Rendus, 1978, 287, 13.
8. Ramirez, F., Tetrahedron, 1968, 24, 5041.
9. Prepared in quantitative yield from (3c)(8) by reaction with phenol under nitrogen in dry methylene chloride over 10 min at room temperature; m.p. 109-112°C ($^{31}P_{\delta}$-29.7 p.p.m.; CH_2Cl_2).
10. Berlin, K.D., and Nagabhushanam, M., J. Org. Chem., 1964 29, 2056.
11. Krimen, L.I., and Cota, D.J., Org. React., 1969, 17, 213.
12. Voncken, W.G., and Buck, H.M., Rec. Trav. Chim., 1974, 93, 14.
13. Pfitzner, K.E., and Moffatt, J.G., J. Am. Chem. Soc., 1965, 87, 5661, 5670.
14. Izumiya, N., and Muraoka, M., J. Am. Chem. Soc., 1969, 91, 2391; Izumiya, N., Muraoka, M., and Aoyagi, H., Bull. Chem. Soc. Japan, 1971, 44, 3391.
15. Separation of diastereomic Gly.Ala.Leu was kindly carried out by Dr R.P. Ambler, Department of Molecular Biology, University of Edinburgh, using a Beckmann 120 C automatic analyser.

RECEIVED July 14, 1981.

9

Synthetic Applications of α-Amino Substituted Phosphine Oxides

A. VAN DER GEN and N. L. J. M. BROEKHOF

Gorlaeus Laboratories, Department of Organic Chemistry, University of Leiden, P.O. Box 9502, 2300 RA Leiden, The Netherlands

Enamines are highly valued intermediates in organic synthesis. Almost invariably they are prepared by reaction of a carbonyl compound with a secondary amine. In principle, another attractive route to enamines could be based on construction of the double bond by a Wittig- or Horner-Wittig reaction. The enamines 1 thus obtained could be easily converted into the corresponding homologous aldehydes, if desired fitted with an extra electrophilic substituent at the original carbonyl carbon atom.

$$>C{=}O \; + \; -\overset{+}{P}{-}\overset{-}{C}H{-}N< \;\longrightarrow\; >C{=}C(H){-}N< \;(\mathbf{1}) \;\xrightarrow{H_3O^+}\; >C(H){-}C(H){=}O$$

$$>C{=}C(H){-}N< \;(\mathbf{1}) \;\xrightarrow{1.\ E^+ \quad 2.\ H_2O}\; >C(E){-}C(H){=}O$$

Earlier attempts to effect carbonyl homologation in this way have met with little success.(1) The necessary presence of additional stabilizing substituents at the α-carbon atom lowered the reactivity of the anions to a level where only reactions with aromatic aldehydes gave reasonable yields. Recently, Martin and coworkers(2) have obtained interesting results with diethyl dialkylaminomethyl phosphonates. In their reaction sequences enamines, although not isolated, appeared to act as intermediates.

We have found that α-aminosubstituted diphenylphosphine oxides are excellent reagents for effecting the conversion of carbonyl compounds into their homologous enamines. For example, with the aid of morpholinomethyldiphenylphosphine oxide (2), aromatic as well as aliphatic and α,β-unsaturated aldehydes, can be converted into homologous enamines 1 in excellent yields. Reaction of 2 with n-BuLi in tetrahydrofuran (THF) at 0° gives a solution of the stable, orange colored anion. Upon reaction at -78° with 1 equiv of an aldehyde and quenching with ammonium chloride, the adducts 3 are obtained in almost quantitative yields. Treatment of these

0097-6156/81/0171-0047$05.00/0

$$\underset{H}{\overset{R}{>}}C{=}O \xrightarrow[\text{2. } NH_4Cl]{\text{1. } Ph_2P(O)\text{-}CH(Li)\text{-}N(CH_2CH_2)_2O} Ph_2P(O)\text{-}CH(N<)\text{-}CH(R)\text{-}OH \;(\underline{3}) \xrightarrow{KH} \underline{1}$$

adducts with potassium hydride leads to completion of the Horner-Wittig reaction to provide the desired enamines 1 in high yields. Some representative results are compiled in Table I.

Table I: Conversion of R^1R^2CO into Morpholino Enamines 1.

R^1	R^2	3 (mp)	1 yield	1 (mp/n_D)
p-BrC_6H_4-	H	176- 7°	93%	137- 8°
p-$CH_3C_6H_4$-	H	155- 8°	96%	68-70°
p-$OCH_3C_6H_4$-	H	146- 7°	91%	84- 6°
C_6H_5-	H	168-70°	99%	73- 5°
(E)-$C_6H_5CH{=}CH$-	H	167- 9°	---	---
(E)-$C_2H_5CH{=}C(CH_3)$-	H	oil	72%	n_D^{23}=1.5057
C_6H_5	C_6H_5	not isol.	90%	78-81°
$CH_3CH(CH_3)CH_2$-	H	149-50°	83%	n_D^{21}=1.4720
$CH_3CH_2CH(CH_2CH_3)$-	H	136- 7°	63%	n_D^{24}=1.4688

The results obtained by reaction of 2 with enolizable ketones are less satisfactory, because in that case the anion also acts as a base, converting part of the carbonyl compound into its lithium enolate. It was anticipated that the use of a less strongly basic, but still sufficiently nucleophilic anion could circumvent this problem. Indeed, anion 4, obtained from the corresponding *N*-methyl-anilinosubstituted phosphine oxide as described before, reacted smoothly with a large variety of carbonyl compounds at -78° to afford, after quenching with ammonium chloride, the adducts 5 in virtually quantitative yields. The much lower basicity of this anion is clearly reflected by the observation that no starting material is recovered, even with highly enolizable ketones such as cyclopentanone and acetophenone. Conversion of the adducts 5 into the *N*-methylanilino enamines 6 is conveniently effected by treatment with KOt-Bu in THF at ambient temperatures. A number or representative examples is compiled in Table II. Because aldehydes already gave good results with the morpholinosubstituted anion, only

$$R^1R^2C{=}O \xrightarrow[2.\ NH_4Cl]{1.\ Ph_2P(O)-CH(Li)-N(CH_3)Ph\ (\underline{4})} \underline{5} \xrightarrow{KOt\text{-}Bu} R^1R^2C{=}C(H_A)-N(CH_3)Ph\ \ \underline{6}$$

Table II: Conversion of R^1R^2CO into Enamines $R^1R^2C{=}C(H_A)N(CH_3)Ph$

R^1R^2CO	6 yield	E/Z ratio	H_A	n_D/mp
Cyclopentanone	86%		5.99 (m)	$n_D^{22}=1.5578$
Cyclohexanone	92%		5.73 (m)	$n_D^{25}=1.5628$
2-Me cyclohexanone	84%	72/28	E 5.75 (m) Z 5.69 (m)	$n_D^{23}=1.5598$
4-t-Bu cyclohexanone	86%		5.73 (m)	mp=66-67°
Adamantanone	85%		5.75 (s)	mp=86-87°
Cycloheptanone	84%		5.80 (m)	$n_D^{21}=1.5608$
Acetophenone	80%	76/24	E 6.42 (q) Z 6.20 (q)	$n_D^{22}=1.6368$
Benzophenone	81%		6.65 (s)	$n_D^{20}=1.6613$
Pentanone-3	87%		5.77 (m)	$n_D^{22}=1.5402$
Butanone-2	85%	50/50	E 5.82 (m) Z 5.74 (m)	$n_D^{21}=1.5418$
Cyclohex-2-enone	84%	95/5	E 5.94 (m)	$n_D^{22}=1.6133$
β-Ionone	81%	53/47	E 6.17 (m) Z 5.98 (m)	$n_D^{22}=1.5902$
2-Ethylbutanal	90%	65/35	E 6.52 (d, 14 Hz) Z 6.04 (d, 8 Hz)	$n_D^{22}=1.5458$

one example is included here. The requisite phosphine oxides are easily accessible by an Arbusov reaction of the *N*-methoxymethyl derivatives of the required amine with chlorodiphenyl phosphine.

$$Ph_2P-Cl + CH_3O-CH_2-N(R^3)(R^4) \longrightarrow Ph_2P(O)-CH_2-N(R^3)(R^4) + CH_3Cl$$

While studying the conversion of the adducts, obtained from aldehydes, it was discovered that upon thermolysis they lose diphenylphosphine oxide (Ph_2POH) to form α-aminomethyl ketones 7. The overall reaction thus results in α-aminomethylation of the parent aldehyde. This Ph_2POH-elimination is probably not an intramolecular *syn* process like the *N*-oxide elimination or the sulfoxide elimination,[3] but more likely follows an *E1*-pattern.

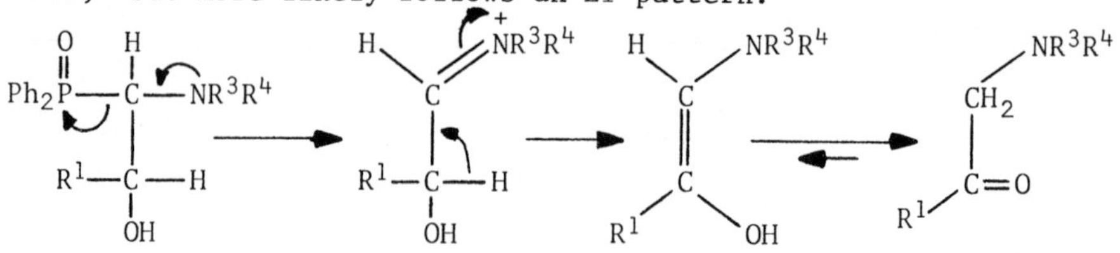

Representative results obtained with this method are summarized in Table III.

Table III: Conversion of R^1CHO into $R^1COCH_2NR^3R^4$

R^1	R^3NR^4	Yield	n_D/mp
C_6H_5	MeNPh	82%	mp=117.5-118°
$p\text{-}CH_3C_6H_4$	MeNPh	86%	mp= 86-87°
$p\text{-}CH_3OC_6H_4$	MeNPh	83%	mp=109-110°
C_6H_5	morpholino	79%	n_D^{21}=1.5661
C_6H_5	piperidino	77%	n_D^{21}=1.5401
$C_6H_5CH_2$	MeNPh	75%	n_D^{21}=1.5914
$n\text{-}C_7H_5$	MeNPh	82%	n_D^{19}=1.5178
$(C_2H_5)_2CH$	MeNPh	82%	n_D^{21}=1.5178

One known application of *N*-methylanilinomethyl ketones is their use in the Bischler indole synthesis.[4] We have found that the α-aminomethyldiphenylphosphine oxides with R^3= phenyl and R^4= methyl or benzyl, are excellent reagents for the synthesis of 3-substituted indoles, starting from aldehydes.

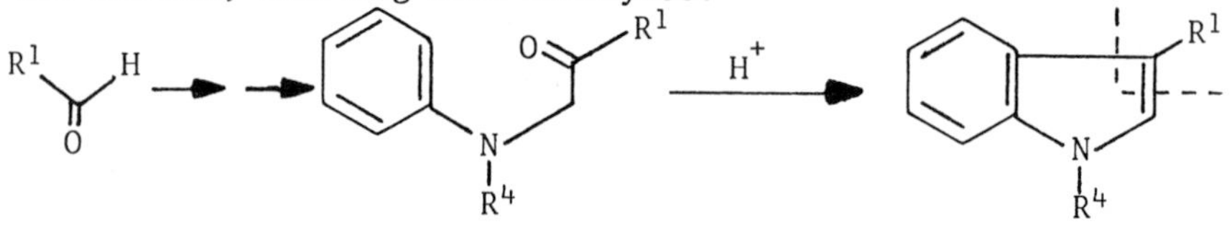

Finally, we wish to report that the Horner-Wittig reaction has been extended to α-alkyl and α-aryl substituted α-aminomethylphosphine oxides. After a facile hydrolysis, the homologous ketones are obtained in excellent yields (85-95%).

REFERENCES

1. H.Gross and W.Bürger, J.Prakt.Chem. 311,395(1969) and ref. cited.
2. S.F.Martin, T.-S.Chou and C.W.Payne, J.Org.Chem. 42,2520(1977).
3. R.D.Bach, D.Andrzejewski and R.Dusold,J.Org.Chem.38,1742(1973).
4. R.K.Brown in "The Chemistry of Heterocyclic Compounds", Vol. 25, Part I, pp 317-384.

RECEIVED July 7, 1981.

10

Reactions of Aziridines, 4-Oxazolines, and Their Derivatives with Alkylidene Phosphoranes and Phosphorus(III) Nucleophiles

M. VAULTIER and R. CARRIÉ

Groupe de Physicochimie Structurale, ERA CNRS N° 389, Université de Rennes, 35042 Rennes Cédex, France

The aziridines 1a, b and the oxazolines 2c, 2d and 2e, bearing at least one electron withdrawing group are thermically in equilibrium with the corresponding azomethine ylids 3 which may be protonated to give the functional iminium salts 4 according to the following scheme (1, 2) :

1a : Z=Ph ; 1b : Z=CO_2Me — 3 — 4

2c : R=Me, Y=CO_2Me
2d : R=Ph, Y=CO_2Me — 3 — 4
2e : R=Ph, Y=CN

0097-6156/81/0171-0051$05.00/0

We have shown that the ylids 3 react with a number of nucleophiles (3). In particular 3a and 3b react quantitatively with the alkylidene phosphoranes 5 to give 3-pyrrolines 6a and 6b as a mixture of diastereoisomers. The cyclisation occurs via an intramolecular Wittig reaction on the carbonyl of an ester group (4). In the same conditions, 2c reacts with 5 to give quantitatively the pyrrolines 7c (two diastereoisomers) as a result of a Wittig reaction on the carbonyl of the keto group in 3c.

However, the reactivity of β-ketophosphoranes 8 with R'=Ph or Me is different from the alkylidene phosphoranes 5 owing to the double bond character of the C-C(=O) bond due to an important contribution of the mesomeric structure 8'. As a consequence,

$$Ph_3P{=}CH{-}\underset{\underset{O}{\|}}{C}{-}R' \;(8) \longleftrightarrow Ph_3\overset{\oplus}{P}{-}CH{=}\underset{\underset{|\underline{O}|^{\ominus}}{|}}{C}{-}R' \;(8') \longleftrightarrow Ph_3\overset{\oplus}{P}{-}\overset{\ominus}{C}H{-}\underset{\underset{O}{\|}}{C}{-}R' \;(8'')$$

certain 1,3 dipoles (azides, nitrile imines, nitrile oxydes, nitrones) add either to the P=C or C=C double bond (5). We have shown that the azomethine ylids 3a and 3c add to the pseudo double bond of the β-ketophosphoranes 8, R' = Ph, R' = Me. This process may be competitive with the nucleophilic addition. This is outlined in the following scheme.

1a : Z=Ph, 1b : Z=CO_2Me

$ZCH—C(CO_2Me)_2$ (aziridine, N—Ph) **1** ⇄ $ZCH \overset{4e}{\frown} C(CO_2Me)_2$ (N—Ph) **3**

+

$Ph_3P{=}CHCOR'$ ⟷ $Ph_3\overset{\oplus}{P}—CH{=}C(—R')—\overline{\underline{O}}|^{\ominus}$

R' = Ph   R' = Me

cycloaddition

a1 → $R'—C(O^{\ominus})—CH(\overset{\oplus}{P}Ph_3)$, ZCH—N(Ph)—$C(CO_2Me)_2$ → $-Ph_3PO$ → $R'—C{=}CH$, ZCH—N(Ph)—$C(CO_2Me)_2$ **9**

a2 → $Ph_3\overset{\oplus}{P}—CH—C(—R')—O^{\ominus}$, ZCH—N(Ph)—$C(CO_2Me)_2$ → $-Ph_3PO$ → $HC{=}C—R'$, ZCH—N(Ph)—$C(CO_2Me)_2$ **10**

(B) nucleophilic addition

→ $Ph_3\overset{\oplus}{P}—CH—COR'$, ZCH—N(Ph)—$\overset{\ominus}{C}(CO_2Me)_2$ → $R'C(=O)—C(=PPh_3)—ZCH—N(Ph)—CHCO_2Me—C(=O)—OMe$ → $R'C(=O)—C(\overset{\oplus}{P}Ph_3)—C(O^{\ominus})(—OMe)$, ZCH—N(Ph)—$CHCO_2Me$ → $R'C(=O)—C{=}C—OMe$, ZCH—N(Ph)—$CHCO_2Me$ **6**

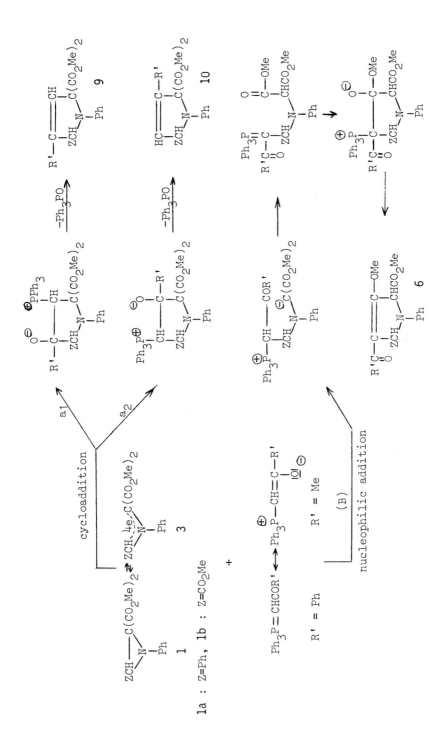

Three routes a_1, a_2 and B are observed leading to the 3-pyrrolines 9, 10 and 6 respectively.

The reaction of functionalized iminium salts 4 derived from aziridines with trimethylphosphite gives aminophosphonates 11 by a Michaelis-Arbuzov rearrangement. This reaction is general(6).

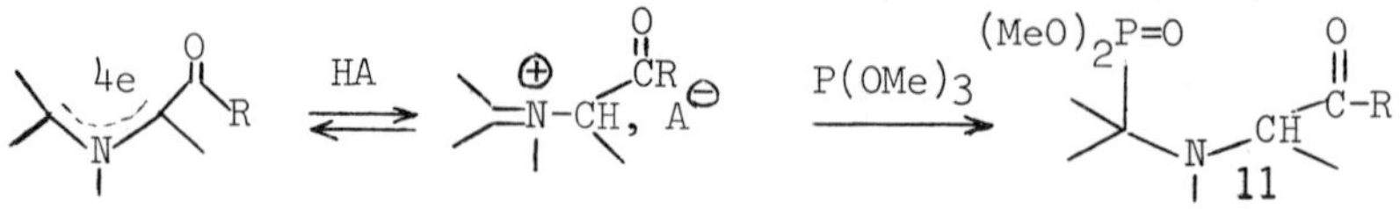

With those derived from 4-oxazolines and orthoaminophenols, depending on the reaction conditions, linear or cyclic aminophosphonates of type 11 or 12 are obtained.

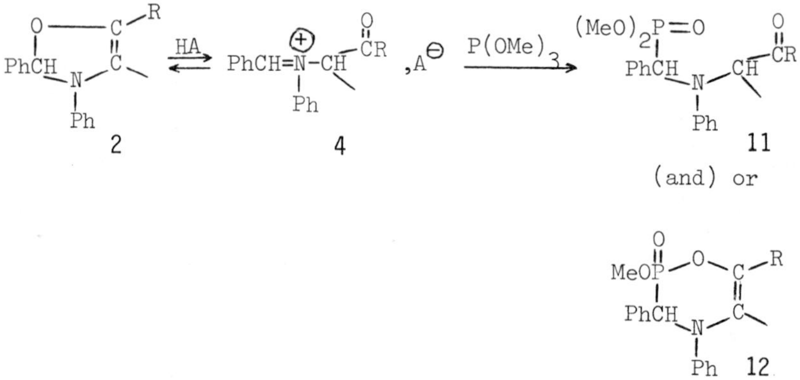

References

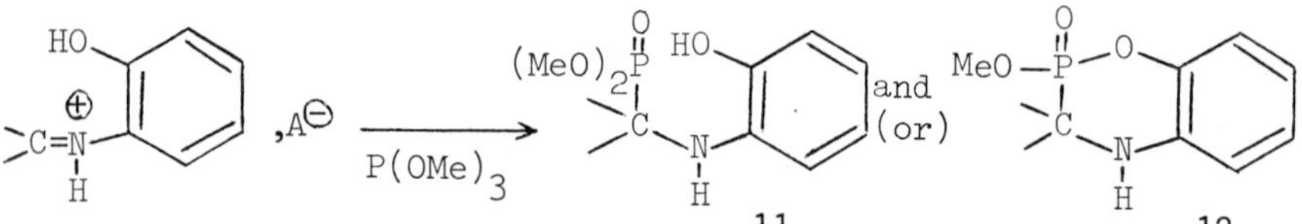

1. Vaultier, M. ; Carrié, R. Tetrahedron, 1979, 35, 1357.
2. Vaultier, M. ; Mullick, G. ; Carrié, R. Can.J.Chem., 1979, 57, 2876.
3. Vaultier, M. ; Danion-Bougot, R. ; Danion, D. ; Hamelin, J. and Carrié, R. J. Org. Chem., 1975, 40, 2990.
4. Vaultier, M. ; Danion-Bougot, R ; Danion, D., Hamelin, J. and Carrié, R. Bull. Soc. Chim., 1976, 1537.
5. Bestmann, H.J. and Kunstmann, R. Chem.Ber., 1969, 102, 1816. Huisgen, R. and Wulff, J. Chem.Ber., 1969, 102, 1841. Ykman, P.; L'Abbé, G. and Smets, G. Tetrahedron, 1973, 29, 195.
6. Vaultier, M. ; Ouali, M.S. and Carrié, R. Bull.Soc.Chim., II, 1980, 633.

RECEIVED July 7, 1981.

Phosphonates Containing Sulfur and Selenium: Synthesis, Reactions, and New Applications

M. MIKOŁAJCZYK, S. GRZEJSZCZAK, W. MIDURA, M. POPIELARCZYK, and J. OMELAŃCZUK

Center of Molecular and Macromolecular Studies, Polish Academy of Sciences, Department of Organic Sulfur Compounds, 90-362 Łódź, Boczna 5, Poland

The ambident behaviour of sulfur in terms of its ability to stabilize adjacent negative and positive charge at carbon has made this element a focal point for new reagents and reactions (1). For this reason our concern has focused on the synthesis of new phosphonates containing sulfur and selenium and the exploration of their application in synthetic and stereochemical studies.

It was found (2,3) that the best synthetic approach to phosphonates containing sulfur and selenium in the α-position is based on the reaction of elemental sulfur and selenium with α-phosphonate carbanions. It is interesting to note that, in contrast to sulfur, selenium may be added in this way to the tertiary carbon atom.

$$(RO)_2\overset{O}{\overset{\|}{P}}-\bar{C}H-R \xrightarrow{1.\,S_8;\ 2.\,R'X} (RO)_2\overset{O}{\overset{\|}{P}}-\underset{}{CH(SR')}-R$$

$$(RO)_2\overset{O}{\overset{\|}{P}}-\bar{C}H-R \xrightarrow{1.\,Se;\ 2.\,R'X} (RO)_2\overset{O}{\overset{\|}{P}}-CH(SeR')-R$$

R = alkyl, aryl
R' = alkyl

In an extension of this work a new route to unsaturated six-membered sulfur heterocycles 1 has been investigated which consists in the internal Horner-Wittig reaction of the appropriate α-phosphoryl sulfoxides or sulfones 2 prepared as shown in Scheme I.

Thus, the sodium salts of α-phosphoryl thiols 3 |R=H(a) R=Me (b)| were reacted with ethylene ketal of 5-chloro-pentanone-2 in ethanol solution under reflux to give the corresponding condensation products 4 in high yields (80-90%). Their oxidation followed by deblocking of the carbonyl group afforded sulfoxides 2,a,b and sulfones 2,c,d which on treatment with n-butyllithium were converted into the desired unsaturated thian-S-oxides or S-dioxides 1.

0097–6156/81/0171–0055$05.00/0

Scheme I

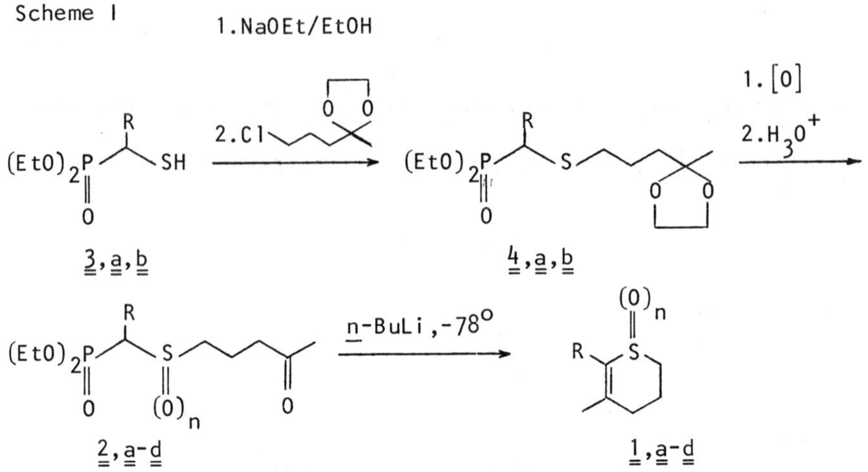

1.NaOEt/EtOH

2.Cl

$(EtO)_2P(O)CH(R)SH$ (**3**, **a**, **b**) → $(EtO)_2P(O)CH(R)S$– (**4**, **a**, **b**)

1.[O]

2.H_3O^+

$(EtO)_2P(O)CH(R)S(O)_n$–(CO) (**2**, **a**–**d**)

n-BuLi, -78°

(O)n, R, S (**1**, **a**–**d**)

		Yield (%)		Yield (%)
a, R=H,	n=1	91	**a**,	60
b, R=Me,	n=1	90	**b**,	68
c, R=H,	n=2	70	**c**,	73
d, R=Me,	n=2	81	**d**,	84

The next part of the present study concerns the oxidation of the phosphonate and thiophosphonate carbanions. Generally, this reaction was found to occur in two directions depending on the structure of the starting phosphonate and the reaction conditions. The lithium salts of thiophosphonate carbanions **5** give on treatment with oxygen the corresponding α-hydroxy thiophosphonates **6**, exclusively.

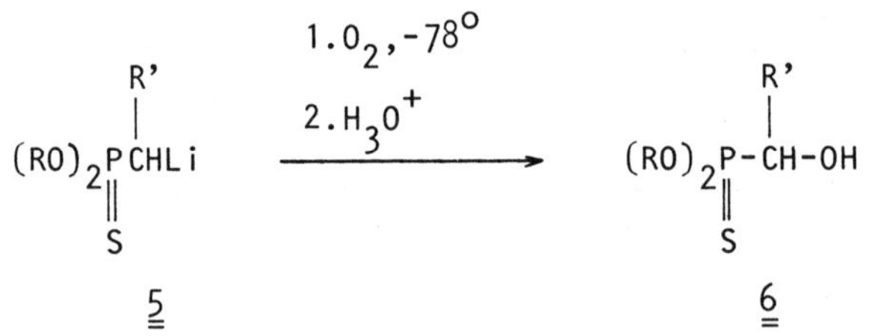

$(RO)_2P(S)CHLi(R')$ (**5**)

1.O_2, -78°

2.H_3O^+

→ $(RO)_2P(S)$-CH(R')-OH (**6**)

		Yield (%)	$\delta_{^{31}P}$
a, R=Me,	R'=H	58	91.2
b, R=Me,	R'=Me	55	99.8
c, R=Et,	R'=Me	67	96.3

As expected, the reaction of the lithium salts of phosphonates 7 with oxygen results in the formation of dialkyl phosphoric acids 8 and carbonyl products i.e. the cleavage of the C-P bond takes place (4,5). However, when the halomagnesium salts of phosphonates 7 were oxidized, α-hydroxyphosphonates 9 were formed.

Scheme II

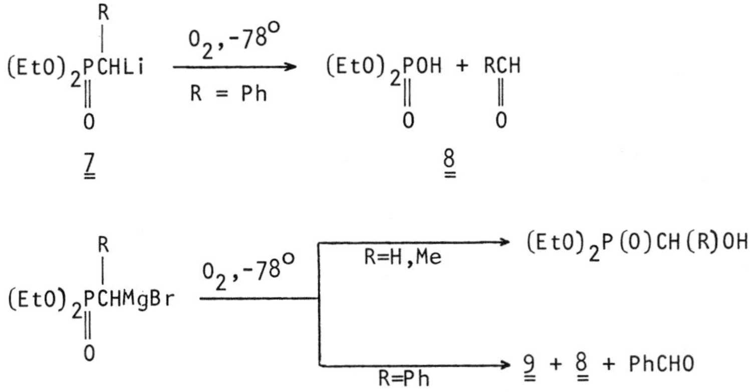

$$\underset{\mathbf{7}}{(EtO)_2P(O)CH(R)Li} \xrightarrow[R\,=\,Ph]{O_2,\,-78^{o}} \underset{\mathbf{8}}{(EtO)_2P(O)OH} + RC(O)H$$

$$(EtO)_2P(O)CH(R)MgBr \xrightarrow{O_2,\,-78^{o}} \begin{cases} \xrightarrow{R=H,Me} (EtO)_2P(O)CH(R)OH \\ \xrightarrow{R=Ph} \mathbf{9} + \mathbf{8} + PhCHO \end{cases}$$

The bidirectional course of the oxidation may be rationalized by assuming that the peroxide intermediate A formed in the first reaction step undergoes the intramolecular decomposition and/or intermolecular reaction as shown in Scheme III.

Scheme III

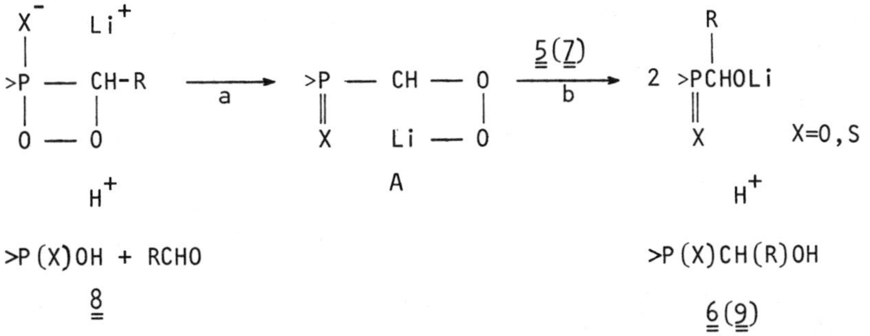

$$>\!P(X^-)(-O-O-)CH\text{-}R\;Li^+ \xrightarrow{a} \underset{\mathbf{A}}{>\!P(=X)-CH-O-O-Li} \xrightarrow[b]{\mathbf{5}(\mathbf{7})} 2\; >\!P(=X)CH(R)OLi \qquad X=O,S$$

$$H^+:\quad \underset{\mathbf{8}}{>\!P(X)OH + RCHO} \qquad\qquad H^+:\quad \underset{\mathbf{6}(\mathbf{9})}{>\!P(X)CH(R)OH}$$

With regard to the phosphorus stereochemistry, a new stereospecific synthesis of chiral O,O-dialkyl thiophosphoric acids (6,7) has been developed which is based on the Horner-Wittig reaction of the optically active phosphonothionate carbanions containing the sulfoxide or dithioacetal moieties. The transformation of (R)-(+) O-methyl O-isopropyl methanephosphonothionate (10) (8) into (R)-(-) O-isopropyl phosphorothioic acid (11) best illustrates this method.

Scheme IV

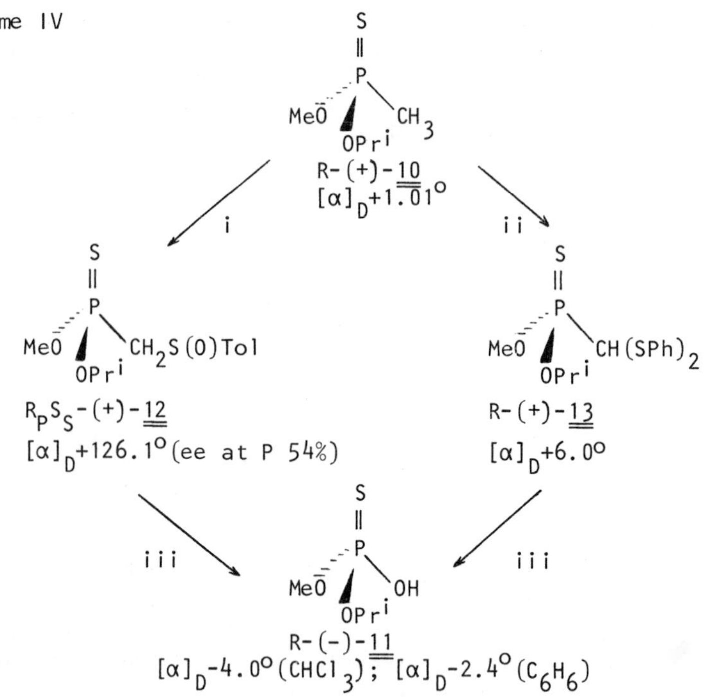

(i) n-BuLi, (-)TolS(O)OMen; (ii) n-BuLi, $(PhS)_2$; (iii) n-BuLi, PhCHO

In a similar way the absolute configuration R at phosphorus was assigned to (-) O-ethyl O-isopropyl phosphorothioic acid (14), (-) O-*n*-butyl O-isopropyl phosphorothioic acid (15) and (+) O-methyl O-naphthyl phosphorothioic acid (16). The correctness of the configurational assignments was checked by X-ray analysis of (+)-S-methyl O-methyl O-naphthyl phosphorothiolate (17) by means of the Bijvoet method.

Literature Cited

1. Block, E. "Reactions of Organosufur Compounds"; Academic Press New York, 1978.
2. Mikołajczyk, M.; Grzejszczak, S.; Chefczyńska, A. and Zatorski, A. J.Org.Chem. 1979, 44, 2967.
3. Mikołajczyk, M.; Grzejszczak, S.; and Korbacz, K. Tetrahedron Letters in press.
4. Davidson, A.H. and Warren, S. J.Chem.Soc.Chem.Commun. 1975, 148.
5. Zimmer, H.; Koenigkramer, R.E.; Cepulic, R.L.; and Nene, D.M. J.Org.Chem., 1980, 45, 2018.
6. Mikołajczyk, M. and Leitloff, M. Russ.Chem.Rev. 1975, 44, 670.
7. Hall, C.R.; Inch, T.D. J.Chem.Soc.Perkin I, 1978, 1104.
8. Mikołajczyk, M.; Omelańczuk, J.; Para M. Tetrahedron, 1972, 3855.

RECEIVED June 30, 1981.

12

Umpolung of α,β-Ethylenic Ketones and Aldehydes by Phosphorus Groups

H. J. CRISTAU, J. P. VORS, Y. BEZIAT, C. NIANGORAN, and H. CHRISTOL

Laboratoire de Chimie Organique ENSCM, (Equipe de Recherche Associée au CNRS N° 610), 8, rue de l'Ecole Normale, 34075 Montpellier, France

Reactivity umpolung and synthetic equivalents are very fruitful concepts in modern organic synthesis, in particular for the multistep preparation of natural compounds. In this area, organophosphorus compounds can be useful tools, particularly for reversible umpolung by heteroatom exchange (1-4).

Our objective was to prepare a phosphorus equivalent of β-acylvinyl anion in order to obtain an anomalous nucleophilic reactivity on the vinylic carbon β to the carbonyl group in α,β-ethylenic ketones and aldehydes. Examples of such umpolung (5-21), including the use of phosphorus groups (22-27), are described in the literature. Except for a special case (25), a Wittig or a Wittig-Horner reaction is used to remove the phosphorus group, in such a way that the carbonyl compound, which functions as an oxidative alkylating reagent, is the only electrophilic counterpart to be used for the β-acylvinyl anion equivalent.

We present here a reversible umpolung of α,β-ethylenic ketones and aldehydes, in which the ylides **3** are synthetic equivalents of β-acylvinylanions in two unlike pathways (i) and (ii) (Scheme I, see next page).

The preparation of the γ-thioacetalated phosphonium salts **2** takes place in a one-pot reaction by addition of triphenylphosphine hydrobromide on the α,β-unsaturated carbonyl compound and subsequent thioacetalation (28, 29) :

$R^1CH{=}CH{-}C(O){-}R^2$ (**1**) $\xrightarrow[\text{2) HS-(CH}_2)_3\text{-SH (-H}_2\text{O) cat. HBr, CHCl}_3\text{, 24h, 60°C}]{\text{1) Ø}_3\text{P, HBr/CHCl}_3\text{, 4h, 20°C}}$ $R^1CH(\overset{+}{P}Ø_3){-}CH_2{-}C(R^2)(S(CH_2)_3S)$ Br^- (**2**)

R^1	H	Me	Ø	Me	Ø	$-(CH_2)_3-$	H	Me	Ø
R^2	Me	Me	Me	Ø	Ø		H	H	H
Yield (%)	75	87	88	63	76	98	69	97	80

0097-6156/81/0171-0059$05.00/0

Scheme I.

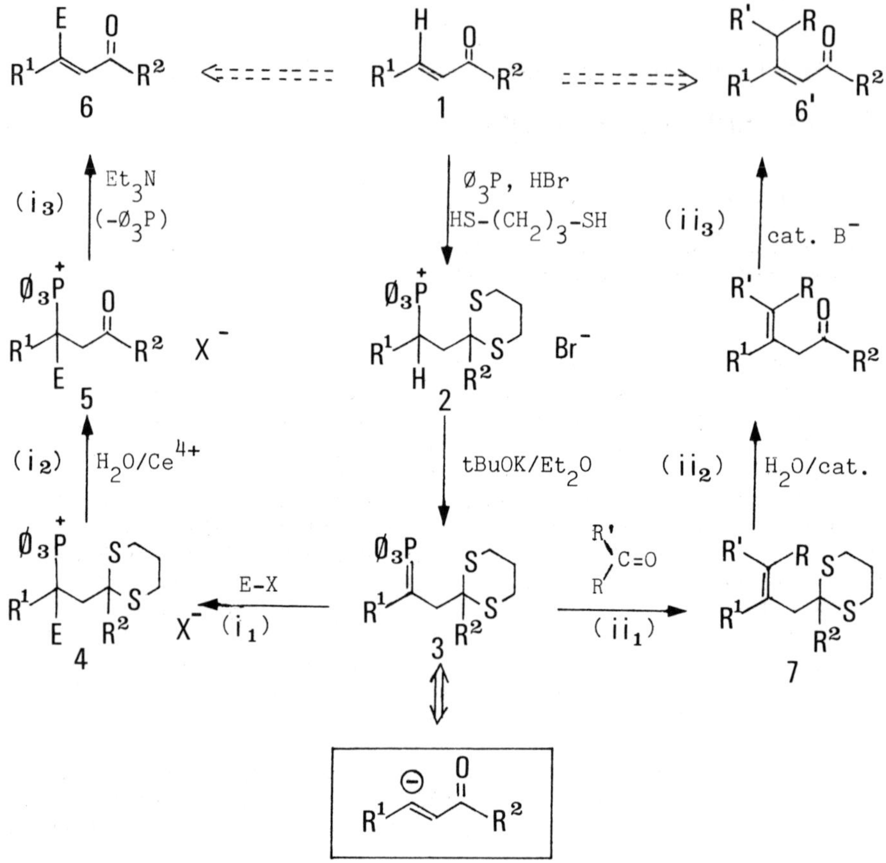

In all cases, the ylides **3** are easily generated from the corresponding phosphonium salts **2** by reaction with one equivalent of potassium tert-butoxide in ether at room temperature.

There are then two different pathways for transformation of these ylides :

First pathway.

In the first step (i_1) the ylides **3** (R^1 = H) can be alkylated with primary alkylhalogenide or with ethyl bromoacetate ; they can also be acylated with acetic anhydride in a reverse addition operation, (In spite of trans-ylidation process, the acylated phosphonium salt can be easily purified by liquid -liquid extraction with bases). Semi-stabilized ylides **3** (R^1 = Ø) are not reactive enough to be alkylated or acylated in this way.

$\emptyset_3P$ ... 3 → $(MeCO)_2O/(Et_2O$, reflux) → 4 (X^-)

(R^1= H, R^2 = Me, **4a**: 68 %)

(R^1= Me, R^2 = Me, **4b** : 47 %)

3 → $(RCH_2X)/(DMSO$, 25°C) → (X^-)

(for example : R^1= H, R^2= Me) ⇒

R	Me	Et	Et	nBu	(CH_2CO_2Me
X	I	Br	I	Br	Br
(%)yield	76	53	72	70	69)

For the thioacetal cleavage in the second step (i_2) a classical cerium(IV)-catalysed hydrolysis gives good yields (> 85 %) with any phosphonium salts we have tested. Subsequent elimination of triphenylphosphine by reaction of the salts 5 with triethylamine was always nearly quantitative.

Second pathway.

With non-stabilized ylides (R^1 = H or Me) and numerous saturated or unsaturated aldehydes and ketones the Wittig reaction (ii_1) gives the corresponding dithianes 7 in good yields (70-93 %). The autoxidation of these ylides gives the expected products 8 and 9 . The semi-stabilized ylides (R^1 = Ø) react only with a reactive aldehyde.

$\emptyset_3P$ ... → R'(R)C=O → 7

R'= H : $4h/Et_2O/20°C$

R'≠ H : 24h/glyme/reflux

→ O_2/N_2 →

R^1= H, R^2= Me (81 %) → 8

R^1= Me, R^2= Me (53 %) → 9

The last steps of the second pathway are well known problems of organic chemistry which have satisfactory solutions. Therefore we have not tested them.

In conclusion, the two pathways of our umpolung method can be satisfactorily applied to the α,β-ethylenic aldehydes or ketones bearing no substituent or an alkyl at the vinylic carbon β to the carbonyl group ; but, when an aryl group substitutes this position, only the second pathway is possible with some reactive aldehydes.

Literature Cited.

1. Seebach, D. ; Enders, D. Angew. Chem. Int. Ed. 1975, 14, 15.
2. Lever, D.W. Tetrahedron. 1976, 32, 1943.
3. Gröbel, B.T. ; Seebach, D. Synthesis. 1977, 357.
4. Seebach, D. Angew. Chem. Int. Ed. Engl. 1979, 18, 239.
5. Piers, E. ; Morton, H.E. J. Org. Chem. 1979, 44, 3437.
6. Hori, I. ; Hayashi, T. ; Midorikawa, M. Synthesis. 1975, 727.
7. Nakai, T.; Shiono, H. ; Okawara, M. Tetrahedron Lett. 1974, 3625.
8. Hirai, K. ; Kishida, Y. Tetrahedron Lett. 1972, 2743.
9. Reich, H.J. J. Org. Chem. 1975, 40, 2570.
10. Corey, E.J. ; Erickson, B.W. ; Noyori, R. J. Am. Chem. Soc. 1971, 93, 1724.
11. Oshima, K. ; Yamamoto, H. ; Nozaki, H. Bull. Chem. Soc. Jpn. 1975, 48, 1567.
12. Carlson, R.G. ; Mardis, W.S. J. Org. Chem. 1975, 40, 817.
13. Cohen, T. ; Bennett, D.A. ; Mura, A.J. J. Org. Chem. 1976, 41, 2506.
14. Grayson, J.I. ; Warren, S. J. Chem. Soc. Perkin I. 1977, 2263.
15. Kondo, K. ; Tunemoto, D. Tetrahedron Lett. 1975, 1007.
16. Julia, M. ; Badet, B. Bull. Soc. Chim. Fr. 1975, 1363.
17. Cooper, G.K. ; Dolby, L.J. ; Farnham, T. J. Org. Chem. 1977, 42, 1349.
18. Debal, A. ; Cuvigny, T. ; Larcheveque, M. Tetrahedron Lett. 1977, 3187.
19. Corey, E.J. ; Ulrich, P. Tetrahedron Lett. 1975, 3685.
20. Leroux, Y. ; Mantione, R. Tetrahedron Lett. 1973, 2586.
21. Clinet, J.C. ; Linstrumelle, G. Tetrahedron Lett. 1978, 1137.
22. Stowell, J.C. ; Keith, D.R. Synthesis. 1979, 132.
23. Bell, A. ; Davidson, A H. ; Earnshaw, C. ; Norrish, H.K. ; Torr, R.S. ; Warren, S. J. Chem. Soc. Comm. 1978, 988.
24. Blatcher, P. ; Grayson, J.I. ; Warren, S. J. Chem. Soc. Comm. 1978, 657.
25. Corbel, B. ; Paugam, J.P. ; Dreux, M. ; Savignac, P. Tetrahedron Lett. 1976, 835.
26. Martin, S.F. ; Garrison, P.J. Tetrahedron Lett. 1977, 3875.
27. Shevchuk, M.I. ; Megera, I.V. ; Burachenko , N.A. ; Dombrovskii, A.V. Zhur. Org. Khim. 1974, 10. 167.
28. Cristau, H.J. ; Vors, J.P. ; Christol, H. Synthesis, 1979, 538.
29. Cristau, H.J. ; Vors, J.P. ; Christol, H. Tetrahedron Lett. 1979, 2377.

RECEIVED June 30, 1981.

BIOCHEMISTRY OF PHOSPHORUS COMPOUNDS

Dedicated to Frank H. Westheimer

13

Monomeric Metaphosphates in Enzymic and in Enzyme-Model Systems

F. H. WESTHEIMER

James Bryant Conant Laboratory of Chemistry, Harvard University, 12 Oxford Street, Cambridge, MA 02138

Much of the chemistry of phosphate esters can best be interpreted in terms of monomeric metaphosphate (1). These unstable intermediates can be generated, among other ways, by the Conant-Swan fragmentation (2,3,4) of β-halophosphonates, e.g.:

$$C_6H_5CBr(CH_3OPO_2^-)-CHBr-CH_3 \xrightarrow{base} C_6H_5CBr=CHCH_3 + Br^- + CH_3OPO_2 \quad (1)$$

$$C_6H_5CBr(PO_3^=)-CHBr-CH_3 \xrightarrow{base} C_6H_5CBr=CHCH_3 + Br^- + PO_3^- \quad (2)$$

In the presence of water or other nucleophiles, monomeric metaphosphates yield orthophosphates

$$CH_3OPO_2 + H_2O \longrightarrow CH_3OPO_3H^- + H^+ \quad (3)$$

$$PO_3^- + H_2O \longrightarrow H_2PO_4^- + H^+ \quad (4)$$

The evidence for the mechanism shown above derives from rate data. Electrically neutral dibromophenylpropylphosphonic acids are stable; further, the dianion (eq. 2) undergoes fragmentation 10^5 times as fast as does the monoester monoanion (eq. 1).

The faster rate for the dianion is consistent with direct fragmentation, which disperses, and therefore is accelerated by, negative charge; the faster rate for the dianion is inconsistent with nucleophilic attack upon the phosphonates, since such reactions (5,6,7) are slowed by factors of 10^2-10^5 by a negative charge.

$$C_6H_5-C(Br)(O=P(O^-)-O^-)-CH(Br)-CH_3 \qquad C_6H_5-C(Br)(H{:}\ddot{O}(H){:}\rightarrow PO_3^=)-CH(Br)-CH_3 \quad (5)$$

0097-6156/81/0171-0065$05.00/0

Enol Phosphates

When the fragmentations yielding monomeric methyl metaphosphate (8,9) or metaphosphate ion (10) are carried out in the presence of 2,2,6,6-tetramethylpiperidine in acetophenone as solvent, the major products are the enol phosphates. Presumably the processes take place by initial attack of the monomeric metaphosphates on the carbonyl group of the ketone.

$$PO_3^- + C_6H_5\text{-}\overset{\overset{O}{\|}}{C}\text{-}CH_3 \rightleftharpoons C_6H_5\text{-}C(=\overset{+}{O}\text{-}PO_3^{=})\text{-}CH_3 \xrightarrow{\text{Base}} C_6H_5\text{-}C(\text{O-}PO_3H^-)\text{=}CH_2 \quad (6)$$

$$CH_3OPO_2 + C_6H_5\text{-}\overset{\overset{O}{\|}}{C}\text{-}CH_3 \rightarrow C_6H_5\text{-}C(=\overset{+}{O}\text{-}PO_2^-\text{-O-}CH_3)\text{-}CH_3 \xrightarrow{\text{Base}} C_6H_5\text{-}C(\text{O-}PO_2^-\text{-O-}CH_3)\text{=}CH_2 \quad (7)$$

Exchange of C=O for C=N

Monomeric metaphosphate ion will also activate the carbonyl group of esters (10). In particular, it promotes the reaction of ethyl acetate with aniline to yield O-ethyl-N-phenylacetimidate. The reaction is complete within seconds, whereas, in the absence of monomeric metaphosphate, the reaction between ethyl acetate and aniline is slow, and yields acetanilide.

$$PO_3^- + CH_3\text{-}C(=O)OC_2H_5 \rightleftharpoons CH_3\text{-}C(=\overset{+}{O}\text{-}PO_3^{=})OC_2H_5 \xrightarrow{C_6H_5NH_2} CH_3\text{-}C(\text{O-}PO_3^{=})(OC_2H_5)(\overset{+}{N}H_2C_6H_5) \xrightarrow{\text{Base}} CH_3\text{-}C(OC_2H_5)\text{=}N\text{-}C_6H_5 + H_2PO_4^- \quad (8)$$

Enzymology

Some of the reactions of PO_3^- parallel enzymatic reactions promoted by adenosine triphosphate (ATP). Pyruvate kinase catalyzes the equilibration of ATP and pyruvate with adenosine diphosphate (ADP) and phosphoenol pyruvate (11,12). In a formal sense, this reaction resembles the preparations of enol phosphate (eqs. 6 and 7). Cytidine triphosphate synthetase catalyzes the reaction of uridine triphosphate with ammonia to yield cytidine triphosphate (13). In a formal sense, this reaction resembles the replacement of the ester carbonyl group of ethyl acetate by the nitrogen of aniline (eq. 8).

Despite these resemblances, questions must necessarily arise as to the detailed mechanisms of the enzymic processes. Does ATP dissociate to yield monomeric metaphosphate? Does ATP (by way of monomeric phosphate or by direct reaction) attack carbonyl groups in pyruvate, or UTP (or other biochemical substrates) to yield activated intermediates similar to the zwitterions postulated in eqs. 6, 7 and 8? Each of the enzymic processes here cited could occur by either of two general pathways shown in eqs. 9 and 10, and might or might not involve monomeric metaphosphate.

$$CH_3-\overset{O}{\overset{\|}{C}}-CO_2^- \rightleftharpoons CH_2=\overset{OH}{\overset{|}{C}}-CO_2^- \underset{ADP}{\overset{ATP}{\rightleftharpoons}} CH_2=\overset{O-PO_3H^-}{\overset{|}{C}}-CO_2^-$$
$$CH_3-\overset{O}{\overset{\|}{C}}-CO_2^- \underset{ATP}{\overset{ADP}{\rightleftharpoons}} CH_3-\overset{^+O-PO_3^=}{\overset{\|}{C}}-CO_2^- \rightleftharpoons CH_2=\overset{O-PO_3H^-}{\overset{|}{C}}-CO_2^- \quad (9)$$

H₂N OH; HN; O; N; R; NH₃; ATP; H₂N OPO₃⁼; HN; O; N; R; NH₂; N; O; N; R; ⁺O-PO₃⁼; ATP; NH₃; HN; O; N; R (10)

In both schemes, the upper pathway represents the formation of an unstable intermediate that is then trapped by reaction with ATP. By contrast, the lower pathways represent activation of a carbonyl group by ATP, followed by loss of a proton (to form enol pyruvate; eq. 9) or reaction with ammonia (to yield an intermediate for the formation of CTP; eq. 10). ATP has long been known to drive metabolic processes; if it phosphorylates carbonyl groups, it will prove to serve a catalytic role as well.

Available evidence (14,15) favors the pathway for pyruvate kinase by way of phosphorylation of pyruvate enol. Furthermore, J. Knowles and his coworkers (16,17), using chiral thiophosphates and chiral ($^{16}O,^{17}O,^{18}O$) phosphate have shown that pyruvate kinase transfers phosphate from phosphoenolpyruvate to ADP with stereochemical inversion at phosphorus. Since monomeric metaphosphate is presumably planar, a chemical reaction by way of that ion should proceed with racemization. In the active site of an enzyme, however, all components might be held so rigidly that racemization need not occur. Furthermore, no information is yet available on the detailed mechanism of reactions catalyzed by cytidine synthetase; our own experiments, designed to distinguish among the mechanisms here discussed, are as yet incomplete.

Acknowledgment

The work here described was supported by the National Science Foundation.

Literature Cited

1. Westheimer, F. H. Chem. Rev. (in press).
2. Conant, J. B.; Pollack, S. M. J. Am. Chem. Soc. 1921, 43, 1665.
3. Conant, J. B.; Jackson, E. L. J. Am. Chem. Soc. 1924, 46, 1003.
4. Maynard, J. A.; Swan, J. M. Aust. J. Chem. 1963, 16, 596.
5. Kumamoto, J.; Cox, Jr., J. R.; Westheimer, F. H. J. Am. Chem. Soc. 1956, 78, 4858.
6. Cox, J. R.; Ramsay, O. B. Chem. Rev. 1964, 64, 317.
7. Gerrard, A. F.; Hamer, N. K. J. Chem. Soc. (B), 1967, 1122.
8. Satterthwait, A. C.; Westheimer, F. H. J. Am. Chem. Soc. 1978, 100, 3197.
9. Satterthwait, A. C.; Westheimer, F. H. J. Am. Chem. Soc. 1980, 102, 4464.
10. Satterthwait, A. C.; Westheimer, F. H. J. Am. Chem. Soc. 1981, 103, 1177.
11. Rose, I. A. J. Biol. Chem. 1960, 235, 1170.
12. Robinson, J. L.; Rose, I. A. J. Biol. Chem. 1972, 247, 1096.
13. Levitzki, A.; Koshland, Jr., D. E. Biochem. 1971, 10, 3365.
14. Midelfort, C. F.; Rose, I. A. J. Biol. Chem. 1976, 251, 5881.
15. Rose, I. A. Adv. Enzymol. 1979, 50, 361.
16. Blättler, W. A.; Knowles, J. R. J. Am. Chem. Soc. 1979, 101, 510.
17. Knowles, J. R. Ann. Rev. Biochem. 1980, 49, 877.

RECEIVED July 7, 1981.

14

Stereoelectronic Effects in Phosphate Esters

D. G. GORENSTEIN, R. ROWELL, and K. TAIRA

University of Illinois, Chemistry Department, P.O. Box 4348, Chicago, IL 60680

The role of orbital orientation in organic and enzymatic reactions has been of considerable current interest. Deslongchamps and coworkers(1) in studying tetracovalent carbon species have recently demonstrated selective cleavage of bonds which are trans, antiperiplanar (app) to lone pairs on directly bonded oxygen and nitrogen atoms.

Molecular orbital calculations have also provided theoretical justification for these stereoelectronic effects in tetracovalent and pentacovalent phosphorus species (2-7). As has been shown in molecular orbital calculations on the $\bar{X}_1$-P-X_2 (X = O,N) structural fragments, the X_1-P bond is strengthened (as indicated by an increase in the Mulliken overlap population) while the P-X_2 bond is weakened when the X_1 atom lone pair is app to the P-X_2 bond. Thus, in the g,t conformation of dimethyl phosphate (Structure **1**) the overlap population for the trans P-O bond is .017 electron lower than the overlap population for the gauche P-O bond. As shown for g,t dimethyl phosphate one lone pair (shaded in **1**) on the gauche bond oxygen is app to the trans bond, while no lone pairs on the trans bond oxygen are app to the gauche bond. Thus, the weakest X_1-P bond has one app lone pair and no lone pairs on X_1 app to the P-X_2 bond.

Further ab initio molecular orbital calculations on the reaction profile for the base-catalyzed hydrolysis of dimethyl phosphate in various ester conformations have provided support for this theory (5,6):

$$(CH_3O)_2\bar{P}O_2^- + {}^-OH \rightarrow (CH_3O)_2PO_3H^{--} \rightarrow (CH_3O)PO_3H^- + CH_3O^-$$

Typically, for the methoxide elimination step the transition state which has an antiperiplanar lone pair to the methoxide leaving group is ca. 11 kcal/mol lower in energy than the transition state without this app lone pair (Fig.1).

In order to experimentally assess the magnitude of the stereoelectronic effect at phosphorus, we have prepared and studied the reactions of the six-membered ring cyclic esters such as **2**-**6**:

Configurational and conformational analysis of these isomeric 2-aryloxy-2-oxo-trans-5,6-tetramethylene-1,3,2-dioxaphosphorinanes

0097-6156/81/0171-0069$05.00/0

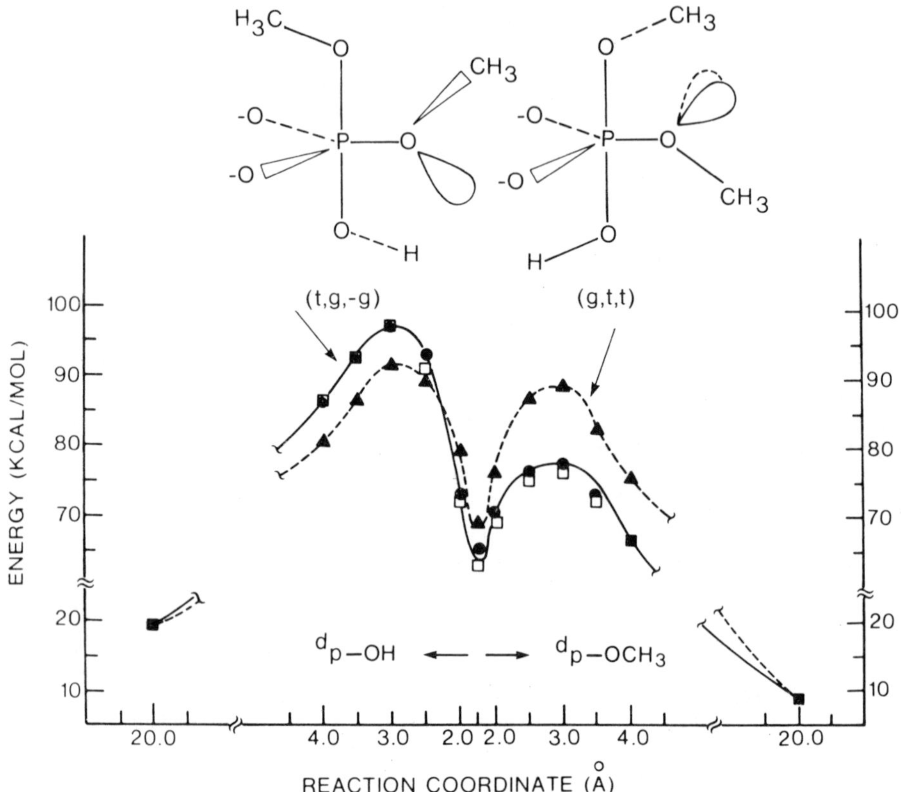

Figure 1. Reaction profiles for hydroxide-catalyzed hydrolysis of dimethyl phosphate (5). Torsional angles about the P–OMe bonds for the phosphorane intermediates are defined by the MeOPOMe structural fragment, and the torsional angle about the P–OH bond is defined by the MeO$_e$POH fragment. Conformers are defined by the following order for the torsional angles: apical ester, equatorial ester, and apical P–OH bonds. The reaction coordinate is defined by the P–OH distance (d_{POH}) and P–O(a)CH_3 distance (d_{POCH_3}) for the attack and displacement steps, respectively. Solid line represents profile for OH^- attack on g,g DMP to yield g,g,–g DMPanes (□) and for OH^- attack on g,t DMP to yield t,g,–g DMPanes (•). Dashed line represents profile for OH^- attack on g,t DMP to yield g,t,t DMPanes (△).

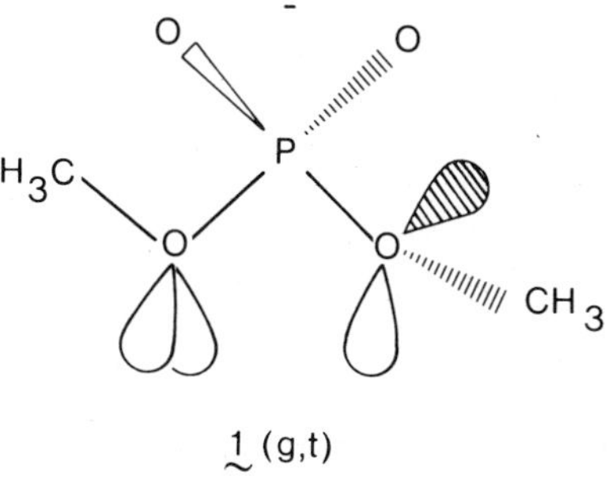

1 (g,t)

2a-5a (R=H); 6a (R=CH_3)

2b-5b (R=H); 6b (R=CH_3)

2-5 (ArO = p-methoxyphenoxy, p-nitrophenoxy, phenoxy, and 2,4-dinitrophenoxy, respectively) and isomeric 2-p-nitrophenoxy-2-oxo-5-methyl (cis and trans)-5,6-tetramethylene-1,3,2-dioxaphosphorinanes, 6 and 7 have been determined by ^{1}H NMR coupling data and P-31 and C-13 NMR and IR spectra. The axial aryloxy isomers 2a-6a of these trans decalin-type six-membered ring phosphorinanes (and cis-decalin 7a) are in chair conformations. However, NMR and IR data support the assignment of a twist boat conformation for the "equatorial" isomer of the 2,4-dinitrophenoxy ester 5b. Mixed chair and twist boat conformations are found for the other aryloxy esters 2b-4b. A pure chair conformation is found for the other equatorial p-nitrophenoxy ester, 6b (R = CH_3) presumably due to the steric interaction of the axial methyl group with the phosphate in any twist conformation. The axial isomers 2a-6a are 1.5-2 kcal/mole lower energy and hydrolyze in base 4-17 times slower than their epimers. Only the twist boat isomer of 2,4-dinitrophenoxy ester, 5b, reacts with 100% inversion of configuration with methoxide. All other compounds react with 4-83% inversion of configuration with the most retention obtained for the poorest leaving groups.

An O-18 isotopic shift on P-31 chemical shifts of epimeric 1,3,2-dioxaphosphorinane esters (10) has been used to investigate the stereochemistry of hydroxide catalyzed hydrolysis in these aryl phosphorinane triesters. As shown in Scheme I the axial epimer of 2-(2,4-dinitrophenoxy)-2-oxo-trans-5,6-tetramethylene-1,3,2-dioxaphosphorinane) 5a, was hydrolyzed in O-18 enriched hydroxide. The cyclic diester, 8, was methylated with diazomethane in methanol. The P-31 NMR spectrum of the O-18 labeled axial methyl phosporinane product, 9, showed three signals representing the unlabeled phosphate, the P—O—18-CH_3 (ester oxygen) labeled phosphate, and the P=O-18 (phosphoryl) labeled phosphate. Analysis of the O-18 distribution by direct integration of the three signals gave 100% P-O aryl bond cleavage and 82% inversion of configuration at phosphorus. The hydroxide catalyzed hydrolysis of the equatorial epimer of the p-methoxyphenoxy-1,3-2-dioxaphosphorinane yielded 59% inversion.

The epimeric substituted aryloxy phosphorinanes 2-5 were hydrolyzed in 30% dioxane/water at 70°C with a variety of nucleophiles (11). The reactivities were sensitive to changes in both the nucleophile and leaving group. The Bronsted β_{nuc}'s for the equatorial leaving groups 2,4-DNP, PNP, and Ph are 0.48, 0.64, and 0.75 respectively. The Bronsted β_{1g}'s are -0.96, -1.04, -0.85, -0.66, -0.64, -0.57, -0.46, and -0.35 for nucleophiles water, methoxyacetate, acetate, phosphate, hexafluoroisopropoxide, carbonate, trifluorethoxide, and hydroxide respectively. These results suggest a concerted $S_N2(P)$ mechanism especially in reactions where β_{1g} is greater than 0.6. However, the hydroxide catalyzed hydrolysis of 5a and 2b yielded substantial retention product at phosphorus. Thus an intermediate is required for attack by very basic nucleophiles in contrast to the Bronsted relationship results. Together, the stereochemical data, multiple structure reactivity relationships, and epimer rate ratios may be reconciled

Scheme I.

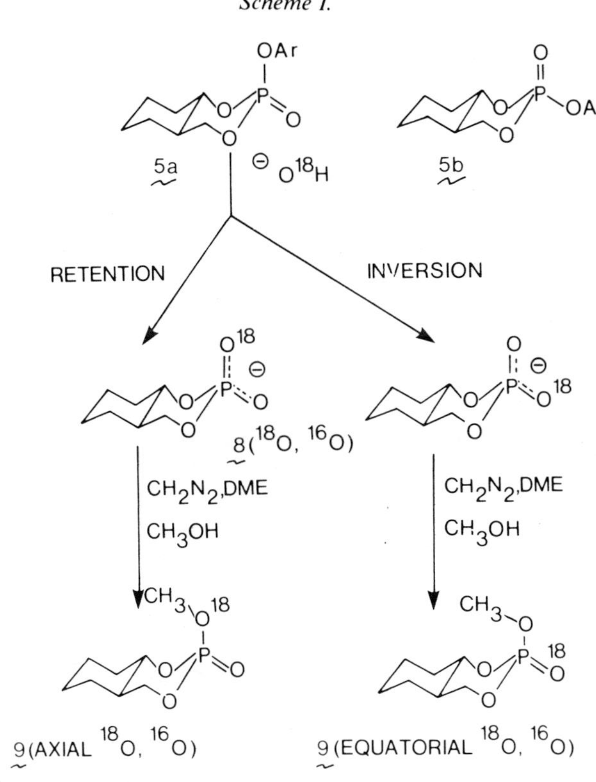
OAr
O
5a
⊖ O18H
5b
RETENTION
INVERSION
O18
8(18O, 16O)
CH2N2,DME
CH3OH
CH3
9(AXIAL 18O, 16O)
9(EQUATORIAL 18O, 16O)

Scheme II.

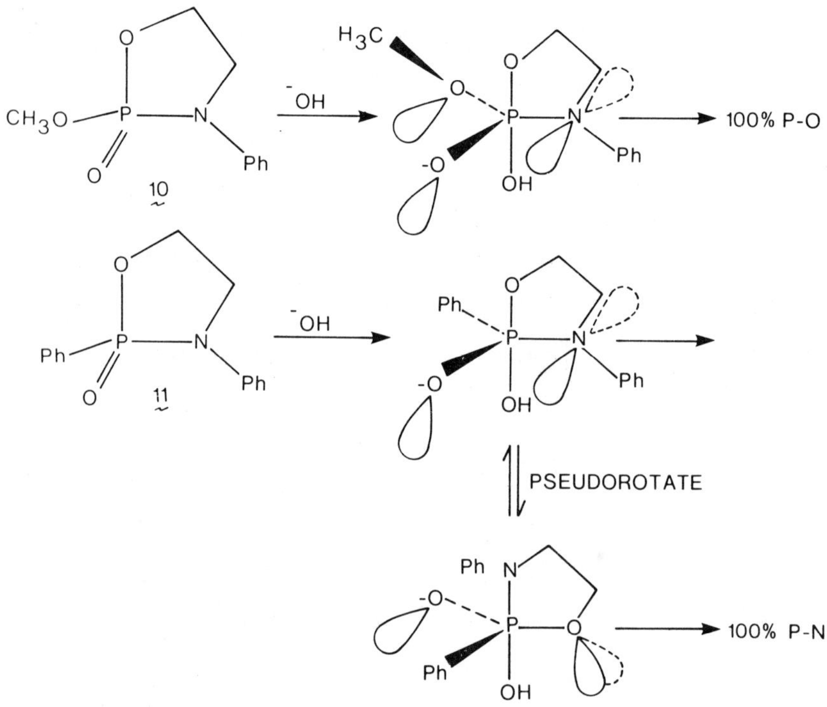

by assuming a continuum of mechanisms for reactions of phosphate triesters, from addition/elimination to concerted $S_N2(P)$.

Unfortunately the unusual conformational flexibility of the esters 2-6 preclude a definitive estimate of the magnitude of the stereoelectronic effect in phosphate ester reactivity. However, we have earlier proposed (4) that a significant fraction of the 10^6 to 10^8-fold rate acceleration in the hydrolysis of strained cyclic five-membered-ring phosphate esters relative to their acyclic counterparts (12) was due to a stereoelectronic effect. In addition, the dramatic difference in P-O bond cleavage (13) in 10 and 11 could arise from a stereoelectronic effect. Thus, as shown in Scheme II the phosphorane formed by hydroxide attack opposite the ring oxygen in 10 has two app lone pairs on the basal oxygen which facilitate ring P-O bond cleavage. The phosphorane formed from hydroxide attack on 11 has only one app lone pair opposite the apical ring oxygen (the nitrogen lone pair will lie in the basal plane) and P-O bond cleavage is not favored. Pseudorotation concomitant with proton reorganization yields a new phosphorane in which the apical nitrogen is app to two oxygen lone pairs, possibly providing an explanation for the 100% P-N cleavage in 11.

LITERATURE CITED

1. Deslongchamps, P. Tetrahedron 1975, 31, 2463.
2. Lehn, J.M.; Wipff, G. J. Chem. Soc. Chem. Comm. 1975, 800.
3. Gorenstein, D.G.; Findlay, J.B.; Luxon, B.A.; Kar, D. J. Am. Chem. Soc. 1977, 99, 3473.
4. Gorenstein, D.G.; Luxon, B.A.; Findlay, J.B.; Momii, R. J. Am. Chem. Soc. 1977, 99, 4170.
5. Gorenstein, D.G.; Luxon, B.A.; Findlay, J.B. J. Am. Chem. Soc. 1977, 99, 8048.
6. Gorenstein, D.G.; Luxon, B.A.; Findlay, J.B. J. Am. Chem. Soc. 1979, 101, 5869.
7. Gorenstein, D.G.; Luxon, B.A.; Goldfield, E. J. Am. Chem. Soc. 1980, 102, 1759.
8. Gorenstein, D.G.; Rowell, R. J. Am. Chem. Soc. 1979, 101, 4929 and refs therein to related studies by other authors.
9. Gorenstein, D.G.; Rowell, R.: Findlay, J.B. J. Am. Chem. Soc. 1980, 102, 5077.
10. Gorenstein, D.G.; Rowell, R. J. Am. Chem. Soc. 1980, 102 6165.
11. Gorenstein, D.D.; Rowell, R. J. Am. Chem. Soc., accepted for publication.
12. Westheimer, F.H. Acc. Chem. Res. 1968, 1, 70.
13. Boudreau, J.A.; Brown, C.; Hudson, R.F. J. Chem. Soc. Chem. Comm. 1975, 679.

RECEIVED September 10, 1981.

15

Stereospecific Synthesis and Assignment of Absolute Configuration at Phosphorus in Nucleoside 3′- and 5′-*O*-Arylphosphorothioates and Nucleoside Cyclic 3′,5′-Phosphorothioates

J. BARANIAK, Z. J. LEŚNIKOWSKI, W. NIEWIAROWSKI, W. S. ZIELIŃSKI and W. J. STEC

Polish Academy of Sciences, Centre of Molecular and Macromolecular Studies, Department of Bioorganic Chemistry, Boczna 5, 90-362 Łódź, Poland

The new class of phosphorylating reagents, asymmetric O,N-diarylphosphoramidochloridates and O,N-diarylphosphoramidochloridothioates, has been reported from this Laboratory */1,2/*. By virtue of the asymmetric phosphorus atom, nucleoside 5'- and 3'-O,N-diarylphosphoramidates (diarylphosphoramidothioates) consist of a mixture of diastereomers which can be separated into individual P-enantiomeric species. These, by the method established in this laboratory */3/*, can be stereospecifically converted by means of NaH/CY_2 (Y=O,S,Se) to the desired nucleoside O-arylphosphates, -phosphorothioates and -phosphoroselenoates */2,4/*, which may be further utilized for the synthesis of oligonucleotides */5/*. Nucleoside 3'- and 5'-O,N-diarylphosphoramidates can be also used for preparation of nucleoside cyclic 3',5'-phosphoranilidates, the key intermediates in the stereospecific synthesis of nucleoside cyclic 3',5'-phosphorothioates */6,7,8/* and -|^{18}O|phosphates */9,10/*. Since the absolute configuration at the P-atom in deoxyadenosine cyclic 3',5'-|R_P|-phosphoranilidate was assigned by means of X-ray crystallography */11/* and this assessment has confirmed the earlier determination made by means of ^{31}P-NMR spectroscopy */7/*, the knowledge of the absolute configuration at the P-atom in diastereomeric deoxyadenosine 3'-O,N-diarylphosphoramidates would answer the question about the stereochemistry of Borden-Smith ring-closure process */12/*. Additionaly, assignment of the absolute configuration at the P-atom in starting nucleoside O,N-diarylphosphoramidates, due to the stereospecificity of P-N→P-X conversion */3/* would disclose the problem of the absolute configuration at the P-atom of diastereomeric nucleoside 3'- or 5'-O-arylphosphorothioates, important tools for investigation of enzymes.

Recently, the absolute configuration and ^{31}P-NMR data of adenosine 5'-|S_P|-O-phosphorothioate O-p-nitrophenyl ester (1) as the product remaining after complete enzymatic digestion of diastereomeric mixture of |R_P|-1 and |S_P|-1, were described by Eckstein et. al. */13/*. This result allowed us to correlate the absolute configuration between 1 and adenosine 5'-O-p-nitrophenyl-N-phenylphosphor-

0097–6156/81/0171–0077$05.00/0

Scheme 1.

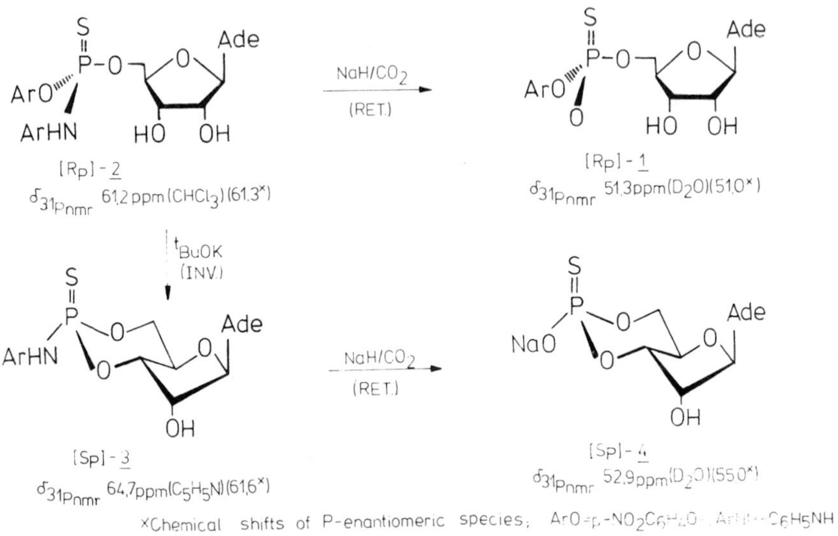

[Rp]-2 $\delta_{^{31}P\,nmr}$ 61.2 ppm ($CHCl_3$) (61.3ˣ)

[Rp]-1 $\delta_{^{31}P\,nmr}$ 51.3 ppm (D_2O) (51.0ˣ)

[Sp]-3 $\delta_{^{31}P\,nmr}$ 64.7 ppm (C_5H_5N) (61.6ˣ)

[Sp]-4 $\delta_{^{31}P\,nmr}$ 52.9 ppm (D_2O) (55.0ˣ)

ˣChemical shifts of P-enantiomeric species; ArO = p-$NO_2C_6H_4O$-, ArHN = C_6H_5NH

amidothioate (2). This last diastereomeric compound under treatment with t-BuOK gives only one diastereomer of adenosine cyclic 3',5'-phosphoranilidothioate (3), which in the reaction of established stereochemistry is converted to cAMPS (4) of known absolute configuration. Correlation of the absolute configuration between 1 and 4 and the knowledge of the stereochemical course of the conversions 2→1 and 3→4 allowed us to establish the inversion of the configuration in the ring closure process 2→3 under conditions originally described by Borden and Smith /12/. The elucidation of the stereochemistry of intramolecular cyclisation was essential in the light of Wadsworth's results concerning the stereochemistry of nucleophilic substitution at phosphoryl and phosphorothioyl P-atom with aryloxy- leaving groups /14/. The conclusion of inversion of configuration at the P-atom during intramolecular substitution (proved independently by Gerlt /15/) was extended on conversion of deoxyribonucleoside 3'-O-aryl-N-phenyl phosphoramidates (5) to deoxyribonucleoside cyclic 3',5'-phosphoranilidates (6). These are stereospecifically converted to diastereomerically pure deoxyribonucleoside cyclic 3',5'-phosphorothioates (7, B=Thy,Ade,Cyt,Gua).
Because the absolute configuration of 6 was assessed earlier on the basis of chemical shift and spin-spin coupling criteria in ^{31}P-NMR spectra, and recently confirmed (B=Ade) by means of X-ray crystallography /11/, the absolute configurations at the P-atom in diastereomers of 7,5 and deoxyribonucleoside 3'-O-arylphosphorothioates (8) are also assigned.

Scheme 2.

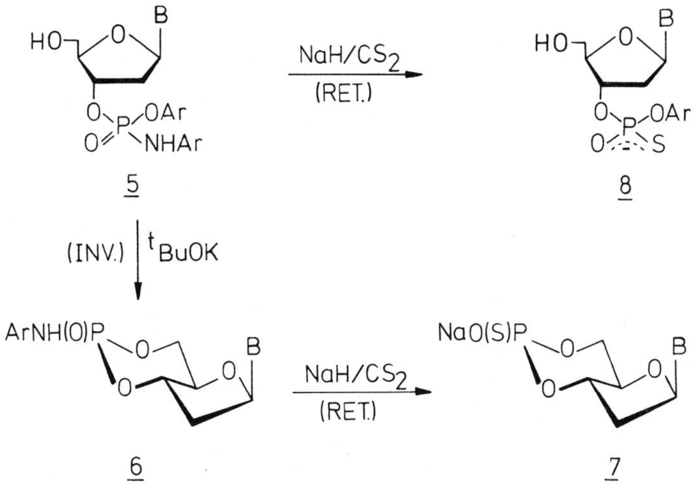

$ArO = o\text{-}Cl\text{-}C_6H_4O-$, $ArNH = C_6H_5NH-$

Scheme 3.

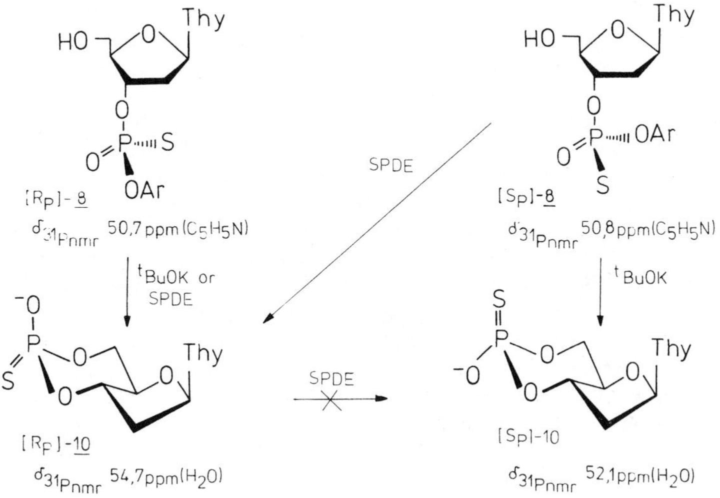

Since it is well established that the P-chiral phosphorothioates serve as effective probes in mechanistic studies for the phosphoryl group transfer enzymes */16/*, we turned our attention to the application of diastereomeric 8 (B=Thy, Ar=$pNO_2C_6H_4$-,*/17/*) to elucidate the mode of action of spleen phosphodiesterase (SPDE, EC 3.1.4.18). This enzyme splits the phosphodiester bonds to yield nucleoside 3'-phosphates. In the case of 8 it was expected that its SPDE-catalyzed hydrolysis in $|^{18}O|H_2O$ medium leads to P-chiral thymidine 3'-$|^{18}O|$phosphorothioate. On the contrary to our expectation, the main product of this reaction was thymidine cyclic 3',5'-$|R_P|$-phosphorothioate (10) */6/*. By treatment of 8 under the same conditions, but in the absence of the enzyme, no trace of 10 was detected.

The hydrolytic action of SPDE towards thymidine 3'-p-nitrophenyl phosphate does not lead to thymidine cyclic 3',5'-phosphate; the only product is thymidine 3'-phosphate. This fact would suggest that hydrolysis of natural nucleotides in the presence of SPDE involves intermediate nucleoside cyclic 3',5'-phosphates which, when bound to active site/s/ of the enzyme, undergo further hydrolysis to nucleoside 3'-phosphates. According to this hypothesis nucleoside phosphorothioates are converted in the presence of SPDE into nucleoside cyclic 3',5'-phosphorothioates, but these are less susceptible to nucleophilic attack and are not further hydrolysed to nucleoside 3'-phosphorothioates. This hypothesis is, however, weakened by the facts that thymidine cyclic 3',5'-phosphate is resistant towards SPDE and it does not show any inhibitory effect towards SPDE-catalyzed hydrolysis of thymidine 3'-O-p-nitrophenyl phosphate.

The fact that both diastereomers of 8 are stereoselectively converted by SPDE to $|R_P|$-10 suggests that an intermediate must be involved on the reaction pathway which is common for both substrates.

A possibility would be monomeric thymidine 3'-metaphosphorothioate, which in the active site allowing the rotation around C-O bond would lead to the same preferred orientation being adopted whatever the stereochemistry of the starting diester and this would then lead to a single diastereomeric product $|R_P|$-10. If this explanation is correct and hydrolysis of thymidine 3'-p-nitrophenyl-$|^{18}O|$phosphate would follow the same mechanistic course, its hydrolysis in $|^{17}O|H_2O$ containing medium in the presence of SPDE should lead to formation of P-racemic thymidine 3'-$|^{16}O,^{17}O,^{18}O|$phosphate. The work on elucidation of this hypothesis is in progress.

Independently on the final result the stereochemical course of SPDE-catalyzed hydrolysis of 8 argue against the presumption, that "all enzymatic reactions at phosphorus proceed with inversion and that, therefore, they occur without pseudorotation at phosphorus" */18/*.

Literature Cited

1. Zieliński, W.S.; Leśnikowski, Z. Synthesis 1976, 185-187.
2. Lesiak, K.; Leśnikowski, Z.J.; Stec, W.J.; Zielińska, B. Pol.J. Chem. 1979, 53, 2041-50.
3. Stec, W.J.; Okruszek,A.; Lesiak, K.; Uznański,B.; Michalski, J. J.Org.Chem. 1976, 41, 227-33.
4. Zieliński, W.S.; Leśnikowski, Z.J.; Stec, W.J.; J.Chem.Soc.,Chem. Commun. 1976, 772-3.
5. Leśnikowski, Z.J.; Stec, W.J.; Zieliński, W.S. Synthesis 1980, 397-400.
6. Zieliński, W.S.; Stec, W.J. J.Am.Chem.Soc. 1977, 99, 8365-66.
7. Leśnikowski, Z.J.; Stec, W.J.; Zieliński, W.S. Nucleic Acids Research, Spec.Publ. No 4, 1978, s49-52.
8. Baraniak, J.; Kinas, R.W.; Lesiak, K.; Stec, W.J. J.Chem.Soc., Chem.Commun. 1979, 940-41.
9. Baraniak, J.; Lesiak, K.; Sochacki, M.; Stec, W.J. J.Am.Chem.Soc. 1980, 102, 4533-34.
10. Gerlt, J.A.; Coderre, J.A. J.Am.Chem.Soc. 1980, 102, 4531-33.
11. Leśnikowski, Z.J.; Stec, W.J.; Zieliński, W.S.; Adamiak, D.; Saenger, W. J.Am.Chem.Soc. 1981, in press.
12. Borden, R.K.; Smith, M. J.Org.Chem. 1966, 31, 3247-53.
13. Burgers, P.M.J.; Sathyanarayana, B.K.; Saenger, W.; Eckstein, F. Eur.J.Biochem. 1979, 100, 585-91.
14. Bauman, M.; Wadsworth, W.S.,Jr. J.Am.Chem.Soc. 1978, 20, 6388-94.
15. Gerlt, J.A.; Mehdi, S.; Coderre, J.A.; Rogers, W.O. Tetrahedron Lett. 1980, 21, 2385-88.
16. Eckstein, F. Acc.Chem.Res. 1979, 12, 204-10.
17. Niewiarowski, W.; Stec, W.J.; Zieliński, W.S. J.Chem.Soc., Chem. Commun. 1980, 524-25.
18. Westheimer, F.H. in "Rearrangements in Ground and Excited State", P. de Mayo, Ed., Academic Press, New York, 1980, Vol.2, p.266.

RECEIVED June 30, 1981.

16

The Stereochemistry of Chiral Cyclic Phosphorus Esters

Do Theories of Bond-Forming and Bond-Breaking Processes Fit the Facts?

THOMAS D. INCH and C. RICHARD HALL

Chemical Defence Establishment, Porton Down, Salisbury, Wiltshire, SP4 OJQ, England

Starting in 1973, 6-membered cyclic phosphorus esters were prepared with the objectives of (a) establishing their configuration at phosphorus (b) stereospecifically opening the ring and detaching the phosphorus from the carbohydrate to afford optically pure phosphorus esters, with configurations established by synthesis. The stereochemical guidelines for synthetic strategy appeared adequate since (a) the concepts of Berry Pseudorotation (BPR) and Turnstile Rotation (TR) were well developed, (b) apicophilicites would suggest the probable direction of attack in substitution reactions and (c) the application of some of these principles to reactions of cyclic esters had been admirably demonstrated notably by Westheimer. During subsequent synthetic stereochemical studies, some of which have given optically pure acyclic phosphorus esters with diverse substituents (1), the complexity of the factors which determine the nature and stereochemistry of the bonds broken have become increasingly apparent. Results to date with acyclic phosphorus esters show that with a few exceptions (notably reactions involving P-SR bond cleavage) acyclic esters undergo reactions with highly stereoselective inversion and are not the kind of results to induce mechanistic speculation. This paper therefore will by reference to results with cyclic phosphorus esters, (2) attempt to illustrate the problems of reconciling experimental data with mechanistic hypothesis.

Firstly attention must be drawn to the importance of ring conformation. In the synthesis of the cyclic methylphosphonates Iax and Ieq from methyl 2,3-di-O-methyl-α-D-glucopyranoside 6-(RS)-methylphosphonofluoridate, Iax is kinetically and Ieq is thermodynamically favoured. The postulate was that ring closure occurred via the non-chair intermediate II. IIa is likely to be preferred over IIb since in IIa the exocylic P-O bond rather than P-Me is pseudoaxial (electronegative substituents at phosphorus in 1,3,2-dioxophosphorinanes prefer an axial orientation viz the 'gauche or anomeric effect'). Structure IIa is the precursor of the kinetic product Iax. If the ring closure had occurred via a chair

0097–6156/81/0171–0083$05.00/0

intermediate the anomeric or gauche effect would promote formation of Ieq as the kinetic product. This reasoning which applies to other ringforming reactions is not easy to validate. Nevertheless, during any considerations of the steric course of reactions at phosphorus in cyclic esters conformational effects must be considered.

For example, conformational effects may be important in reactions with Grignard reagents of Iax, Ieq and their P=S analogues IIIax and IIIeq. With PhMgI, Iax and Ieq gave preponderant cleavage with inversion of the P-O6 bond to give about 50% yields of the major products. In contrast the main product (61%) from IIIeq-PhMgI was from P-O6 cleavage with retention. Further with IIIax-PhMgI both P-O6 cleavage with inversion (16%) and P-O4 cleavage with retention (28%) occurred. In these and other reactions there is a slight tendency for axial phosphonates to react more readily and give products with inversion whereas equatorial phosphonates react more slowly and give products with retention. These examples are representative of many results of cyclic esters with Grignard reagents and alkyllithiums. The results are all complex with no overriding trends and no completely convincing explanations. In arriving at any explanations it will be necessary to establish whether differences in the course of the various reactions are due primarily to the effect of the configuration and conformation of the starting esters on the direction of attack by the incoming nucleophile or whether the relative ease with which non-chair conformations can be adopted can affect overall reaction rates and the relative rates of direct displacement and reorganisation processes.

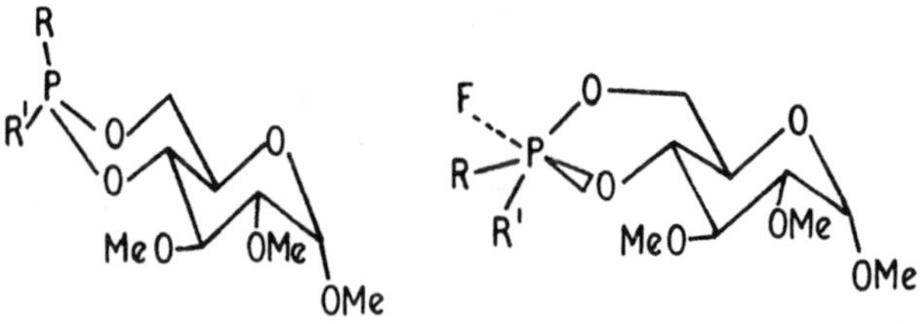

Iax R=Me, R'=O
Ieq R=O, R'=Me
IIIax R=Me, R'=S
IIIeq R=S, R'=Me

IIa R=Me, R'=O
IIb R=O, R'=Me

Studies of ring opening reactions of 6-membered cyclic esters with hydroxide and with alkoxides have in addition to providing some support for the possible importance of conformational effects, more importantly provided new information about the ability of trigonal bipyramidal intermediates (TBP's) to undergo ligand reorganisation processes. The scheme shows the results (with possible TBP's) of ring opening the cyclic ethylphosphates IVax and IVeq with methoxide. The reactions were not stereospecific but gave preponderant products with inversion. For example the ratios of V:VI was 4:1 from IVax and 3:7 from IVeq. However VII was produced stereospecifically from IVax. What was most important however was that the migration of VI to VII was stereospecific with retention and that the much slower reverse migration of VII to VI was also stereospecific with retention. These results appear to suggest that the mixed products from IVax and IVeq result not from multiple BPR or TR processes following initial attack opposite O4 or O6 but rather that the mixed products arise from attack opposite both O4 and O6 with ring opening occurring following direct displacement or displacement following a single BRP or TR process. (Results from all other stereochemical studies are consistent with this). Migration sequences involving TBP's which undergo more than a single BPR are precluded by the requirements that the 6-membered ring spans only equatorial-equatorial or apical-equatorial positions in the TBP and that the P-O group should remain in an equatorial position. The consequences are that migration can only occur with retention (except where exchange processes involving bond breaking may operate). Multiple TR or TR switch mechanisms, which would circumvent any restraints to ligand reorganisation imposed by the ring system, can in theory allow migration with "racemisation". In the migrations VI to VII and VII to VI where all the ligands attached to phosphorus are oxygen, it is reasonable to assume many of the TBP's will be of similar energy (excluding those causing the 6-membered ring to span apical positions, etc), thereby facilitating their interconversion causing racemisation of the migrating phosphorus ester. The fact that no such "racemisation" is observed makes it extremely unlikely that multiple TR processes will make a significant contribution to any reactions involving TBP intermediates.

The title of this paper asks if theories fit the facts. So far it has been argued that for cyclic esters, conformational effects make it exceptionally difficult to control the variables to test any theory but by showing that ligand reorganisation processes in reaction intermediates (as opposed to stable phosphoranes) are limited then at least some simplification of mechanisms is possible. It is pertinent now to examine the one assumption that was so valuable in explaining the differences in hydrolysis patterns of 5-membered phosphates and phosphonates, viz, the C-P bond provides a barrier to pseudorotation.

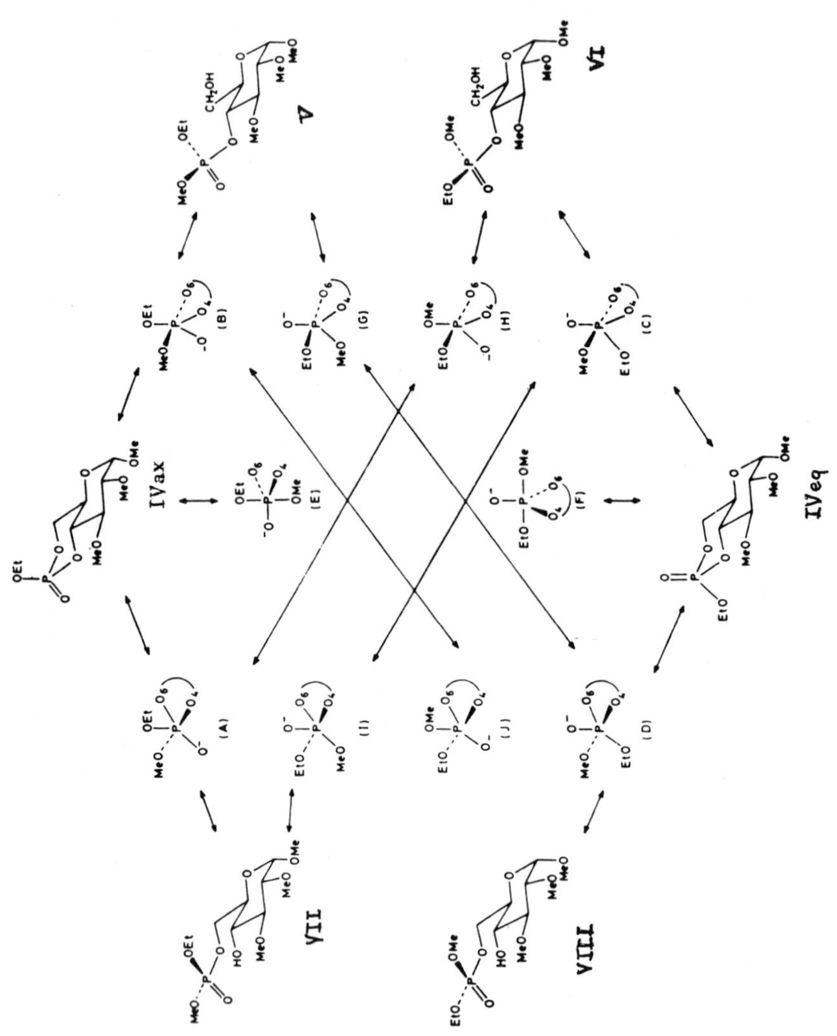
V
VI
IVax
IVeq
VII
VIII
(A)
(B)
(C)
(D)
(E)
(F)
(G)
(H)
(I)
(J)

In I and III although ring opening occurs usually with exclusive inversion, in some derivatives traces of retention products are observed. Thus compared with the ring opening of IV there is a barrier to pseudorotation. This is further exemplified by the fact that migration of alkyl methylphosphonate groups between positions 4 and 6 on glucose do not occur by pseudorotation of a TBP intermediate as for the migration VI to VII but by reformation of the cyclic ester (eg I) followed by re-ring opening to establish the equilibrium product mixture. However in a recent stereochemical examination of ring opening by acid or base of 2-methyl 1,3,2-dioxaphospholan-2-thiones although an inversion pathway preponderated there was a significant retention pathway. Also, although migration occurred by ring re-formation and re-opening again there was competition from a pseudorotation process.

These results and of course some of the Grignard results where retention mechanisms in phosphonates are preponderant show that caution must be taken when even a foundation stone of modern phosphorus stereochemistry is being considered.

Literature Cited

1. Hall, C. R.; Inch, T. D. Phosphorus and Sulphur. 1979, 7 171.
2. Hall, C. R.; Inch, T. D. Tetrahedron. 1980, 36, 2059.

RECEIVED June 30, 1981.

17

The Stereochemical Course of the Alkaline Hydrolysis of 1,3,2-Oxazaphospholidine-2-thiones

C. RICHARD HALL and THOMAS D. INCH

Chemical Defence Establishment, Porton Down, Salisbury, Wiltshire, SP4 OJQ, England

The basic hydrolysis of acyclic phosphoramidates occurs with a mixture of C-O and P-O bond cleavage (1, 2) however similar treatment of 1,3,2-oxazaphospholidines can result in substantial P-N cleavage (1, 3). 1,3,2-Oxazaphospholidines derived from (-)-ephedrine react with solutions of alkoxides in alcohol to give only the products of P-N cleavage and with inversion of configuration at phosphorus. Scheme I (4, 5).

We wish to report that on treatment with a solution of sodium hydroxide in aqueous dioxan, (A) was converted by endocyclic P-O cleavage into (C). Alkylation of (C) with methyl iodide followed by acetylation with acetic anhydride/pyridine resulted in (D), which reacted with a dilute solution of anhydrous hydrogen chloride in methanol to give R-(+)-ethyl O,S-dimethyl phosphorothioate (E). Since both S-alkylation and O-acetylation involve no reaction at phosphorus, and available evidence suggests that under the conditions used acid catalysed P-N cleavage in N,N-dialkyl phosphoramidothioates occurs with inversion (6), the observed sequence (Scheme 2) implies that endocyclic P-O cleavage occurs with retention of configuration at phosphorus.

A similar procedure starting with (B) [the phosphono analogue of (A)] again results in predominant endocyclic P-O cleavage but in this case the reaction is not stereospecific (See Table for the observed stereochemistry as a function of solvent) and a minor product resulting from P-N cleavage with inversion is also formed (Scheme 3).

We believe (5) that pseudorotations, or other ligand reorganisation processes, in TBP reaction intermediates are limited to the pairwise exchange of apical and equatorial ligands, with the constraint that P-X $^{\ominus}$ remains equatorial. Thus if nucleophilic attack occurs opposite the eventual leaving group substitution occurs with inversion; where this is not the case substitution occurs, after pseudorotation, with retention. When mixed stereochemistry is observed we believe it results from competitive attack opposite more than one substrate ligand. The factors that determine the direction of nucleophilic attack are not yet

0097-6156/81/0171-0089$05.00/0

understood. It is not always correct to assume that attack will occur opposite the most apicophilic group, because apicophilicity is a thermodynamic term that relates the propensity of ligands to occupy apical positions in stable TBP's. For reactions under kinetic control we suggest the concept of ligand apical potentiality in order to relate the likelihood, during nucleophilic attack at tetracoordinate phosphorus, of a ligand being in-line with the nucleophile and therefore of occupying an apical position in the first formed TBP.

Although oxygen is expected to be much more apicophilic than nitrogen (7) the results reported here are most conveniently explained if, in this ring system, nitrogen has a higher apical potentiality than oxygen. Thus, in the phosphoro series, attack of the nucleophile occurs opposite nitrogen to give the TBP (F). When the nucleophile is alkoxide P-N cleaves to give the observed product with inversion. When the nucleophile is hydroxide pseudorotation to the new TBP (G) occurs. P-O cleavage in (G) results in the observed retention. That pseudorotation of (F, nuc = OH) is fast compared to the rate of P-N cleavage can be explained by suggesting that the apical OH is rapidly deprotonated (by base or the adjacent $P-S^{\ominus}$) and the resulting unfavourable apical $P-O^{\ominus}$ provides the driving force (8).

In the phosphono series the first formed TBP is (H). In this case pseudorotation of (H) to (I) will be retarded because of the low apicophilicity of the methyl group, therefore both P-N cleavage in (H) and pseudorotation followed by P-O cleavage in (I) can compete. That the major product is usually formed by P-O cleavage with inversion (Table) suggests apical attack of hydroxide opposite endocyclic oxygen to give (J) is also competitive. These three processes are affected to different extents by changes in solvent.

One possible explanation of why in this ring system nitrogen has a higher apical potentiality than oxygen is that attack opposite oxygen to form (J) is sterically hindered by the fixed bulk of the methyl substituent on nitrogen. If this were the case an increase in this bulk should make the formation of (J) even less favourable and thus, particularly in the phosphono series, result in P-O cleavage with a greater percentage retention of configuration. This does not occur (Table). An alternative explanation may be provided by stereoelectronic considerations. Correctly oriented (trans-antiperiplanar, app) lone pairs of electrons on ligands that are, or are about to become, equatorial in a TBP may be able to overlap with anti-bonding orbitals in an adjacent bond thereby weakening and lengthening it. Since this is what is required of the bond to one ligand as it makes the transition from tetrahedral to TBP geometry, nucleophilic attack might be expected to be directed opposite the ligand that is, or is potentially, app to the greatest number of lone pairs. In (H) endocyclic oxygen has two lone pairs partially app to the apical P-N bond and this, it has been suggested (9), results in a greater net overlap than from one completely app lone pair. The equatorial nitrogen in (J) has only one lone pair.

Scheme 1.

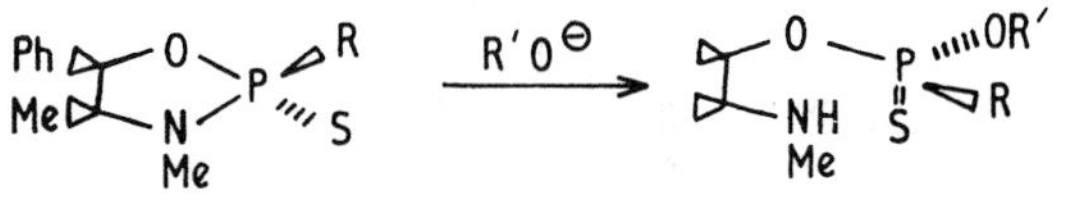

Ph Me O N Me P R S $\xrightarrow{R'O^{\ominus}}$ O NH Me P OR' R S

(A) R=OEt, (B) R=Me

Scheme 2.

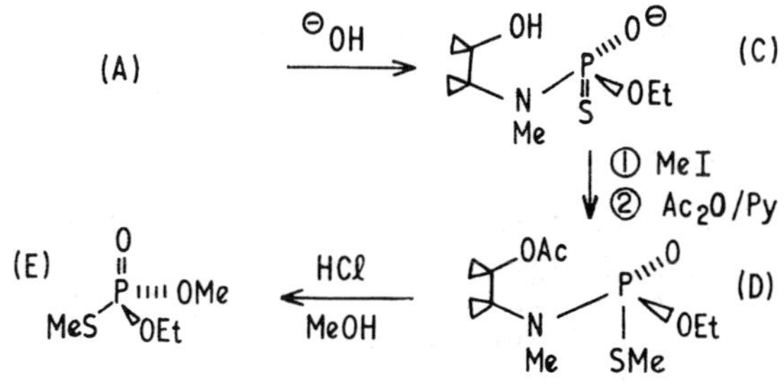

(A) $\xrightarrow{^{\ominus}OH}$ (C)

(C) ① MeI ② Ac_2O/Py → (D)

(D) $\xrightarrow{HCl / MeOH}$ (E)

Scheme 3.

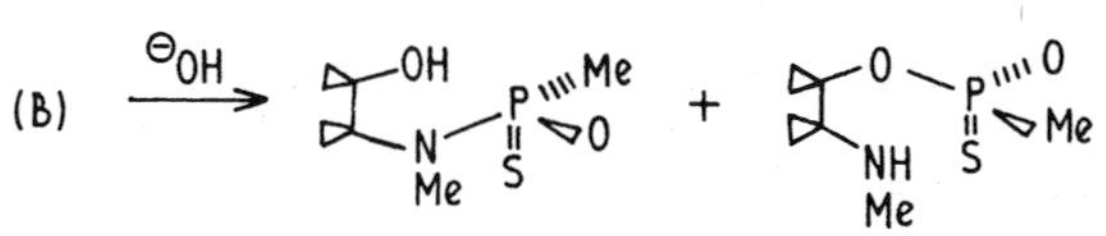

(B) $\xrightarrow{^{\ominus}OH}$ OH N Me P Me O S + O NH Me P O Me S

Scheme 3

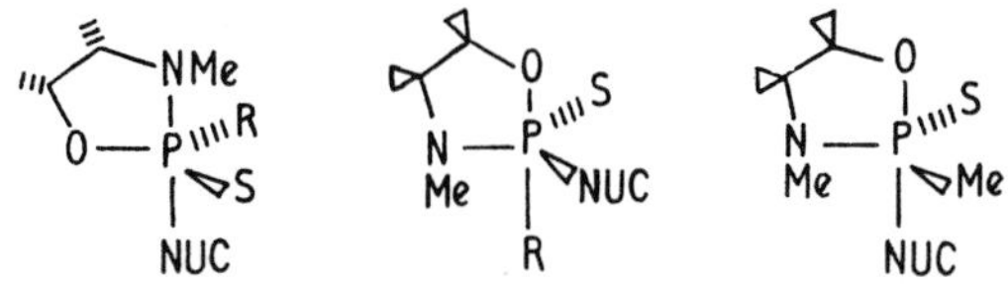

(F) R=OEt
(H) R=Me

(G) R=OEt
(I) R=Me

(J)

TABLE

Basic Hydrolysis of 2-Methyl-1,3,2-Oxazaphospholidine-2-thiones

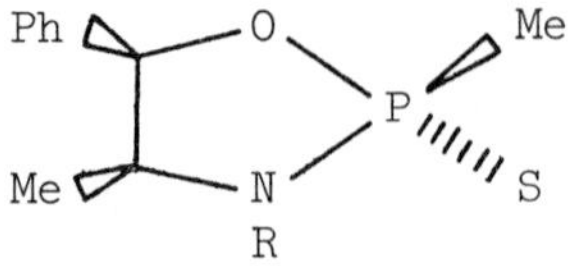

Solvent	Yield % (Stereochemistry)		
	R = Me		R = i-Pr
	P-N Cleavage	P-O Cleavage	P-O Cleavage
EtOH/H_2O	17 (I)	83 (80% I)	(≥ 95% I)
Dioxan/H_2O	5 (I)	95 (60% I)	(≥ 95% I)
Benzene/18-crown-6		98 (50% I)	
CH_3CN		95 (85% R)	(75% I)

I = Inversion, R = Retention of configuration at phosphorus

Literature Cited

1. Brown, C.; Boudreau, J. A; Hewitson, B.; Hudson, R. F. J.C.S. Chem. Comm. 1975, 504.
2. Hamer, N. K; Tack, R. D. J.C.S. Perkin II. 1974, 1184.
3. Brown, C.; Boudreau, J. A; Hewitson, B.; Hudson, R. F. J.C.S. Perkin II. 1976, 888.
4. Cooper, D. B.; Hall, C. R.; Harrison, J. M.; Inch, T. D. J.C.S. Perkin I. 1977, 1969.
5. Hall, C. R.; Inch, T. D. Tetrahedron. 1980, 36, 2059.
6. Hall, C. R.; Inch, T. D. J.C.S. Perkin I. 1979, 1646.
7. Trippett, S. Pure Appl. Chem. 1974, 40, 595.
8. Kluger, R.; Covitz, F.; Dennis, E.; Williams, D.; Westheimer, F. H. J. Amer. Chem. Soc. 1969, 91, 6066.
9. Gorenstein, D. G.; Luxon, B. A.; Findlay, J. B.; Momii, R. J. Amer. Chem. Soc. 1977, 99, 4170.

RECEIVED June 30, 1981.

18

Hydrolysis of Adenosine 5′-Triphosphate

An Isotope-Labeling Study

SEYMOUR MEYERSON and EUGENE S. KUHN

Research Department, Standard Oil Company (Indiana), Naperville, IL 60566

FAUSTO RAMIREZ and JAMES F. MARECEK

Department of Chemistry, State University of New York, Stony Brook, NY 11794

The key role played by adenosine 5'-triphosphate (ATP) in biophosphorus chemistry has prompted numerous investigations of its mechanism of hydrolysis(1-6), but many questions remain unanswered(7). In principle the nonenzymatic hydrolysis of ATP can occur *via* four primary pathways:

$$H_2O + ATP(P_\gamma) \rightleftharpoons ADP + P_i \quad (1)$$
$$H_2O + ATP(P_\beta) \rightleftharpoons P_i + ADP \quad (2)$$
$$H_2O + ATP(P_\beta) \rightleftharpoons AMP + PP_i \quad (3)$$
$$H_2O + ATP(P_\alpha) \rightleftharpoons PP_i + AMP \quad (4)$$

Further hydrolysis of the ADP can proceed *via* two different pathways:

$$H_2O + ADP(P\beta) \rightleftharpoons AMP + P_i \quad (5)$$
$$H_2O + ADP(P\alpha) \rightleftharpoons P_i + AMP \quad (6)$$

A recent paper(8) reported a study of the hydrolysis of ATP in 0.01 M aqueous solutions at 70°C in the pH range 0-10. Representative values of the pseudo-first order rate constants are: (a) in 1 N HCl, $k = 13.9 \times 10^{-2}$ min^{-1}, $t_{½}$ = 5 min; (b) in 0.1 N HCl, $k = 1.98 \times 10^{-2}$ min^{-1}, $t_{½}$ = 35 min; (c) at pH 8.30, $k = 11.5 \times 10^{-5}$ min^{-1}, $t_{½}$ = 100.5 hr. Hydrolysis rates were determined using a liquid-chromatographic (LC) technique that also permitted examination of the various stages of the reaction. In addition, acetonitrile solutions of tetra-*n*-butylammonium (M^+) salts of the tetra- and trianions, $ATP^{4-}4M^+$ and $ATPH^{3-}3M^+$, produced *t*-butyl and isopropyl phosphates at approximately equal rates when the solutions contained *t*-butyl and isopropyl alcohols, respectively(8). These results led to the conclusion that the tetra- and trianions of ATP undergo hydrolysis by an elimination-addition sequence *via* the monomeric metaphosphate anion, in accord with a previous suggestion(9), while the hydrolysis of the acid, $ATPH_4$, and the monoanion, $ATPH_3^-$, proceeds by addition-elimination, presumably *via* an oxyphosphorane intermediate. The mechanism of hydrolysis of the dianion, $ATPH_2^{2-}$, remained unsettled.

We have now supplemented LC with ^{18}O labeling as a probe of the relative contributions of hydrolytic attack on the three

0097-6156/81/0171-0093$05.00/0

phosphorus atoms of ATP and on the two phosphorus atoms of ADP in nonenzymatic hydrolyses. We have also compared the results of these hydrolyses with those carried out under catalysis by myosin CaATPase. This enzymatic reaction has been extensively utilized as a control to establish the isotopic distribution in [γ-^{18}O]ATP samples used in investigations of intermediate oxygen exchange during actomyosin MgATPase action in muscle contraction(10, 11).

We hydrolyzed ATP and ADP in 1 N and 0.1 N HCl and in buffered solutions at pH 4 and 8 in which the hydrolysis medium was variously enriched in ^{18}O to either ~10% or ~20%. To assess the isotopic enrichment of each such solution for use in the nucleotide hydrolysis experiments, we hydrolyzed PCl_5 in the solution, esterified the resultant phosphoric acid/inorganic phosphate (P_i) by reaction with diazomethane, and determined the isotopic distribution of the trimethyl phosphate (TMPO) by mass spectrometry. The 1 N and 0.1 N HCl hydrolyses were allowed to proceed for 45 min and 10 hr, respectively, at 70°, insuring complete conversion of ATP into AMP + $2P_i$. The pH 8 hydrolyses were allowed to proceed for 36 hr at 70° to a point (20-25% completion) at which the ratio of ADP to AMP established that 96% and 4%, respectively, of the P_i released had arisen by the primary and secondary hydrolysis steps, namely, ATP → ADP + P_i and ADP → AMP + P_i. The pH 4 hydrolyses were allowed to proceed for 24 hr, also at 70°, to ~40% completion.

In a parallel set of experiments, we hydrolyzed highly enriched [γ-^{18}O]ATP in 1 N and 0.1 N HCl and at pH 8.0 and, as above, converted the resulting P_i to TMPO and determined its isotopic composition by mass spectrometry. The [γ-^{18}O]ATP was prepared *via* reaction of ADP-morpholidate intermediate with enriched tri-*n*-butylammonium orthophosphate(12), which in turn was prepared by hydrolyzing PCl_5 in a 50% excess of $H_2{}^{18}O$ of 99% isotopic purity.

The isotopic analysis requires a procedure of sufficient specificity and sensitivity to analyze the ^{18}O-enriched TMPO as a ~1% solution in a 90:10 mixture of methanol/water. We have utilized two different instrumental procedures. One employs a gas chromatographic column directly coupled to the mass spectrometer; the other, probe microdistillation in the ion source in conjunction with high-resolution mass measurement. The two procedures have consistently yielded nearly identical analyses.

Isotopic analysis of the [^{18}O] P_i obtained from the reaction of PCl_5 with $H_2{}^{18}O$ shows that the label is incorporated quantitatively into the P_i, and confirms the assumption that the incorporation is a random process. This assumption furnishes the basis for translating the isotopic composition of the P_i obtained from hydrolysis of [γ-^{18}O]ATP into the isotopic composition of the ATP and the contributions of the various reaction paths that participate in the hydrolysis. Thus, we have typically obtained an isotopic enrichment of 96 atom % ^{18}O in the γ-phosphoryl group of the [γ-^{18}O] ATP synthesized from [^{18}O]P_i of 99 atom %

enrichment. Further, isotopic analysis of the P_i obtained from the complete hydrolysis of this typical [γ-^{18}O]ATP to AMP + 2P_i in 1 N and 0.1 N HCl reveals the inadvertent incorporation of ~4% of enriched phosphoryl groups in the β position. In some of our earlier experiments, this value reached 9%; in a few instances, it dropped as low as 2%.

The combined labeling and kinetic studies have enabled us to reach the following conclusions:

(i) In 1 N and 0.1 N HCl the hydrolysis of ATP by addition-elimination occurs with initial attack 93% on P_γ and 7% on P_β; both lead only to ADP + P_i. In the subsequent hydrolysis of the ADP to AMP + P_i, attack is 83% on P_β and 17% on P_α.

(ii) We have detected no exchange of oxygen atoms between water and ATP, ADP, or P_i at any pH. Such exchange would have constituted positive evidence for the formation of an oxyphosphorane intermediate in the addition-elimination mechanism, but the absence of exchange does not rule out such an intermediate. The extent of exchange--and, in particular, its occurrence at a level great enough to allow detection--depends upon the relative rates of exchange and collapse of the intermediate to products. The hypothetical oxyphosphorane intermediates corresponding to P_γ and P_α attack by water on ATP are pictured in formulas **1** and **4** of Figure 1. Those corresponding to the two possible reaction paths associated with P_β attack are pictured in formulas **2** and **3**. The LC results have effectively ruled out the pathways implied by intermediates **3** and **4**. The 7% P_β attack indicated by the labeling results supports the participation of intermediate **2**, although the bulk of the product appears to form *via* **1**.

(iii) The labeling results for ATP hydrolysis at pH 8 indicate a clean process in which the P_i stemming from the primary reaction contains exclusively three oxygen atoms derived from P_γ and one from water. This finding constitutes independent evidence, supplementing the kinetic studies, in support of the elimination-addition mechanism at this pH.

(iv) The 1 N and 0.1 N HCl and the pH 8 hydrolyses of [γ-^{18}O]ATP consistently lead to nearly identical derived values for the isotopic enrichment in the ATP P_γ group. Comparison between the ATP P_γ isotopic composition so determined and that of the same sample arrived at by the myosin-CaATPase procedure(11) shows that even under very carefully selected conditions the myosin-CaATPase hydrolysis may be subject to some 2% exchange of P_γ oxygen atoms with the water. Under less nearly optimum conditions, the use of such calcium-mediated enzymatic hydrolysis as a control is subject to greater uncertainties.

(v) There are basic differences in the course of the nonenzymatic and enzymatic ATP hydrolyses. With respect to the attack by water at the P_β atom of ATP in acid medium, no enzyme is known that catalyzes such a reaction, i. e., there is no known "P_β-ATPase." Three enzymes have been reported that catalyze

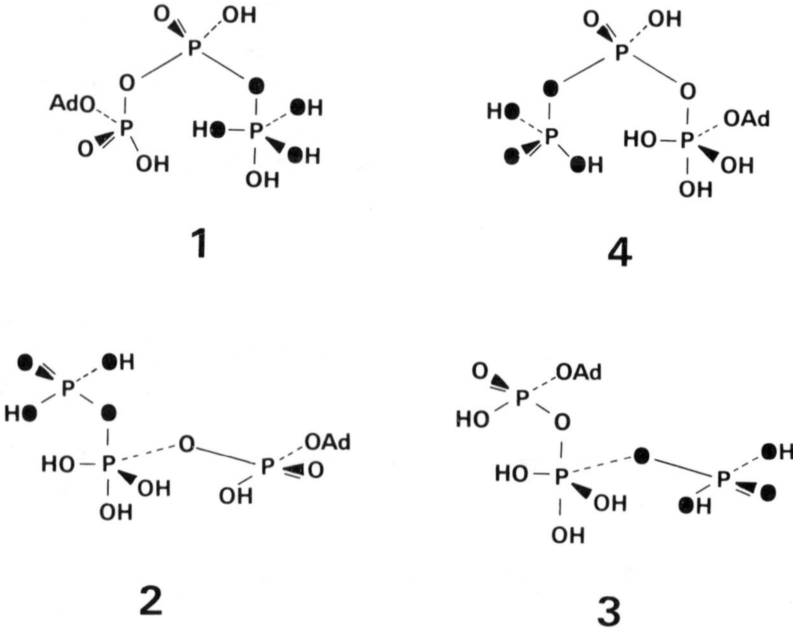

Figure 1. Oxyphosphorane intermediates corresponding to possible primary reaction pathways (1)–(4) in hydrolysis of [γ-^{18}O] *ATP in 1*N *and 0.1*N *HCl by the addition–elimination mechanism. Key: 1 denotes attack by water on* P_γ*; 2 and 3, on* P_β*; 4, on* P_α*;* • *denotes the heavy isotope* ^{18}O.

attack by an alcohol group at P_β of ATP, but in all cases the products are AMP and the product of pyrophosphoryl transfer to the alcohol(13). Thus, the directions of cleavage in nonenzymatic and enzymatic nucleophilic attacks at P_β of ATP are different. In the operation of actomyosin MgATPase in muscle contraction, extensive intermediate oxygen exchange takes place between water in the medium and species bound to the active site. This process generates P_i species that have suffered one, two, and three exchanges as well as no exchange at all(11). In the absence of actin, or when proteolysis of the myosin is carried out to certain enzymatically active subfragments (e.g., the myosin "head" or S1 subfragment), most of the P_i released to the medium has undergone three exchanges, the maximum number possible. The differences in enzymatic vs nonenzymatic ATP hydrolyses have been attributed(10, 11, 14) to effects associated with the tight binding of the MgATP to the protein. This binding results not only from interactions of adenine, ribose, and the polyphosphate chain with the protein, but also from ligation of amino acid residues, notably histidine(10, 14), to the magnesium ion.

Literature Cited

1. Friess, S. L. J. Am. Chem. Soc. 1953, 75, 323.
2. Tetas, M.; Lowenstein, J. M. Biochem. 1963, 2, 350.
3. Schneider, P. W.; Brintzinger, H. Helv. Chim. Acta 1964, 47, 1717.
4. Buisson, D. H.; Sigel, H. Biochim. Biophys. Acta 1974, 343, 45.
5. Taqui Khan, M. M.; Srinivas Mohan, M. J. Inorg. Nucl. Chem. 1974, 36, 707.
6. Taqui Khan, M. M.; Srinivas Mohan, M. Indian J. Chem. 1976, 14A, 945, 951.
7. Satchell, D. P. N.; Satchell, R. S. Annu. Rep. Prog. Chem, Sect. A 1978 1979, 75, 41.
8. Ramirez, F.; Marecek, J. F.; Szamosi, J. J. Org. Chem. 1980, 45, 4748.
9. Miller, D. L.; Westheimer, F. H. J. Am. Chem. Soc. 1966, 88, 1507.
10. Levy, H. M.; Ramirez, F.; Shukla, K. K. J. Theor. Biol. 1979, 81, 327.
11. Shukla, K. K.; Levy, H. M.; Ramirez, F.; Marecek, J. F.; Meyerson, S.; Kuhn, E. S. J. Biol. Chem. 1980, 255, 11344.
12. Wehrli, W. E.; Verheyden, D. L. M.; Moffatt, J. G. J. Am. Chem. Soc. 1965, 87, 2265.
13. Switzer, R. L., in "The Enzymes"; Boyer, P. D., Ed.; Academic Press: New York, 1974; Chap 19.
14. Sarma, R.; Ramirez, F.; Narayanan, P.; McKeever, B.; Marecek, J. F. J. Am. Chem. Soc. 1979, 101, 5015.

RECEIVED July 7, 1981.

19

Nucleoside Phosphorothioates for the Study of Enzyme Mechanisms

F. ECKSTEIN

Max-Planck-Institut für Experimentelle Medizin, Abteiling Chemie, Hermann-Rein-Strasse 3, D-3400 Göttingen, FRG

In recent years nucleotides have become available which can be applied to the study of two problems associated with enzymatic phosphoryl and nucleotidyl transfer reactions: the stereochemical course of the reaction as well as the structure of the active metal nucleotide complex. Isotopically labelled ATP has been developed for the determination of the stereochemistry of such reactions (1,2) and exchange inert Cr^{III} and Co^{III} complexes of ATP for the elucidation of the metal substrate complexes (3,4). A third class of compounds, the nucleoside phosphorothioates, are suitable for investigations of both these questions (5,6). This versatility becomes apparent when one considers that exchange of a non-bridging oxygen at either the α- or ß-phosphorus of ATP by sulfur renders the particular phosphorus chiral which gives rise to the existence of a pair of diastereomers each of ATPαS as well as of ATPßS. Although such a substitution of α-phosphorus does not create a new centre of chirality, the additional stereospecific exchange of one ^{16}O for one ^{18}O does (7,8). All these diastereomers of ATPαS, ATPßS and ATPγS have been synthesized and their absolute configurations determined. Thus the stage is set for investigations of the stereochemical courses of transfer reactions involving any of these three phosphorus as long as the absolute configuration of the products of such reactions can be assigned.

That compounds like the diastereomers of ATPαS and ATPßS can also be used to study the metal coordination to the triphosphate in an enzymatic reaction is based on the observation that the selectivity of an enzyme for one of the two diastereomers of ATPαS or ATPßS depends on whether a hard (Mg^{2+}) or soft (Cd^{2+}) metal is used in the reaction (9). The explanation for this is that metal ions such as Mg^{2+} preferentially coordinate to oxygen and Cd^{2+} to sulfur. Thus, with the same diastereomer Mg^{2+} forms a complex with Δ (3) and with Cd^{2+} with Λ screw sense, for example. Since the active site of an enzyme is asymmetric it will have a preference for the Δ or Λ screw sense isomer of the metal-nucleotide complex. It follows that if e.g. the S_p-isomer is active in the presence of Mg^{2+}, the R_p-isomer is expected to be active in the presence of

0097-6156/81/0171-0099$05.00/0

Cd^{2+} as long as the metal ion is really coordinated to that phosphorothioate group.

Using two enzymes, a mammalian adenylate cyclase and myosin ATPase, as examples the application of phosphorothioate analogues to the study of the mechanism of nucleotidyl and phosphoryl transfer will be described.

Adenylate Cyclase

Mammalian adenylate cyclase is a membrane bound highly regulated enzyme. Since it has not been purified to homogeneity yet, many questions concerning the mechanism of cyclisation are so far unsolved. One of these is whether the reaction proceeds via a covalent enzyme intermediate. We have applied the diastereomers of $[^{35}S]$ATPαS to investigate this problem (10). It was found that only the S_p-isomer is a substrate. The analysis of the product $[^{35}S]$cAMPS was performed by high performance liquid chromatography where the two diastereomers whose synthesis has been described (11), can be separated. Radioactivity was found to be associated with the R_p-isomer only. Thus, the reaction has proceeded with inversion of configuration. The simplest interpretation of this is that it is the result of a single S_N2 reaction. It indicates that cyclisation has proceeded without going through a covalent enzyme intermediate since this would involve two nucleophilic substitution steps, resulting in overall retention of configuration. Thus, the 3-OH group directly attacks the α-phosphorus.

The question arises whether the result obtained with the phosphorothioate analogue is of any relevance for the reaction with normal ATP or whether the enzyme alters its mechanism when encountering a sulfur instead of an oxygen. That this is not the case is shown by an investigation of the reverse reaction with a bacterial enzyme using $[^{18}O]$dcAMP as substrate (12). This reaction also proceeds with inversion of configuration. Other enzymes such as glycerol kinase (13), cyclic phosphodiesterase (14,15), hexokinase, pyruvate kinase and snake venom phosphodiesterase (16) have also been investigated with the phosphorothioates as well as with isotopically labelled ATP and in all cases the stereochemistry observed is the same. Thus, apparently enzymes do not change at least the stereochemical course of the reaction although the rate of the chemical step can be considerably reduced (17).

Myosin ATPase

The subfragment 1 of myosin was investigated with the diastereomers of ATPαS as well as ATPβS in the presence of various cations, particularly Mg^{2+}, Cd^{2+}, NH_4^+ and Tl^+ (18). The S_p/R_p activity ratios obtained with the isomers of ATPαS were largely independent of the cation. The simplest explanation for this result is that the cations are not coordinated to the α-phosphate group. Additional support for this contention comes from experiments employing the diastereomers of ATPαS which are methylated at sulfur making, thus, metal coordination to the α-phosphate group extremely unlikely. These compounds are still substrates for the myosin ATPase.

With ATPßS essentially only the S_p-isomer is a substrate with a S_p/R_p activity ratio in presence of Mg^{2+} of > 3000, in presence of Cd^{2+} of 0.2. With NH_4^+ a ratio of 8 and in presence of Tl^+ of 0.04 was found. This data indicate that both mono- and divalent cations coordinate to the ß-phosphate group of the nucleoside triphosphate substrate. The combined results obtained with ATPαS and ATPßS suggest that myosin uses the metal α,ß,γ-bidentate nucleotide chelate as substrate. This information together with the known stereochemical course of the ATPase reaction (19) allows a description of the events at the acitve site in stereochemical terms.

These two examples illustrate the usefulness of the phosphorothioate analogues of nucleotides as tools for the elucidation of certain aspects of enzymatic nucleotidyl and phosphoryl transfer reactions.

Literature Cited

1. Knowles, J.R. Ann. Rev. Biochem. 1980, 49, 877.
2. Buchwald, S.L., Hansen, D.E., Hassett, A., Knowles, J.R. Methods in Enzymology, in press.
3. Dunaway-Mariano, D., Cleland, W.W. Biochemistry 1980, 19, 1496.
4. Dunaway-Mariano, D., Cleland, W.W. Biochemistry 1980, 19, 1506.
5. Eckstein, F. Acc. Chem. Res. 1979, 12, 204.
6. Eckstein, F. Trends in Biochem. Sci. 1980, 157.
7. Orr, G.A., Simon, J., Jones, S.R., Chin, G.J., Knowles, J.R. Proc. Natl. Acad. Sci. USA 1978, 75 , 2230.
8. Richard, J.P., Ho, H.T., Frey, P.A. J. Amer. Chem. Soc. 1978, 100, 7756.
9. Jaffe, E.K., Cohn, M. J. Biol. Chem. 1978, 253, 4823.
10. Eckstein, F., Romaniuk, P.J., Heideman, W., Storm, D.R. J. Biol. Chem. in press.
11. Baraniak, J., Kinas, R.W., Lesiak, K., Stec, W.J. J.C.S. Chem. Com. 1979, 940.
12. Coderre, J.A., Gerlt, J.A. J. Amer. Chem. Soc. 1980, 102, 6594.
13. Pliura, D.H., Schomberg, D., Richard, J.P., Frey, P.A., Knowles, J.R. Biochemistry 1980, 19, 325.
14. Jarvest, R.L., Lowe, G., Potter, B.V.L. J.C.S. Chem. Com. 1980, 1142.
15. Cullis, P.M., Jarvest, R.L., Lowe, G., Potter, B.V.L. J.C.S. Chem. Com. 1981, 245.
16. Lowe, G., Cullis, P.M., Jarvest, R.L., Potter, B.V.L., Sproat, B.S. Phil. Trans. Roy. Soc. London, in press.
17. Bagshaw, C.R., Eccleston, J.F., Trentham, D.R., Yales, D.W., Goody, R.S. Cold Spring Harbor Symp. Quant. Biol. 1972, 37, 127.
18. Connolly, B., Eckstein, F. J. Biol. Chem., submitted.
19. Webb, M.R., Trentham, D.R. J. Biol. Chem. 1980, 255, 8629.

RECEIVED July 7, 1981.

Chiral [^{16}O, ^{17}O, ^{18}O] Phosphate Monoesters for Determining the Stereochemical Course of Phosphokinases

GORDON LOWE, PAUL M. CULLIS, RICHARD L. JARVEST, and BARRY V. L. POTTER

The Dyson Perrins Laboratory, Oxford University, South Parks Road, Oxford OX1 3QY England

Phosphokinases catalyse a large and important group of reactions which play a key role in intermediary metabolism and its regulation. It appears that in general they follow a sequential pathway, that is, the phosphoryl transfer step takes place only when both phosphoryl donor and phosphoryl acceptor are bound to the enzyme. Kinetic criteria however provide no evidence concerning the mechanism of the phosphoryl transfer step itself. It is commonly observed that phosphokinases catalyse partial reactions, such as ATP-ADP exchange in the absence of co-substrate, albeit at rates significantly slower than the overall reaction. Such results imply the intervention of a phosphoryl-enzyme intermediate in the exchange reaction, and it is always possible to invoke substrate synergism to suggest that the putative phosphoryl-enzyme could be a kinetically competent intermediate when all substrates are present. Such a suggestion however would be insecurely founded.

In order to solve this mechanistic problem, a method for analysing the stereochemical course of phosphokinases has been developed using chiral[^{16}O,^{17}O,^{18}O]phosphate esters. Stereochemical analysis should allow a clear mechanistic distinction to be made, since inversion of configuration at phosphorus implies a direct 'in-line' phosphoryl transfer mechanism whereas retention of configuration suggests a double-displacement mechanism with a phosphoryl-enzyme intermediate on the reaction pathway.

The general route developed for the synthesis of chiral [^{16}O,^{17}O,^{18}O]phosphate monoesters of known absolute configuration is shown in Scheme 1(1).

The stereochemical analysis depends on converting a chiral[^{16}O,^{17}O,^{18}O]phosphate monoester into two diastereoisomeric conformationally locked six membered cyclic phosphate triesters (2). In the cyclization step any one of the peripheral oxygen isotopes will be lost with equal probability, and the residual oxygen isotopes will take up axial and equatorial positions. Methylation of the isotopically labelled cyclic phosphate

0097–6156/81/0171–0103$05.00/0

Scheme 1. Synthesis of [(S)-^{16}O, ^{17}O, ^{18}O] *phosphate monoesters: Reagents i, PhLi; ii, $HOCH_2CH_2OH$, p-$MeC_6H_4SO_3H$; iii, $H_2{}^{18}O$, dioxan, p-$MeC_6H_4SO_3H$; iv, $LiAlH_4$; v, $P^{17}OCl_3$, C_5H_5N; ROH, C_5H_5N; vi, H_2, Pd/C.*

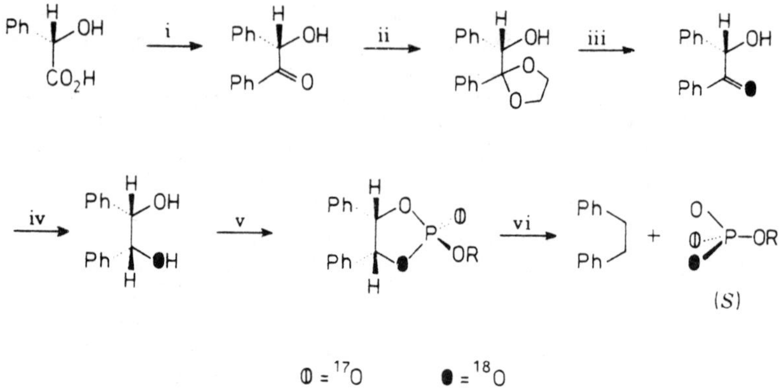

diester should give the diastereoisomeric axial and equatorial triesters. D-Glucose 6[(*S*)-$^{16}O,^{17}O,^{18}O$]phosphate is used in Scheme 2 to illustrate the method, but adenosine 5'[(*S*)-$^{16}O,^{17}O,^{18}O$]phosphate has also been cyclized and methylated in essentially the same way. Species 1 - 6 would be formed if cyclization occurred with retention of configuration at phosphorus, whereas 7 - 12 would be formed if cyclization occurred with inversion of configuration. If all the sites were fully enriched as indicated, only 1 and 4 or 7 and 10 would be observed in the ^{31}P NMR spectrum since the nuclear electric quadrupole moment of ^{17}O when directly bonded to phosphorus, broadens and effectively obliterates the phosphorus resonance. Moreover ^{18}O when directly bonded to phosphorus causes an isotope shift which is smaller when singly bonded than when doubly bonded to phosphorus, so that 1 and 4 can be distinguished from 7 and 10(3). In practice, the sites labelled ^{17}O are not fully enriched, so that species containing ^{16}O and ^{18}O at such sites will also be present and will be observed in the ^{31}P NMR spectrum. This in fact is an advantage since the [$^{16}O_2$]- and [$^{18}O_2$]-triesters arising in this way will provide the necessary internal references, and the ratio of the [$^{16}O_{ax}$, $^{18}O_{eq}$]- to [$^{18}O_{ax}$, $^{16}O_{eq}$]-triesters will provide the stereochemical information. Thus in practice, 1 and 4 or 7 and 10 will predominate in the ^{31}P NMR spectrum but will not be the exclusive resonances.

D-Glucose 6[(*S*)-$^{16}O,^{17}O,^{18}O$]phosphate was cyclized with diphenylphosphorochloridate and potassium t-butoxide and esterified with methyl iodide. The cyclization occurs with inversion of configuration at phosphorus since 7 and 10 predominated in the ^{31}P NMR spectrum; it was possible to calculate that the stereospecificity of the reaction was in excess of 94%. Adenosine 5'[(*S*)-$^{16}O,^{17}O,^{18}O$]phosphate was similarly cyclized and esterified, the cyclization occurring with inversion of configuration in excess of 94% (2).

Adenosine 5'[γ(*S*)-$^{16}O,^{17}O,^{18}O$]triphosphate, prepared by the general method of synthesis was incubated with D-glucose and yeast hexokinase. The D-glucose 6[$^{16}O,^{17}O,^{18}O$]phosphate was shown to have the (*R*)-configuration at phosphorus by ^{31}P NMR spectroscopy after cyclization and esterification (4). Incubation of adenosine 5'[γ(*S*)-$^{16}O,^{17}O,^{18}O$]triphosphate with adenosine 3'-phosphate and polynucleotide kinase (from T_4 infected *E. coli* gave adenosine 3',5'[$^{16}O,^{17}O,^{18}O$]bisphosphate which was converted into adenosine 5'[$^{16}O,^{17}O,^{18}O$]phosphate with nuclease P_1. After cyclization and esterification the ^{31}P NMR spectrum showed the adenosine 5'[$^{16}O,^{17}O,^{18}O$]phosphate to have the (*R*)-configuration at phosphorus (5). Thus both yeast hexokinase and T_4 infected *E.coli* polynucleotide kinase catalyse phosphoryl transfer with inversion of configuration at phosphorus.

Since ATP is a common substrate for all phosphokinases, it is now possible to analyse the stereochemical course of any

Scheme 2. Cyclization of D*-glucose 6[(S)-^{16}O, ^{17}O, ^{18}O]phosphate with retention of configuration at phosphorus followed by methylation should give 1–6, whereas cyclization with inversion of configuration at phosphorus should give 7–12 after methylation.*

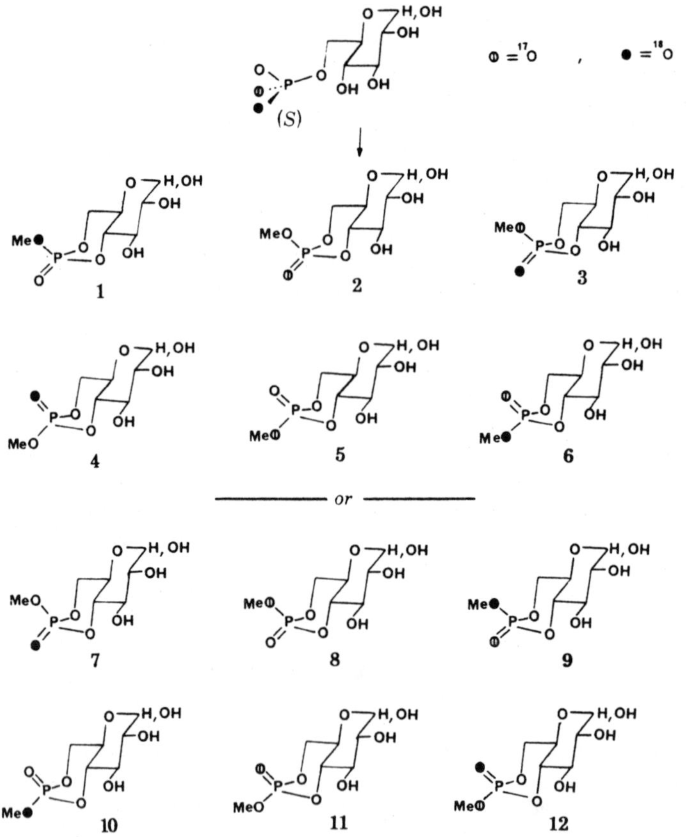

enzyme of this class by preparing the appropriate [(*S*)-$^{16}O,^{17}O,^{18}O$]phosphate ester and enzymically transferring the chiral phosphoryl group to ADP. The ATP[γ-$^{16}O,^{17}O,^{18}O$]triphosphate can then be used as a substrate for hexokinase and the D-glucose 6[$^{16}O,^{17}O,^{18}O$]phosphate so formed, analysed for chirality at phosphorus as before.

In this way we have shown that phosphoryl transfer catalysed by *Bacillus stearothermophilus* and rabbit skeletal muscle phosphofructokinase (6), and rabbit skeletal muscle pyruvate kinase occurs with inversion of configuration at phosphorus (7). The simplest interpretation of these stereochemical results is that phosphoryl transfer occurs by an 'in-line' mechanism in the enzyme substrate ternary complexes. Stereochemical analysis is thus proving to be of considerable importance for delineating the mechanism adopted by phosphokinases.

Literature Cited

1. Cullis, P.M. and Lowe, G., J. Chem. Soc. Perkin Trans. 1, 1981 in press.
2. Jarvest, R.L., Lowe, G. and Potter, B.V.L., J. Chem. Soc. Perkin 1, 1981 in press.
3. Lowe, G., Potter, B.V.L., Sproat, B.S. and Hull, W.E., J. Chem. Soc. Chem. Comm., 1979, 733.
4. Lowe, G. and Potter, B.V.L., Biochem. J., submitted for publication.
5. Jarvest, R.L. and Lowe, G., Biochem. J., submitted for publication.
6. Jarvest, R.L., Lowe, G. and Potter, B.V.L., Biochem. J., submitted for publication.
7. Lowe, G., Cullis, P.M., Jarvest, R.L., Potter, B.V.L. and Sproat, B.S., Phil. Trans. R. Soc. Lond. 1981, B293, 75.

RECEIVED June 30, 1981.

Syntheses and Configurational Assignments of Thymidine 3′- and 5′-(4-Nitrophenyl [^{17}O,^{18}O] Phosphates)

SHUJAATH MEHDI, JEFFREY A. CODERRE,[1] and JOHN A. GERLT

Department of Chemistry, Yale University, New Haven, CT 06511

Determination of the stereochemical course of nucleophilic displacement reactions at phosphorus in enzymology is the most direct experimental method for determining if an enzyme catalyzed reaction involves the formation of a covalent adduct between the enzyme and substrate. Whereas the classical approach to this problem has been to use chiral phosphorothioate esters which are analogs of the natural substrates (1,2), recent work reported by a number of laboratories has demonstrated that the syntheses and configurational assignments of oxygen chiral phosphate mono- and diesters are technically feasible (3-6). These recent advances allow the stereochemical course of phosphoryl and nucleotidyl transfer reactions to be investigated with oxygen chiral substrates; such studies are desirable because the results obtained with the phosphorothioate analogs are mechanistically ambiguous due to the low rates at which these substrate analogs are processed by enzymes. At present, stereochemical experiments employing both the phosphorothioate and oxygen chiral approaches have been performed on three enzymes: glycerol kinase from yeast (7,8), adenylate cyclase from *Brevibacterium liquefaciens* (9,10), and cyclic nucleotide phosphodiesterase from bovine heart (11,12). In each case, results were obtained which demonstrated that sulfur substitution does not alter the stereochemical course of the enzymatic reaction.

To facilitate determination of the stereochemical course of both enzymatic and nonenzymatic hydrolyses of phosphodiesters, we have decided to prepare phosphodiesters which are [^{17}O, ^{18}]-chiral so that the hydrolysis reactions can be performed in $H_2{}^{16}O$. We have previously reported the syntheses and configurational analyses of the diastereomers of cyclic 2'-deoxyadenosine 3',5'-[^{17}O, ^{18}O]monophosphate (13) and the stereochemical course of the hydrolysis of one of the diastereomers catalyzed by the cyclic nucleotide phosphodiesterase from bovine heart (12). In this communication we report the syntheses and configurational assign-

[1] Current address: Department of Biochemistry and Biophysics, University of California, San Francisco, CA 94143

0097-6156/81/0171-0109$05.00/0

ments of the diastereomers of thymidine 3'- and 5'-(4-nitrophenyl [^{17}O, ^{18}O]phosphates); these oxygen chiral phosphodiesters will be used to determine the stereochemical course of reactions catalyzed by nucleases.

The diastereomeric [^{17}O]-enriched 4-nitrophenyl esters of the P-anilidates of 3'-monomethoxytrityl thymidine and of 5'-monomethoxytrityl thymidine were prepared according to the procedures developed in this laboratory for the preparation of unlabelled materials (14), except that [^{17}O]$POCl_3$ was used to prepare the required 4-nitrophenyl N-phenyl phosphoramidic chloride; the [^{17}O]-enrichment of the labelled $POCl_3$ was 51%. The diastereomers were separated by short column chromatography on Merck silica gel 60H. The [^{17}O]-P-anilidates were reacted separately with a 10-fold excess of 99% enriched $C^{18}O_2$ and after removal of the monomethoxytrityl groups, the products were purified by chromatography on Amberlite XAD-2. The chemistry is summarized in the Figures. The chiral [^{17}O, ^{18}O]-phosphodiesters which were obtained were identical with authentic thymidine 4-nitrophenyl phosphates using the criteria of TLC and ^{1}H NMR at 270 MHz; the ^{31}P spectra at 32 MHz and at 81 MHz revealed the expected ratio of [^{16}O, ^{18}O]- and [^{18}O, ^{18}O]-resonances.

Although the configurations of the unlabelled P-anilidates have been assigned (14) and the reaction of cyclic P-anilidates with $C^{18}O_2$ (5) or [^{18}O]benzaldehyde (6) has been shown to proceed with the anticipated retention of configuration at phosphorus, we considered it necessary to determine the configurations of the acyclic [^{17}O,^{18}O]-chiral phosphodiesters. The required configurational assignments have been accomplished by two independent methods.

The most straightforward method of configurational assignment is to determine the configuration of the cyclic thymidine 3',5'-[^{17}O, ^{18}O]monophosphate which can be obtained by tert-butoxide induced cyclization of the 4-nitrophenyl ester; Borden and Smith have reported that this is a facile method for the preparation of 3',5'-cyclic nucleotides (15). The accepted principles of nucleophilic displacement reactions at phosphorus allow the prediction that this cyclization reaction should proceed with inversion of configuration at phosphorus (16). Accordingly, we have converted each of the [^{17}O,^{18}O]-chiral acyclic nucleotide esters to cyclic thymidine 3',5'-[^{17}O, ^{18}O]monophosphates using the conditions described by Borden and Smith (15); cyclic nucleotides were obtained in yields of 70% following purification by chromatography on DEAE-Sephadex A-25.

The configurations of the cyclic thymidine 3',5'-[^{17}O, ^{18}O]-monophosphates can be conveniently assigned by ^{17}O NMR spectroscopy. We have previously reported that in aqueous solution at 95° C the ^{17}O NMR chemical shifts of the phosphoryl oxygens of cyclic 2'-deoxyadenosine 3',5'-[^{17}O, ^{18}O]monophosphate are sufficiently different at 36.6 MHz such that two resonances can be resolved in a racemic mixture of the diastereomers, with the down-

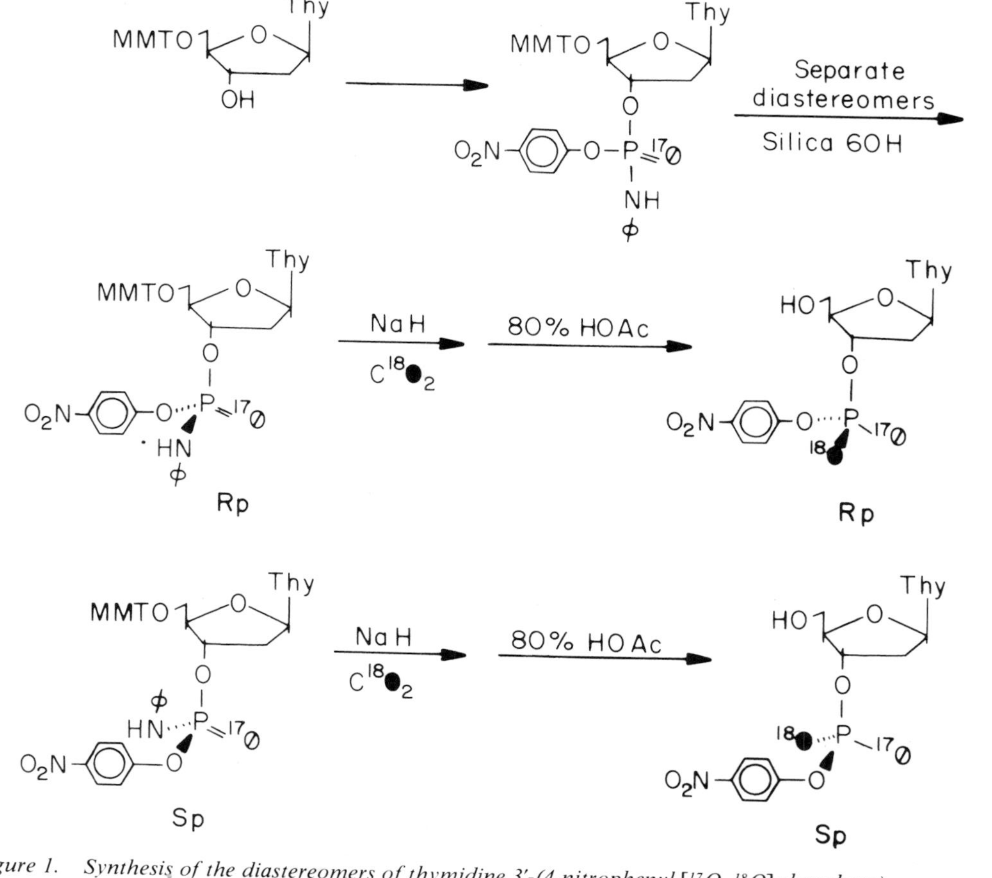

Figure 1. Synthesis of the diastereomers of thymidine 3′-(4-nitrophenyl [^{17}O, ^{18}O]phosphate).

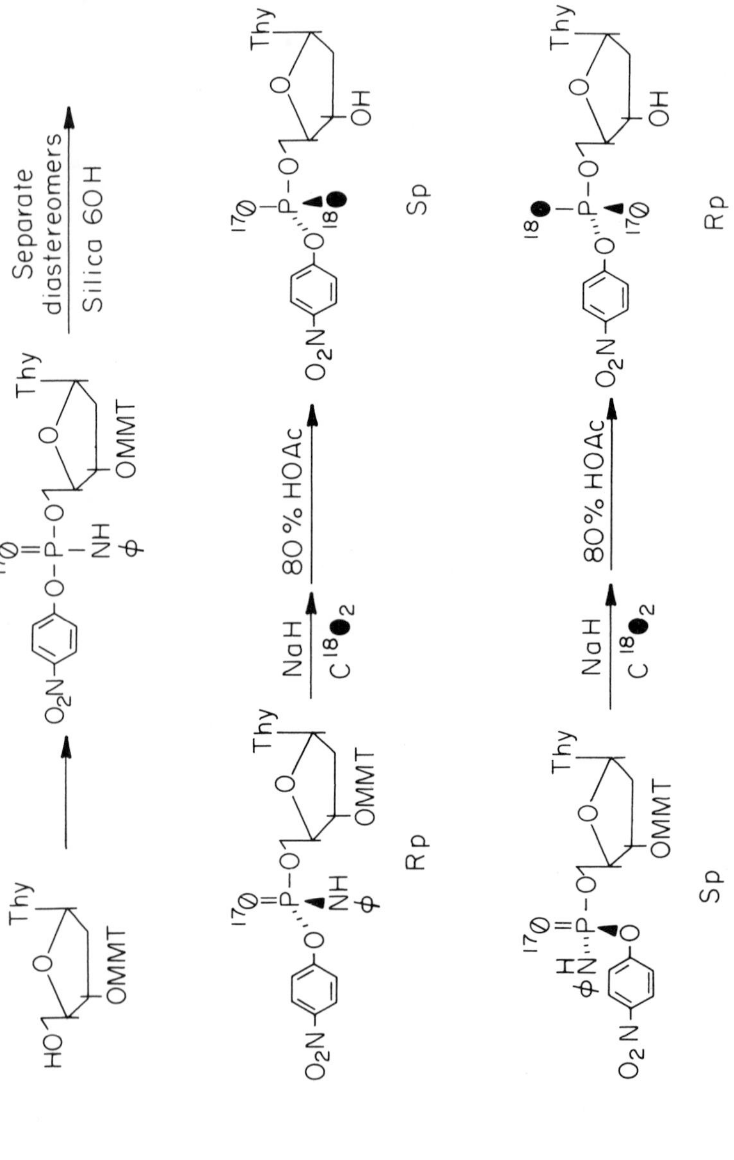

Figure 2. Synthesis of the diastereomers of thymidine 5'-(4-nitrophenyl [^{17}O, ^{18}O] phosphate).

field resonance being associated with the axially positioned [^{17}O]-nucleus (13); it is unlikely that the identity of the heterocyclic base would influence the chemical shifts of the phosphoryl oxygens. One major resonance was observed in the ^{17}O NMR spectrum of each cyclic thymidine 3',5'-[^{17}O, ^{18}O]monophosphate; the chemical shifts were those expected if the reaction of the acyclic P-anilidates with $C^{18}O_2$ proceeded with the predicted retention of configuration at phosphorus. The configurations of the chiral [^{17}O, ^{18}O]-cyclic phosphodiesters were verified by conversion to a mixture of axial and equatorial methyl esters and measurement of the [^{18}O]-perturbations on the ^{31}P NMR chemical shifts, as previously described (12).

Because the configurational assignments derived from ^{17}O NMR chemical shifts rely on a reaction whose stereochemical course had not been firmly established (tert-butoxide induced cyclization), we have also assigned the configurations of the acyclic 4-nitrophenyl esters in an unambiguous fashion. Ethanolic solutions of the acyclic esters were hydrogenolyzed in the presence of Adam's catalyst and an excess of HCl. The reduction proceeds via C-O bond cleavage and results in the formation of [^{16}O, ^{17}O, ^{18}O]-chiral phosphate monoesters. Configurational analysis of these chiral monoesters can be accomplished by cyclization using reactions of known stereochemical course to yield a mixture of three types of chiral cyclic thymidine 3',5'-monophosphate, i.e., [^{16}O, ^{17}O]-, [^{17}O, ^{18}O]-, and [^{16}O, ^{18}O]-labelled; the configuration of the [^{16}O, ^{18}O]-chiral cyclic ester present in the mixture can be accomplished by measurements of the [^{18}O]-perturbations on the ^{31}P chemical shifts of the axial and equatorial methyl esters. In the absence of appropriate enzymes (12), this procedure must be carried out chemically. We have selected activation by diphenylphosphorochloridate followed by tert-butoxide induced cyclization to carry out the required ring closure, since recent studies have demonstrated that this reaction sequence is accompanied by the predicted inversion of configuration (16) at the chiral phosphorus atom (4,12). Following hydrogenolysis, chemical activation and cyclization, and methylation, measurements of the [^{18}O]-perturbations on the ^{31}P chemical shifts of the methyl esters revealed that the configurations of the acyclic 4-nitrophenyl phosphates were those predicted if the reaction of the acyclic P-anilidates with $C^{18}O_2$ proceeded with retention of configuration at phosphorus. Thus, the configurations of the acyclic [^{17}O, ^{18}O]-chiral 4-nitrophenyl esters can be considered to be firmly established.

Determination of the stereochemical course of the reactions catalyzed by the exonucleases from snake venom and bovine spleen and by Staphylococcal nuclease is in progress.

Acknowledgements

This research was supported by a grant from the National Institutes of Health (GM-22350). The high field NMR spectrometers used in this research are supported by a grant from the National Science Foundation (CHE-791620). J.A.G. is the recipient of a Research Career Development Award from the National Institutes of Health (CA-00499, 1978-83) and of a fellowship from the Alfred P. Sloan Foundation (1981-83).

Literature Cited

1. Eckstein, F. Acc. Chem. Res. 1979, 12, 204.
2. Knowles, J. R. Annu. Rev. Biochem. 1980, 49, 877.
3. Abbott, S. J.; Jones, S. R.; Weinman, S. A.; Bockhoff, F. M.; McLafferty, F. W.; Knowles, J. R. J. Am. Chem. Soc. 1979, 101, 4323.
4. Cullis, P. M.; Jarvest, R. L.; Lowe, G.; Potter, B. V. L. J. Chem. Soc., Chem. Commun. 1981, 245.
5. Gerlt, J. A.; Coderre, J. A. J. Am. Chem. Soc. 1980, 102, 4531.
6. Baraniak, J.; Lesiak, K.; Sochacki, M.; Stec, W. J. J. Am. Chem. Soc. 1980, 102, 4533.
7. Blättler, W. A.; Knowles, J. R. J. Am. Chem. Soc. 1980, 102, 510.
8. Pliura, D. H.; Schomburg, D.; Richard, J. P.; Frey, P. A.; Knowles, J. R. Biochemistry 1980, 19, 325.
9. Gerlt, J. A.; Coderre, J. A.; Wolin, M. S. J. Biol. Chem. 1980, 255, 331.
10. Coderre, J. A.; Gerlt, J. A. J. Am. Chem. Soc. 1980, 102, 6594.
11. Burgers, P. M. J.; Eckstein, F.; Hunneman, D. H.; Baraniak, J.; Kinas, R. W.; Lesiak, K.; Stec, W. J. J. Biol. Chem. 1979, 254, 9959.
12. Coderre, J. A.; Mehdi, S.; Gerlt, J. A. J. Am. Chem. Soc. 1981, 103, 1872.
13. Coderre, J. A.; Mehdi, S.; Demou, P. C.; Weber, R.; Traficante, D. D.; Gerlt, J. A. J. Am. Chem. Soc. 1981, 103, 1870.
14. Gerlt, J. A.; Mehdi, S.; Coderre, J. A.; Rogers, W. O. Tetrahedron Lett. 1980, 2385.
15. Borden, R. K.; Smith, M. J. Org. Chem. 1966, 31, 3247.
16. Westheimer, F. H. in "Rearrangements in Ground and Excited States;" DeMayo, P., Ed.; Academic Press, New York, 1980; Vol. II, p. 229.

RECEIVED June 30, 1981.

22

The Mechanism of Aldehyde-Induced ATPase Activities of Kinases

W. W. CLELAND and ALAN R. RENDINA

Department of Biochemistry, University of Wisconsin, Madison, WI 53706

Kinases are enzymes which transfer the γ-phosphate of MgATP to various acceptors. While much is known about the stereochemistry (1), we know little about the chemical mechanism except that the acid-base catalyst which accepts the proton from the alcohol in the hexokinase (2) and fructokinase (3) reactions is a carboxyl group, while that for creatine kinase is a histidine (4). The ATPase activity induced by suitable aldehydes is thus a reaction of considerable interest. Such activity was first seen with glycerokinase (5), which phosphorylates L-glyceraldehyde at the 3 position, but in the presence of D-glyceraldehyde splits MgATP to MgADP and P_i with no evidence of any intermediates being formed. It was thought at the time that the ATPase activity involved phosphorylation of the aldehyde hydrate to give the phosphate adduct of the aldehyde, which promptly decomposed. We have now found that this is not correct, and in this paper we will detail studies on 3 enzymes which show ATPase activites in the presence of aldehydes and on two which do not, and discuss the possible chemistry of the reaction.

While checking a sample of 2,5-anhydromannose-6-P for fructose-6-P by incubating it with phosphofructokinase and MgATP, we discovered that this aldehyde, which is sterically hindered from forming an internal hemiacetal, induced an ATPase activity (6). Since aldehyde hydration shows a large inverse equilibrium isotope effect of 0.73 when the hydrogen on the carbonyl carbon is replaced by deuterium (7,8), 2,5-anhydromannose-6-P-1-d will be 60% hydrated, compared to 52% hydration of the unlabeled aldehyde. If the free aldehyde were the activator, 48% of the unlabeled and 40% of the deuterated compound would be active, and a normal deuterium isotope effect of 0.48/0.40 = 1.2 would be seen on V/K (the apparent first order rate constant) for the activator, while if the hydrate were the active form, an inverse isotope effect of 0.52/0.60 = 0.87 would be seen. The observed value of 1.23 ± 0.03 showed that the free aldehyde and not the hydrate was the activator (6).

0097–6156/81/0171–0115$05.00/0

The absence of an isotope effect on the V_{max} for the ATPase reaction, however, means that the hydrate does not bind appreciably in the active site as a competitive inhibitor (when a competitive inhibitor is present in the variable substrate, there is no effect on V/K, but V_{max} is decreased). A normal isotope effect on V_{max} of 1.2 would be seen if the hydrate and free aldehyde had the same affinity for the enzyme. The failure of the hydrate to bind presumably reflects steric restraints in the active site. Since the free aldehyde and a water molecule would take up more space than the hydrate, these data suggest that the aldehyde-induced ATPase does not result from induction by the aldehyde of the proper conformation change to permit reaction of MgATP with a bound water molecule.

When we attempted to induce ATPase activity by acetate kinase with redistilled acetaldehyde, a slow release of ADP from MgATP was seen. However, acetaldehyde freshly passed over a column of Dowex-1-Cl, or condensed from a stream of nitrogen passing over an aqueous solution at neutral pH did not show this activity, and we conclude that the initial observation was due to traces of acetic acid. Acetaldehyde did competitively inhibit acetate kinase, however, with a K_i of 57 mM for unlabeled and 49 mM for deuterated acetaldehyde. This inverse isotope effect of 0.87 shows that the hydrate is the inhibitory species (acetaldehyde is 60% hydrated and acetaldehyde-1-d will be 67% hydrated, so that the K_i of the deuterated species should be lower by the ratio 0.60/0.67 = 0.89). Since the active site of acetate kinase has room for both oxygens of acetate, it is not surprising to find that the hydrate of acetaldehyde is bound in the active site. The failure of the hydrate to be phosphorylated (which would result in ATPase activity) could be due to incorrect geometry, or to the lack of an acid-base catalyst to accept the proton of the hydroxyl group being phosphorylated. With 3-P-glycerate kinase, which also catalyzes phosphorylation of a carboxyl group by MgATP, D-glyceraldehyde-3-P did not induce ATPase activity.

2,5-Anhydromannose is phosporylated at the 6-hydroxyl by fructokinase as well as by hexokinase (9). When we used fructokinase and hexokinase to check the concentrations of our 2,5-anhydromannose preparations (from ADP monitored by a pyruvate kinase, lactate dehydrogenase couple), more ADP was produced with fructokinase, and the excess ADP was accompanied by an equal amount of inorganic phosphate. 2,5-Anhydromannose is thus inducing an ATPase activity by fructokinase as well as becoming phosphorylated. Because of the symmetry of 2,5-anhydromannose, when C-6 is adsorbed in the active site, phosphorylation occurs, while when C-1 is adsorbed, MgATP is split to MgADP and P_i. Since the two reactions are competitive, the product ratio equals the ratio of V/K values for the two reactions. Thus: $(V/K)_{ATPase}/(V/K)_{kinase} = [ADP]_{FK}/[ADP]_{HK} - 1$. With purified 2,5-anhydromannose the (ATPase)/(kinase) ratio was 0.13, and was constant from pH 5.5 to 9.4. Since V/K for the kinase reaction decreases below a pK of 6 (3), the ATPase reaction must show the same pH dependence.

The (ATPase)/(kinase) ratio was higher for unlabeled 2,5-anhydromannose than for the deuterated compound by a factor of 1.22, while for the two compounds as substrates for fructokinase (ADP production followed) the V/K isotope effect was 1.04 ± 0.02, and there was no V_{max} isotope effect. If only the free aldehyde is an activator for the ATPase activity, while the free aldehyde and the hydrate have equal V/K values for the kinase reaction, the predicted isotope effect on the (ATPase)/(kinase) ratio is 1.19, while that on the V/K for ADP production is 1.024. Thus, the aldehyde is the activator for fructokinase as well as for phosphofructokinase.

With glycerokinase, the ATPase activity induced by D-glyceraldehyde shows the same pH profile for the V/K of MgATP as the kinase activity, decreasing at low pH as the group which is presumably the acid-base catalyst becomes protonated. Since the most likely chemical mechanisms for the aldehyde-induced ATPase reactions appeared to be metaphosphate cleavage of MgATP or direct phosphorylation of the aldehyde to give an unstable oxycarbonium ion, we decided to run the ATPase reaction in the presence of methanol, which should react with metaphosphate to give methyl phosphate, or with a phosphorylated aldehyde to give a phosphoryl methyl acetal. When [^{32}P]-ATP was incubated with D-glyceraldehyde and glycerokinase in 35% [^{14}C]-methanol and the products chromatographed on Dowex-1 with a borate gradient no compounds containing both ^{14}C and ^{32}P were detected. In a similar reaction mixture, inorganic phosphate and ADP were formed at equal rates, and thus it appears that if metaphosphate or a phosphorylated aldehyde were formed, that they reacted with water (either trapped in the active site, or being the first molecule to have access to the active site after reaction) in total preference to methanol.

The failure to isolate a stable methanol and phosphate-containing compound from ATPase reactions run in the presence of methanol leaves the chemical mechanism of the ATPase reaction in doubt. The requirement for the proper protonation state of the acid-base catalyst implies either that the aldehyde is phosphorylated to an oxycarbonium ion, with the negative charge on the acid-base group stabilizing the positive charge on the oxycarbonium ion, or that the conformation change which produces ATP cleavage requires ionization of the acid-base catalyst. In the latter case, ATP could either cleave to give metaphosphate or transfer a phosphoryl group to a bound water molecule, although the failure of the aldehyde hydrates to bind makes the binding of water in the active site less likely, and such bound water would be expected to interfere with the normal kinase reaction. Perhaps this water is bound in such a way that it has access to the reactants only after reaction, but before they are released into solution.

It is possible to test for phosphorylation of the aldehyde by seeing if ^{18}O is transferred from the aldehyde to phosphate during the ATPase reaction. ^{18}O should have a half life in the carbonyl group of D-glyceraldehyde in water at 20° of about 50 seconds (10), so by running the reaction at low temperature with a high level of enzyme and adding the aldehyde in a small volume of $H_2[^{18}O]$ it should be possible to carry out such an experiment, and we hope to do so in the near future.

Acknowledgement

This work was supported by NIH grant GM 18938.

Literated Cited

1. Knowles, J. R. Annu. Rev. Biochem. 1980, 49, 877.
2. Viola, R. E.; Cleland, W. W. Biochemistry 1978, 17, 4111.
3. Raushel, F. M.; Cleland, W. W. Biochemistry 1977, 16, 2176.
4. Cook, P. F.; Kenyon, G. L.; Cleland, W. W. Biochemistry 1981, 20, 1204.
5. Janson, C. A.; Cleland, W. W. J. Biol. Chem. 1974, 249, 2562.
6. Viola, R. E.; Cleland, W. W. Biochemistry 1980, 19, 1861.
7. Hill, E. A.; Milosevich, S. A. Tetrahedron Lett. 1976, 50, 4553.
8. Lewis, C. A.; Wolfenden, R. Biochemistry 1977, 16, 4886.
9. Raushel, F. M.; Cleland, W. W. J. Biol. Chem. 1973, 248, 8174.
10. Trentham, D. R.; McMurray, C. H.; Pogson, C. I. Biochem. J. 1969, 114, 19.

RECEIVED July 12, 1981.

23

Kinetic and Thermodynamic Studies of Yeast Inorganic Pyrophosphatase

BARRY S. COOPERMAN

Department of Chemistry, University of Pennsylvania, Philadelphia, PA 19104

Despite their widespread distribution and central importance to cellular metabolism, phosphoryl-transfer enzymes remain incompletely understood with respect to their detailed mechanisms, especially when compared with what is known about more well studied enzymes, such as the serine proteases. Our goal is to obtain a detailed understanding of enzymatic catalysis of phosphoryl transfer and toward this end we have chosen to study yeast inorganic pyrophosphatase (PPase) as an example of a phosphoryl transfer enzyme. PPase is a dimer made up of identical subunits of molecular weight 32,000 daltons (1). Its covalent structure has been determined (2), two research groups have reported low-resolution crystal structures (3,4), and a high resolution structure determination is underway (D. Voet, personal communication). PPase catalyzes three different reactions, inorganic pyrophosphate (PPi) hydrolysis, H_2O-inorganic phosphate (Pi) oxygen exchange, and, considerably more slowly, PPi:Pi equilibration (5,6). PPase requires divalent metal ions for activity. The highest activity is conferred by Mg^{2+}, although substantial activity (>5% of that found with Mg^{2+}) is also found in the presence of $Zn^{2+} > Co^{2+} \simeq Mn^{2+}$ (7).

In this paper we report on recent findings of ours which have 1) demonstrated that PPase activity requires three divalent metal ions per subunit and 2) allowed formulation of a minimal kinetic scheme for PPase catalysis which accounts quantitatively for the three activities it manifests.

Binding Studies

It had been previously shown by Rapoport et al. (8) that native PPase binds two divalent metal ions (Mg^{2+}, Co^{2+}, Mn^{2+}) per subunit. We have now used equilibrium dialysis to extend these studies, by measuring the effect of added Pi on Mn^{2+} and Co^{2+} binding (9). The results (Figure 1) demonstrate that in the presence of Pi a third divalent metal ion is bound per subunit. For Mn^{2+}, one intrinsic dissociation constant characterizes the bind-

0097-6156/81/0171-0119$05.00/0

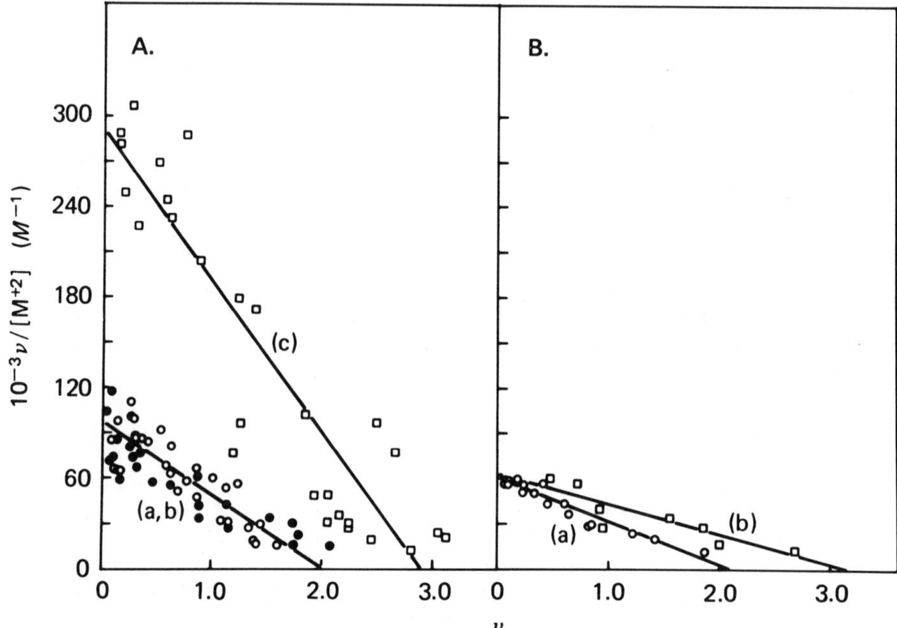

Figure 1. Scatchard plots of metal ion binding to PPase as measured by equilibrium dialysis. A: Mn^{2+} binding (a) no added Pi, enzyme concentration 7-42μM (○); (b) in the presence of 50μM Pi, enzyme concentration 7-36μM (●); (c) in the presence of 4mM Pi, enzyme concentration 8-116μM (□). B: Co^{2+} binding (a) no added Pi, enzyme concentration 8-96μM (○); (b) in the presence of 2.5mM Pi, enzyme concentration 30–35μM (□) (see Ref. 9).

ing in the absence of Pi, and another, lower by about three-fold, characterizes the binding in the presence of Pi. For Co^{2+}, essentially the same intrinsic dissociation constant characterizes the binding in both the presence and absence of Pi. These studies, along with parallel studies on divalent metal ion effects on Pi binding (as measured by equilibrium dialysis protection of activity against chemical modification, ^{31}P and water proton relaxation rates) have allowed evaluation of all of the equilibrium constants in Scheme I describing divalent metal ion (Mn^{2+} or Co^{2+}) and Pi binding to PPase.

Kinetic and Thermodynamic Studies

We have combined the results of three different types of measurement, enzyme-bound PPi, rates of H_2O-Pi oxygen exchange, and rates of PPi hydrolysis, to formulate the minimal scheme for PPase catalysis shown in equation (1), and to evaluate the rate constants contained within it (10). We note that all constants are apparent for pH 7.0

$$Mg_2E + MgPPi \underset{6s^{-1}\ (k_2)}{\overset{(k_1)\ 3 \times 10^7 M^{-1}s^{-1}}{\rightleftharpoons}} Mg_2EMgPPi \underset{222s^{-1}\ (k_4)}{\overset{H_2O^*\ (k_3)\ 1070s^{-1}}{\rightleftharpoons}} MgE(MgPi)_2$$

$$\underset{1.6 \times 10^5 M^{-1}s^{-1}\ (k_6)}{\overset{(k_5)\ 740s^{-1}}{\rightleftharpoons}} MgEMgPi + MgPi^* \underset{9 \times 10^5 M^{-1}s^{-1}\ (k_8)}{\overset{(k_7)\ 464s^{-1}}{\rightleftharpoons}} Mg_2E + Pi \qquad (1)$$

(pH 7.0, 25°C)

and further that the complexes shown in equation (1) are defined with respect to stoichiometry but not with respect to relative positioning. Thus, for example, $Mg_2EMgPPi$ refers to a complex with three Mg and one PPi, and is not meant to imply that MgPPi is necessarily bound as a complex to Mg_2E.

The major features of equation (1) are: 1) the implied requirement for three bound metal ions per active subunit; 2) the release of the electrophile Pi (i.e., the phosphoryl group attacked nucleophilically by H_2O in the forward direction) prior to the release of the leaving group Pi; and 3) the numerical evaluation of the rate constants. We now discuss each of these features in turn.

Boyer and his co-workers have recently presented evidence that H_2O-Pi exchange proceeds via enzyme-bound PPi (denoted EPPi) formation from Pi (6,11). Extending their results, we have investigated EPPi formation as a function of MgPi concentration at dif-

Scheme I. Relevant equilibria in solutions of enzyme, divalent metal ion, and Pi (see Ref. 9).

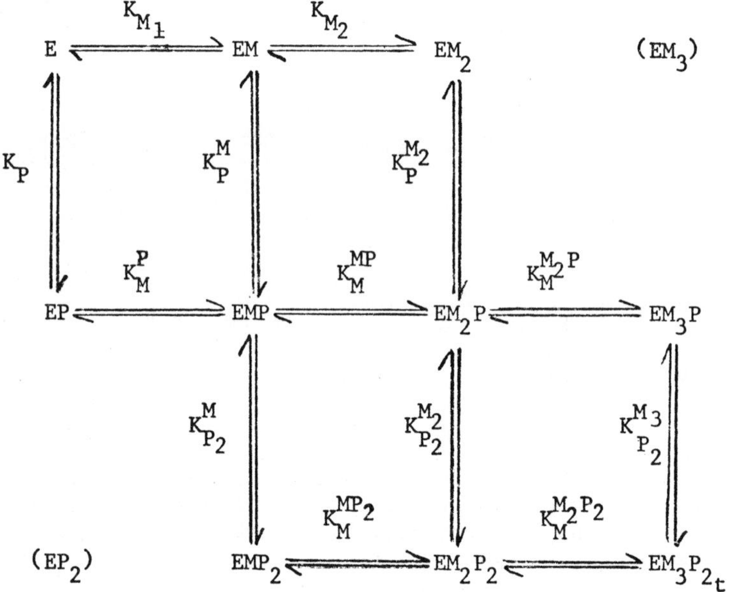

ferent fixed levels of $[Mg]_{free}$. These measurements required development of two methods for quantitatively measuring PPi in the presence of a 10^3-10^4 molar excess of Pi. The first involves the selective extraction into isobutanol from water of ^{32}Pi as a phosphomolybdate complex leaving behind ^{32}PPi in the aqueous solution. The second involves the separation of PPi from Pi by two-dimensional tlc on polyethyleneimine plates. We found that the dependence of EPPi formation on [MgPi] obeyed equation (2), where $[EPPi]_t$ and $[E]_T$ represent the total concentrations of EPPi and enzyme, respectively,

$$\frac{[E]_T}{[EPPi]_t} = \frac{A}{[MgPi]^2} + \frac{B}{[MgPi]} + C \qquad (2)$$

and that whereas the empirical parameters B and C were independent of $[Mg^{2+}]$ over the range studied (10-30mM), parameter A was approximately proportional to $[Mg^{2+}]$. The qualitative significance of this result was that enzyme forms containing either two bound Pi's or one bound PPi required one additional Mg^{2+} compared with enzyme forms containing one or no Pi. Thus, the third metal ion which is bound on Pi addition (Figure 1) appears necessary for EPPi formation. Rates of H_2O-Pi oxygen exchange were determined by measuring ^{18}O release from ^{18}O-labeled Pi using the nmr method of Cohn and Hu (12), which resolves the ^{31}P peaks due to the five ^{18}O-labeled and unlabeled species. We found that the exchange rate has essentially the same dependence on [MgPi] as does EPPi formation, thus providing confirmatory evidence for the intermediacy of EPPi in oxygen exchange. Further, we found, in agreement with the results of Hackney (13), that the value of the partition coefficient for H_2O-Pi oxygen exchange, Pc, defined as the rate at which enzyme-bound Pi loses water in the exchange step divided by the sum of this rate and the rate of release of Pi to the medium, was much less than one (we find 0.23; Hackney reports 0.30) and essentially independent of MgPi concentration over a wide range. This result requires that the Pi containing the oxygen from H_2O be released first, since, were it released second, it is clear from equation (1) that Pc should increase with increasing MgPi until it attained a limiting value of 1.0.

The eight rate constants in equation (1) were evaluated using the following eight equations. The measured parameters A,B, and C in equation (2) allow evaluation of K_3 (k_3/k_4), K_5 (k_5/k_6), and K_7 (k_7/k_8). Knowing these constants and previously determined values for the solution equilibrium of PPi$\rightleftharpoons$2Pi and for Mg^{2+} binding to PPi and Pi allows calculation of K_1 (k_1/k_2), by closing a thermodynamic loop. The remaining 4 equations (3-6) were provided by measuring Pc and enzyme-catalyzed rates of oxygen exchange and of PPi hydrolysis. An independent check on the validity of the calculation is provided by the good agreement

$$Pc = \frac{k_4}{k_4 + k_5} \quad (3) \qquad k_{cat,ex} = \frac{k_5 K_3}{K_3 + 1} \times \frac{4Pc}{4 - 3Pc} \quad (4)$$

$$k_{cat,hyd} = \frac{k_3 k_5 k_7}{k_3 k_5 + k_7 (k_3 + k_4 + k_5)} \quad (5) \qquad K_{m,hyd} = \frac{k_7 [k_3 k_5 + k_2 (k_4 + k_5)]}{k_1 [k_3 k_5 + k_7 (k_3 + k_4 + k_5)]} \quad (6)$$

of our calculated value for k_2 ($6s^{-1}$) and the value of $5s^{-1}$ which can be estimated from a measured value of PPi release from enzyme (6). The rate constant values shown in equation (1) lead to the following conclusions regarding the three reactions catalyzed by PPase:

(1) for PPi hydrolysis all three steps following PPi binding, steps 3,5, and 7, are partially rate-determining;

(2) for H_2O-Pi exchange, PPi synthesis, step 4, is almost exclusively rate-determining;

(3) for PPi:Pi equilibration, PPi release, step 2, is exclusively rate-determining.

Literature Cited

1. Heinrikson, R. L.; Sterner, R.; Noyes, C.; Cooperman, B. S.; Bruckmann, R. H. J. Biol. Chem. 1973, 248, 2521-8.
2. Cohen, S. A.; Sterner, R.; Keim, P. S.; Heinrikson, R. L.; J. Biol. Chem. 1978, 253, 889-97.
3. Bunick, G.; McKenna, G. P.; Scarbrough, F. E.; Uberbacher, E. C.; Voet, D. Acta Crystallog, Sect. B. 1978, 34, 3210-5.
4. Makhaldiani, V. V.; Smirnova, E. A.; Voronova, A. A.; Kuranova, I. P.; Arutyunyun, E. G.; Vainshtein, B. K.; Höhne, W. E.; Binwald, B.; Hansen, G. Dokl. Akad. Nauk SSSR 1978, 240 1478-81.
5. Cohn, M. J. Biol. Chem. 1958, 230, 369-79.
6. Janson, C. A.; Degani, C.; Boyer, P. D. J. Biol. Chem. 1979 254, 3743-9.
7. Butler, L. G.; Sperow, J. W. Bioinorg. Chem. 1977, 7, 141-50.
8. Rapoport, T. A.; Höhne, W. E.; Heitmann, P.; Rapoport, S. M. Eur. J. Biochem. 1973, 33, 341-7.
9. Cooperman, B. S.; Panackal, A.; Springs, B.; Hamm, D. J. Biochemistry 1981, 20, in press.
10. Springs, B.; Welsh, K. M.; Cooperman, B. S. Biochemistry 1981, 20, in press.
11. Hackney, D. D.; Boyer, P. D. Proc. Natl. Acad. Sci. USA 1978, 75, 3133-7.
12. Cohn, M.; Hu, A. Proc. Natl. Acad. Sci. USA 1978, 75, 200-3.
13. Hackney, D. D. J. Biol. Chem. 1980, 255, 5320-8.

RECEIVED June 30, 1981.

24

The Role of Histidine Residues and the Conformation of Bound ATP on ATP-Utilizing Enzymes

P. R. ROSEVEAR[1], G. M. SMITH, S. MESHITSUKA, and A. S. MILDVAN[1]

Institute for Cancer Research, Fox Chase Cancer Center, Philadelphia, PA 19111

P. DESMEULES and G. L. KENYON

Departments of Pharmaceutical Chemistry and Biochemistry and Biophysics, University of California San Francisco, San Francisco, CA 94143

Histidine residues often play important roles in the functioning of proteins. This point is cogently illustrated by the fact that viruses, the most efficient organisms known, rarely if ever incorporate histidine into their coat proteins, but generally incorporate it into their enzymatic proteins (1). Proton NMR has long provided a method for monitoring the role of individual histidine residues in small proteins (2). Improvements in instrumentation (3) have prompted us to study by NMR the role of histidines in large phosphotransferase enzymes, a field to which Professor Westheimer has made profound mechanistic contributions. The enzymes we have studied are adenylate kinase (M_r 22,000), creatine kinase (M_r 82,000) and pyruvate kinase (M_r 237,000), and the roles detected for histidine are summarized in Figure 1.

Creatine kinase was purified from rabbit muscle by the method of Kuby et al. (4). Rabbit muscle pyruvate kinase was purchased from Boehringer. Porcine muscle adenylate kinase was purchased from Sigma, and was further purified by gel filtration on Sephadex G-50. The enzymes were homogeneous as judged by their specific activities and by their migration as single components in sodium dodecyl sulfate gel electrophoresis. Proton NMR spectra at 250 MHz of 0.5-2.0 mM enzyme sites in ^2H_2O solution were obtained with a Bruker WM 250 MHz pulse FT spectrometer at 25°. At least 256 transients were accumulated over 8192 data points using 16 bit A/D conversion. Relaxation rates and histidine pK' values were determined by standard NMR methods (5, 6).

Creatine Kinase. Six of the sixteen imidazole C-2 proton resonances and one imidazole C-4 proton resonance per subunit of this dimeric enzyme (M_r = 82,000) were detected. Titrations measuring their chemical shifts as a function of pH* yielded pK' values of 7.0, 7.1, 5.9, and 5.2 for his(2), (3), (4) and (6), respectively, and permitted the assignment of the C-4 resonance to his(3) (7). The pK' of his(2) was unaffected by saturation of the enzyme with creatine but was increased by 0.6 - 0.7 units on saturation with the phosphorylated substrates phosphocreatine or MgATP, in

[1]Current address: Department of Physiological Chemistry, The Johns Hopkins Medical School, Baltimore, MD 21205.

0097-6156/81/0171-0125$05.00/0

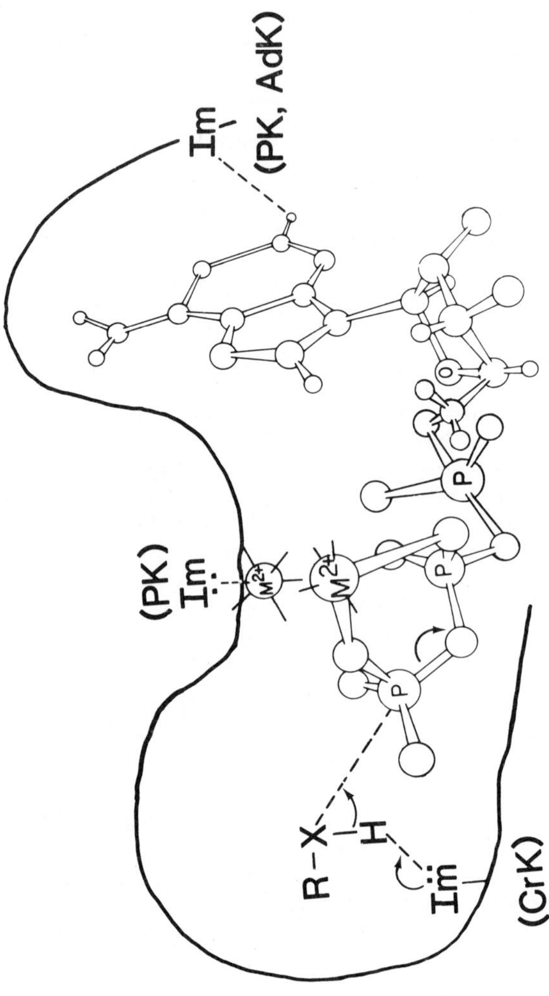

Figure 1. Composite diagram summarizing the roles of histidine imidazole groups at the active sites of the phosphotransferase enzymes creatine kinase (CrK), pyruvate kinase (PK), and adenylate kinase (AdK).

quantitative agreement with the results of a kinetic study of the creatine kinase reaction as a function of pH (8) indicating that his(2) is the general acid/base catalyst which deprotonates the guanidinium group of creatine as it is phosphorylated by MgATP. Because of the pitfalls in pH-rate studies alone, as incisively pointed out by Westheimer and co-workers (9, 10) independent investigations are necessary to identify acid/base catalysts on enzymes.

The pK' values of his(4) and his(6) of creatine kinase, while too low to fit the kinetic data as the general acid/base catalyst, also increased (by 0.4 units) in response to the binding of the phosphorylated substrates. Titrations of creatine kinase with substrates at constant pH* (6.8) monitoring the chemical shifts of his(2) or his(6) yielded dissociation constants for phosphocreatine (8.7 mM) and MgADP (110 μM) consistent with those derived from kinetic data (8, 11, 12), indicating active site binding. Direct evidence for the presence of his(2), his(3) and his(6) at or near the active site was provided by the paramagnetic effects of the substrate analog β,γ-bidentate Cr^{3+}ATP on the relaxation rates of their imidazole protons. The $1/T_1$ values yielded distances of 12 ± 0.5 Å from Cr^{3+} to the C-2 protons of his(2) and his(6), consistent with his(2) functioning as the general acid/base catalyst (Figure 2), and with his(6) interacting electrostatically with the substrates. His(3) is somewhat farther from Cr^{3+}ATP and is so positioned that its C-4 proton is oriented toward the Cr^{3+}, at a distance of 14 Å.

The presence of a general acid/base catalyst at the creatine binding site, together with a recent measurement of a short (∿ 6 Å) distance from Cr^{3+}ADP to the phosphorus of P-creatine (13), imply an associative mechanism for the creatine kinase reaction.

Pyruvate Kinase. Six of the fourteen imidazole C-2 proton resonances and three imidazole C-4 proton resonances of this tetrameric enzyme (M_r 237,000) were detected and their pK' values determined. The substrate P-enolpyruvate selectively decreased the pK' of one histidine (his(3)) by 0.4 units, from a value of 6.2 to 5.8, only in the presence of the cation activators Mg^{2+} and K^+ (3). The metal activators Mg^{2+} and K^+ alone produced smaller decreases in the pK' value of his(3) suggesting that the binding of P-enolpyruvate lowers the pK' of his(3) by strengthening its interaction with the metal activators. Direct evidence for the proximity of the divalent cation to his(3) was obtained by the paramagnetic effects of the activator Ni^{2+} on $1/T_1$ and $1/T_2$ values of the C-2 proton of his(3) in the pyruvate kinase-K^+, Ni^{2+} complex at pH* 6.0. The calculated Ni^{2+}-^1H distance of 6 Å is consistent with a second sphere imidazole complex. The presence of P-enolpyruvate increases the paramagnetic effect of Ni^{2+} on $1/T_2$ of this proton by a factor of 8.7, suggesting direct coordination of his(3) by Ni^{2+} in the substrate complex (14).

The paramagnetic substrate, β,γ-bidentate Cr^{3+}ATP broadens all of the histidine C-2 resonances. Specific displacement of Cr^{3+}ATP from the active site by P-enolpyruvate resulted in the selective narrowing not only of his(3) as expected, but also of his(2),

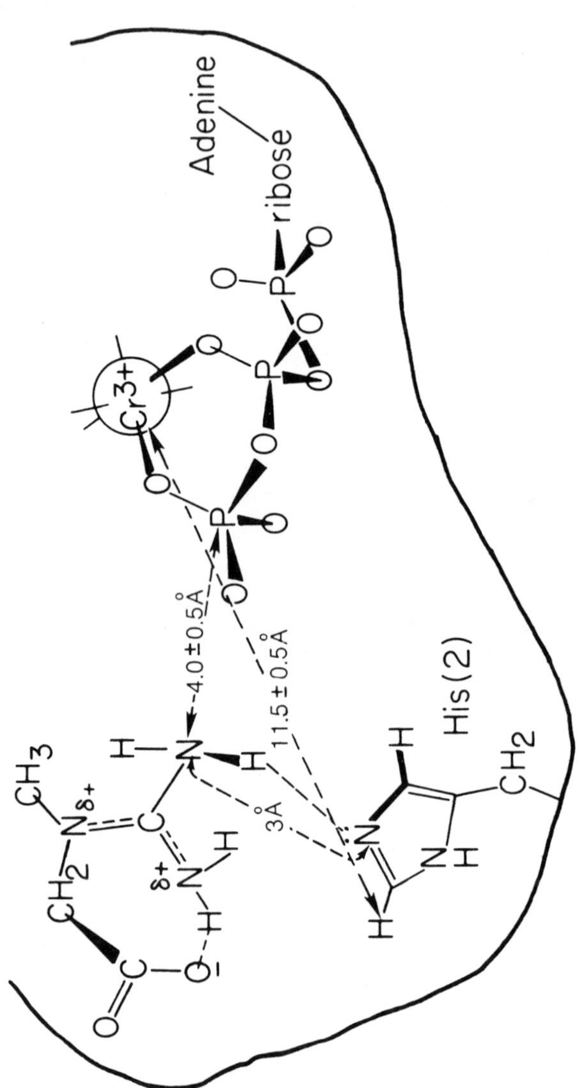

Figure 2. Model of the creatine kinase–Cr^{3+} ATP–creatine complex showing his(2) functioning as a general base, with the upper limit distance from Cr^{3+} to the C-2 proton of his(2), based on other reasonable distances.

demonstrating the presence of both his(3) and his(2) at the active site. An intermolecular negative nuclear Overhauser effect (NOE) was detected between the C-2 proton of his(2) and the adenine H-2 proton of MgATP in the pyruvate kinase-Mg^{2+}-ATP-Mg^{2+} complex indicating the proximity of his(2) to the adenine H-2 proton (Figure 1). Independent evidence for the presence of his(2) at the nucleotide binding site was provided by the selective chemical modification of this residue with diethyl pyrocarbonate as monitored by NMR and by kinetic studies (14).

Conformation of ATP on Pyruvate Kinase, Creatine Kinase, and Protein Kinase. A fundamental problem in the determination of the conformation of flexible enzyme-bound substrates such as ATP by distance measurements from a paramagnetic reference point is that a weighted root-mean-sixth average conformation is detected (5,15). From the average conformation alone it is not possible to determine the number and nature of the individual conformations that give rise to this average. In the case of the pyruvate kinase-metal-ATP complex we have solved this problem by determining the average conformation of bound ATP by two independent methods with differing locations of the reference points and with different observation frequencies (14, 16). Intramolecular negative NOE's were observed on the adenine H-8 proton of enzyme-bound MgATP upon the pre-irradiation of certain of the ribose protons of the bound nucleotide. The magnitude of the NOE's decreased in the order $H_2' > H_1' \sim H_3' >> H_4'$, H_5', indicating correspondingly increasing distances between the adenine H-8 and these ribose protons. These effects are consistent with a previous determination of the glycosidic torsional angle of pyruvate kinase-bound ATP based on the paramagnetic effects of enzyme-bound Mn^{2+} on the T_1 of the protons of ATP, ($\chi = 30 \pm 10°$ (16)) provided a 2'-endo ribose conformation is assumed. This agreement in the nucleotide conformations determined by the two methods, can be explained only by the existence of a unique average conformation about the glycosidic bond of ATP on pyruvate kinase (14). Preliminary observations have also been made, with creatine kinase and bovine heart protein kinase, of negative NOE's on the adenine H-8 proton of bound metal-ATP, upon pre-irradiation of the ribose protons (17). The magnitude of the NOE's decreased in the order $H_2' > H_3' > H_1'$, H_4', H_5' suggesting a 2'-endo ribose pucker and a high anti-conformation, differing in the glycosidic torsional angle from that on pyruvate kinase. The results with protein kinase are consistent with an independent determination of the conformation of bound metal-ATP based on distances from Mn^{2+} ($\chi = 84 \pm 10°$ (18)), suggesting a unique average nucleotide conformation on this enzyme.

Adenylate Kinase. As in the pyruvate kinase-ATP complex, an intermolecular negative NOE is detected on adenylate kinase between the C-2 proton of his(36) and the adenine H-2 proton of bound MgATP, indicating that his(36) is very near the adenine H-2 proton of the bound MgATP substrate (Figure 1). His(36) had previously been assigned (19). Consistent with this finding, the paramagnetic substrate analog α,β,γ-tridentate Cr^{3+}ATP selectively increases $1/T_1$ and $1/T_2$ of the C-2 proton of his(36) yielding a Cr^{3+}-^1H distance of 11 ± 1 Å.

Acknowledgement. It is a great pleasure for the authors to dedicate this paper to Frank Westheimer. Though most of us have not had the privilege of working in his laboratory, we are all his students nevertheless, since we have learned so much from the crisp clarity and profound wisdom of his publications, his lectures, and his comments on science and life.

This work was supported by N.I.H. Grants AM-13351, CA-06927, RR-05539, CA-22780, RR-542, and AM-17323, N.S.F. Grant PCM-79-23154, and an appropriation from the Commonwealth of Pennsylvania.

Literature Cited

1. Dayhoff, M. O. "Atlas of Protein Sequence and Structure"; Nat. Biomed. Res. Fdn.: Silver Spring, MD 1969; D-173.
2. Roberts, G. C. K.; Jardetzky, O. Adv. Prot. Chem. 1970, 24, 447.
3. Meshitsuka, S.; Smith, G. M.; Mildvan, A. S. J.B.C. 1981, 256, 4460.
4. Kuby, S. A.; Noda, L.; Lardy, H. A. J.B.C. 1954, 209, 191.
5. Mildvan, A. S.; Gupta, R. K. Methods Enzymol. 1978, 49G, 322.
6. Meadows, D. L. Methods Enzymol. 1972, 26, 638.
7. Rosevear, P. R.; Desmeules, P.; Kenyon, G. L.; Mildvan, A. S. Fed. Proc. 1981, 40, 1872.
8. Cook, P. F.; Kenyon, G. L.; Cleland, W. W. Biochem. 1980, 20, 1204.
9. Frey, P. A.; Kokesh, F. C.; Westheimer, F. H. J. Am. Chem. Soc. 1971, 93, 7266.
10. Kokesh, F. C.; Westheimer, F. H. J. Am. Chem. Soc. 1971, 93, 7270.
11. Morrison, J. F.; James, E. Biochem. J. 1965, 97, 37.
12. Schimmerlik, M. I.; Cleland, W. W. J.B.C. 1973, 248, 8418.
13. Gupta, R. K. Biophys. J. 1980, 32, 225.
14. Meshitsuka, S.; Smith, G. M.; Mildvan, A. S. Int. J. Quantum Chem. 1981, (in press).
15. Jardetzky, O. Biochem. Biophys. Acta. 1980, 621, 227.
16. Sloan, D. L.; Mildvan, A. S. J.B.C. 1976, 251, 2412.
17. Rosevear, P. R.; Desmeules, P.; Kenyon, G. L.; Bramson, H. N.; Kaiser, E. T.; Mildvan, A. S. Abs. A.C.S. Mtg., Biol. Div. 1981.
18. Granot, J.; Kondo, H.; Armstrong, R. N.; Mildvan, A. S.; Kaiser, E. T. Biochem. 1979, 18, 2339.
19. McDonald, G. G.; Cohn, M.; Noda, L. J.B.C. 1975, 250, 6947.

RECEIVED September 16, 1981.

25

[$^{18}O/^{16}O$]^{31}P-NMR Studies of Phosphoryl Transfer Enzymes

J. J. VILLAFRANCA, F. M. RAUSHEL, R. P. PILLAI, M. S. BALAKRISHNAN, C. DEBROSSE, and T. D. MEEK

Department of Chemistry, The Pennsylvania State University, University Park, PA 16802

In 1978 Cohn and Hu (1) demonstrated the isotopic effect by ^{18}O on the ^{31}P-nmr spectrum of inorganic phosphate. For each ^{18}O that replaces an ^{16}O, an upfield shift of 0.021 ppm results. Thus the chemical shift for $HP^{16}O_4^{=}$ is 0.084 ppm downfield from $HP^{18}O_4^{=}$. Based on this observation it is obvious that any chemical event that potentially involves substitution or exchange of an ^{18}O for ^{16}O in phosphorus containing compounds can be followed by monitoring the ^{31}P-nmr spectrum throughout the course of the reaction. This article describes experiments from my laboratory on two enzymes that catalyze phosphoryl transfer reactions and our use of the [$^{18}O/^{16}O$]^{31}P-nmr methodology to detect intermediates in the enzymic reactions.

Glutamine synthetase catalyzes the formation of glutamine from ATP, gluatmate, and ammonia. The other products are ADP and P_i. The reaction mechanism is though to proceed through a γ-glutamyl phosphate intermediate. We have shown that incubation of ADP, glutamine, and [^{18}O]P_i with glutamine synthetase resulted in the loss of ^{18}O from the P_i (2). Analysis of the data showed that only one ^{18}O was lost per encounter and that the rate constant for exchange was 5-7 times faster than net turnover of products. This was also demonstrated by Stokes and Boyer (3) using mass spectrometry.

H●–P(=●)(–●⁻)(–●⁻) + R–C(=O)–NH₂ ⇌ R–C(=O)–●–P(=●)(–●⁻)(–●⁻) + NH₃
(I)

R–C(=O)–●–P(=●)(–●⁻)(–●⁻) + ADP ⇌ R–C(=O)–●⁻ + ADP–O–P(=●)(–●⁻)(–●⁻)
(I) (II)

0097–6156/81/0171–0131$05.00/0

This exchange reaction is explained by assuming that the above reactions are occurring on the enzyme surface and that the γ-carboxyl of glutamate (II) is free to rotate, thus allowing, upon reversal of the reaction, the formation of γ-glutamyl phosphate (I) with an ^{16}O in the bridge position. Reaction of I with NH_3 would then produce P_i with only 3 atoms of ^{18}O instead of the original 4. Since a random distribution of ^{18}O is maintained at all times, only one ^{18}O is lost per encounter with the enzyme and thus P_i must be dissociating from the enzyme faster than glutamine.

Recently Midelfort and Rose (4) introduced a positional isotope exchange technique that is also suitable for study by the ^{18}O chemical shift technique. Briefly, this technique follows the exchange of label from one part of a substrate to another due to rotational equivalence of some intermediate. The method was first applied to glutamine synthetase in the reaction:

$$\text{Ado-O-}\overset{\text{O}}{\overset{\|}{\underset{\text{O}^-}{\underset{|}{\text{P}}}}}\text{-O-}\overset{\text{O}}{\overset{\|}{\underset{\text{O}^-}{\underset{|}{\text{P}}}}}\text{-}\bullet\text{-}\overset{\text{O}}{\overset{\|}{\underset{\text{O}^-}{\underset{|}{\text{P}}}}}\text{-O}^- + \text{Glu} \rightleftharpoons \text{Ado-O-}\overset{\text{O}}{\overset{\|}{\underset{\text{O}^-}{\underset{|}{\text{P}}}}}\text{-O}\}\overset{\text{O}}{\overset{\|}{\underset{\text{O}^-}{\underset{|}{\text{P}}}}}\text{-}\bullet + \gamma\text{-Glutamyl-P}$$

"rotation"

$$\text{Ado-O-}\overset{\text{O}}{\overset{\|}{\underset{\text{O}^-}{\underset{|}{\text{P}}}}}\text{-O-}\overset{\bullet}{\overset{\|}{\underset{\text{O}^-}{\underset{|}{\text{P}}}}}\text{-O}^- + \gamma\text{-Glutamyl-P} \rightleftharpoons \text{Ado-O-}\overset{\text{O}}{\overset{\|}{\underset{\text{O}^-}{\underset{|}{\text{P}}}}}\text{-O-}\overset{\bullet}{\overset{\|}{\underset{\text{O}^-}{\underset{|}{\text{P}}}}}\text{-O-}\overset{\text{O}}{\overset{\|}{\underset{\text{O}^-}{\underset{|}{\text{P}}}}}\text{-O}^- + \text{Glu}$$

ATP is synthesized with ^{18}O in the β-γ bridge position. In phosphate-containing species such as ATP where the phosphorus (α, β, and γ) and oxygen (bridge and nonbridge) atoms are in different environments the effect of ^{18}O substitution on the ^{31}P chemical shift is dependent on the oxygen environment. The shift is largest for those oxygens with most double bond character. For the β-P, the shift per ^{18}O atom is 0.016 ppm for the β-γ bridge position and 0.028 ppm for the β nonbridge position.

Presuming glutamine synthetase catalyzed the formation of γ-glutamyl-P from ATP and glutamate in the absence of NH_3, then ADP would be formed; because of rotational equivalence it would allow, upon the reversal of the above reactions, the formation of ATP with ^{18}O in the nonbridge position of the β-P. Midelfort and Rose found this to be the case for glutamine synthetase by mass spectral analysis of the $^{18}O/^{16}O$ exchange and we have followed this reaction by nmr. We detected the loss of ^{18}O from the γ-P signal (the starting material had ^{18}O in all four oxygens of the γ-P) and the shift of signal from the β-γ bridge to β nonbridge position as predicted above. Thus all lines of evidence point to formation of γ-glutamyl-P as a kinetically competent intermediate in catalysis.

Carbamyl phosphate synthetase catalyzes the synthesis of carbamyl-P from HCO_3^-, glutamine, and 2 moles of ATP. The enzyme also catalyzes the HCO_3^--dependent hydrolysis of ATP. Raushel and Villafranca (5) followed the exchange of ^{18}O from the bridge to the nonbridge position of [γ-^{18}O]ATP after incubation with enzyme and bicarbonate. The exchange rate was 0.4 times the rate of ADP formation. These results support the formation of carboxy phosphate as the first intermediate in the catalytic sequence. The rate of formation of intermediates in the reaction was also studied using rapid reaction techniques by Raushel and Villafranca (6) and the data agree with the ^{31}P-nmr studies presented above.

The positional isotope exchange has also been measured with ^{31}P-nmr in the reverse reaction of carbamyl phosphate synthetase:

$$\text{Carbamyl-P} + \text{ADP} \rightarrow \text{ATP} + NH_3 + HCO_3^-.$$

Carbamyl phosphate was synthesized with ^{18}O in all oxygens except the carbonyl oxygen of carbon. The exchange of the bridge oxygen into the carbonyl oxygen was followed by nmr and was evidence for the following series of reactions

$$NH_2-\overset{\overset{O}{\|}}{C}-\bullet-\underset{\underset{\bullet^-}{|}}{\overset{\overset{\bullet}{\|}}{P}}-\bullet^- + \text{ADP} \rightleftharpoons NH_2-\overset{\overset{O}{\|}}{C}-\bullet^- + \text{ATP}$$

$$\Updownarrow$$

$$NH_2-\overset{\overset{\bullet}{\|}}{C}-O-\underset{\underset{\bullet^-}{|}}{\overset{\overset{\bullet}{\|}}{P}}-\bullet^- + \text{ADP} \rightleftharpoons NH_2-\overset{\overset{\bullet}{\|}}{C}-O^- + \text{ATP}$$

These data support the formation of a second intermediate in the reaction pathway, viz, carbamate ($NH_2CO_2^-$) and its formation is ~4 times faster than ATP formation for the reverse reaction outlined above.

In conclusion the [$^{18}O/^{16}O$]^{31}P-nmr method is a powerful technique for the study of reaction intermediates in phosphoryl transfer reactions. Both the nature of the reactive species as well as the rate of formation and breakdown of the reactive species on the enzyme can be established.

Work supported by NIH grants AM-21785, AM-05996, GM-23529 and NSF Grant PCM-7807845.

Literature Cited

1. Cohn, M.; Hu, A. Proc. Natl. Acad. Sci. USA 1978, 75, 200-3.

2. Balakrishnan, M. S.; Sharp, T. R.; Villafranca, J. J. Biochem. Biophys. Res. Commun. 1978, 85, 991-8.
3. Stokes, B. O.; Boyer, P. D. J. Biol. Chem. 1976, 251, 5558-64.
4. Midelfort, C. F.; Rose, I. A. J. Biol. Chem. 1976, 251, 5881-7.
5. Raushel, F. M.; Villafranca, J. J. Biochemistry 1980, 19, 3170-4.
6. Raushel, F. M.; Villafranca, J. J. Biochemistry 1979, 18, 3424-9.

RECEIVED June 30, 1981.

26

Potential Antiviral Nucleotides

D. W. HUTCHINSON

Department of Chemistry and Molecular Sciences, University of Warwick, Coventry, CV4 7AL, United Kingdom

In contrast to antibacterial agents, few clinically useful antiviral agents are in use at the present time. One very promising antiviral agent is the glycoprotein, interferon, and biochemical events which take place when interferon interacts with susceptible cells are beginning to be understood. For example, one result of the interaction of interferon with cells is the increased synthesis of the 5'-triphosphate of adenylyl(2'-5') adenylyl(2'-5')adenosine (2-5A) together with the tetramer and higher oligomers (1). These unusual oligonucleotides are very effective inhibitors of protein synthesis in cell-free systems and can also activate a nuclease which degrades viral mRNAs.

Many syntheses of 2-5A have been reported but most of these are multistage preparations involving the selective use of protecting groups and hence the final yields of 2-5A are usually very low. We find that the metal ion-catalysed oligomerisation of adenosine 5'-phosphoroimidazolidate (2) is a rapid and efficient route to the 5'-phosphate of the 'core' trinucleoside diphosphate of 2-5A. This synthetic method arose from studies on the formation of oligonucleotides under prebiotic conditions when it was observed that nucleoside 5'-phosphoroimidazolidates oligomerise in the presence of divalent metal ions. Thus, adenosine 5'-phosphoroimidazolidate in the presence of lead(II) ions gives oligomers with mainly the thermodynamically more stable 2'-5' links. On the other hand, guanosine 5'-phosphoroimidazolidate gives 2'-5' linked oligomers in the presence of lead(II) ions but 3'-5' linked oligomers in the presence of zinc(II) ions (3). Presumably, the differences in linkage arise from differences in the structures of the nucleotide-metal ion complexes. In our hands, chromatographic separation of the oligomers from the lead(II) catalysed oligomerisation followed by removal of 3'-5' linked oligomers by digestion with RNase T_2 and rechromatography is a rapid, convenient, method of obtaining the 5'-phosphate of the 'core' oligonucleotide of 2-5A. The 5'-phosphate can then be converted into the triphosphate by standard methods. The enzyme which converts ATP into 2-5A has been found in a variety

0097-6156/81/0171-0135$05.00/0

of cells and it has been suggested that 2-5A may have hormonal properties (4). We are currently investigating the effect of the 'core' oligonucleotide on cell growth and virus replication.

The use of 2-5A as an antiviral agent in animals does not seem to be feasible at present as the highly charged oligonucleotide is not taken up by cells to any extent and is broken down rapidly in interferon treated cells. We are looking at the mode of action of another class of stable phosphorus-containing antivirals which appear to be readily taken up by cells. These antiviral compounds are phosphonoacetic (PAA) and phosphonoformic (PFA) acids, analogues of pyrophosphoric acid.

The antiviral properties of PAA against Herpes Simplex virus were discovered during routine screening (5), and these compounds also inhibit influenza virus replication (6). While the precise mode of action of PAA and PFA are not known at present, they will inhibit the polymerase enzymes in the two viruses. Two mechanisms for this inhibition appear plausible. As enzymic reactions are in theory reversible, an analogue of a nucleoside triphosphate (e.g. I or II) may be formed in which the β,γ-phosphoryl residues of the triphosphate moiety are replaced by PAA or PFA residues. We have prepared the ATP analogue of PAA. This compound has, not unexpectedly, no effect on enzymes such as hexokinase which transfer the γ-phosphoryl residue of ATP to a substrate. This ATP analogue is also not a substrate RNA polymerase from *E. coli* or from influenza virus. Furthermore, the dTTP analogue is not a substrate for the DNA polymerase of HSV (7).

We have been unable to prepare the ATP analogue of PFA(II) by the phosphoromorpholidate route. When we attempt this preparation all we observe by TLC is the rapid formation of AMP. We believe that the ATP analogue is formed but it undergoes rapid intramolecular breakdown to give a highly reactive formyl phosphate. The difference in stability between (I) and (II) is striking, but it has been observed that diethyl 2-carboxymethylphenylphosphonate (III) undergoes hydrolysis by an intramolecular route 10^5 times more slowly than diethyl 2-carboxyphenylphosphonate(IV) at pH 3.0 and 79.5^o (8).

While we have not investigated the effect of PAA and PFA analogues of nucleoside triphosphate in detail, we do not think that the inhibitory effect of PAA and PFA is due to the formation of these analogues. Rather we believe that PAA and PFA act by complexing with a pyrophosphate-binding site in enzymes, probably by co-ordinating with an essential metal ion such as zinc. DNA and RNA polymerases are zinc-requiring enzymes (9), and it is relevant that reverse transcriptase, another zinc-requiring enzyme (10), is inhibited by PFA (11). The stability constants of some metal complexes of PAA are known (12). Unlike pyrophosphate, PAA does not form strong complexes with magnesium ion but does form strong complexes with zinc ion. Bathocuproin(V) and 2-acetylpyridine thiosemicarbazone(VI), both good chelating agents for

AdOP(O)(HO)–N(imidazole) —Pb(II), 7 days, R° → oligo A —i) separate, ii) RNase T_2, iii) separate → 2-5 A

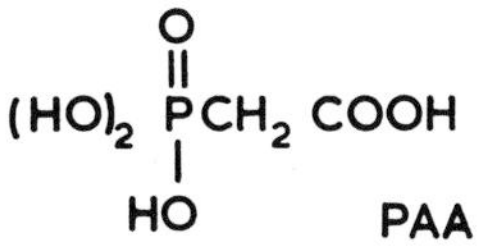

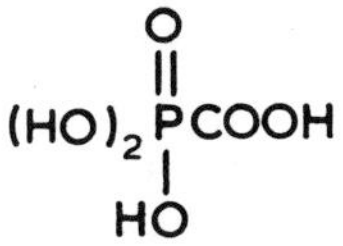

$(HO)_2P(O)CH_2COOH$ PAA

$(HO)_2P(O)COOH$ PFA

AdOP(O)(HO)–N(morpholine) —PAA→ AdO P(O)(HO)–O–P(O)(OH)CH_2COOH (I)

—PFA→ (II) → AdO P(O)($^-$O) O C(O) P(O)(O^-)$_2$ —H_2O→ AMP + PFA

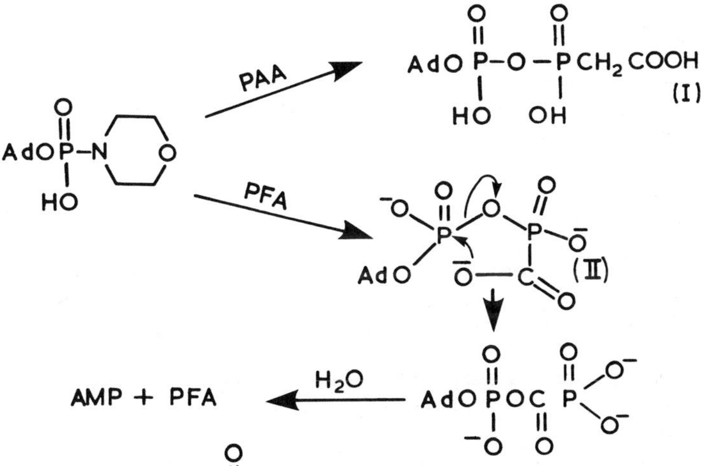

(III) $P(O)(OEt)_2$ / COOH

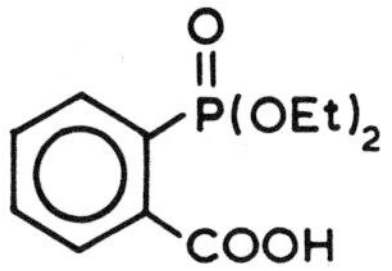

(IV) $P(O)(OEt)_2$ / CH_2COOH

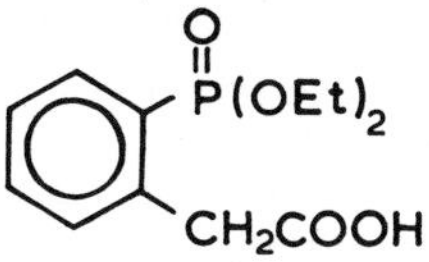

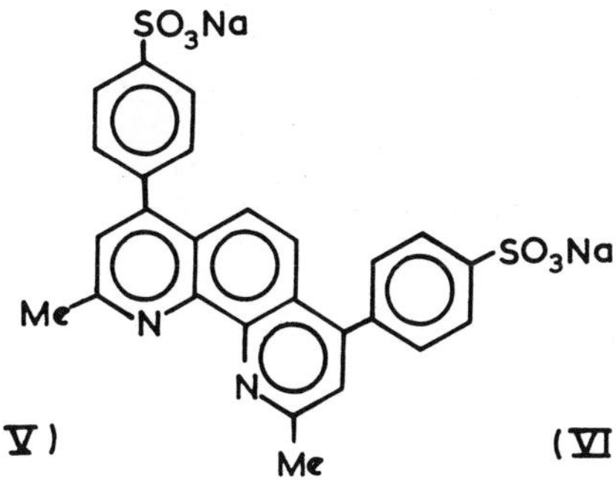

(V) SO_3Na, SO_3Na, Me, N, N, Me

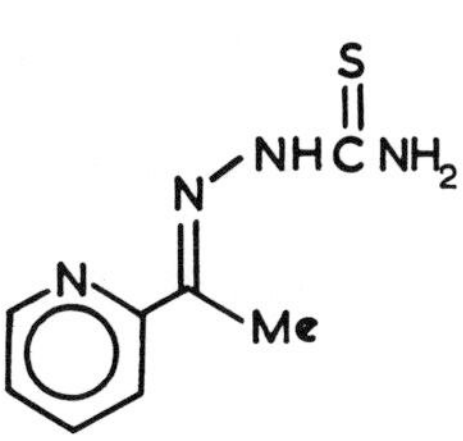

(VI) S, NHCNH$_2$, N, N, Me

'soft' metal ions such as zinc, inhibit influenza virus replication by inhibiting the RNA-dependent RNA polymerase of the virus (13). Reverse transcriptase is also inhibited by o-phenanthroline, an analogue of bathocurpoin (14).

Neither PAA nor PFA are in clinical use as the former is concentrated in bones when taken systemically. There is some hope that PFA, which is less irritant than PAA, may be useful topically against HSV(II). We are in the process of synthesising analogues of PAA and PFA in the hope of obtaining compounds which chelate with zinc ion but do not have unwanted side effects in animals.

Literature Cited

1. Williams, B. R. G.; Golgher, R. R.; Brown, R. E.; Gilbert, C. S.; Kerr, I. M. Nature (London) 1979, 282, 582
2. Sawai, H.; Shibata, T.; Ohno, M. Tetrahedron Letters 1979, 4573
3. van Roode, J. H. G.; Orgel, L. E. J. Mol. Biol. 1980, 144, 579
4. Etienne-Smekens, M.; Vassart, G.; Content, J.; Dumont, J. E. F.E.B.S. Letters 1981, 125, 146
5. Boezi, J. A. Pharmac. Ther. 1979, 4, 231
6. Stridh, S.; Helgstrand, E.; Lannerö, B.; Misorny, A.; Stenig, G.; Öberg, B. Arch. Virol. 1979, 61, 245
7. Leinbach, S. S. Ph.D. Thesis, Michigan State University 1976; Diss.Abs. 1977, 37B, 6098
8. Blackburn, G. M.; Brown, M. J. J. Amer. Chem. Soc. 1969, 91, 525
9. Mildvan, A. S.; Leob, L. A. Crit. Rev. Biochem. 1979, 6, 219
10. Poesz, B. J.; Battula, N.; Leob, L. A. Biochem. Biophys. Res. Comm. 1974, 56, 959
11. Sundquist, B.; Öberg, B. J. gen. Virol. 1979, 45, 273
12. Stünzi, H.; Perrin, D. D. J. Inorg. Biochem. 1979, 10, 309
13. Oxford, J. S. Postgrad. Med. J. 1979, 55, 105
14. Valenzuela, P.; Morris, R. W.; Faras, A.; Levinson, W.; Rutter, W. J. Biochem.Biophys.Res.Comm. 1973, 53, 1036

RECEIVED July 7, 1981.

NEW SYNTHETIC METHODS FOR PHOSPHORUS COMPOUNDS

Phosphinomethanes: Synthesis and Reactivity

H. H. KARSCH

Technische Universität München, Anorganisch-chemisches Institut, Lichtenbergstrasse 4, D-8046 Garching, FRG

Diphosphinomethanes are distinguished by two nucleophilic centers close together and an additional electrophilic site (acidic H).

$$\begin{array}{ccccc} & & H\leftarrow|N & & \\ >\underline{P} & - & C & - & \underline{P}< \\ \downarrow & & H & & \downarrow \\ E & & & & E' \end{array} \qquad \begin{array}{l} E,E' = H^+, R^+, (M), O, S \\ N = R^- \end{array}$$

We have synthesized $Me_2PCH_2PR_2$ (R=Ph,t-Bu),$MeP(CH_2PMe_2)_2$, $P(CH_2PMe_2)_3$ and mixed Cl/t-Bu and Me/CH_2SiMe_3 diphosphinomethanes by various methods.

Because of its simplicity and low steric requirements, we chose the methyl derivate $Me_2PCH_2PMe_2$ as model for these compounds.

Examples for nucleophilic reactions of the phosphinomethanes are (1):

$$Me_2PCH_2PMe_2 \xrightarrow{E} Me_2PCH_2\overset{E}{P}Me_2 \xrightarrow{E} Me_2\overset{E}{P}CH_2\overset{E}{P}Me_2 \qquad (1)$$

$$E=H^+, Me^+, t\text{-}Bu^+, O, S...$$

$$Me_2PCH_2PMe_2 + 2\ Me_2PCl \rightleftharpoons [Me_2PPMe_2CH_2PMe_2PMe_2]Cl_2 \qquad (2)$$

$$\uparrow$$

$$2\ Me_2PPMe_2 + CH_2Cl_2$$

Reaction (2) offers a new route to the preparation of $Me_2PCH_2PMe_2$, since the equilibrium may be shifted to the left by hydrolysis of the salt.

The ligand properties are very similar to those of PMe_3 in electronic respects, but sterically it is much more favourable. Thus, whereas $(PMe_3)_5Fe$ is not known as an isolable complex,(2),$(Me_3P)_3(Me_2PCH_2PMe_2)Fe$ is a stable compound (no C-H-cleavage occurs!) The

0097–6156/81/0171–0141$05.00/0

complex may be compared with the isoelectronic Co(I)- and Ni(II)-complexes, which are less stable in respect to pentacoordination. The cationic cobalt(I)-complex $[(Me_3P)(Me_2PCH_2PMe_2)_2Co]Cl$ decomposes rather rapidly in THF, yielding $(PMe_3)_2(Me_2PCH_2PMe_2)CoCl_2$ and a novel, paramagnetic complex $[(Me_3P)(Me_2PCH_2PMe_2)Co]_2PMe_2$ which contains a unique Co_2^{1+}-entity within a triple-bridging framework. Other recently isolated $Me_2PCH_2PMe_2$-complexes contain Cr(O),Fe(II),Co(II,III) or Ag_2^{2+}-moieties.

The coordination behaviour of $MeP(CH_2PMe_2)_2$ is comparable to that of PMe_3, as is shown in reaction (3).

$$"(Me_3P)_4Fe" + 2\ MeP(CH_2PMe_2)_2 \longrightarrow "[MeP(CH_2PMe_2)_2]_2Fe" \quad (3)$$

No 4- or 5- coordinated Fe(O)-complex is obtained, C-H-cleavage of a PCH_3-group as in the starting complex [cf.(6)] being observed instead. The resulting H-Fe(II)-complex is unique by its structural features: one six-membered ($\widehat{FePCPCP}$) ring and a novel metallabicyclo-system. In contrast, the lithiation of the free ligand occurs at the bridging carbon atom (4):

$$MeP(CH_2PMe_2)_2 + LiCH_2PMe_2 \xrightarrow[-PMe_3]{} Li[Me_2PCHP(Me)CH_2PMe_2] \quad (4)$$

This reaction (4) also demonstrates the enhanced acidity according to the following sequence:

$$R_2PCH_3 < (R_2P)_2CH_2 < (R_2P)_3CH$$

The metalated phosphinomethanes $Li[CH_{3-n}(PMe_2)_n]$ (n=1-3) represent examples for a rare class of compounds, where two nucleophilic centers, which may compete for electrophiles, are linked together directly.

$$Me_2\bar{P}-\overset{\ominus}{\bar{C}}< \qquad Me_2\overset{\ominus}{\bar{P}}=C<$$

We have shown (3), that the reactions with MeI proceed via attack at phosphorus, e.g.(5).

$$Li[C(PMe_2)_3] + 3\ MeI \longrightarrow [(Me_3P)_3C]I_2 \quad (5)$$

$$\left.\begin{array}{c} L_3(H)\widehat{Fe(CH_2PMe_2)} \\ \Updownarrow \\ L_4Fe \end{array}\right\} + MeI \begin{cases} \xrightarrow[-60^{o}C]{THF} L_3(Me_2EtP)(H)FeI \\ \xrightarrow[0^{o}C]{pentane} L_4(Me)FeI \\ \xrightarrow[+20^{o}C]{THF/Mg} L_3(Me)\widehat{Fe(CH_2PMe_2)} \end{cases} \quad (6)$$

$L=PMe_3$

In contrast, MeI yields different products in the reaction with $(Me_3P)_4Fe$ [which exists in equilibrium with $(Me_3P)_3(H)Fe(CH_2PMe_2)$ (2)], depending on the reaction conditions, but in neither case attack at phosphorus is observed (6).

With diorganochlorphosphines as electrophiles, C- and/ or P- attack may occur, depending on the steric bulk of the substituents, (4), e.g.

$$Li[CH_2PMe_2] + R_2PCl \longrightarrow Me_2PCH_2PR_2 \qquad (7)$$

$$Li[CH(PMe_2)_2] + R_2PCl \longrightarrow \begin{cases} (Me_2P)_2(R_2P)CH \\ R_2P\text{-}PMe_2\text{=}CH\text{-}PMe_2 \end{cases} \qquad (8)$$

$$Li[C(PMe_2)_3] + Me_2PCl \longrightarrow \begin{cases} \not\rightarrow (Me_2P)_4C \\ Me_2P\text{-}PMe_2\text{=}C(PMe_2)_2 \end{cases} \qquad (9)$$

This ylid, formed in reaction (9), belongs to a rare class of compounds, distinguished by three phosphorus atoms attached to a ylidic carbon atom. Their ^{31}P-NMR-data are given in Table I. The parent compound of this type may be obtained according to (10) by 3 independent routes.

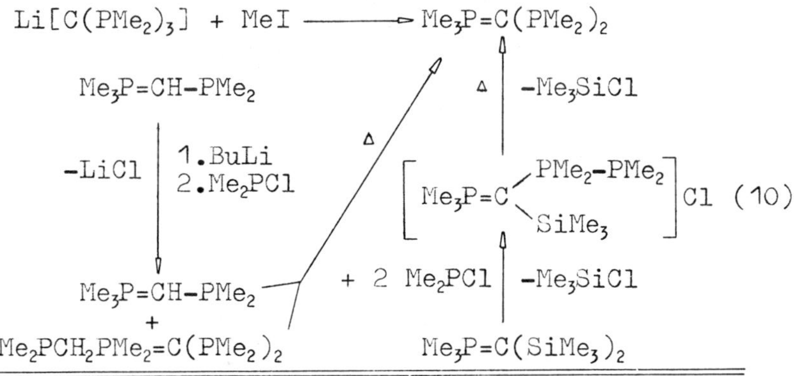

Table I: ^{31}P-NMR-data for $RMe_2P_A\text{=}C(P_BMe_2)_2$ (C_6D_6, H_3PO_4 ext.)

R=	δP_A	δP_B	δP_R	$J(P_AP_B)$	$J(P_AP_R)$	$J(P_BP_R)$
Me	+ 8,15	-43,20		123,4		
PMe_2	+14,15	-33,96	-58,87	140,4	213,6	12,2
CH_2PMe_2	+13,06	-39,30	-57,63	138,8	46,5	7,6
$CH_2^{\ominus}$	+26,22	-48,83		85,5		

These ylids may be looked at as the neutral counterparts of the diphosphinomethanides.

$$Me_2\underline{P}\text{—}\overset{\oplus PR_3}{\underset{\ominus}{C}}\text{—}\underline{P}Me_2 \qquad\qquad Me_2\underline{P}\text{—}\overset{H}{\underset{\ominus}{C}}\text{—}\underline{P}Me_2$$

In fact, they act as powerful ligands in metal complexes, e.g. in $\{(Me_3P)_3[Me_3PC(PMe_2)_2]Co\}_2(PF_6)_2$.

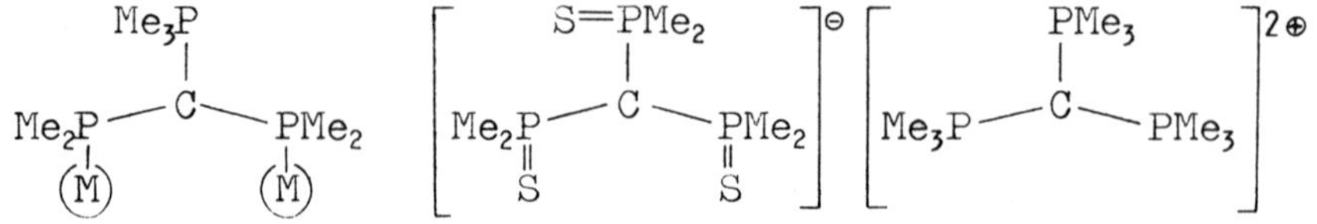

Me_3P–C(Me_2P–Ⓜ)(PMe_2–Ⓜ) — A

$[(S{=}PMe_2)C(Me_2P{=}S)(PMe_2{=}S)]^{\ominus}$ — B

$[(PMe_3)C(Me_3P)(PMe_3)]^{2\oplus}$ — C

The mesomeric stabilisation in this system (A) is comparable to the di-cationic system (C) [see reaction (5)] and the anionic system (B), which my be obtained according to (11).

$$(Me_2PS)_3CH + Me_3P{=}CH_2 \rightleftharpoons [Me_4P][(Me_2PS)_3C] \quad (11)$$

The diphosphinomethanes should also be suitable for preparing PCPCPC six-ring systems for the first time. In fact, we obtained a colourless solid from reaction (12), and from the ^{31}P-NMR-spectrum we believe, that it contains a six-ring-dication, along with some polymer, from which separation could not be achieved, however.

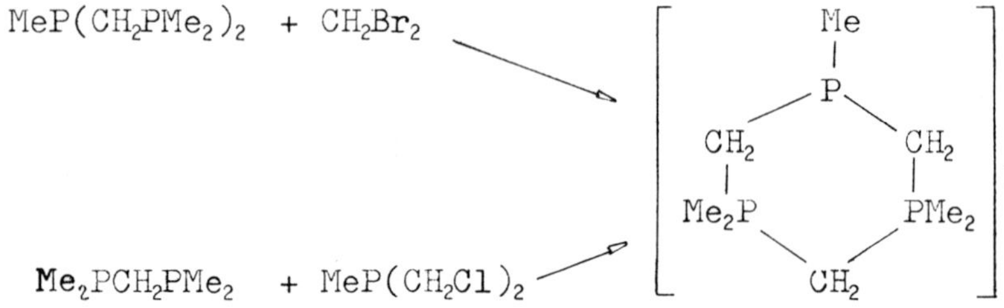

$$MeP(CH_2PMe_2)_2 + CH_2Br_2 \rightarrow \left[\text{cyclo-}MeP(CH_2)_2(PMe_2)_2CH_2\right]X_2 \quad (12)$$

$$Me_2PCH_2PMe_2 + MeP(CH_2Cl)_2 \rightarrow \left[\text{cyclo-}MeP(CH_2)_2(PMe_2)_2CH_2\right]X_2 \quad (13)$$

The same product is obtained from reaction (13). [^{31}P-NMR (CD_3OD, H_3PO_4 ext.): $\delta P|$: -49,14; δP^+: +25,63; J(PP)= 30Hz]. We could not achieve further reaction with MeI and neither with $[MeP(CH_2PMe_3)_2]I_2$. [^{31}P-NMR(CD_3OD, H_3PO_4 ext): $\delta P|$: -50,65; δP^+: +25,92; J(PP)=64Hz]. In contrast, $RP(CH_2SiMe_3)_2$ (R=Me,t-Bu) is readily converted by MeI to the corresponding phosphonium salts.

From these and related findings, we conclude, that the simplicity (low steric requirements), synthetic utility, and powerful coordination ability of substituted phosphinomethanes and their anions create a wide, prospering field of organophosphorus and coordination chemistry to be cultivated in the future.

1. Karsch, H.H. Z. Naturforsch. 1979, 34b, 31
2. Karsch, H.H. Inorg. Synth. VolXX, Wiley Interscience, New York, 1980; p.69 and refernces therein
3. Karsch, H.H. Z. Naturforsch. 1979, 34b, 1178
4. Karsch, H.H. Z. Naturforsch. 1979, 34b, 1171

RECEIVED July 7, 1981.

28

Preparation and Synthetic Reactions of α- Alkoxyallyl Phosphorus Ylides

M. MALEKI and J. A. MILLER

Dundee University, Dundee, Scotland, United Kingdom

O. W. LEVER, JR.

Burroughs Wellcome Company, Research Triangle Park, NC 27709

Allylic phosphonium ylids have long had a valued place in the synthesis of 1,3-dienes, and similarly α-heterosubstituted ylids have been used in the preparation of vinyl derivatives (1). This paper is concerned with the preparation and ylid chemistry of a series of α-alkoxyallylic phosphorus derivatives which possess both the structural features alluded to above.

There are many examples of Wittig-type reactions in the literature, in which the behavior of phosphonium ylids, phosphine oxide ylids, and phosphonate ylids is significantly different. This is particularly so with α-heterosubstituted derivatives, as shown by a number of recent examples in which oxide derived ylids have been found to give vinyl derivatives in circumstances where the corresponding phosphonate derived ylids do not (2-5). It was therefore decided to prepare the parent (unsubstituted) α-methoxyallyl phosphorus compounds 1, 2 and 3, in order to investigate their Wittig-type chemistry.

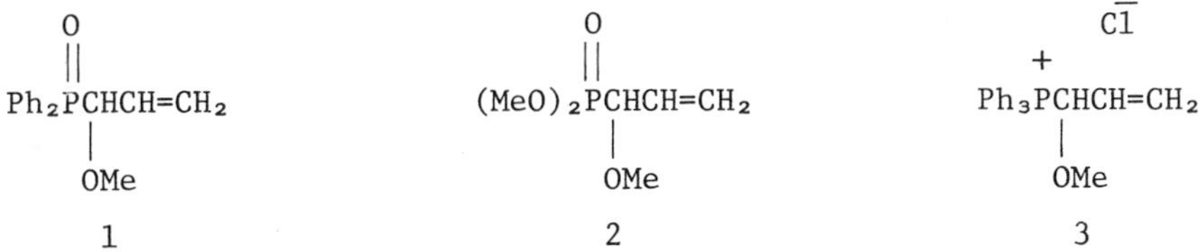

The basic method used to prepare 1, 2, and 3 was the known reaction (6,7) of phosphorus (III) chlorides with acetals. This was first applied to the preparation of the α-methoxyallylphosphine oxide 1, as shown in equation (a), and subsequently extended (8) to a wide range of aryl and alkyl derivatives of 1 .

$$Ph_2PCl \;+\; (MeO)_2CH\text{-}CH{=}CH_2 \longrightarrow Ph_2\overset{O}{\overset{\|}{P}}CH(OMe)CH{=}CH_2 \quad \text{(a)}$$

1

0097–6156/81/0171–0145$05.00/0

When this type of reaction was applied to the preparation of the ester 2, phosphorus trichloride was used as the phosphorus (III) halide. Treatment of this with one mole equivalent of 3,3-dimethoxyprop-1-ene yields mainly the enol ether 4, with smaller amounts of the isomeric α-chloroallyl methyl ether 5. This mixture is not very stable and has to be treated with trimethyl phosphite rapidly, in order to provide 2 in good yield, as shown in Scheme 1. The significance of this sequence is that it reveals that since 2 is the only product, the reactions leading to 2 via 4 must proceed by two allylic inversions, whilst those via 5 involve direct substitution twice at the original acetal carbon.

$$PCl_3 + (MeO)_2CHCH{=}CH_2 \longrightarrow MeOPCl_2 + \underset{4}{ClCH_2CH{=}CHOMe} + \underset{5}{CH_2{=}CH\text{-}CH(Cl)OMe}$$

$$\text{4 or 5} + (MeO)_3P \longrightarrow \underset{2}{(MeO)_2P(O)CH(OMe)CH{=}CH_2} + MeCl$$

Scheme 1

Another unexpected feature of these α-methoxyallylation sequences was revealed by the preparation of the phosphonium salt 3. This was achieved by adding triphenylphosphine to the mixture of ethers 4 and 5, prepared as above, in dry toluene. The salt 3 precipitates from the reaction mixture at low temperature (-30 to +10°C). In contrast, when quaternization is carried out at room temperature under homogeneous conditions (e.g. in dry chloroform), the product is the γ-methoxyallylphosphonium salt 6. It was subsequently shown that the salt 3 isomerizes readily in solution to give its isomer 6, and these are clearly the kinetic and thermodynamic products, respectively, of quaternization of the ethers 4 and 5, as shown in Scheme 2.

$$\underset{3}{Ph_3\overset{+}{P}CH(OMe)CH{=}CH_2\ \bar{Cl}} \xleftarrow[<20°C]{Ph_3P} \left\{ \underset{4}{ClCH_2CH{=}CHOMe} + \underset{5}{CH_2{=}CHCH(Cl)OMe} \right\} \xrightarrow[>20°C]{Ph_3P} \underset{6}{Ph_3\overset{+}{P}CH_2CH{=}CHOMe\ \bar{Cl}}$$

Scheme 2

To our knowledge, such a facile 1,3-migration of a phosphonium group has not been reported before, although one example of apparently related behavior has been observed (9). The phosphonium salt 6 has previously been reported, as its bromide, and used as a three carbon chain extension reagent for aldehydes (10).

Unfortunately the phosphonate 2, and the salt 3, are very hygroscopic, and this led us to select the oxide 1 for scrutiny as precursor of unusual Wittig reagents. Thus a range of ylids have been prepared from 1 and its substituted derivatives using standard conditions (n-BuLi, i-Pr_2NH, in dry THF at -78°C). In general, these ambident ylids can react at either the α-carbon or at the γ-carbon with electrophiles, and examples of both types of behavior will be published elsewhere (11).

When the ylid derived from the oxide 7 was treated with benzaldehyde, a 2-methoxy-1,3-diene was formed. This led us to attempt a synthesis of (±) ar-turmerone by a sequence (Scheme 3) in which 7 effectively acts as an acyl anion equivalent. The required aldehyde 8 was synthesized in good yield using initially diphenyl(methoxymethyl)phosphine oxide (3), and then chlorotrimethylsilane-sodium iodide (12) to cleave the enol ether.

O i,ii,iii OMe iv CHO 8(70%)

O Ph$_2$P OMe 7 v,vi 9(8%) OMe vii O ar-turmerone

Scheme 3 i $Ph_2\overset{O}{\overset{\|}{P}}CHLi(OMe)$; ii H_2O; iii NaH, Δ; iv Me_3SiCl, NaI v LDA; vi 8, Δ; vii aq. acid.

The very low yield in the key Wittig-Horner step, leading to the dienol ether 9, was not due to the lack of reactivity of the lithiated derivative of 7, since the parent oxide was not recovered from the reaction. The major reaction product was the diphosphine dioxide 10, which, it was subsequently established, could also be obtained in high yield (75%) if solutions of the anion were made at low temperatures, and then warmed to 20°C in the absence of any added electrophile, before quenching with water.

A rationalization for the formation of 10 is presented in Scheme 4. The crucial feature is that lithiation of the oxide 7 occurs at two different sites, in both of which internal coordination to lithium is important. When lithiation occurs on the δ-carbon, giving 13, formation of the dienylphosphine oxide 11 results, and hence anion 12 arises. Upon quenching with water the dimeric dioxide 10 is formed. In a separate experiment, addition of benzaldehyde led to isolation of a trienylphosphine oxide whose formation we take to be further evidence for the anion 12.

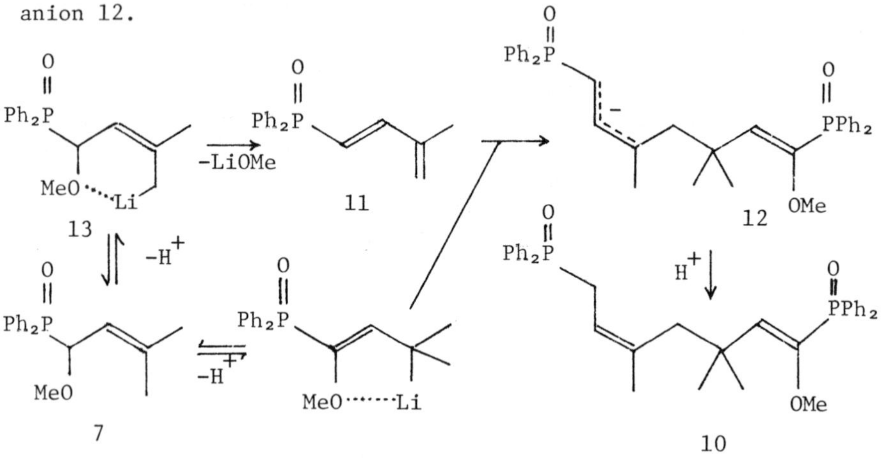

Scheme 4

LITERATURE CITED

1. Gosney, I.G.; Rowley, A.G.; Walker, B.J., in "Organophosphorus Reagents in Organic Synthesis, ed. J.I.G. Cadogan, Academic Press, London, 1979, Chapters 2 and 3.
2. Schlosser, M.; Tuong, H.B., Chimia, 1976, 30, 197.
3. Earnshaw, C.; Wallis, C.J.; Warren, S., J. Chem. Soc. Perkin Trans. I, 1979, 3099.
4. Boutagy, J.; Thomas, R.; Chem. Rev., 1974, 74, 37.
5. Kluge, A.F.; Cloudsdale, I.F., J. Org. Chem., 1979, 44, 4847.
6. Gazizov, M.B.; Sultanova, D.B.; Razumov, A.I.; Zykova, T.V.; Anoshina, N.A.; Salakhutdinov, R.A.; J. Gen. Chem., USSR, 1975, 45, 1670.
7. Dietsche, W.; Annalen, 1968, 712, 21.
8. Maleki, M.; Miller, A.; Lever, O.W., Tetrahedron Lett., 1981, 22, 365.
9. Font, J.; March, P., Tetrahedron Lett., 1978, 3601.
10. Martin, S.F.; Garrison, P.J., Tetrahedron Lett., 1977, 3875.
11. Maleki, M.; Miller, J. A.; Lever, O. W., submitted for publication.
12. Kosarych, Z.; Cohen, T., Tetrahedron Lett., 1980, 21, 3959.

RECEIVED July 7, 1981.

α-Phosphorylated Carbanions: Synthetic Features

G. STURTZ, B. CORBEL, M. BABOULÈNE, and J. J. YAOUANC

Laboratoire de Chimie Hétéroorganique—ERA 612, Faculté des Sciences et Techniques, 6 Avenue Le Gorgeu, 29283 Brest Cédex, France

The study of the chemical reactivity of some phosphates and more specifically, the various possibilities of ester bond rupture, can to some extent be of aid in understanding a few of the mechanisms involved in metabolism effects, while allowing the implementation of new reactions in organic synthesis.

For these reasons it seemed of interest to attempt to create an α-carbanion from a phosphoric acid derivatives 1

$$(A)_2\underset{\underset{O}{\|}}{P} - O - \overset{\ominus}{C} - \xrightarrow{?} \Box\overset{\ominus}{C} -$$

1

A = RO or R_2N -

The following question could, for example, be put forward as to whether a carbenic or carbenoic entity would be yielded.

By selecting certain terpenic phosphates as a substratum, biosynthesis advantages could be brought out by avowing the "natural" possibility of creating this carbanion in a terpenic pyrophosphate, for example ; this to our knowledge has never been brought up in reactions with respect to these compounds.

First (1), we have shown that O-benzylic phosphoric acid derivatives 2 treated with a strong base produced phosphate carbanions which immediately rearranged into α-hydroxy phosphonate oxanions 3 .

$$A_2P(=O)-O-CH_2-C_6H_4X \;(\mathbf{2}) \xrightarrow{n\,BuLi} A_2P(=O)-O-\overset{\ominus}{C}H-C_6H_4X \longrightarrow A_2P(=O)-CH(O^{\ominus})-C_6H_4X \;(\mathbf{3})$$

0097–6156/81/0171–0149$05.00/0

Using two equivalents of n butyl lithium one can isolate benzoine.

The presence of propionic acid in small quantity beside the α-hydroxy phosphonamide 5 when O-allylic phosphoramide 4 is treated by one equivalent of n-Butyllithium has required the analysis of the conditions of its formation (2).

$$[(CH_3)_2N]_2 \underset{\underset{O}{\|}}{P}-OCH_2-CH=CH_2 \xrightarrow[2)\ H_3O^+]{1)\ n\text{-Bu Li THF, 70°C}} [(CH_3)_2N]_2 \underset{\underset{O}{\|}}{P}-\underset{\underset{OH}{|}}{CH}-CH=CH_2$$

4 5 (60 %)

$+ CH_3CH_2CO_2H$ 10%

The mesomeric form 6 B explains the obtention of this acid :

$$[(CH_3)_2N]_2 \underset{\underset{O}{\|}}{P}-O-CH_2-CH=CH_2 \xrightarrow{n\text{-Buli THF -70°C}} [(CH_3)_2N]_2 \underset{\underset{O}{\|}}{P}-\overset{H}{\underset{\underset{O^{\ominus}}{|}}{C}}-CH=CH_2$$

$$\xrightarrow{nC_4H_9Li} [(CH_3)_2N]_2 \underset{\underset{O}{\|}}{P} - \overset{\ominus}{\underset{\underset{O}{|}}{C}} - CH = CH_2 \quad \text{6 A}$$

$$\updownarrow$$

$$[(CH_3)_2N]_2 \underset{\underset{O}{\|}}{P} - \underset{\underset{O^{\ominus}}{|}}{C} = CH - \overset{\ominus}{CH_2} \quad \text{6 B}$$

$$\xrightarrow{H_3O^+} [(CH_3)_2N]_2 \underset{\underset{O}{\|}}{P} - \underset{\underset{O}{\|}}{C} - CH_2 - CH_3 \longrightarrow CH_3CH_2CO_2H$$

unstable 7

From that interpretation we have developed synthetic interest in this bianion 6 B (carbanion in β position of a potential carboxylic acid) towards different electrophilic reagents (3).

Alkyl halogenides were condensed on our bianion. This method may turn out to be useful for setting non-saturation in fatty-acid chains.

With the carbonyl derivatives, butyrolactones are obtained in one step.

For example the O-tetrahydropyranyl derivative of 3β- hydroxy 17-oxo 5-androstene 8 reacted on the bianion and give after methanolysis of protective group the 3β, 17β- dihydoxy 5-androstene 17α- propanoïc acid γ-lactone 9' (4).

ThpO 8

$+2[(CH_3)_2N]_2P(=O)-C(R)=C(O^-)-CH_2^-$

HO 9

With Oxiranes the valerolactones are prepared.

The stabilization of the carbanionic form of some thiolophosphates 10 seems to be possible through the electron withdrawing mesomeric effect of the carbethoxy group. Condensations of these carbanions with derivatives lead to the corresponding α,β-unsaturated esters (5).

$$(R^1O)_2P(=O)-S-CH(R^3)-CO_2R^2 \xrightarrow[\text{2) } R^4COR^5]{\text{1) LDA, -78°C, THF}} R^4R^5C=C(R^3)CO_2R^2$$

10

A mechanism, via a β-hydroxy phosphorothiolate β-mercaptophosphate rearrangement, will be shown, and these new synthons will be compared with the phosphonate analogs (Wittig -Horner reagent).

Our work is in fact supported by Tanaka's results who sometimes after us proposed a synthesis of trisubstituted α,β-ethylenic esters from this same molecule (6).

The extension of this reaction to carbanions of β-ketophosphorothiolates 11 which could have led, by analogy, to a new α,β-ethylenic ketone synthesis, has brought about a different reaction path.

The delocalization of enolate allows a transposition into an enol mercapto phosphate which can be quite easily alkylated. Then enol phosphate thioethers 12 can be obtained.

$$(R^1O)_2\underset{\underset{O}{\|}}{P} - S - CH^{\ominus} - \overset{O}{\overset{\|}{C}} - R^2 \quad \underline{11}$$

$$\updownarrow$$

$$(R^1O)_2\overset{O}{\overset{\|}{P}} - S - CH = \underset{O^{\ominus}}{C} - R^2 \longrightarrow (R^1O)_2\underset{O}{\overset{\|}{P}} - O - \overset{R^2}{C} = CH - S^{\ominus}$$

$$\downarrow R^3I$$

$$(R^1O)_2\underset{\underset{O}{\|}}{P} - O - \overset{R^2}{C} = CH - SR^3 \quad \underline{12}$$

Indeed in the same experimental conditions the geranyl phosphate brings the formation of an α-hydroxy phosphonate which can give the citral.

By using a double quantity of base, we get citronellic acid and we know that this type of compound is found in some plants. So we may consider that in the field of a biology enzymatic system would exist, which could lead from geranyl pyrophosphate to such transpositions. This hypothesis would explain the presence of these compounds into some plants.

LITERATURE CITED

1. STURTZ, G. and CORBEL, B. C.R. Acad. Sc. Paris, série C t. 276, 1973, p 1807.
2. STURTZ, G. and CORBEL, B. C.R. Acad. Sc. Paris, série C t. 277, 1973, p 395.
3. STURTZ, G., CORBEL, B. and PAUGAM, J.P. Tétrahedron Letters, 1976, p 47.
4. STURTZ, G., YAOUANC, J.J., KRAUSZ, F. and LABEEUW, B. Synthesis, 1980, p 289.
5. BABOULENE, M. and STURTZ, G. Journal of Organometallic Chemistry, 1979, p 27.
6. TANAKA, K., UNEME, H., ONO, N. and KAJI, A. Chemistry Letters 1979, p 1039.

RECEIVED July 7, 1981.

A New Synthesis of Indoles

M. LE CORRE, A. HERCOUET, and H. LE BARON

Laboratoire de Synthèse Organique, Université de Rennes I, Avenue du Général Leclerc, 35042 Rennes Cédex, France

Indoles were prepared (80-95%) from *o*-acylaminobenzyltriphenylphosphoranes by intramolecular Wittig reaction.

Madelung synthesis of indoles[1] is initiated only by strong bases at elevated temperatures. Prior deprotonation of the amide group greatly reduces the electrophilicity of the carbonyl group, and the useful scope of this synthesis is therefore limited to molecules which can survive very drastic conditions (scheme 1)

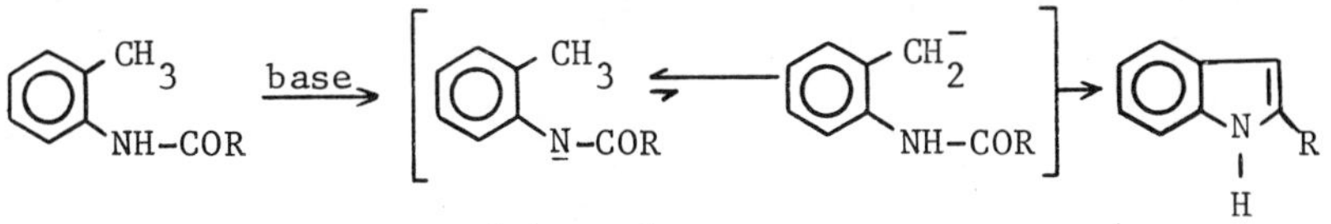

Scheme 1

We now wish to report a new simple method for the synthesis of indoles which is based on the intramolecular reaction of amide with alkylidenephosphoranes[2](scheme 2)

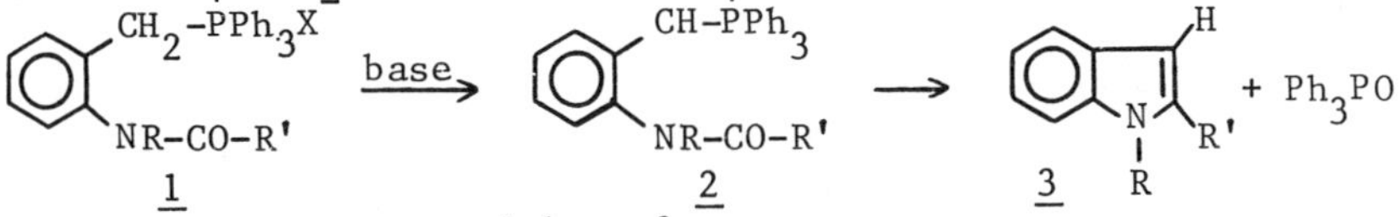

Scheme 2

In a general procedure, 1.1 equivalent of base (CH_3ONa, *t*-BuOk or *t*-AmONa) was added to a stirred suspension of 1 equivalent of dry phosphonium salt 1 in boiling toluene. After being stirred for 10 min, the solution was filtered and evaporated to dryness. The crude material was essentially the desired indole with triphenylphosphine oxide. The yields for the reaction in scheme 2 are for isolated pure products obtained by preparative

0097–6156/81/0171–0153$05.00/0

layer chromatography on silica gel (1 : 1 ether-benzene) or by crystallization from ethanol.

As shown in Table I, the yields of substituted indoles, obtained through the use of this procedure were quite good. Not only is this reaction compatible with a wide variety of substituents at the 2 position, but our procedure can also be used to prepare 1-substituted indoles.

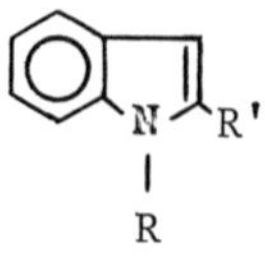

R	R'	yield(%)	mp (°C)
H	CH_3	96	61 (a)
H	$-C(CH_3)=CH_2$	93	118-120
H	C_6H_5	97	189-190 (b)
H	$p-NO_2-C_6H_4$	93	249-251 (c)
H	$p-CH_3O-C_6H_4$	85	228-230 (d)
C_6H_5	CH_3	80	liq.

(a) lit. [4], m.p. 62°C ; (b) lit. [5], m.p. 188-189°C ; (c) lit. [6], m.p. 251-252°C ; (d) lit. [6], m.p. 230-230,5°C.

The *o*-acylaminobenzylphosphonium salts are available from two ways : (1) reduction of *o*-nitrobenzylphosphonium salt[7] with zinc in hydrobromic acid-ethanol (yield 95%). The reductive cleavage of the carbon-phosphorus bond[8] can be avoided by using only two equivalents of reagent (2) condensation of *o*-aminobenzyl alcohols with triphenylphosphine and hydrobromic acid.

These salts are then readily acylated by acyl chlorides in the presence of two equivalents of triethylamine or pyridine to give the desired *o*-acylaminobenzylphosphonium salts (85-96% yield).

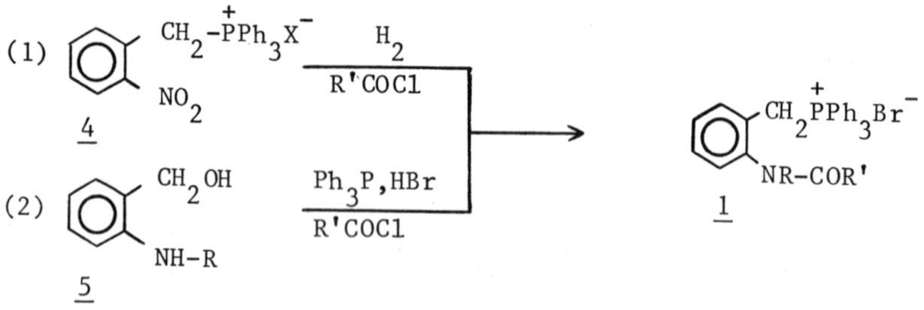

(1) $CH_2-\overset{+}{P}Ph_3X^-$, NO_2 (4) — H_2 / R'COCl →

(2) CH_2OH, NH-R (5) — Ph_3P,HBr / R'COCl →

→ $CH_2\overset{+}{P}Ph_3Br^-$, NR-COR' (1)

Scheme 3

1 - W. MADELUNG, *Chem. Ber.*, 1912, 25, 1128.
2 - To our knowledge, amides failed to react with phosphorus ylids the only published exception is the reaction of $Ph_3P{=}CH{-}CO_2Et$ with a lactam[3].
3 - W. FLITSCH, H. PETERS, *Tetrahedron Letters*, 1969, 1161.
4 - L. MARION, C.W. OLDFIELD, *Canad. J. Res.*, 1947, 25B, 1.
5 - J.I.G. CADOGAN, M. CAMERON-WOOD, *Proc.Chem.Soc.*, 1962, 361.
6 - C.E. BLADES, WILDS, *J. Org. Chem.*, 1956, 21, 1031.
7 - A.A. ARDANAKI, N. MALEKI, M.R. SAADEIN, *J.Org.Chem.*, 1978, 43 4128.
8 - A. SHONBERG, K.H. BROSOWSKI, E. SINGER, *Chem. Ber.*, 1962, 95, 2984.

RECEIVED July 7, 1981.

Alkylation by Way of Monomeric and Polymeric Alkoxyphosphonium Salts

DONALD W. HAMP and EDWARD S. LEWIS

Department of Chemistry, Rice University, P.O. Box 1892, Houston, TX 77001

There are numerous examples of the conversion of alcohols to alkylating agents using phosphorus compounds (1-5). We became interested in the use of alcohols as general alkylating agents using methyltriphenoxyphosphonium triflate according to eq. (1) and (2) (5).

$$ROH + MeP(OPh)_3 \rightarrow Me(PhO)_2POR + PhOH \qquad (1)$$

$$Me(PhO)_2POR + Nu^- \rightarrow Me(PhO)_2PO + RNu \qquad (2)$$

This procedure, although rather general and susceptible to further variation has not easily given good yields, and the desired product is contaminated with PhOH and $Me(PhO)_2PO$. To avoid the latter problems we prepared a polymer version of the reagent in which one of the phenyl groups is incorporated in polystyrene. The polymeric reagent, prepared by Scheme 1, does indeed show some of the desired reactions, but undesired side reactions, difficulty of polymer recycling and yield problems in the monomeric model system have led us to study a simpler system.

Scheme 1

styrene + p-acetoxystyrene + DVB + toluene $\xrightarrow{(a)}$

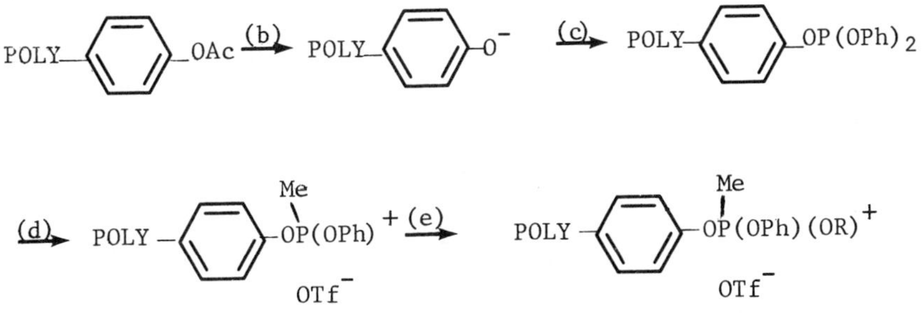

POLY–C$_6$H$_4$–OAc $\xrightarrow{(b)}$ POLY–C$_6$H$_4$–O$^-$ $\xrightarrow{(c)}$ POLY–C$_6$H$_4$–OP(OPh)$_2$

$\xrightarrow{(d)}$ POLY–C$_6$H$_4$–OP(Me)(OPh)$^+$ OTf$^-$ $\xrightarrow{(e)}$ POLY–C$_6$H$_4$–OP(Me)(OPh)(OR)$^+$ OTf$^-$

0097–6156/81/0171–0157$05.00/0

a) H_2O, polyvinyl alcohol benzoyl peroxide, 80° (6). b) H_4N_2, dioxane. c) $(PhO)_3P$, 1% $(PhO)_2POH$ (7). d) MeOTf, CH_2Cl_2. e) ROH, CH_3CN.

The polymer bound triphenylphosphine oxide, 1b, is readily converted to the ditriflate, 2b, following Hendrickson and Schwartzman's procedure for the synthesis of triflyltriphenylphosphonium triflate, 2a, (8). Much of our work has been the exploration of the monomeric system, first the conversion of the alcohol to $ROP(Ph)_3{}^+OTf^-$ and the subsequent reactions of these with various nucleophiles.

Scheme 2

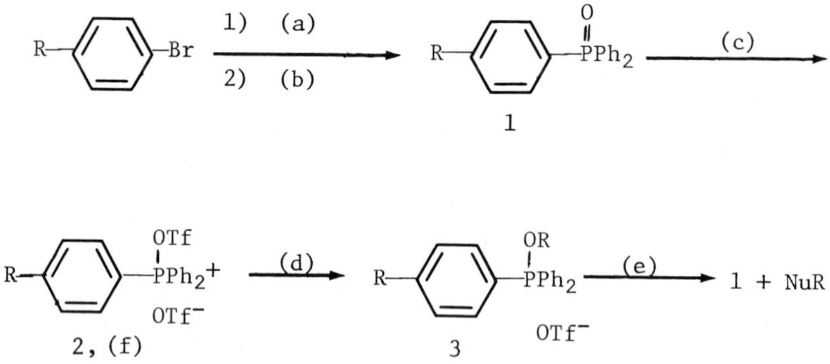

R—C6H4—Br → 1) (a), 2) (b) → R—C6H4—P(O)Ph2 (1) → (c) →

R—C6H4—P(OTf)Ph2+ OTf− (2, (f)) → (d) → R—C6H4—P(OR)Ph2 OTf− (3) → (e) → 1 + NuR

1a, 2a, 3a, R=H. 1b, 2b, 3b, R=polymer

a) $LiPPh_2$, THF. b) H_2O_2, acetone. c) Tf_2O, CH_2Cl_2, -78°. d) ROH, 2,6-lutidine, CH_2Cl_2, -78°. e) Table I solvents, reflux. f) 2a is monomeric in solution, but crystallization produces $Ph_3{}^+PO^+PPh_3{}^-(OTf)_2$ (9).

A variety of the salts, 3a, were prepared in solution and characterized by proton nmr (these were virtually predictable using normal values of J_{P-H}) and proton decoupled ^{31}P nmr. The following were prepared in solution, with P chemical shifts in parenthesis: R=CH_3 (64.0), R=CH_3CF_2 (68), R=$CH(CH_3)_2$ (59.4), R=$(CH_3)_2$-$CH_1CH_2CH_2$ (61.7), R=$(CH_3)_3CCH_2CH_2$ (61.6), cyclohexyl (59.2), benzyl (60.6), octyl (60.87) $ClCH_2CH_2OCH_2CH_2$ (62.8), C_6H_5 (65.6). The last was isolated as a crystalline solid, and further characterized: Mp 171°, ^{19}F nmr singlet, +2.4 ppm relative to CF_3COOD, the high resolution mass spectrum showed the intact cation. The yields of the salts depend upon the solvent; rigorously dried CH_2Cl_2 and CH_3CN give near quantitative yields, THF much lower yields (even at -78°) to suppress the triflic anhydride THF reaction), and liquid SO_2 yields no detectable product.

Toward iodide, bromide and thiocyanate ions the phosphonium salts reacted to give alkylated products in fair yields. Yields, based on starting triflic anhydride which was allowed to react with a slight excess of phosphine oxide, are given in Table I.

Table I

Yields of $ROPPh_3^+OTf^- + Nu^- \rightarrow RNu$ Reaction[a]

R	solvent[b]	Nu	product	(yield)
1-octyl	THF	LiBr	RBr	(71)
"	CH_3CN	"	"	(61)
"	THF	LiCl	RCl	(5)
2-(2-chloroethoxy)-ethyl	CH_2Cl_2	LiBr	RBr	(55)
3,3-dimethylbutyl	CH_2Cl_2	NMe_4Br	RBr	(43)
3-methylbutyl	CH_2Cl_2	NaI	RI	(70)
isopropyl	$CDCl_3$	NaI	RI	(42)
tertbutyl	$CDCl_3$	NaI	RI	(52)
1-octyl	CH_3CN	KSCN[c]	RSCN[d]	(87)
1-cyclohexyl	$CHCl_3$	Et_3NCH_2PhCl	RCl	(0)

a) Yields are based on Tf_2O and are determined by gc with an internal standard. b) This is the solvent used in the second step in Scheme 1, the solvent for the first step was sometimes left, sometimes removed. c) 18-crown-6 was used in this experiment, but its function was not explored. d) No RNCS was seen in the gc traces.

Some other reaction courses were also seen. For example, 3a, R=octyl with butyllithium gave no significant yield of dodecane. The ^{31}P chemical shift of an important component suggested an ylide, perhaps derived from butyltriphenylphosphonium from attack at P rather than C. Ethyl magnesium bromide with 3a, R=octyl gave mostly 1-bromooctane, but with butylmagnesium chloride, about 20% yield of dodecane was observed with no 1-chlorooctane.

The polymer bound triphenylphosphine oxide was synthesized using established techniques from a macro reticular DVB/styrene copolymer, Amberlite XE-305 (10). The phosphorus content was determined gravimetrically using the Kjeldahl method (11). Various polymers contained between 30 and 50% of the aromatic rings functionalized.

Reactions of the bound alkoxyphosphonium triflate (formed in an identical manner to that of the unbound analog 3a), with the nucleophiles shown in Table II were followed by ir and the yields determined by gc. Our few preliminary results demonstrate that their alkylating abilities are comparable to that of the unanchored alkoxyphosphoniums. The triphenylphosphine oxide polymer produced can be recycled back to the ditriflate, with triflic anhydride.

Table II

$$POL-\overset{+}{P}(OR)Ph_2 + Nu^- \rightarrow POL-P(O)Ph_2 + NuR$$

R	solvent	Nu	product	(yield)
3-methylbutyl	CH_2Cl	NaI	RBr	(60)
1-octyl	CH_3CN	LiBr	RBr	(60)
2-phenylethyl	CH_3CN	LiBr	RBr	(52)
2-(2-chloroethoxy) ethyl	CH_3CN	LiBr	RBr	(20)
"	HMPA	LiBr	RBr	(0)

Acknowledgement: Support of this work by the Robert A. Welch Foundation is gratefully acknowledged.

Literature Cited

1. Landauer, S. R.; Rydon, H. N. J. Chem. Soc. 1953, 2224.
2. Castro, B.; Selve, C. Bull. Soc. Chim. Fr. 1971, 12, 4368-4372.
3. Appel, R. Angew. Chem. Int. Ed. Engl. 1975, 14, 801-811.
4. Hendrickson, J. B.; Schwartzman, S. M. Tet. Lett. 1975, 4, 277-280.
5. Lewis, E. S.; Walker, B. J.; Ziurys, L. M. Chem. Comm. 1978, 424-425.
6. Arshady, R.; Kenner, V. W. J. Poly. Sci. Poly. Ed. 1974, 12, 2017-2025.
7. U. S. patent 3,375,304, 1968; Chem. Abst. 1968, 68, 965595.
8. Schwartzman, S. M. Ph.D. thesis, 1975, Brandeis University.
9. Aaberg, B.; Gramstad, T.' Husebye, S. Tet. Lett. 1979, 24,
10. Delles, H. W.; Schwarz, R. W. J. Am. Chem. Soc. 1974, 96, 6469-6480.
11. Steyermark, A. "Quantitative Organic Microanalysis 2nd Ed." 1961, New York, Academic Press; p 354.

RECEIVED June 30, 1981.

Some Preparative and Mechanistic Aspects of the Chemistry of Phosphoric Acid and Thiophosphoric Acid Chloride Betaines

M. MEISEL, C. DONATH, and H. GRUNZE

Zentralinstitut für Anorganische Chemie der Akademie der Wissenschaften der DDR, DDR-1199 Berlin-Adlershof, Rudower Chaussee, GDR

Phosphoric acid chloride and thiophosphoric acid chloride pyrididium betaines (I) which are easily prepared by reacting P_4O_{10}, $P_4O_6S_4$ and P_4S_{10}, respectively, with phosphorus oxychloride or phosphorus thiochloride in the presence of pyridine (1), are of interest for the synthesis of substituted phosphates and thiophosphates.

$$\mathrm{C_5H_5N^{\oplus}-P(=X)(-X^{\ominus})-Cl} \qquad \mathrm{I}$$

Ia X = O

Ib X = O and S

Ic X = S

During nucleophilic attack on the P-atom of I both the P-Cl and the P-N bond may be broken and three possible reaction types may result:

$$\text{type A:}\quad Nu^- + Py{\cdot}PX_2Cl \longrightarrow Py{\cdot}PX_2Nu + Cl^- \qquad (1)$$

$$\text{type B:}\quad Nu^- + Py{\cdot}PX_2Cl \longrightarrow NuPX_2Cl^- + Py \qquad (2)$$

$$\text{type C:}\quad 2Nu^- + Py{\cdot}PX_2Cl \longrightarrow NuPX_2Nu^- + Cl^- + Py \qquad (3)$$

Nu = nucleophile; Py = pyridine

A typical example for a type A reaction is the "halosilane condensation", i.e. the reaction of the betaines I with trimethylsilyl derivatives with liberation of trimethylsilyl chloride. This reaction is especially smooth with Ic:

$$Py{\cdot}PS_2Cl + Me_3SiY \longrightarrow Py{\cdot}PS_2Y + Me_3SiCl \qquad (4)$$

(Y = e.g. NCS, OR, NRR')

The amido derivatives can also be prepared by reacting $Py{\cdot}PS_2Cl$ with amines at a molar ratio of Ic : amine = 1:2.

0097–6156/81/0171–0161$05.00/0

$Py{\cdot}PS_2Cl$ (Ic) reacts according to type A with NaF in dry acetonitrile yielding $Py{\cdot}PS_2F$ (2). The analogous reaction with $Py{\cdot}PO_2Cl$ (Ia) does not take place. Only when stronger fluorinating agents like AsF_3 are employed, the Cl-F exchange occurs and the formation of the fluoride betain $Py{\cdot}PO_2F$ is observed (3).

The intermediary formation of a "monometaphosphate" (cf. (4)) in a type A reaction is indicated by the hydrolysis of $Py{\cdot}PO_2Cl$. The reaction of Ia with equimolar amounts of water in nitromethane yields a polyphosphate mixture, whereas in acetonitrile cyclization of the $[PO_3]^-$ - anions to trimetaphosphate is observed. The reaction can be formulated as an $S_N1(P)$ (elimination - addition) as follows:

$$Py^{\oplus}-P(=O)(Cl)-O^{\ominus} + H_2O \longrightarrow \left\{ Py^{\oplus}-P(=O)(-O\cdots H)-O^{\ominus} \right\} + HCl$$

$$\longrightarrow \left\{ [PyH]^+ \left[O-P{\lt}^{O}_{O} \right]^- \right\} \longrightarrow \frac{1}{n} \; [PyH]_n[PO_3]_n \qquad (5)$$

In contrast, during the reaction of $Py{\cdot}PS_2Cl$ (Ic) with water polymeric thiophosphates are never found.

The reaction with saturated aqueous NH_4F also indicates the different behaviour of Ia and Ic (5). For $Py{\cdot}PO_2Cl$ (Ia) the product consists of monofluorophosphate (75 %) and monophosphate (25 %). It is probable that initially formed $[PO_3]^-$ (cf. eq. (5)) rapidly reacts with F^- to $[PO_3F]^{2-}$ ($S_N1(P)$).

The reaction of $Py{\cdot}PS_2Cl$ (Ic) with aqueous NH_4F, however, leads to difluorodithiophosphate (40 %), monofluorodithiophosphate (5 %) and monothiophosphate (35 %). Here the first step most likely is an attack of the two competing nucleophiles which should yield the intermediates $Py{\cdot}PS_2F$ and $Py{\cdot}PS_2(OH)$. The subsequent steps of the reaction probably correspond to an $S_N2(P)$ mechanism.

Reactions of the betaines I with initial cleavage of the P-N bond (type B) are rarely observed. One example is the reaction with hydrogen halides leading to dihalogeno phosphates and -thiophosphates:

$$Py{\cdot}PX_2Cl + H[Hal] \longrightarrow [PyH][HalPX_2Cl] \qquad (6)$$

(Hal = halide)

This type of reaction was first described for $Py \cdot PS_2F$ which reacts with HCl to fluoro chloro dithiophosphate (2). The chloride betaines Ia and Ic react with HCl quantitatively to the respective dichlorophosphates (6). $Py \cdot POSCl$ (Ib) reacts, as has been found, too, with other nucleophiles, under partial dismutation, and a mixture of $[PO_2Cl_2]^-$, $[POSCl_2]^-$ and $[PS_2Cl_2]^-$ is obtained. The reaction of $Py \cdot PS_2Cl$ (Ic) with HBr certainly at first yields $[BrPS_2Cl]^-$ according to (6). With an excess of HBr also Cl-Br exchange under formation of $[PS_2Br_2]^-$ takes place.

The reactions of chloride betaines of type C proceed easily and with high yields. Aliphatic alcohols, phenols, ammonia and primary or secondary amines can be used as the corresponding nucleophiles (7).

Studies of the reaction of $Py \cdot PS_2Cl$ with alcohols in different molar ratios show that in contrast to what has been reported (2) this reaction does not proceed under primary cleavage of the P-N bond (type B) via the intermediate $[ROPS_2Cl]^-$, but rather via $Py \cdot PS_2OR$ (type A). However, in most cases, the second step, i.e. cleavage of the P-N bond, is so fast, that a simultaneous reaction of Ic corresponding to type C results.

An interesting example for the type C reaction is the simultaneous reaction of I with alcohol/amine mixtures. In the case of Ic the reaction can be controlled in such a way, that practically exclusively the esteramide of the dithiophosphoric acid is formed (8):

$$Py \cdot PS_2Cl + ROH + 3\ RR'NH \longrightarrow [RR'NH_2][ROPS_2(NRR')] + [RR'NH_2]Cl + Py \qquad (7)$$

A study of the reaction (7) at different alcohol/amine ratios has revealed that apparently at first the alcohol reacts under formation of the ester betaine $Py \cdot PS_2OR$. Thus, at a molar ratio of Ic : amine = 1:1 in the presence of ROH only dialkyldithiophosphate is produced, while at a ratio of Ic : amine = 1:2 a mixture of esteramide and dialkylester results. Only at a molar ratio of Ic : amine = 1:3 the esteramide is formed almost exclusively (cf. eq. (7)).

It was found that $Py \cdot PO_2Cl$ is an excellent phosphorylation agent for nucleotide syntheses (9). When Ia reacts with an OH group of a nucleoside an ester betaine is formed according to type A as demonstrated by NMR. This intermediate betaine rapidly reacts with nucleophilic agents like $[PO_4]^{3-}$, $[P_2O_7]^{4-}$, R'OH and produces in relatively high yields the respective nucleoside phosphates and dinucleoside phosphates:

$$Py{\cdot}PO_2Cl + ROH \xrightarrow{Py} Py{\cdot}PO_2OR \begin{cases} \xrightarrow{[P_2O_7]^{4-}} [RO\text{-}P_3O_9]^{4-} \\ \xrightarrow[R'OH]{} RO\text{-}PO_2\text{-}OR' \end{cases} \quad (8)$$

ROH = protected and unprotected nucleosides

R'OH= ROH; other alcohols

This synthesis can be performed as an one-pot reaction.

The reactions of the betaines I demonstrate the great variety of applications of these compounds for the preparation of substituted phosphoric and thiophosphoric acid derivatives, and they open new routes for the synthesis of new or difficult to prepare compounds.

Literature Cited

1. Meisel, M.; Grunze, H. Z.anorg.allg.Chem. 1968, 360, 277 and unpublished results respectively.
2. Fluck, E.; Retuert, P.J.; Binder, H. Z.anorg.allg. Chem. 1973, 397, 225.
3. Donath, Ch.; Cernik, M.; Meisel, M.; Dostal, K.; Grunze, H. unpublished results.
4. Ramirez, F.; Marecek, J.F. Pure and Appl.Chem. 1980, 52, 1021.
5. Meisel, M.; Donath, Ch. Chemiedozententagung, Freiberg, 1979.
6. Meisel, M.; Donath, Ch., to be published.
7. Meisel, M.; Donath, Ch.; Wolf, G.-U. DDR Pat. WP C 07 F/213 416 (6.6.79).
8. Meisel, M.; Donath, Ch. DDR Pat. WP C 07 F/205 560 (31.10.78)
9. Bauschke, E.; Meisel, M.; Brankoff, T. DDR Pat. WP C 07 H/223 311 (14.8.80).

RECEIVED July 7, 1981.

"Activated" Phosphoranes for the Cyclodehydration and Chlorination of Simple Diols

S. WOODY BASS, CAREY N. BARRY, PHILIP L. ROBINSON, and SLAYTON A. EVANS, JR.

The William R. Kenan, Jr. Laboratories of Chemistry, The University of North Carolina, Chapel Hill, NC 27514

The scope and synthetic utility of dioxyphosphoranes, and particularly, diethoxytriphenylphosphorane (DTPP), as useful cyclodehydrating "reagents" for the conversion of diols to cyclic ethers have received only superficial attention (1). The DTPP-diol → ether route has several unique advantages over existing methods: (a) the reaction conditions are effectively neutral and mild, (b) the stereoselectivity in the closure of both unsymmetrical and symmetrical diols to cyclic ethers is high, and (c) the isolation of the product(s) from triphenylphosphine oxide (TPPO) is convenient.

We have systematically examined the facility with which DTPP promotes the cyclodehydration of simple diols to cyclic ethers: 1,3-propanediol (1) → oxetane (2) (2-5%); 1,4-butanediol (3) → tetrahydrofuran (4) (85%); 1,5-pentanediol (5) → tetrahydropyran (6) (72%); 1,6-hexanediol (7) → oxepane (8) (55-68%). Increased alkyl substitution at the carbinol carbon significantly diminishes the facility for cyclic ether formation. For example, a mixture of meso- and d,l-2,6-heptanediol gave only 6-10% of the cis- and trans-2,6-dimethyltetrahydropyrans when treated with DTPP. While diol 1 resists cyclodehydration with DTPP to oxetane, some 2,2-disubstituted 1,3-propanediols are readily converted to the appropriate oxetanes [e.g., 2-ethyl-2-phenyl-1,3-propanediol → 3-ethyl-3-phenyloxetane (78%)].

Treatment of diol 9 with DTPP gives starting material and ethyl ether 10 (32%) but no bicyclic oxetane. trans-2-Hydroxycyclohexyl 2-hydroxyethyl sulfide (11) reacts with DTPP affording 63% of trans-1,4-oxathiadecalin (12) and none of the corresponding cis isomer. These results indicate that the primary hydroxyl group undergoes preferential activation by DTPP and subsequent displacement as TPPO.

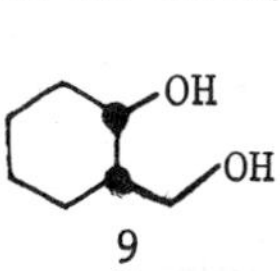

OH OH — 9; OH O — 10; OH S OH — 11; O S — 12

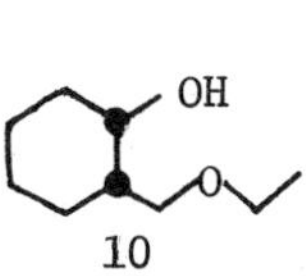

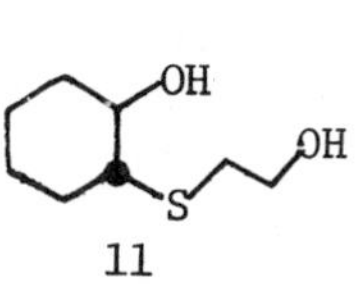

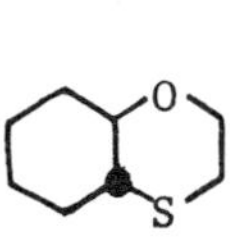

0097–6156/81/0171–0165$05.00/0

We have also determined that the reaction of (Z)-2-butene-1,4-diol with DTPP in refluxing dichloromethane (CH_2Cl_2) affords 2,5-dihydrofuran (85-87% GLC and 60% isolated yield). On the other hand, treatment of (E)-2-butene-1,4-diol with DTPP in chloroform (61°, 18 h) gave a distilled material (42%) whose ^1H NMR spectrum was completely superimposable on an authentic sample of 3,4-epoxy-1-butene (**13**). This result is in agreement with the ring closure predictions of Baldwin where the "3-exo-trig" cyclization is predicted to be favored (2).

$$HOCH_2C(H)=C(H)CH_2OH \xrightarrow[61°]{DTPP} \underset{42\%}{\text{epoxy}(CH_2)\text{-}CH\text{-}CH=CH_2} \quad \mathbf{13}$$

Stereochemical information on the mode of cyclodehydration of unsymmetrical diols to cyclic ethers could obviously have important consequences regarding useful, preparative routes to chiral cyclic ethers of high enantiomeric purity. For example, dioxyphosphorane promoted cyclodehydration of a chiral diol can, in principle, give the enantiomeric ethers by either of two stereochemically distinct routes. Separate stepwise decomposition of oxyphosphonium betaines, A and B, although proceeded by a number of equilibria could ultimately afford a nonracemic mixture of cyclic ethers.

$$Ph_3P(OEt)_2 + HO(CH_2)_n\overset{*}{C}H(OH)R \xrightarrow{-\ 2\ EtOH} A \text{ or } B \quad (n=1\text{-}3;\ R=Me,Ph)$$

A: $Ph_3\overset{+}{P}$–O–$\overset{*}{C}H(R)$–$(CH_2)_n$–O^-

B: $Ph_3\overset{+}{P}$–O–$(CH_2)_n$–$\overset{*}{C}HR$–O^-

$$A, B \longrightarrow \text{cyclic ether } O\text{–}(CH_2)_n\text{–}\overset{*}{C}HR + Ph_3PO$$

The results (Table I) indicate that the collapse of betaine B affording largely retention of stereochemistry at the carbinol carbon (91.8-96.4%) is highly favored. While the length of the hydrocarbon chain does not appear to influence the stereochemical course of the cyclodehydration, the % retention at the chiral carbon is diminished slightly when a phenyl group replaces a methyl group (Entry 1 and 4). This may imply formation of a benzylic carbocation with the expected loss of stereochemical integrity. We have excluded pathways which might involve concerted decomposition of dioxyphosphoranes to cyclic ethers with retention of stereochemistry at least for symmetrical 1,2-diols by examining the reaction of d,l-2,3-butanediol with DTPP. The ^{13}C NMR spectrum of the reaction mixture is consistent only with the cis epoxide exhibiting resonances at δ 12.9 and 52.4 ppm.

Table I. Stereochemistry of Cyclodehydration of Diols with DTPP

Entry	Diol	%Optical Purity	Cyclic Ether	% Ret
1	(S)-(+)-1,2-Propanediol	71.4	2-Methyloxirane	96.4
2	(R)-(-)-1,3-Butanediol	100	2-Methyloxetane	95.6
3	(R)-(-)-1,4-Pentanediol	100	2-Methyloxolane	93.8
4	(S)-(+)-1-Phenyl-1,2-ethanediol	97.5	1-Phenyloxirane	91.8

During the course of our work with DTPP, we became interested in developing other useful and effective cyclodehydrating "media" for diols and triols, with particular emphasis on TPP and tetrachloromethane (CCl_4).

When trans-1,2-cyclohexanediol is treated with equimolar TPP in excess CCl_4, a 88% yield of trans-2-chlorocyclohexanol can be realized with no evidence (^1H, ^{13}C NMR, GLC) for trans-1,2-dichlorocyclohexane or cis-2-chlorocyclohexanol. Since the trans chlorohydrin could not arise from simple displacement of TPPO by chloride ion with retention of stereochemistry (3), we suspected the intermediacy of cyclohexene oxide which could subsequently undergo ring opening by the hydrochloric acid (HCl) generated in solution. This was easily proven by repeating the reaction in the presence of solid potassium carbonate and realizing a 86% yield of cyclohexene oxide and no cis or trans chlorohydrins.

Treatment of diol 1 with TPP-CCl_4 in CH_3CN solvent gives predominantly 3-chloropropanol (75%) and 1,3-dichloropropane (16%) but no oxetane. It is unlikely that some 3-chloropropanol is a consequence of ring opening of 2 with HCl since repeating the reaction in the presence of K_2CO_3, an HCl scavenger, gave identical results.

We have observed that diol 3, cis-2-butene-1,4-diol, and cis-1,2-bis(hydroxymethyl)cyclohexane react smoothly with TPP-CCl_4 to afford 4 (78%), 2,5-dihydrofuran (65%), and cis-8-oxabicyclo[4.3.0]-nonane (84%). Reaction of diol 5 with TPP-CCl_4 in CH_3CN gives 52% of 5-chloropentanol, 6 (11%), and 1,5-dichloropentane (25%) while diol 7 affords 6-chlorohexanol (48%) and 1,6-dichlorohexane (39%). Comparisons of the ether:chlorohydrin:dichloride product distributions arising from these simple diols reveal a trend for efficiency of chain closure to 3 → 7 membered rings where the formation of cyclic ethers appear to decrease in order of the following ring size: $3 \simeq 5 > 6 > 4 \simeq 7$.

In an attempt to prepare bicyclic ether 12 by cyclodehydrating diol 11 with one equiv of TPP in CCl_4, we discovered a regiospecific chlorination of the primary hydroxyl group affording trans-2-hydroxycyclohexyl 2-chloroethylsulfide (**14**) in 70% yield. Formation of **14** may result from chloride displacement of TPPO or chloride capture of a thiiranium ion formed by neighboring sulfenyl sulfur displacement of TPPO. The relatively small amount of **12** (6%) formed in this reaction may result from (a) intramolecular cyclization of 14, capture of a thiiranium ion by the C_1 hydroxyl group, or (c) betaine (e.g., B) collapse to ether **12** and TPPO.

TPP, ex CCl_4, 77°, 24 h: **11** → **14** + **15** + **12**

	14	15	12
1 eq TPP	70%	0%	6%
2 eq TPP	39%	48%	13%

When diol **11** is allowed to react with 2 equiv of TPP-CCl_4, chlorohydrin 14 and **12** are formed as well as trans-2-chlorocyclohexyl 2-chloroethyl sulfide (**15**) in 48%. The evidence seems firm that formation of **15** must arise from the intermediacy of a thiiranium ion which allows for retention of configuration during the HO → Cl conversion at C_1. This is further supported by the fact that trans-2-thiomethyl cyclohexanol under similar reaction conditions (1 eq TPP in CCl_4) gives only trans-2-thiomethylcyclohexyl chloride (62% by ^{13}C NMR). By contrast, the reaction of trans-2-methoxycyclohexanol with TPP-CCl_4 gives exclusively cis-2-methoxycyclohexylchloride arising from chloride displacement of TPPO with complete inversion of stereochemistry at C_1. Our present findings corroborate the results of Billington and Golding where it was determined that the reaction between $CH_3SCH_2{*}CH_2OH$ and TPP-CCl_4 gave a mixture (1:1) of $CH_3S{*}CH_2CH_2Cl$ and $CH_3SCH_2{*}CH_2Cl$ (${*}CH_2$= ^{13}C enriched methylene carbon)(4).
TPP=triphenylphosphine; DTPP=diethoxytriphenylphosphorane; TPPO= triphenylphosphine oxide; GLC= gas-liquid chromatography.

Acknowledgement is made to the National Science Foundation (CHE 78-05921) and Research Corporation for support of this research.

Literature Cited

1. Chang, B. C.; Conrad, W.; Denney, D. B.; Denney, D. Z.; Edelman, R.; Powell, R. L.;White, D. W. J. Am. Chem. Soc. 1971, 93, 4004.
2. Baldwin, J. E. J. C. S. Chem. Commun.1976, 734.
3. Jones, L. A.; Sumner, C. E.; Franzus, B.; Haung, T. T.-S.; Snyder, E. I. J. Org. Chem. 1978, 43, 2821.
4. Billington, D. C.; Golding, B. T. J. C. S. Chem. Commun. 1978, 208.

RECEIVED June 30, 1981.

N-Alkylation of Organophosphorus Amides

A New, Convenient Route to Primary and Secondary Amines

A. ZWIERZAK

Institute of Organic Chemistry, Technical University (Politechnika), Zwirki 36, 90-924 Łódź 40, Poland

Gabriel synthesis, which is most often considered as a classical approach to primary amines, can be generalized as monoalkylation of a suitably protected ammonia derivative with subsequent removal of the phthaloyl group from nitrogen. Despite its wide applicability this procedure suffers, however, from several drawbacks: (i) hydrazinolysis is unsuitable for deprotection in the presence of some functional groups (i.e. carbonyl or carboalkoxyl group); (ii) deprotection cannot be carried out under non-hydrolytic conditions; (iii) direct synthesis of secondary amines is not possible.

It is well documented that the phosphorus-nitrogen bond in organophosphorus amides can be easily and effectively cleaved under acidolytic conditions using gaseous hydrogen chloride in benzene |1| or tetrahydrofuran |2|:

$$>P(O)N \xrightarrow[\text{r.t.}]{\text{HCl/THF (benzene)}} >P(O)Cl + \overset{+}{N}H_3\ Cl^-$$

This particular feature of P-N bond containing compounds can be utilized for synthetic purposes by using "phosphoryl protection" in the synthesis of amines based on alkylative procedures. The solution of this problem introducing two new reagents, i.e. diphenylphosphinic amide (I) and sodium N-(t-butyloxycarbonyl) diethyl phosphoroamidate (II) as useful synthetic equivalents of an amino moiety is the subject of this communication.

$$\underset{(I)}{Ph_2P(O)NH_2} \qquad \underset{(II)}{(EtO)_2P(O)N(Na)COOBu^t}$$

Diphenylphosphinic amide (I) can be readily prepared in 70% yield from diphenylphosphinic chloride by a modified procedure described by Russian workers |3|.
Amide (I) is a crystalline material (m.p. 166-167°), perfectly stable at ambient temperature for indefinite periods of time. Sodium hydride in benzene easily converts amide (I) into its sodium

0097-6156/81/0171-0169$05.00/0

derivative which can be then directly alkylated in boiling benzene. Alkylation is not selective, however, even when an equimolar amount of the corresponding halide is slowly added to the Na-salt of (I). Owing to transmetalation and subsequent alkylation of the N-alkyl derivatives (III) mixtures containing 25-50% of N,N-dialkyldiphenylphosphinic amides (IV) are always formed:

$$\underset{(I)}{Ph_2P(O)NH_2} \xrightarrow[2.\ R\text{-}X]{1.\ NaH/benzene,80^{o}} \underset{(III)}{Ph_2P(O)NHR} + \underset{(IV)}{Ph_2P(O)NR_2}$$

Recently it was found that diphenylphosphinic amide(I) can be selectively mono- or dialkylated under the conditions of phase-transfer catalysis. Monoalkylation is feasible in a two-phase system consisting of 50% aqueous sodium hydroxide and refluxing benzene in the presence of 5 mol-% of tetra-n-butylammonium hydrogen sulfate as catalyst. Alkyl bromides are the alkylating agents of choice. Dialkylation or further alkylation of monoalkyl derivative (III) can be readily accomplished under similar conditions when benzene is replaced by boiling toluene. The following scheme illustrates both possibilities:

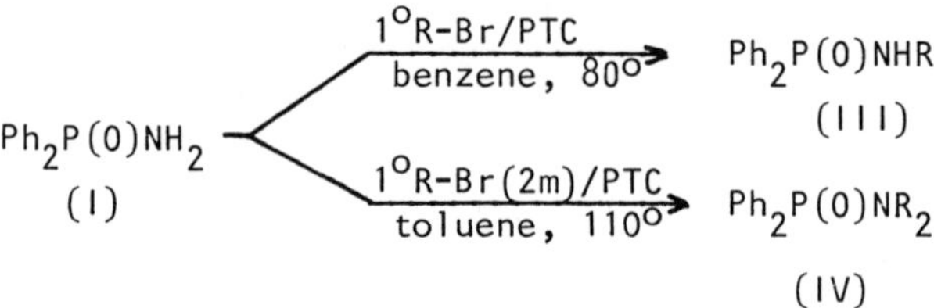

When primary alkyl bromides are used as alkylating agents the yields of (III) and (IV) are high (80-95%). All crude products are spectroscopically homogeneous (^{31}P-NMR) and free from any undesirable impurities. Monoalkylation of (I) can be also accomplished by means of secondary alkyl bromides when conventional liquid-liquid PTC system is replaced by the solid-liquid one. Powdered sodium hydroxide-potassium carbonate in boiling benzene in the presence of 10 mol-% of tetra-n-butylammonium hydrogen sulfate as catalyst was found to be the system of choice for the preparation of N-alkyldiphenylphosphinic amides (V) containing secondary alkyl groups linked to nitrogen:

$$\underset{(I)}{Ph_2P(O)NH_2} \xrightarrow[benzene,\ 80^{o}]{2^{o}R\text{-}Br/solid\text{-}liquid\ PTC} \underset{(V)}{Ph_2P(O)NHR\ (2^{o})}$$

Although the monoalkyl derivatives (V) are formed in reasonable yields (40-60%) they cannot be subjected to further alkylation by the same general procedure involving the use of secondary or primary alkyl bromides. This markedly reduced reactivity is possibly due to steric reasons.

Due to its remarkable acid lability the protecting diphenylphosphinic group can be almost quantitatively removed from the alkylation products (III), (IV) or (V) by treatment of the latter with gaseous HCl in tetrahydrofuran at room temperature |2|:

$$\underset{(III),(V)}{Ph_2P(O)NHR} \xrightarrow[\text{r.t.; 12h}]{\text{HCl/THF}} \underset{(VI)}{R\text{-}NH_3^+ \; Cl^-}$$

$$\underset{(IV)}{Ph_2P(O)NR_2} \xrightarrow[\text{r.t.; 12h}]{\text{HCl/THF}} \underset{(VII)}{R_2NH_2^+ \; Cl^-}$$

The alkylation - deprotection two-step procedure can be utilized for the preparation of hydrochlorides of primary (VI) and secondary (VII) amines thus offering a substantial extension to conventional variants of Gabriel synthesis.

Phase-transfer catalysed N-alkylation of diphenylphosphinic amide (I) is totally unsuitable in the case of easily hydrolysable organic halides, especially those containing additional functional groups (i.e. carbonyl or carboalkoxyl) which decompose readily in strongly alkaline medium. To circumvent the difficulty connected with simultaneous mono- and dialkylation of (I) under anhydrous conditions, N-(t-butyloxycarbonyl) diethyl phosphoroamidate (X), a doubly-protected ammonia derivative, was devised as a useful and superior substitute of phthalimide in the Gabriel-type synthesis of amines.

Compound (X) is readily available by a simple two-step procedure involving transformation of diethyl phosphoroamidate (VIII) into the corresponding isocyanate |4| (IX) followed by addition of t-butanol to the carbonyl group:

$$\underset{(VIII)}{(EtO)_2P(O)NH_2} \xrightarrow[65\%]{(COCl)_2/CCl_4;\ -5\text{-}0^o} \underset{(IX)}{(EtO)_2P(O)NCO} \xrightarrow[\text{r.t.; 100\%}]{Bu^tOH/CCl_4}$$

$$\rightarrow \underset{(X)}{(EtO)_2P(O)NHCOOBu^t} \xrightarrow[100\%]{MeONa/MeOH} \underset{(II)}{(EtO)_2P(O)N(Na)COOBu^t}$$

By the action of an equivalent amount of sodium methoxide in methanol (X) can be quantitatively converted into its sodium salt (II), which is non-hygroscopic and perfectly stable at ambient temperature.

It was found that sodium salt (II) can be easily and cleanly monoalkylated with a variety of polyfunctional organic halides under strictly anhydrous conditions. The reactions are generally carried out in refluxing benzene in the presence of 10 mol-% of tetra-n-butylammonium bromide (TBAB) as phase-transfer catalyst:

$$(EtO)_2P(O)N(Na)COOBu^t \xrightarrow[\text{benzene, } 80^o]{1^oR\text{-}X/TBAB} (EtO)_2P(O)N(R)COOBu^t$$

(II) (XI)

N-Alkylated derivatives (XI) are formed in high yields (80-95%) and need not any further purification. The reaction is restricted to primary halides but according to our experience this is the only severe limitation. Crude (XI) can be almost quantitatively deprotected by treatment with gaseous HCl in benzene or tetrahydrofuran at room temperature to give the corresponding amine hydrochlorides (XII). Selective removal of Boc group is also possible by means of an excess of trifluoroacetic acid at 0°.
It affords N-alkyl diethyl phosphoroamidates (XIII) which are potential starting materials for the preparation of secondary amines |5|.

$$(EtO)_2P(O)N(R)COOBu^t \text{ (XI)} \xrightarrow[\text{r.t.; 12h}]{\text{HCl/benzene (THF)}} R\text{-}NH_3^+\ Cl^- \text{ (XII)}$$

$$(EtO)_2P(O)N(R)COOBu^t \text{ (XI)} \xrightarrow[0^o]{CF_3COOH} (EtO)_2P(O)NHR \text{ (XIII)}$$

The use of sodium salt (II) is strongly recommended for nucleophilic amination of primary halides under anhydrous conditions. Both reagents (I) and (II) can be considered as useful substitutes of potassium phthalimide, free from well known preparative inconveniances involving application of the latter.

LITERATURE CITED

1. Skrowaczewska, Z.; Mastalerz, P. Roczniki Chem. 1955, 29, 415.
2. Zwierzak, A.; Osowska, K. Angew.Chem.Int.Ed.Engl. 1976, 15, 302.
3. Zhmurova, I.N.; Voitsekhovskaya, I.Y.; Kirsanov, A.V. Zh.Obshch. Khim. 1959, 29, 2083.
4. Samaraj, L.I.; Derkatsch, G.I. Zh.Obshch.Khim. 1966, 36, 1433.
5. Zwierzak, A.; Brylikowska-Piotrowicz, J. Angew.Chem. Int. Ed. Engl. 1977, 16 107.

RECEIVED June 30, 1981.

Phosphoric Amide Reagents

ERIK B. PEDERSEN

Department of Chemistry, Odense University, DK-5230 Odense M, Denmark

In 1935 Fricker showed that N-substituted amides could be prepared by treating pyridinecarboxylic acids with a mixture of secondary amines and phosphorus pentoxide [1]. The first report in the literature on the reaction of hexamethylphosphoric triamide (HMPT) with organic molecules was given by Heider [2]. Carboxylic acids form stable (1:1) complexes with HMPT. They decompose at 200 °C and transamidation products were isolated in high yields. By heating a series of potential hydroxy heterocyclic compounds in HMPT, the corresponding dimethylamino-heterocycles were produced. The yields for polycyclic compounds were generally above 50%, as exemplified here by the synthesis of 2-dimethylaminoquinoline in 79% yield [3]. It is generally found

$(Me_2N)_3P=O$ → 79%

that oxo groups in heterocyclic compounds can be transformed directly into an amino-group by heating the oxo compound to 200 °C with an appropriate phosphoramide reagent [4,5]. However, the phosphoramides are not generally commercially available. So for laboratory preparations the method will be of minor importance only.

The phosphorus pentoxide-amine mixtures, which can readily be prepared in an exothermic reaction, were found to react analogously to phosphoramides.

0097–6156/81/0171–0173$05.00/0

4-pyridone was used as a model compound for demonstrating the possibility of making a direct conversion of an oxo group into an amino group by heating at 210 °C [6]. 2-Dialkylaminopyridines were similarly optained in 16-39% yield.

OH

P_2O_5, HNR_2 →

N

NR_2

N

23-49%

The corresponding reagents prepared from primary amines and phosphorus pentoxide were not easy to handle. Instead, a reagent mixture which was prepared from phosphorus pentoxide, primary amine hydrochloride, and N,N-dimethylcyclohexylamine was useful at 250 °C for the synthesis of N-substituted 2-aminoquinolines from the corresponding oxo compound [7].

CH_3

N O

H

RNH_3Cl, P_2O_5 →

$C_6H_{11}N(CH_3)_2$

CH_3

N NHR

31-86%

The corresponding 4-aminoquinolines were obtained in 42-98% yield. In the synthesis of 4-methyl-2-quinolinamine which was prepared in 49% yield, using N,N-dimethylcyclohexylamine as the tertiary amine component of the reaction mixture, N-mono- and N,N-dimethylated quinolinamines could also be isolated in 16% and 3% yield, respectively. If tributylamine was used as the tertiary amine component, the yield of 4-methyl-2-quinolinamine increased to 68% and N-butyl-4-methyl-2-quinolinamine was only formed in 6% yield. In the latter reaction pure 4-methyl-2-quinolinamine was easily obtained by recrystallizing the raw material.

The most striking similarity between HMPT and mixtures of phosphorus pentoxide and dialkylamines was found in a new quinoline synthesis. 2-Dialkylaminoquinolines were synthesized in 23-61% yield by heating acetanilides and N,N-diethylformamide in mixtures of phosphorus pentoxide and dialkylamines [8]. Similar reactions were obtained using

HCONEt$_2$; P_2O_5, HNR_2 → 2-(NR_2)quinoline, 23-61%

DMF; $(Me_2N)_3P{=}O$ → 2-$N(CH_3)_2$-quinoline, 40-76%

phenyl N,N,N',N'-tetraethylphosphorodiamidate, instead of a mixture of phosphorus pentoxide and diethylamine [9]. In refluxing DMF and HMPT, acetanilides afforded 2-dimethylaminoquinolines in 40-76% yield [10].

We have also shown that HMPT can be used as a ring closure reagent in the synthesis of 9-dimethylamino-1,2,3,4-tetrahydroacridine from cyclohexanone and the methyl ester of anthranilic acid, with polyphosphoric acid as catalyst [11]. Heating of esters of N-acetylanthranilic acid in HMPT afford 2,4-bis-(dimethylamino)quinoline [12]. However, if each nitrogen atom of the phosphoric amide reagent were substituted with only one alkyl group, esters of N-acetylanthranilic acid reacted quite differently. With phenyl N,N'-dimethylphosphorodiamidate as reagent, 2,3-dimethyl-4(3H)-quinazolinone was isolated [13]. The disadvantage of the latter reaction is the synthesis of the phosphordiamidate reagent. Recently it was found that a mixture of phosphorus pentoxide, amine hydrochloride and N,N-dimethylcyclohexylamine similarly could be used, and quinazolinones were synthesized in 35-92% yield [14].

$COOCH_3$, $NHCR^1$ (C=O) → P_2O_5, R^2NH_3Cl; $C_6H_{11}NMe_2$; 180°C, 45 min. → quinazolinone (R^1, R^2), 35-92%

The parallelism in reactions with HMPT and mixtures of phosphorus pentoxide and amines can best be realized by looking at the mechanism of HMPT reactions [15]. In the reaction of p-methoxybenzylalcohol with HMPT which produce the corresponding N,N-dimethylbenzyl amine, p-methoxybenzyl phosphate was identified in the ^{31}P NMR spectrum of the reaction mixture. Since pyrophosphate was also observed during the reaction of the benzyl alcohol with HMPT, the intermediate formation of the metaphosphate anion can be assumed. Addition of the benzyl alcohol to that anion then produce the benzyl phosphate. Phosphate ions are known to be good leaving groups so that a benzylamine can easily be produced from the phosphate by reaction with dimethylamine released from the dimethylammonium ion. The phosphoric acid produced during the reaction is believed to react with HMPT, so that more dihydrogen-bis(dimethylammonium) pyrophosphate is formed.

1. Fricker, K. Fr. 791, 783; Chem. Abstr. 1936, 30, 4178[8].
2. Heider, R. L. US. 2,603,660; Chem. Abstr. 1953, 47, 4900.
3. Pedersen, E. B. and Lawesson, S.-O Tetrahedron, 1974, 30, 875.
4. Arutyunyan, E. A., Gunar, V. I., Gracheva, E. P., and Zav'talov, S. I. Izv. Akad. Nauk SSSR, Ser. Khim. 1969, 655.
5. Eck, H., Kinzel, P., Müller, F Ger. Offen 2, 403,165; Chem. Abstr. 1975, 83, 159165 t.
6. Pedersen, E. B., and Carlsen, D. Synthesis 1978, 844.
7. Pedersen, E. B., and Carlsen, D. Chem. Scr. (in press).
8. Hansen, B. W., and Pedersen, E. B. Liebigs Ann. Chem. (in press).
9. Pedersen, E. B., and Carlsen, D. Synthesis 1977 890.
10. Pedersen, E. B., and Lawesson, S.-O. Acta Chem. Scand. B 1974, 28,1045.
11. Osbirk, A., and Pedersen, E. B. Ibid, 1979 33, 313.
12. Pedersen, E. B. Tetrahedron 1977, 33, 217.
13. Pedersen, E. B. Synthesis 1977, 180.
14. Nielsen, K. E., and Pedersen, E. B. Acta Chem. Scand. B 1980, 34, 637.
15. Pedersen, E. B., and Jacobsen, J. P. J. C. S. Perkin II 1979, 1477.

RECEIVED June 30, 1981.

BIOLOGICALLY IMPORTANT PHOSPHORUS COMPOUNDS, NATURAL AND SYNTHETIC

Reversible Masking of Acetylcholinesterase by Covalent Phosphorylation in the Presence of a Novel Cyclic Phosphate Ester

H. LEADER, L. RAVEH, R. BRUKSTEIN, M. SPIEGELSTEIN, and Y. ASHANI

Israel Institute for Biological Research, P.O. Box 19, Nes Ziona 70450, Israel

In recent studies, it was shown that eel acetylcholinesterase (AChE, EC 3.1.1.7) previously inhibited with 1,3,2-dioxaphosphorinane 2-oxides (I) undergoes spontaneous reactivation with $t_{1/2}$ = 12 minutes at pH 7.0, in marked contrast to enzyme inhibited

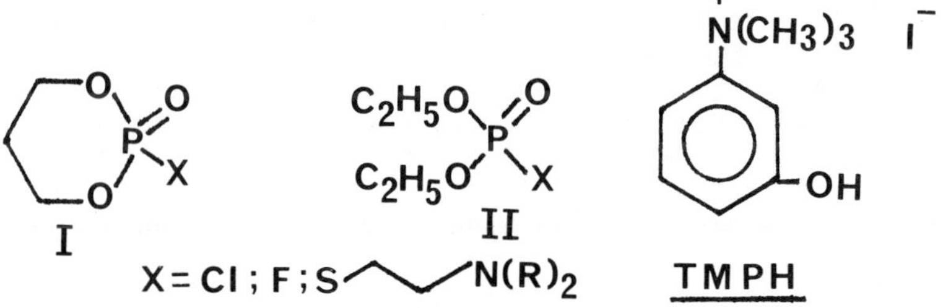

I II TMPH

X = Cl ; F ; $S\text{-}CH_2CH_2\text{-}N(R)_2$

TMPH: $\overset{+}{N}(CH_3)_3$ I^-, OH

with O,O-diethylphosphoryl derivate (II) (1,2).

The formation of unstable phosphoryl-AChE conjugate of type I (where X=AChE) prompted us to design a molecule which may exhibit special effects on cholinergic mechanisms. It was assumed that the rate of return of enzyme activity after the inhibition by I should be the same, irrespective of the leaving group X.

We wish to report here on the synthesis and biological properties of O-[m-N,N,N-trimethylammoniophenyl] 1,3,2-dioxaphosphorinane 2-oxide iodide (TDPI; I, X = TMPH).

The rationale behind the molecular combination of cyclic phosphate ester (I) and potential leaving group such as m(N,N,N-trimethylammonio) phenyl iodide (TMPH) may be summarized as follows:

a. Introducing quaternary nitrogen to I should increase the rate of inhibition while restricting the phosphorylation of AChE *in vivo* to peripheral sites.

b. Reversible phosphorylation of AChE by I may be controlled efficiently by specific reactivators (oximes) in contrast to carbamoyl-AChE conjugate.

0097-6156/81/0171-0179$05.00/0

c. The corresponding leaving group, TMPH is a structural analog of the extremely potent reversible inhibitor of AChE, namely edrophonium bromide, where the ethyl group was substituted for one of the nitrogen methyls (3).

d. Phenyl esters of type I are predicted to be very stable under physiological conditions.

TDPI was prepared by reacting 2-chloro-1,3,2-dioxaphosphorinane 2-oxide (I, X=Cl) with m(N,N-dimethylamino) phenol followed by methyl iodide to produce pale yellow crystals. ^{31}P and ^{1}H nmr spectroscopy, tlc and C,H,P,N analyses verified the structure and homogeneity of TDPI. UV-absorption spectroscopy indicates that free phenol (TMPH), if present, is less than 0.2%.

When AChE from either electric eel or rat-brain homogenate was incubated with TDPI, the decrease in enzyme activity approached a steady state. The initial rate of release of TMPH was found to be in good agreement with the rate constant obtained from inhibition measurements. The inhibited enzyme recovered spontaneously upon extensive dilution at a rate similar to the rate constants calculated from the inhibition studies (Scheme 1).

2-Pyridiniumaldoxime methiodide (2-PAM) accelerated significantly the spontaneous reactivation whereas the 3-analog (3-PAM) did not enhance the regeneration of enzyme activity. On the basis of these results and comparative evaluation with previous studies, where different cyclic esters (I) were utilized (1,2), we suggest the following scheme for the inhibition of AChE by TDPI.

Scheme 1

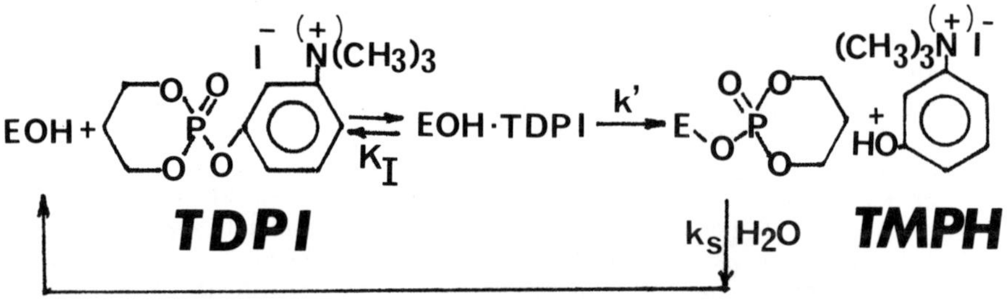

$$\mathrm{EOH} + \mathbf{TDPI} \underset{K_I}{\rightleftharpoons} \mathrm{EOH{\cdot}TDPI} \xrightarrow{k'} \mathrm{E\text{-}O\text{-}P(O)(cyclic)} + \mathbf{TMPH}$$

$$\mathrm{E\text{-}O\text{-}P(O)(cyclic)} \xrightarrow[k_s]{H_2O} \mathrm{EOH}$$

Butyrylcholinesterase (BuChE EC 3.1.1.8) from either horse or mice serum displayed different profiles. A steady state was not developed, although the rate constants of inhibition decreased with time. Since the presence of multiforms of serum BuChE has been established, it is likely that the first-order plot represents more than one exponent. The inhibited enzyme did not regenerate as fast as AChE-TDPI conjugate. However, 2-PAM enhanced the reactivation of horse-serum BuChE after inhibition with TDPI. The various rate constants were computed from the initial slopes of the inhibition and reactivation of

BuChE in contrast to AChE where the individual constants were computed from equations derived on the basis of Scheme 1. We note that aging was observed only in the case of BuChE. Table I summarizes several kinetic parameters associated with the inhibition of AChE and BuChE by TDPI:

Table I: Kinetic Parameters for TDPI at pH 7.0

Enzyme	$k_i (M^{-1} min^{-1})^a$	$k_S (min^{-1})^b$	$k_r (min^{-1})^c$	%-aging[d]
AChE, eel	8.4×10^3	0.07	0.38	2
BuChE, horse	1.8×10^4	0.01	0.05	69

a) $k_i = k'/K_I$ (Scheme 1); b) From direct measurements;
c) 0.5 mM 2-PAM. k_S subtracted; d) After 20 hr in presence of 0.05 mM TDPI.

The formation of an unstable,covalent phosphoryl-AChE conjugate is evidently accompanied by a non-covalent reversible complex (Michaelis complex) between AChE and TDPI (AChE · TDPI). Although TMPH (K_I = 0.25 μM) was found to be 100 fold more powerful than TDPI (K_I = 0.02 mM) in terms of concentration required to achieve similar amount of Michaelis complex, TDPI provided better protection of AChE against irreversible phosphorylation by II. In this set of experiments, the efficiency of TDPI and TMPH were compared on the basis of equal concentration/affinity ratio were I/K_I = 10.

When TDPI was added to an isolated rat-hemidiaphragm, AChE activity decreased at a rate similar to AChE from either rat-brain homogenate or electric eel. Removal of TDPI by washing restored the activity of the enzyme in rat-hemidiaphragm to the original level within 15-30 minutes. TDPI protected the rat-hemidiaphragm from irreversible inhibition of tetanic tension, caused by the powerful anticholinesterase 1,2,2-trimethylpropyl methylphosphonofluoridate (Soman). In hemidiaphragms that were not pre-treated with TDPI tetanic activity was abolished completely and could not be restored upon removal of the inhibitor. In addition, TDPI protected mice (in conjunction with atropine) against 5 x LD_{50} soman and 22 x LD_{50} paraoxon (II, X = p-nitrophenoxy).

The correlation between in vitro findings and in vivo observations suggest that the excellent protection of AChE provided by TDPI is due to the formation of a reversible covalent-phosphoryl conjugate and reversible non-covalent Michaelis complexes, AChE · TDPI and AChE ·TMPH.

The toxicity of TDPI in terms of LD_{50} was found to be 450 mg/kg (sc) in mice. It is believed that TDPI is a potential non-toxic drug, that may be applied for treatments of cholinergic impairments.

References
1. Ashani, Y; Snyder, S.L; Wilson, I.B. Biochemistry 1972, 11, 3518.
2. Ashani, Y; Leader, H. Biochem. J.1979, 177, 781.
3. Wilson, I.B; Quan, C. Arch. Biochem. Biophys.1958, 73, 131.

RECEIVED July 7, 1981.

37

α-Aminophosphonous Acids: A New Class of Biologically Active Amino Acid Analogs

E. K. BAYLIS, C. D. CAMPBELL, J. G. DINGWALL, and W. PICKLES

Central Research Laboratories, Ciba-Geigy (UK) Limited, Tenax Road, Manchester M17 1WT, England

Phosphorus analogues of the important naturally occurring amino acids have received considerable attention in the past ten years in both academic and industrial laboratories and some have shown useful biological activity: aminomethanephosphonic acid 1 and phosphinothrycin 2, a naturally occurring glutamic acid analogue, are plant growth regulants/ herbicides while the dipeptide 3 is an antibacterial agent.

The possibility of phosphorus analogues of amino-acids acting as false substrates and so interfering with biological mechanisms led us to consider the α-amino-phosphonous acids 4 as perhaps the closest analogues.

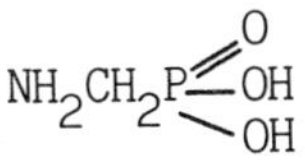

$NH_2CH_2P(=O)(OH)OH$

1

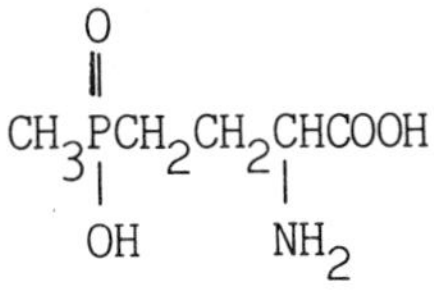

$CH_3P(=O)(OH)CH_2CH_2CH(NH_2)COOH$

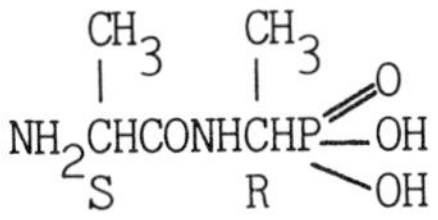

2

$NH_2CH(CH_3)CONHCH(CH_3)P(=O)(OH)OH$ (S, R)

3

$RCH(NH_2)P(=O)(H)OH$

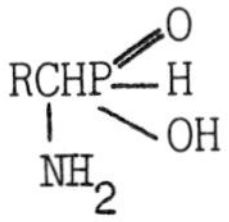

4

Only one α-aminophosphonous acid, the glycine analogue, was reported in the literature at the outset of this work. It was prepared[1] by the ammonolysis of chloromethanephosphonous acid, a method not applicable to other α-amino acid analogues. This paper describes a general synthesis of the α-aminophosphonous acids and some of their physical, chemical and biological properties.

0097-6156/81/0171-0183$05.00/0

Among the many methods available for the synthesis of α-aminophosphonic acids that of Tyka[2], involving the addition of dialkylphosphite to benzylimines, followed by hydrogenolysis, appeared the most adaptable to our purposes, since the addition of hypophosphorous acid to imines to give N-substituted α-aminophosphonous acids was well documented[3]. Thus addition of hypophosphorous acid to the benzylamine imines 5 readily gave the N-benzylaminophosphonous acids 6. However, all attempts to hydrogenolyse 6 failed because of P-H catalyst poisoning, and under more severe conditions with higher catalyst loadings carbon-phosphorus bond cleavage occurred.

$$\underset{\mathbf{5}}{RCH{=}NCH_2Ph} \xrightarrow[EtOH]{H_3PO_2} \underset{\mathbf{6}}{RCH(NHCH_2Ph)P(=O)(H)OH} \qquad R = Ph,\ Pr^i$$

Needing instead an acid labile protecting group we turned to diphenylmethylimines. Hypophosphorous acid added to imines derived from diphenylmethylamine in ethanol (Method A) to give the diphenylmethylaminophosphonous acids 7. Alternatively, reaction of the diphenylmethylamine salt of hypophosphorous acid with aldehydes and ketones in refluxing ethanol (Method B) or preferably dioxan (Method C) gave the same product.

$$RR'C{=}N{-}CH(Ph)_2 + H_3PO_2 \longrightarrow \underset{\mathbf{7}}{RR'C(NHCH(Ph)_2){-}P(=O)(H)OH}$$

Method A (Ethanol)

$$RR'C{=}O + (Ph)_2CHNH_3^+\,H_2PO_2^- \longrightarrow \mathbf{7}$$

Method B (Ethanol)
Method C (Dioxan)

$$\mathbf{7} \xrightarrow[\text{ii) propylene oxide}]{\text{i) } H^+} \underset{\mathbf{8}}{RR'C(NH_2){-}P(=O)(H)OH}$$

Acid cleavage of the diphenylmethyl group could be achieved under a variety of conditions e.g. i) 49% HBr, 100°, 45 min., ii) TFA/Anisole, reflux, 30 min. and the free aminophosphonous acid 8 then obtained by treatment of the hydrobromide salt with propylene oxide in ethanol, or simply by heating the trifluoroacetates in ethanol.

The following amino acid analogues were prepared:
Amino acid [M.Pt., Method (yield %)]

Alanine [223-4°C, C(45)]
Valine [201-5°C, A (69), B (49)]
Leucine [222-3°C, A (47), B(78)]
Phenylalanine [227-8°C, C (36)]
Methionine [231°C, C (25)]
Glutamic acid [154°C, C (38)]

The analogues of serine and tyrosine were prepared from suitably protected hydroxy aldehydes and the tryptophan analogue from indolepyruvic acid. A wide selection of other α-aminophosphonous acids was also prepared from aliphatic, aromatic and heterocyclic aldehydes and aliphatic ketones.

The α-aminophosphonous acids can be resolved by classical procedures using the α-methylbenzylamine salts of N-benzyloxycarbonyl derivatives. We have resolved three α-aminophosphonous acids (alanine, valine and methionine analogues) and obtained both <u>d</u>- and <u>l</u>-enantiomers with >99% optical purity. A crystal structure determination of a protected dipeptide derivative of (-)1-aminoethanephosphonous acid showed the (-)aminophosphonous acid to have R stereochemistry.

Di- and oligo-peptides with a terminal α-aminophosphonous acid residue have been prepared by coupling with N-hydroxy succinimide esters of N-benzyloxycarbonylamino acids or peptides.

$$\text{ZNH-}\underset{}{\overset{R}{\overset{|}{\text{CH}}}}\text{-COONSu} + \text{H}_2\text{N}\overset{R'}{\overset{|}{\text{CH}}}\text{P}(=\text{O})(\text{H})(\text{OH}) \xrightarrow{\text{i) ii) iii)}} \text{H}_2\text{N}\overset{R}{\overset{|}{\text{CH}}}\text{CONH}\overset{R'}{\overset{|}{\text{CH}}}\text{P}(=\text{O})(\text{H})(\text{OH})$$

i) aq. $NaHCO_3$ ii) HBr/HOAc iii) propylene oxide

As well as peptide formation the α-aminophosphonous acids undergo the typical amine reactions of amino acids. Further, they can be oxidised to the corresponding aminophosphonic acids without racemisation. For example, oxidation of the (S, R) dipeptide 9 with mercuric chloride gave a phosphonic dipeptide identical to Alaphosphin in quantitative yield, and served to confirm the stereochemical assignments.

```
   CH3    CH3                           CH3    CH3
   |      |    //O       HgCl2          |      |    //O
H2NCHCONHCHP—H        -------->    H2NCHCONHCHP—OH
   (S)    (R)  \OH        H2O           (S)    (R)  \OH

       (9)                             ALAPHOSPHIN R
```

When our work on synthesis of α-aminophosphonous acids was essentially complete and a patent filed[4], a brief report appeared[5] describing the synthesis of three α-aminophosphonous acids by addition of hypophosphorous acid to oximes.

In biological studies within CIBA-GEIGY the alanine, valine and methionine analogues and derived peptides show significant antibacterial activity when tested in a minimal agar medium. The valine and methionine analogues and the dipeptide 9 possess in vivo activity and all three are effective against experimental infections in the mouse. The antibacterial activity of the valine analogue is antagonised by valine, and this and other preliminary studies indicate that the mode of action of both the valine analogue and the dipeptide 9 is by transport into the bacterial cell followed by inhibition of protein synthesis. This is in contrast to the phosphonic dipeptide Alaphosphin which acts by transport into the bacterial cell and intra-cellular release of the alanine mimectic which interferes with bacterial cell wall synthesis. The parent aminophosphonic acid is itself inactive and clearly not transported into the bacterial cell[6].

The phosphonous analogue of alanine and peptides derived from it also affect the growth of plants at low concentrations.

REFERENCES

1. U.S. Patent 3,160,632, 1964; Chem. Abstr. 62, 4053F
2. Tyka, R., Tetrahedron Lett. 1970, 677
3. Schmidt, H., Chem. Ber. 1948, 81, 477
4. Ger. Offen. 2,722,162, 1977; Chem. Abstr. 88, 105559
5. Khomutov, R.M., Osipova, T.I , Izv. Akad. Nauk SSSR, Ser. Khim. 1978, 1951
6. Atherton, F.R., Hall, M.J., Hassall, C.H., Lambert, R.W., Ringrose, P.S., Antimicrobial Agents and Chemotherapy, 1979, 696.

RECEIVED July 7, 1981.

Phosphonodipeptides

LIDIA KUPCZYK-SUBOTKOWSKA, PAWEL KAFARSKI, JANUSZ KOWALIK, BARBARA LEJCZAK, PRZEMYSLAW MASTALERZ, and JÓZEF OLEKSYSZYN

Institute of Organic and Physical Chemistry, Technical University of Wrocław, 50-370 Wrocław, Poland

JERZY SZEWCZYK

Department of Organic Chemistry, Technical University of Gdańsk, 80-952 Gdańsk, Poland

Peptides of aminoalkanephosphonic acids (Phosphonopeptides) 4, are now receiving considerable interest because some representatives of this class were found to repress bacterial growth at very low concentration levels [1-3]. Literature data on the synthesis of dipeptides containing aminoalkylphosphonic acids are scarce and very little information is avaiable on blocking and deblocking of the phosphonate moiety [3-9].

In this communication we report synthesis of dipeptides containing P-terminal aminoalkylphosphonic acids related structurally to glycine, alanine, nor-valine, valine, leucine, phenylglycine and phenylalanine, while the N-terminal residues include glycine, alanine, valine, leucine, proline, methionine, beta-alanine, phenylalanine, tyrosine, lysine and glutamine.

The peptides 3 were prepared by condensation of phtaloyl, N-carbobenzoxy, N-t-butoxycarbonyl or N-formylamino acids with 1-aminoalkanephosphonic acids as well as with their dialkyl or diphenyl esters /Scheme/. Special attention was payed on the efectiveness of the peptide bond formation and on the methods for selective and total removal of the blocking groups.

The literature data on the preparation of phosphonodipeptides from 1-aminoalkanephosphonic acids showed [3-5] that the yields of condensation reactions are usually small or moderate. Moreover, the use of bulky N-blocked amino acids drastically decreased the reaction yield. Thus following Martell's method [5] we wre unsuccesful in the preparation of dipeptides containing N-terminal valine or leucine, while peptides of phenylalanine were obtained in 5-10% yield. Also the active ester method appeared to give small yields of the desired products. Our studies using p-nitrophenyl- and cyanomethyl esters of N-phtaloyl amino acids confirmed these observations.

0097–6156/81/0171–0187$05.00/0

$$X\text{-}NH\text{-}\underset{R}{CH}\text{-}COOH \;+\; H_2N\text{-}\overset{R^2}{\underset{R^1}{C}}\text{-}PO_3Y_2$$

$\underline{1}$ $\underline{2}$

$\downarrow$ coupling

$$X\text{-}NH\text{-}\underset{R}{CH}\text{-}\underset{O}{\overset{}{C}}\text{-}NH\text{-}\overset{R^2}{\underset{R^1}{C}}\text{-}PO_3Y_2$$

$\underline{3}$ $\downarrow$ deblocking

$$H_2N\text{-}\underset{R}{CH}\text{-}\underset{O}{C}\text{-}NH\text{—}\overset{R^2}{\underset{R^1}{C}}\text{-}PO_3H_2$$

$\underline{4}$

X = OHC- , $C_6H_5CH_2OCO$- , $/CH_3/_3COCO$- , phtaloyl

Y = Na, Mg, CH_3, CH_3CH_2, $/CH_3/_3CH$, C_6H_5

R = H, CH_3, $CH_3CH_2CH_2$, $/CH_3/_2CH$, $/CH_3/_2CHCH_2$, $C_6H_5CH_2$, p-HO-$C_6H_4CH_2$, $CH_3SCH_2CH_2$, H_2N-$/CH_2/_5$-, $CH_2CH_2CONH_2$

R^1 = H, CH_3

R^2 = H, CH_3, CH_3CH_2, $/CH_3/_2CH$, $/CH_3/_2CHCH_2$, C_6H_5, $C_6H_5CH_2$ and 2-aminoethanephosphonic acid [6]

Although esters of amino acids are much more widely used in peptide synthesis than the corresponding acids, the use of dialkyl 1-aminoalkanephosphonates was strongly limited because of lack of simple method for the esterification of these acids.Also the methods for preparation of these esters given in the literature are rather complex and their yields are low.

The simple method for the preparation of dialkyl 1-aminoalkanephosphonates discovered in our laboratory[10] enabled us to solve that problem.Phosphonopeptides **3** were prepared by condensation of N-blocked amino acids with dialkyl 1-aminoalkanephosphonates by means of dicyclohexylcarbodiimide (DCC) method,or preferably using the mixed carboxylic-carbonic anhydride (MCA) method.

All the blocking groups were removed from the peptides **3** by acidolysis with 45% hydrogen bromide in glacial acetic acid solution in the case of N-carbobenzoxy-,N-formyl or N-t-butoxycarbonyl groups .The selective removal of N-blocking groups was also carried out.Thus N-carbobenzoxy- group was removed with hydrogen gas on palladium catalyst,N-phtaloyl by the action of hydrazine,while N-formyl or N-t-butoxycarbonyl with etheral or methanolic solutions of hydrogen chloride gas.Dialkyl esters of phosphonodipeptides obtained in this manner can be further used for oligopeptide synthesis.

If the racemic form of aminophosphonates were used mixtures of diastereoisomers were obtained.We invented a method for their separation by column ion-exchange chromatography (Dowex 50W X8,H^+ form) .The chromatographically pure diastereoisomeric phosphonodipeptides were obtained in this manner.

The other way for obtaining diastereoisomeric dipeptides was the use of optically active dialkyl 1-aminoalkanephosphonated.They were prepared by the separation of dibenzoyl-L-tartaric salts of these esters.

The use of dialkyl aminoalkanephosphonates seems to be the most preferable method for the synthesis of phosphonodipeptides so far.

Diphenyl 1-aminoalkanephosphonates (**2**,Y=C_6H_5) were not used for the phosphonopeptide synthesis yet.They appeare to be very exciting substrates because these esters can be very easily prepared by the method invented in our laboratory[11].

Coupling of N-carbobenzoxyamino acids with diphenyl 1-aminoalkanephosphonates by the DCC method resulted in the formation of mixtures of diastereoisomers enriched in one of diastereoisomers,i.e. they appeare to form in nonequimolar quantities.It is probably due to kinetic control of the reaction.On the other hand the MCA method gave an equimolar mixtures of diastereoisomers.

The main problem here is removing the phenyl ester groups from the phosphonic moiety. We have found that convenient routes for deblocking are: hydrogenation on Adams' catalyst and transesterification methods followed by hydrogen bromide in glacial acetic acid treatment. The best results ewre obtained if transesterification was carried out using potassium fluoride-crown ether-methanol system.

References:

1./E.Bayer,K.H.Gugel,H.Hägele,H.Hagenmaier,S.Jessipow, W.A.Konig,H.Zähner,Helv.Chim.Acta,55,224/1972/
2./B.K.Park,A.Mirota,H.Sakai,Agric.Biol,Chem.,40,1905 /1976/
3./J.G.Allen,F.R.Atherton,M.J.Hall,C.H.Hassall, S.W.Holmes, .W.Lambert,N.J.Nisbet,P.S.Ringrose, Nature 272,56/1978/
4./M.Hariharan,R.J.Motekaitis,A.E.Martell,J.Org.Chem., 40,470/1975/
5./F.R.Atherton,H.J.Hall,C.H.Hassall,R.W.Lambert, P.S.Ringrose,Ger.Offen.,26.02.193 /1976/
6./P.Kafarski,P.Mastalerz,Roczniki Chemii,51,433/1977/
7./K.P.Poroshin,V.K.Buritchenko,Dokl.Akad.Nauk SSSR, 168,386/1964/
8./W.F.Gilmore,M.A.McBride,J.Pharm.Sci.,63,1087/1974/
9./K.Yamauchi,M.Kinoshita,M.Imoto,Bull Soc.Chem.Jap., 45,2528/1972/
10./J.Kowalik,L.Kupczyk-Subotkowska,P.Mastalerz,Synthesis,in press
11./J.Oleksyszyn,L.Kupczyk-Subotkowska,Synthesis ,985 /1979/

RECEIVED June 30, 1981.

Some Aspects of the Chemical Synthesis of Oligodeoxyribonucleotides

COLIN B. REESE and LYDIA VALENTE

Department of Chemistry, King's College, Strand, London WC2R 2LS, England

We report some recent studies on the synthesis of oligodeoxyribonucleotides. We have prepared three dodecamer restriction site adaptors which were required by Dr. Peter Rigby of the Department of Biochemistry, Imperial College of Science and Technology, University of London, in connection with his studies on the cloning of cDNA (complementary deoxyribonucleic acids). The nucleotide sequences are d[AATTCGGTACCG], d[AATTCGAGCTCG] and d[AATTCGTCGACG] which are Eco RI → Kpn I, Eco RI → Sst I and Eco RI → Pst I adaptors, respectively. These three dodecamers were prepared in satisfactory yields from four fully-protected hexamers (d[Dbmb-*ApAp*Tp*TpCpG*-Px], d[Dbmb-*Gp*Tp*ApCpCpG*-Px], d[Dbmb-*ApCpCp*Tp*CpG*-Px], and d[Dbmb-T*pCpGpApCpG*-Px]. The system of abbreviations for protected oligodeoxynucleotides has been described previously (1). [o-Dibromomethylbenzoyl (2) and 9-phenylxanthen-9-yl (3) are abbreviated to Dbmb and Px, respectively. 6-N-(p-t-Butylbenzoyl)-2'-deoxyadenosine, 4-N-benzoyl-2'-deoxycytidine and 2-N-(p-t-butylphenylacetyl)-2'-deoxyguanosine residues are represented by *A*, *C* and *G*, respectively; phosphate residues which are protected by o-chlorophenyl groups are represented by *p*.]

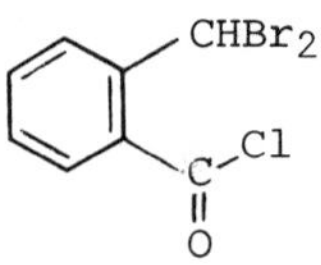

$CHBr_2$ / C(=O)Cl

(*1*) [Dbmb-Cl]

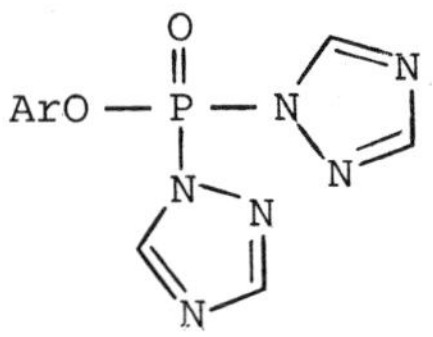

ArO — P(=O) — N (triazolyl)

(*2*) Ar = $2\text{-}ClC_6H_4$

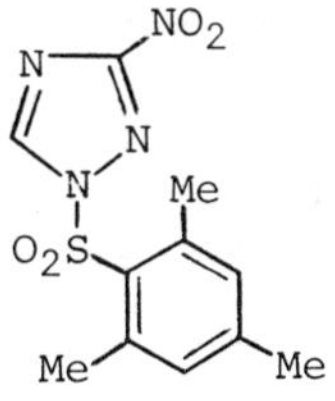

NO_2 / N / N / N / O_2S / Me / Me / Me

(*3*) [MSNT]

Ph Cl / O

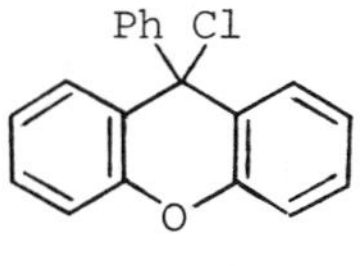

(*4*) [PxCl]

0097–6156/81/0171–0191$05.00/0

Scheme 1

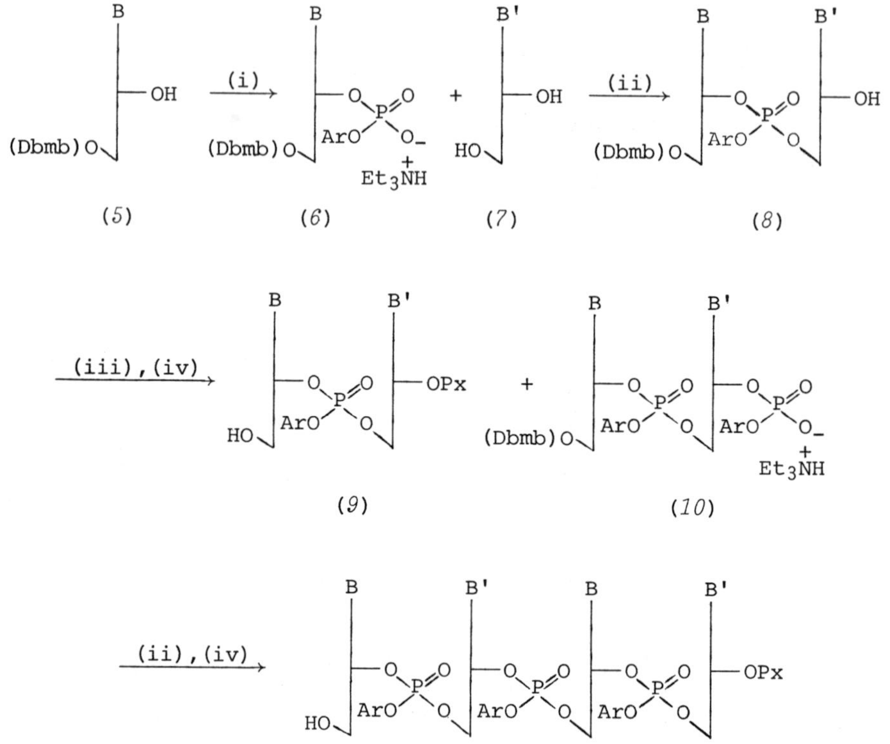

(5) (Dbmb)O–[B]–OH $\xrightarrow{(i)}$ (6) (Dbmb)O–[B]–O–P(=O)(OAr)–O^- $Et_3\overset{+}{N}H$ + (7) HO–[B']–OH $\xrightarrow{(ii)}$ (8) (Dbmb)O–[B]–O–P(=O)(OAr)–O–[B']–OH

$\xrightarrow{(iii),(iv)}$ (9) HO–[B]–O–P(=O)(OAr)–O–[B']–OPx + (10) (Dbmb)O–[B]–O–P(=O)(OAr)–O–[B']–O–P(=O)(OAr)–O^- $Et_3\overset{+}{N}H$

$\xrightarrow{(ii),(iv)}$ (11) HO–[B]–O–P(=O)(OAr)–O–[B']–O–P(=O)(OAr)–O–[B]–O–P(=O)(OAr)–O–[B']–OPx

(*11*) Ar = 2-ClC_6H_4

Scheme 2

(a) d[Dbmb-*ApAp*Tp*TpCpG*-Px] $\xrightarrow{(v),(i)}$ d[Dbmb-*ApAp*Tp*TpCpGp*] (*12*)

(b) d[Dbmb-*GpTpApCpCpG*-Px] $\xrightarrow{(iv)}$ d[HO-*GpTpApCpCpG*-Px] (*13*)

(c) (*12*) + (*13*) $\xrightarrow{(ii)}$ d[Dbmb-*ApAp*Tp*TpCpGpGpTpApCpCpG*-Px] (*14*)

(d) (*14*) $\xrightarrow{(vi),(vii),(viii)}$ d[AATTCGGTACCG] (*15*)

Reagents: (i) (a) (*2*)/acetonitrile-pyridine, (b) Et_3N-H_2O; (ii) (*3*)/pyridine; (iii) (*4*)/pyridine; (iv) (a) silver perchlorate-2,4,6-collidine/acetone-water (98:2 v/v), (b) morpholine; (v) toluene-p-sulfonic acid/$CHCl_3$-MeOH (95:5 v/v); (vi) N^1,N^1,N^3,N^3-tetramethylguanidinium syn-4-nitrobenzaldoximate in dioxan-water (1:1 v/v), 20°C, 24 hr; (vii) aqueous ammonia (d 0.88), 20°C, 48 hr; (viii) acetic acid-water (4:1 v/v), 20°C, 15 min.

The procedure used for the preparation of the fully-protected hexamers is indicated in Scheme 1. The nucleoside building blocks required (*5*) are prepared (1,2) by treating thymidine or appropriate N-acyl derivatives of 2'-deoxyribonucleosides (*7*) with o-dibromomethylbenzoyl chloride (*1*) in acetonitrile-pyridine solution. The resulting Dbmb-derivatives (*5*), which are obtained as crystalline solids in good isolated yields, are first treated with a two- to three-fold excess of o-chlorophenyl phosphorodi-(1,2,4-triazolide) (*2*) (1,4) in acetonitrile-pyridine and the products hydrolyzed with aqueous triethylamine to give the corresponding 3'-(o-chlorophenyl) phosphates (*6*) in very high (usually $\nless$ 90%) yields. The latter mononucleotide building blocks (*6*) are isolated as pure solid triethylammonium salts (1,4). No symmetrical 3'→3'-dinucleoside phosphates may be detected in the products.

When the mononucleotide blocks (*6*) are allowed to react (ca. 20 min, room temperature) with thymidine or the appropriate N-acyl-2'-deoxyribonucleoside (*7*) in the presence of an excess (ca. 3 molecular equivalents) of 1-(mesitylene-2-sulfonyl)-3-nitro-1,2,4-triazole (*3*, MSNT) (5,6) in pyridine solution, the corresponding 3'→5'-partially-protected dinucleoside phosphates (*8*) are obtained in good yields (usually 60-70%). Phosphorylation occurs on the 5'-hydroxy function of (*7*) with a very high degree of regioselectivity and the small quantities of the isomeric 3'→3'-dinucleoside phosphates sometimes obtained are less polar and may readily be removed from the desired products (*8*) by short column chromatography (7) on silica gel.

A partially-protected dinucleoside phosphate (d[HO-*CpG*-Px]) with a free 5'-hydroxy and a protected 3'-hydroxy function is also required. Treatment of the 5'-protected dinucleoside phosphates (*8*) with 9-phenyl-9-xanthenyl chloride (*4*, pixyl chloride) (3) in pyridine solution gives their 5'-O-pixyl derivatives in virtually quantitative yields. When the latter are treated first with silver perchlorate and 2,4,6-collidine in acetone-water (98:2 v/v) for ca. 1 hr at room temperature and the products then treated with morpholine, the 5'-Dbmb group (2) is removed and the desired dinucleoside phosphates (*9*) with protected 3'-terminal hydroxy functions are obtained in good yields. The appropriate 5'-protected dinucleoside phosphates (*8*) are then phosphorylated with o-chlorophenyl phosphorodi-(1,2,4-triazolide) (*2*) and thereby converted into their 3'-(o-chlorophenyl) phosphates (*10*) in yields of ca. 90%. The latter (*10*) are then condensed, in the presence of an excess of MSNT (*3*) with the 3'-protected dinucleoside phosphates (*9*) to give fully-protected tetranucleoside triphosphates usually in good yields (ca. 70%). The corresponding partially-protected tetranucleoside triphosphates (*11*) are then obtained by removing the 5'-Dbmb protecting groups by the procedure described above. The desired fully-protected hexanucleoside pentaphosphates are then prepared in the same way by condensing dinucleotide blocks (*10*) with the appropriate partially-protected tetranucleoside triphosphates (*11*). MSNT (*3*) is again used as the condensing agent and satisfactory yields (ca. 60%) are usually obtained.

The preparation of one of the completely unblocked dodecamers (*15*; the Eco RI → Kpn I adaptor) is illustrated in Scheme 2. The final steps in the preparation and unblocking of the other two dodecamers correspond exactly. In the first place, the hexamer block (d[Dbmb-*ApApTpTpCpG*-Px]) is treated (Scheme 2a) with toluene-p-sulfonic acid in chloroform-methanol (95:5 v/v) and the resulting hexamer with a free 3'-hydroxy function is phosphorylated with o-chlorophenyl phosphorodi-(1,2,4-triazolide) to give the corresponding hexanucleotide block (*12*). The latter is then condensed, in the presence of MSNT (*3*), with the product (*13*, Scheme 2b) obtained by removing the 5'-Dbmb protecting group from the hexamer block (d[Dbmb-*GpTpApCpCpG*-Px]), to give the fully-protected dodecamer (*14*, Scheme 2c). The yield in this final condensation step is good but the purification of the dodecamers by short column chromatography on silica gel proved sometimes to be difficult. The removal of the protecting groups from (*14*) is effected in a three-step process. The fully-protected oligonucleotide is first treated with a large excess (ca. 10 molecular equivalents per phosphotriester group) of a 0.3 M-solution of the N^1,N^1,N^3,N^3-tetramethylguanidinium salt of syn-4-nitrobenzaldoxime (5) in dioxan-water (1:1 v/v) at 20°C for 24 hr to unblock the internucleotide linkages. We now believe (8) that this process is complete in a much shorter time. An ammonolysis step completes the removal of the N-acyl and 5'-O-Dbmb protecting groups. The fully-unblocked oligonucleotide is obtained following the removal of the pixyl group by acidic hydrolysis.

All three of the fully-unblocked dodecamers underwent complete digestion to give their monomeric components when they were treated with Crotalus adamanteus snake venom and spleen phosphodiesterases. Their structures were further confirmed in the usual way.

We thank the Science Research Council for generous financial support.

Literature Cited

1. Chattopadhyaya, J. B.; Reese, C. B. Nucleic Acids Res. 1980, 8, 2039.
2. Chattopadhyaya, J. B.; Reese, C. B.; Todd, A. H. J. Chem. Soc. Chem., Commun. 1979, 987.
3. Chattopadhyaya, J. B.; Reese, C. B. J. Chem. Soc., Chem. Commun. 1978, 639.
4. Chattopadhyaya, J. B.; Reese, C. B. Tetrahedron Lett. 1979, 5059.
5. Reese, C. B.; Titmas, R. C.; Yau, L. Tetrahedron Lett. 1978, 2727.
6. Jones, S. S.; Rayner, B.; Reese, C. B.; Ubasawa, A.; Ubasawa, M. Tetrahedron 1980, 36, 3075.
7. Hunt, B. J.; Rigby, W. Chem. Ind. (London) 1967, 1868.
8. Reese, C. B.; Valente, L., unpublished observations.

RECEIVED July 7, 1981.

Coupling of Fatty Diazomethylketones with Organophosphorus Acids

An Approach to Glycerophospholipid Analog Synthesis

DAVID A. MARSH[1] and JOSEPH G. TURCOTTE

Department of Medicinal Chemistry, University of Rhode Island, Kingston, RI 02881

Intense interest in cell membrane architecture, chemistry, and function as a current frontier of molecular biology research has generated a concomitant interest in, and need for, individual molecular species of glycerophospholipids. Since naturally occurring phospholipids are multispecies and generally "non-separable ", and the chemical synthesis of individual molecular species is multistep, stereospecific, and low yield (overall), new appraoches to the synthesis of natural glycerophospholipids and/or analogs of glycerophospholipids suitable for biochemical, biophysical, and pharmacologic studies are needed. That derivatives of naturally occurring glycerophospholipids can exhibit significant physicochemical properties and pharmacologic activities, can be seen from studies on pulmonary surfactant (1), antihypertensive (2), and anticancer (3,4,5) phospholipid and lysophospholipid analogs.

As an entry to an alternative approach to the synthesis of the spectrum of classes of glycerophospholipids and structurally related analogs, lipid dihydroxyacetone phosphate and phosphonate derivatives were considered. Published routes (6-9) to such derivatives are long and low yielding. For example, 1-palmitoyloxyhydroxyacetone phosphate [III, $CH_3(CH_2)_{14}COOCH_2COCH_2OP(O)(OH)_2$] can be synthesized (10) by direct phosphorylation ($POCl_3$) of 1-palmitoyloxy-hydroxyacetone [II, $CH_3(CH_2)_{14}COOCH_2COCH_2OH$] , obtained (11) from the diazomethylketone [I, $CH_3(CH_2)_{14}COOCH_2COCHN_2$] precursor, the yield of III from II was 25% and from I, would be less than 15%. Alternatively, Hajra and Agranoff (10) found that the diazomethylketone I could be reacted directly with phosphoric acid to give a 60% yield of 1-palmitoyloxyhydroxyacetone phosphate (III) in a single step. This direct coupling procedure could be a promising approach to glycerophospholipid synthesis provided that: the direct phosphorylation or phosphonylation of lipid diazomethylketones structurally related to molecules such as I has general applicability; facile reduction (12) of product ketones to corresponding secondary alcohols (racemic and/or optically active) can be accom-

[1] Current address: Department of Medicinal Chemistry, Massachusetts College of Pharmacy, Hampden Campus, Springfield, MA 01119.

0097–6156/81/0171–0195$05.00/0

$$R_1COCHN_2 \xrightarrow[\underline{VI}]{HOPO(R_2)R_3} R_1COCH_2OPO(R_2)R_3$$

$$\underline{IV} \downarrow$$

$$\underset{\underline{V}}{R_1COCH_2OH} \xrightarrow[\underline{VII}]{ClPO(R_2)R_3} \underset{\underline{VIII}}{R_1COCH_2OPO(R_2)R_3}$$

plished; subsequent acylation of such alcohols can be controlled to give esters esterified at what would be comparable to the sn-2 position of natural glycerophospholipids.

Because of the potential synthetic utility of the direct coupling reaction, a preliminary investigation of its scope was made using 1-diazo-2-heneicosanone [IV, $R_1 = CH_3(CH_2)_{17}CH_2$] as a model compound; little attention has been given to the direct synthesis (10,12,13,14) of α-phospho- or α-phosphonoketones from reaction of organophosphorus acids with α-diazoketones. The results of the study are summarized in Table I. It was found that 1-diazo-2-heneicosanone reacts with acids such as 2-phthalimidoethylphosphonic acid, 2-chloroethylphosphonic acid, chloromethylphosphonic acid, 3-bromopropylphosphonic acid, dioctadecyphosphoric acid, and dibenzylphosphoric acid to give fair to good yields of the corresponding derivatives; yields (Table I) have not been optimized. Most of the reactions were completed within a few hours and in each case the products (VIIIa-VIIIf) were able to be purified readily by column chromatography (SilicAR CC-7); thin-layer chromatography of crude reaction mixtures usually showed the products significantly different in R_f values from both starting materials and side-products.

Diacidic phosphonates (VIa-VId), i.e., those with two protons available for substitution, had a tendency to yield significant amounts of bis-substituted phosphonates [$(R_1COCH_2O)_2PO(R_2)$] upon reaction with IV. However, it was found that careful addition (Table I) of the α-diazoketone to the phosphonate at reaction temperature minimized the formation of this by-product; moreover, the bis-substituted phosphonates could be converted to the corresponding monoacids by refluxing the former with sodium iodide in methyl ethyl ketone. In this manner higher yields of phosphonates (e.g., VIIIa-VIIId) could be obtained.

Although the yields of α-phospho- and α-phosphonoketones were not optimized, the direct synthesis (IV → VIII) of these ketones from diazomethylketones and organophosphorus acids can be expect-

IX	X	XI	XII
CH_2OCOR_1	CH_2R_1	CH_2OCOR_1	CH_2R_1
*CHOH	CHOH	*$CHOCOR_2$	$CHOCOR_2$
$CH_2OPO(X)OH$	$CH_2OPO(X)OH$	$CH_2OPO(X)OH$	$CH_2OPO(X)OH$

R_1, R_2 = saturated or unsaturated fatty chains; X = OH, choline, ethanolamine, serine, etc., or corresponding phosphonate isosteres; *S-configuration.

TABLE I

PHOSPHONATES AND PHOSPHATES DERIVED FROM 1-DIAZO-2-HENEICOSANONE

$$\underset{\underline{IV}^{a}}{CH_3(CH_2)_{18}COCHN_2} + \underset{\underline{VIa}-\underline{VIf}}{HOPO(R_1)R_2} \longrightarrow \underset{\underline{VIIIa}-\underline{VIIIf}}{CH_3(CH_2)_{18}COCH_2OPO(R_1)R_2}$$

Acid	R_1	R_2	VIa-VIf[b]	IV[b]	Solvent	Temperature[c]	% Yield[d]	Product[e]
VIa	OH	$CH_2CH_2N(CO)_2C_6H_4$	28.8	14.9	Dioxane	reflux	57.8	VIIIa
VIb	OH	CH_2CH_2Cl	2.6	0.6	CH_2Cl_2	25°C	45.0	VIIIb
VIc	OH	CH_2Cl	10.0	2.5	EtAc	25°C	66.1	VIIIc
VId	OH	$CH_2CH_2CH_2Br$	0.8	0.3	Dioxane	reflux	49.0	VIIId
VIe	$O(CH_2)_{17}CH_3$	$OCH_2(CH_2)_{17}CH_3$	0.1	0.1	Dioxane	reflux	30.0	VIIIe
VIf	$OCH_2C_6H_5$	$OCH_2C_6H_5$	3.0	3.0	EtAc	reflux	71.4	VIIIf

[a]Prepared by treatment of eicosanoic acid with oxalyl chloride followed by reaction of the resultant acid chloride with diazomethane. [b]Mmol. [c]Additions of solutions of reactants usually made at ambient temperature. [d]Per cent yields based on eicosanoic acid $[CH_3(CH_2)_{18}COOH]$ as starting material. [e]In a typical reaction VIa was dissolved in 900 ml of dioxane (reflux) and IV dissolved in 200 ml of dioxane was added dropwise. After reflux for an additional 1 hr, the solvent was removed in vacuo and $CHCl_3$ added. The precipitate was filtered, the filtrate concentrated, and the product chromatographed on SilicAR CC-7 (400 g): $CHCl_3$; $CHCl_3$/MeOH (19:1); $CHCl_3$/MeOH (10:1); $CHCl_3$/MeOH (5:1); $CHCl_3$/MeOH (1:1). The product eluted in the 1:1 $CHCl_3$/MeOH fraction. Anal. Calcd. for $C_{32}H_{52}NO_6P$: C,66.05; H, 8.94; N, 2.48; P, 5.49. Found: C, 65.81; H, 8.84; N, 2.33; P, 5.49. Structures of all reaction products were confirmed by spectroscopy (^1H NMR,IR) and elemental analysis.

ed to have advantages over the alternative synthetic route (IV → V → VIII). For example, in the application of this reaction to glycerophospholipid and glycerophosphonolipid analog synthesis the α-phosphonoketone VIIIa was synthesized in 58% overall yield from eicosanoic acid (Table I) starting material; by comparison, the alternative 4-step route afforded VIIIa in 26% overall yield and required two chromatographic (column) purifications. Therefore, starting with any readily available fatty acid, upon generation of the acid chloride and diazomethylketone *in situ*, reaction of the latter class of molecules with respective organophosphorus acids, affords an essentially one-chamber preparative route for lipid α-phospho- and α-phosphonoketones. This simple sequence provides a synthesis of analogs (e.g., X, XII, respectively) of glycerolyso-phospholipids (IX) and glycerophospholipids (XI), and a related approach to the synthesis of classes of glycerophospholipids from intermediates such as III.

Literature Cited

1. Turcotte, J. G.; Sacco, A. M.; Steim, J. M.; Tabak, S. A., Notter, R. H. *Biochim. Biophys. Acta* 1977, *488*, 235.
2. Turcotte, J. G. et al. *J. Med. Chem.* 1975, *18*, 1184; Sen, S., Smeby, R. R., Bumpus, F. M., Turcotte, J. G., *Hypertension* 1979, *1*, 427.
3. Turcotte, J. G., Srivastava, S. P., Meresak, W. A., Rizkalla, B. H., and Wunz, T. P. *Biochim. Biophys. Acta* 1980, *619*, 604.
4. Turcotte, J. G. et al. *Biochim. Biophys. Acta* 1980, *619*, 619.
5. Raetz, C. H. R., Chu, M. Y., Srivastava, S. P., Turcotte, J. G. *Science* 1977, *196*, 303
6. Piantadosi, C.; Ishaq, K. S.; Wykle, R. L.; Snyder, F. *Biochem.* 1971, *10*, 1417.
7. Piantadosi, C.; Chae, K.; Ishaq, K. S.; Snyder, F. *J. Pharm. Sci.* 1972, *61*, 971.
8. Piantadosi, C.; Ishaq, K. S.; Snyder, S. *J. Pharm. Sci.* 1970, *59*, 1201.
9. Snyder, F; Blank, M. L.; Malone, B.; Wykle, R. L. *J. Biol. Chem.* 1970, *245*, 1800.
10. Hajra, A. K.; Agranoff, B. W. *J. Biol. Chem.* 1968,*243*, 1617.
11. Schlenk, H.; Lamp, B. G.; DeHaas, B. W. *J. Am. Chem. Soc.* 1952, *74*, 2550.
12. Turcotte, J. G.; Pavanaram, S. K.; Yu, C-S.; Lee, H; Boyd, R. O.; Dadbhawala, K. R.; Marsh, D. A. *Fed. Proc.* (Abs) 1973, *32*, 690; Marsh, D. A. Ph. D. Dissertation, University of Rhode Island, 1976.
13. Rosenthal, A. F.; Chodsky, S. V. *Chem. Commun.* *1968*, 1504.
14. Silverman, J. B.; Babiarz, P. S.; Mahajan, K. P.; Buschek, J.; Fondy, T. P. *Biochem.* 1975, *14*, 2252.

RECEIVED July 7, 1981.

Design of Organophosphorus Reagents for Peptide Synthesis

R. RAMAGE, B. ATRASH, and M. J. PARROTT

Department of Chemistry, The University of Manchester Institute of Science and Technology, Sackville Street, Manchester M60 1QD, England

Two crucial aspects of the stepwise synthesis of peptides are the temporary protection of α-amino functions and activation of carboxylic acids to enable facile formation of amide bonds. The amino protecting groups should be stable, except under specific mild cleavage conditions, and must not lead to diminished stereochemical integrity of the protected α-amino acid during activation. Criteria for the choice of activation procedure adopted for the carboxylic acid function are no less stringent requiring rapid, highly efficient amide formation during the repetitive steps leading to the synthesis of polypeptides.

One of the most successful classes of amino protecting groups is that based on the t-butyl urethane which may be cleaved by mild acid. Structural variation gives rise to groups susceptible to deprotection over a range of acid conditions. A disadvantage of this type of protection is the formation of carbenium ions during the deprotection process which can react with side chain functionality of cysteine, methionine, tryptophan or tyrosine leading to alkylated products. Although this can be mitigated by use of scavengers it was thought desirable to design another series of protecting groups which have the same propensity towards acid cleavage, but which occasion no deleterious side reactions during deprotection. It was decided to investigate for this purpose the utility of the remarkable acid lability of the P-N bond in phosphinamidates.[1] Careful mechanistic researches have led to results which would suggest that acid-catalysed solvolysis of phosphinamidates can proceed via trigonal bipyramidal intermediates thus producing no reactive intermediates capable of entering side reactions. In order to maximise the effect of substituents it was judged that phosphinamidates would be capable of a wider range of reactivity towards acid hydrolysis than phosphoramidates which would also suffer from the disadvantages of offering two modes of fragmentation of the trigonal bipyramidal intermediate during solvolysis. Preliminary work[3] showed that the $Ph_2PO.NHR$ group is more acid-labile than $Bu^tO.CONHR$, therefore a series of protected amino acids were prepared using the readily accessible $Ph_2PO.Cl$

0097-6156/81/0171-0199$05.00/0

and these have now been incorporated into a programme aimed towards the synthesis of prohormones. Recent researches directed to the application of phosphinamidates in this area have involved the mechanistic study of the acid-catalysed methanolysis of a series of phosphinamidates derived from β-phenylethyl amine incorporating substituents on phosphorus which were selected in order to define the optimum reagent for use in peptide synthesis.

$$R_2P(O).NHCH_2CH_2Ph \xrightarrow{HCl/MeOH} R_2P(O)OMe + H_3\overset{+}{N}CH_2CH_2Ph\ Cl^-$$

Kinetic results for a series of such reactions are given in Table 1 and show an interesting combination of steric and electronic effects of the substituent R. From X-ray diffraction data on $Ph_2P(O).N(Me)CH_2CH_2Ph$ it could be seen that the geometry of substituents at the nitrogen atom is non-planar and that only one phenyl ring was oriented to allow interaction with the P=O bond. Comparison of rate data in Table 1 for the phosphinamidates R_2=Ph_2, R_2=Me_2 and R_2=Me/Ph shows the latter to have the optimum balance of steric effect (Me) and electronic effect (Ph) for facile hydrolysis. The rapid onset of steric retardation may be seen from comparison of the rates of hydrolysis of dimethylphosphinamidates with the higher dialkyl analogues. Unfortunately the dimethyl series proved too hygroscopic to be useful in peptide synthesis, however the diethylphosphinamides show promise for side-chain amino protection which requires relatively greater acid stability.

Carboxylic mixed anhydrides are very important for the rapid synthesis of peptides by the stepwise procedure, however the use of carboxylic mixed anhydrides, e.g. those derived from pivalic acid and a protected amino acid (1), suffers from two disadvantages. Firstly, regiospecificity of attack at the desired carboxyl function is largely determined by steric effects and will not be 100% for all coupling reactions. Secondly, such mixed anhydrides have a propensity towards disproportionation to symmetric anhydrides which is highly undesirable in terms of reaction efficiency. This latter process can be depressed by operation of the reaction at -15 °C, but with the concurrent decrease in reaction rate and, on large scale manufacture, increased costs.

$$\underset{(1)}{X.NH.\overset{R^2}{CH}.CO.O.COBu^t} \qquad \underset{(2)}{X.NH.\overset{R^2}{CH}.CO.O.P(O)R_2} \;+\; \underset{(3)}{H_2N\overset{R^1}{CH}COOMe} \rightarrow X.NH.\overset{R^2}{CH}.CONH.\overset{R^1}{CH}.COOMe \;+\; R_2P(O)OH$$

With these considerations in mind it was decided to investigate phosphinic-carboxylic mixed anhydrides in peptide methodology[4]. Mechanistic consideration of the reactants (2) and (3) and products shown above would suggest regiospecific nucleophilic attack by the amine component due to the formation of an amide bond with concomitant generation of a new P-O bond. As in the study of phosophinamidates discussed above, a series of phosphinic acids were selected for preparation of the mixed anhydrides (2) because of the intimate steric and electronic

Table 1

Rate Constants ($s^{-1} \times 10^5$) and Half-lives (min) for Acid Catalysed Methanolysis of Phosphinamidates $R_2P(O)NHCH_2CH_2Ph$

R_2	25°		30°		37°		45°	
	k	$T_{1/2}$	k	$T_{1/2}$	k	$T_{1/2}$	k	$T_{1/2}$
$(PhCH_2)_2$	3.1	373	4.0	287	5.9	197	8.0	144
Et_2	6.8	169	9.7	118	14.8	78	26.1	44
$\underline{n}$-Bu_2	6.8	169	10.4	111	15.4	75	27.0	43
Me_2	145.0	8	201.0	6	326.0	4	520.0	2
Ph_2	41.0	28	53.6	21	68.7	17	94.7	12
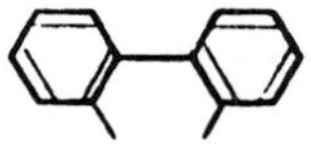	36.0	32	52.8	22	88.0	13	153.5	8
Me,Ph	too fast to measure by HPLC							

Table 2

Rate Constants ($l.mol^{-1}.s^{-1} \times 10^5$) for the Disproportionation of Diphenylphosphinic-Amino Acid Anhydrides

Amino Acid	0 °	30 °	40 °	50 °	Time 10% (min)
Z-Gly	3.3	15.5	49.2	138	74
Z-Ala	1.4	22.0	62.4	113	143
Z-Leu	0.5	7.4	18.1	40	390
Z-Phe	1.4	6.7	26.0	75	170
Z-Val	11.7	43.0	67.3	109	17
Z-Ply	-	52.2	135.0	248	56

Table 3

Rate Constants (l.mol^{-1}.s^{-1} x 10^{5}) for the Disproportionation of Z-Valine-Phosphinic Acid Anhydrides

Phosphinic acid Substituents	0°	30°	40°	50°	60°	Time 10% (min)
Me	1.4	7.4	15.2	21.2	-	138
Et	1.7	4.6	10.8	22.8	-	130
n-Bu	0.7	5.6	17.0	20.1	-	240
i-Bu	2.3	6.4	-	17.1	21.1	85
$PhCH_2$	20.4	119.4	163.0	277.0	-	9
Ph	11.7	43.0	67.3	109.0	-	17
[structure: 2,2'-biphenylyl]	7.7	96.7	128.2	240.3	-	23

effects which substituent R on phosphorus can contribute directly to reactivity towards nucleophilic attack at the P=O.Thus it was decided to investigate the thermal disproportionation of the series of mixed anhydrides shown in Tables 2 and 3 in order to design the most appropriate reagent for peptide synthesis.
The kinetic study involved ^{31}P NMR measurements on the anhydrides (2) in EtOAc/$CDCl_3$(and also DMF/$CDCl_3$)solution at a range of temperatures.In the first study the phosphinic acid moiety was kept constant,where R=Ph,in order to evaulate the effect of varying the nature of the α-amino acid on the rate of disproportionation.With the exception of valine (Table 2) mixed diphenylphosphinic anhydrides showed good thermal stability and, indeed,all were completely regiospecific with respect to ammonolysis at the carbonyl group.Concurrently with this study, rates of acylation were measured using the anhydrides (2,R=Ph)and from the data it could be concluded that the rates of disproportionation were insignificant from a preparative aspect compared with the desired amide formation at 0 °C.In order to evaluate the effect of the substituents R on phosphorus on the rate of disportionation it was decided to employ Z-valine as the protected α-amino acid because it is known to be sterically hindered,and therefore a case where disproportionation could be in serious competition to amide formation.From Table 3 it can be deduced that phosphinic acids having dialkyl substituents are optimum for phosphinic-carboxylic mixed anhydride formation in terms of thermal stability.This is a necessary requirement for mixed anhydride utilisation in solid phase peptide synthesis.

A combination of phosphinamidate NH_2-protection and phosphinic-carboxylic mixed anhydride activation has been applied to the synthesis of the C-terminal tetrapeptide (4) of gastrin and interesting pentapeptide analogues of (Met)-enkephalin (5). These examples contain amino acid units which would preclude either the use of hydrogenolysis to remove N-benzyloxycarbonyl protecting groups or necessitate the use of scavengers if acid labile protection of the t-butyloxycarbonyl type was used.

Trp.Met.Asp.Phe.NH_2 (4)   Tyr.Gly.Gly.Phe.Met (5)

1. G.Tomaschewski and G.Kühn,J.Prakt.Chem.,1968,38,222. D.A.Tyssee,L.P.Bausher and P.Haake,J.Amer.Chem.Soc.,1973,95, 8066;T.Koizumi and P.Haake,J.Amer.Chem.Soc.,1973,85,8073. M.Kenn,W.Jansen and O.Schmitz-DuMont,Z.anorg.allg.Chem., 1979,452,176.
2. M.J.P.Harger,A.J.MacPherson and D.Pickering,Tetrahedron Letters 1975,1797;M.J.P.Harger,J.Chem.Soc.Perkin I,1977,605; M.J.P.Harger,J.Chem.Soc.Perkin I, 1979, 1294.
3. G.W.Kenner,G.A.Moore and R.Ramage,Tetrahedron Letters,1976, 3623.
4. A.G.Jackson,G.W.Kenner,G.A.Moore,R.Ramage and W.D.Thorpe, Tetrahedron Letters, 1976, 3627.

RECEIVED June 30, 1981.

42

The Nature of the Energy Transduction Links in Mitochondrial Oxidative Phosphorylation

FAUSTO RAMIREZ, SHU-I TU[1], PRABHA R. CHATTERJI, HIROSHI OKAZAKI, JAMES F. MARECEK, and BRIAN McKEEVER

Department of Chemistry, State University of New York at Stony Brook, Stony Brook, NY 11794

The membrane-associated oxidative phosphorylation in intact mitochondria depends on the close cooperation between the ATPase, the electron transfer chain, and the metabolite transport systems. It is generally accepted that the hydrolysis of MgATP and the electron transfer catalyzed by the mitochondrial inner membrane generate a transmembrane proton electrochemical potential ($\Delta\mu_{H^+}$), but the precise role of proton translocation in oxidative phosphorylation has remained a subject of research and debate (1). According to Mitchell's chemiosmotic hypothesis, the $\Delta\mu_{H^+}$ generated by redox events is the direct and exclusive driving force for ATP synthesis. This hypothesis postulates that H^+ movement is directly linked to electron flow, and that there is no need for contact between electron transfer enzymes and the ATPase during the synthesis of ATP. It now appears, however, that $\Delta\mu_{H^+}$ may not be an obligatory intermediary state between electron transfer reactions and ATP synthesis (2,3).

It has recently been shown that fluorescamine (FL), a compound which specifically forms covalent derivatives with primary amines, exerts significant effects on mitochondrial activities (4). The data suggested that the H^+ movement and its associated energy-yielding step, i.e., the redox events or the hydrolysis of MgATP, are only indirectly linked. Subsequent work was carried out with compounds of types **1** and **2**, derived from the reaction of primary amines with FL (5,6). The amine-FL compounds do not form covalent derivatives with components of the membrane. Yet, the acyclic form, **1**, of these compounds modify the H^+ movements associated with the ATPase and the respiratory activities to a greater extent than the respective energy-yielding processes in intact rat liver mitochondria. This type of modification is not observed with the lactone form, **2**, of the compounds (7). The acyclic modifiers, **1**, have no effect on the kinetics of the H^+-leak, and on the influx of H^+ across the inner membrane that is induced by conventional uncouplers such as 2,4-dinitrophenol (DNP).

[1]Current Address: USDA SEA-AR Eastern Regional Research Center.

0097-6156/81/0171-0205$05.00/0

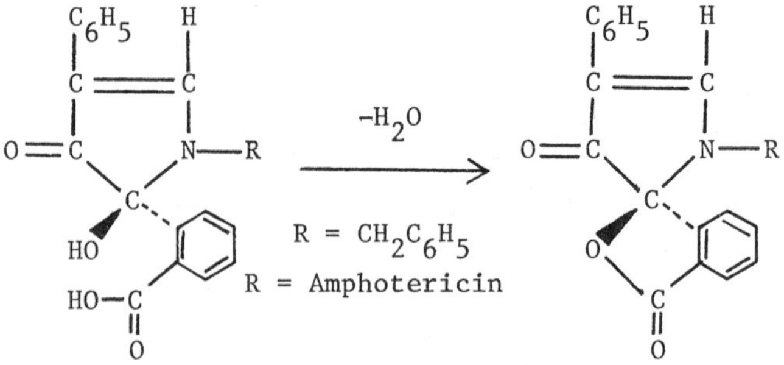

$-H_2O$

R = $CH_2C_6H_5$

R = Amphotericin

1, Acyclic **2**, Lactone

These results strengthen the "indirect-link" concept, and further suggest that the acyclic modifiers, **1**, specifically inhibit the activation of proton pumps associated with the ATPase and the respiratory activities as a result of an implantation process at the corresponding protogenic pump proteins. It is likely that this inhibition is achieved by hydrogen bonding of carboxyl, hydroxyl, keto and amino groups in the acyclic modifier, **1**, with certain membrane proteins.

The activity of the ATPase can be affected by inhibitors of the electron transfer chain (rotenone, antimycin A, cyanide) in intact mitochondria (8,9). We have confirmed this effect (7). These observations have been taken as an indication that there is a direct interaction between redox enzymes and the ATPase, other than $\Delta\mu_{H^+}$, and that this interaction may also serve as the regulatory mechanism in oxidative phosphorylation. The mutual regulation between the ATPase and the respiratory activities is hindered by the presence of the acyclic modifiers, **1**, and we have proposed that this regulation requires the direct interaction between intact, energized proton pumps associated with the ATPase system and the respiratory activities. The present Communication reports new data in support of this "pump-pump interaction" regulatory mechanism. We have studied the kinetics of ATP synthesis in intact mitochondria in the presence of the acyclic modifier **1a**, and have compared the results with those obtained in the presence of oligomycin or of the uncoupler DNP. In addition, we have obtained some information on possible sites of implantation of the acyclic modifiers, **1**, utilizing the radioactive photoaffinity probe made from the reaction of FL with tritiated 1-amino-3-(2'-nitro-4'-azidophenyl)aminopropane, or NAZA [R = $\overset{*}{C}H_2CH_2\overset{*}{C}H_2NHC_6H_3\cdot(2'\text{-}NO_2)(4'\text{-}N_3)$ in **1c**].

Rat liver mitochondria were isolated as described (4). The initial rate of ATP synthesis associated with the oxidation of succinate was followed by monitoring fluorometrically the ATP-linked NADPH production in the presence of hexokinase and glucose-6-phosphate dehydrogenase (10). Control experiments showed no interference from unexpected reduction of $NADP^+$ or from electron backflow. Possible ATP formation via mitochondrial adenylate kinase was inhibited by P^1,P^5-di(adenosine-5')-pentaphosphate (AP_5A) (11). The results are given in Figure 1. It can be seen that the acyclic benzylamine-FL compound, 1a, is a more effective inhibitor of oxidative phosphorylation than oligomycin, which is a specific ATPase agent, and than 2,4-dinitrophenol, which enhances the membrane H^+ conductance.

We interpret these and previous results as follows. If the interaction between the normally operating proton pumps associated with the respiratory chain and the ATPase system is responsible, not only for the mutual regulation between these two enzyme systems, but also for the synthesis of ATP, the modifiers FL, 1, should be relatively effective inhibitors of oxidative phosphorylation, since 1 appear to interact directly with both types of proton pumps. Oligomycins presumably interfere only with the activation of the ATP-linked proton pump possibly by preventing the H^+ flow through a specific channel. Uncouplers (DNP) dissipate the $\Delta\mu_{H^+}$ resulting from the activation of the proton pumps. A greater inhibition of oxidative phosphorylation by amine-FL compounds than by oligomycin and DNP would be expected in this picture, in agreement with our observations. Accordingly, we propose the following hypothesis. In normal mitochondria, the energy released from the minimal hydrolysis of MgATP catalyzed by the ATPase is utilized to raise the enzyme to a higher energy conformation $(ATPase)^*$ and to activate its proton pump $(P_{ATP})^*$. These events generate a $\Delta\mu_{H^+}$ to transport metabolites (redox substrates, ADP-ATP exchange, P_i) needed in subsequent processes. The transported redox substrates initiate the electron transfer of the respiratory chain. The energy released at each coupling site is again stored, partly as the high energy conformation of the transfer enzyme (E^*, e.g. cytochrome oxidase), and partly in the form of $\Delta\mu_{H^+}$ through the activation of respiratory proton pumps $(P_E)^*$. The direct contact between P_{ATP}^* and P_E^* allows the interaction between $ATPase^*$ and E^*. The conformational energy is then released in the condensation of the tightly bound ADP (present in the ATPase) with P_i (present in its appropriate region) to form ATP. In this hypothesis, $\Delta\mu_{H^+}$ is still an indispensable component of the energy transduction since it is required for the initiation and continuation of phosphorylation.

The synthesis of the photoaffinity probe, NAZA-FL, 1c, will be described elsewhere. The effects of this acyclic modifier on mitochondrial activities were assayed as previously described (5,6). The results are summarized in the Table. The effects, and the effective concentration range, of compounds 1c and 1a are similar in the dark. With the photolabile modifier, 1c, illumination with strong light resulted in an irreversible preferential inhibition of the energy-dependent proton movements; this effect was not relieved by phosphatidylcholine vesicles. Hence, the modifier became covalently bonded in the vicinity of its implantation. Additional experiments were performed with phosphate-washed (12) membrane preparations (ATPase-enriched inner membrane frac-

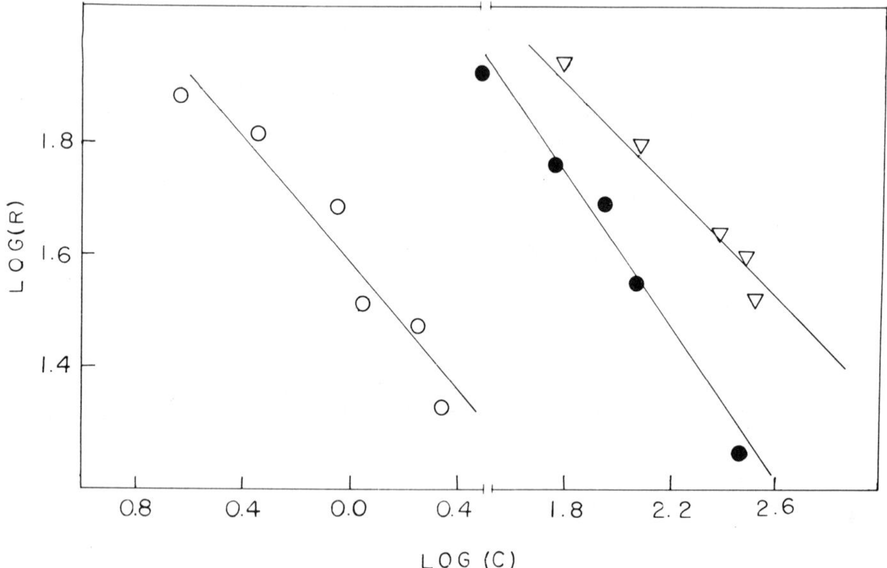

Figure 1. Effect of oligomycin (O), benzylamine–fluorescamine compound (•), and 2,4-dinitrophenol (△) on oxidative phosphorylation. C = concentration of inhibitor in nmoles/mg of protein, R = initial rate of inhibited ATP synthesis expressed as percent of the uninhibited rate (110.4 nmoles of ATP min^{-1} mg^{-1} of protein).

Freshly isolated rat liver mitochondria (100–200μg) were incubated for 2 min in 3 mL of a medium (pH 7.5) containing 170mM sucrose, 50mM tris, 5mM succinate, 8mM P_i, 3mM $MgCl_2$, 20mM glucose, 12.5 units of hexokinase, 0.122M $NADP^+$, 7 units of dehydrogenase, and 60μM AP_5A. After incubation, oligomycin, the benzylamine–fluorescamine compound, 1a, or 2,4-dinitrophenol was added, and incubation was continued for 60 s. The ATP synthesis was initiated by addition of 0.1 μmol of ADP. The rate of NADPH formation was monitored fluorometrically (excitation at 340 nm, emission at 460 nm) using a Perkin Elmer MPF 44A fluorescence spectrophotometer. Quantitative estimation allowing for internal quenching was obtained by injecting standard NADPH samples.

tion). Preliminary labeling pattern analyses (SDS-acrylamide gel electrophoresis) revealed that the acyclic modifier 1c labeled three peptides, MW ∿ 16,000, 13,000 and 2,000. The NAZA precursor labeled only one peptide, MW ∿ 2,000. When the membrane was treated with oligomycin before the labeling step with 1c, labeling of the MW ∿ 16,000 peptide was abolished. Further work with the probe 1c is in progress.

Effects of NAZA-Fluorescamine, 1c, on Mitochondrial Activities[a]

Activity	-hν	+hν
Electron transfer-linked	Preferential inhibition of H^+ movement	Same, but more pronounced
ATP hydrolysis-linked	Enhances P_i formation but inhibits H^+ movement	Same, but more pronounced
Membrane Permeabilities[d]	No effect	No effect

[a]The amine precursor, NAZA, has no effect. [b]Mitochondrial suspension were incubated for 5 min in the presence of modifier, 1c, ranging from 0 to 1.3 x 10^{-4} M. The activities were assayed under dim light. [c]A separate but identical suspension containing 1c was illuminated for 2 min before the activity assays. [d]Including membrane H^+ leak and secondary ion movements.

This work was supported by NHLBI grant #HL 23126.

Literature Cited

1. Boyer, P. D.; Chance, B.; Ernster, L.; Mitchell, P.; Racker, E.; Slater, E. C. Ann. Rev. Biochem. 1977, 46, 955.
2. Villalobo, A.; Lehninger, A. L. J. Biol. Chem. 1979, 254, 4352.
3. Holian, A.; Wilson, D. F. Biochemistry 1980, 19, 4213.
4. Lam, E.; Shiuan, D.; Tu, S. Arch. Biochem. Biophys. 1980, 201, 330.
5. Ramirez, F.; Shiuan, D.; Tu, S.; Marecek, J. F. Biochemistry 1980, 19, 1928.
6. Tu, S.; Lam, E.; Ramirez, F.; Marecek, J. F. Euro. J. Biochem. 1981, 113, 391.
7. Tu, S.; Okazaki, H.; Ramirez, F.; Lam, E.; Marecek, J. F. Arch. Biochem. Biophys. 1981, In Press.
8. Weiner, M. W.; Lardy, H. Arch. Biochem. Biophys. 1974, 162, 568.
9. Alexandre, A.; Rossi, C. R.; Carignani, G.; Rossi, C. S. FEBS Lett. 1975, 52, 107.
10. Gomez-Puyou, A.; Tuena de Gomez-Puyou, M.; Ernster, L. Biochim. Biophys. Acta 1979, 547, 252.
11. Melnick, R. L.; Rubenstein, C. P.; Motzkin, S. M. Anal. Biochem. 1979, 96, 7.
12. Soper, J. W.; Pedersen, P. L. "Methods in Enzymology", Academic Press Inc.: New York, 1979; 55, 329.

RECEIVED July 7, 1981.

ADP Hydrolysis Promoted by Cobalt(III)

MARKUS HEDIGER and RONALD M. MILBURN

Department of Chemistry, Boston University, Boston MA 02215

In recent work we have shown that interaction of $tn_2Co^{III}(aq)$ with pyrophosphate (1) and with ATP (2) in aqueous solution can lead to large rate enhancements (up to 10^5-fold in neutral media) for hydrolysis within a P-O-P linkage [tn = trimethylenediamine; (aq) = $(H_2O)_2$ or $(H_2O)(OH)$ depending on pH; charges omitted for simplicity]. These results indicated that a 1:1 cobalt(III) to ADP complex would be unreactive toward hydrolysis, while a 2:1 species would exhibit markedly enhanced reactivity. The present study confirms this expectation, and provides additional insight into the nature of these metal ion catalyzed processes.

Simple additions of such labile aquo ions as Mg^{2+}, Ca^{2+}, Zn^{2+} and Mn^{2+}, which are of importance in enzymic phosphoryl transfer, have resulted in only very modest catalytic effects for reactions of phosphate species (3). The much more effective tn_2Co^{III} unit possesses the special advantage that it remains intact for long periods, while trans/cis isomerization and substitution in the fifth and sixth coordination sites proceed at moderately rapid and generally convenient rates. These characteristics make it particularly suitable for use in model studies (4).

Figure 1a shows the ^{31}P NMR spectrum (24.2 MHz, positive chemical shifts downfield from 1 M H_3PO_4) for a solution of ADP (0.10 M) and $tn_2Co^{III}(aq)$ (0.10 M), recorded within 10 minutes of adjustment to pH 7.0 (by addition of NaOH, 5 M) after equilibrating the solution for 2.5 hr at pH 4.0 in a pH-stat apparatus. (For these and other experiments the temperature was 20°C and Cl^- was a counter ion; ClO_4^- gave similar results.) The doublets at about -10 and -6 ppm represent the free ADP which remains (28%). At +0.5 and +3.5 ppm are the P_α and P_β resonances characteristic for the six-membered chelate 1 (64% of the original ADP). These results are in agreement with observations on the $(NH_3)_4CoADP$ system (5). The remaining 8% of the original ADP has hydrolyzed to AMP and P_i, represented by the signals further downfield for the free and complexed monophosphate species.

Figure 1b displays the spectrum for a solution which was 0.36 M in $tn_2Co^{III}(aq)$ and 0.12 M in ADP and which was equilibrated at

0097–6156/81/0171–0211$05.00/0

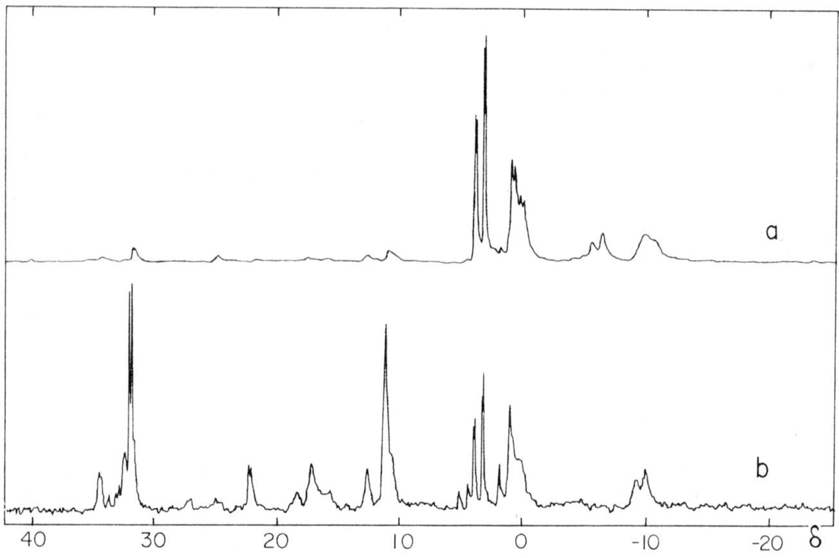

Figure 1. ^{31}P *NMR spectra at pH 7.0 for (a) a solution of 0.10*M *$tn_2Co_{(III)}(aq)$ and 0.10*M *ADP (200 scans; equilibration 2.5 h, pH 4.0, and 20.0° C) and (b) a solution of 0.36*M *$tn_2Co_{(III)}(aq)$ and 0.12*M *ADP (100 scans; equilibration 1 h, pH 4.0, and 20.0° C).*

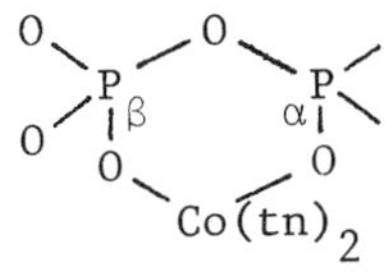

$Co(tn)_2$ — 1 $(tn)_2CoOH_2$ — 2 $(tn)_2CoOH_2$ 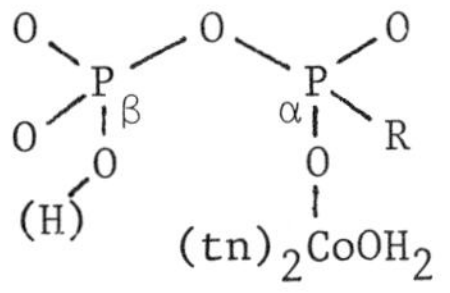 — 3

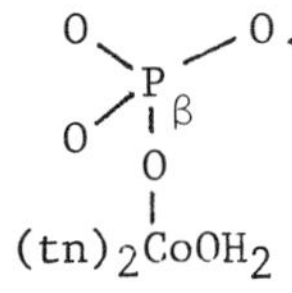

pH 4.0 for 1 hr prior to adjustment to pH 7.0 (spectrum taken within 6 minutes). By comparison with Figure 1a it is obvious that there has been much more hydrolysis. Here we have 4% free ADP and 4% in the form of 2. The signal for P_β of 2 is shifted 10.8 ppm downfield and is represented by the small doublet at 4.8 ppm. The signal for P_α of 2 is scarcely shifted and appears as a doublet with the P_α signal for free ADP. At +0.5 and +3.5, as before, are the P signals for the six-membered chelate 1. A small amount (2%) of 3 is indicated by a P_α signal at +2 (overlapping with the P_α of 1). The signals for complexed AMP (3%) are at 12.7 ppm, and at 11.5 ppm overlapping with complexed P_i. The other signals downfield from 3.5 ppm have been discussed (1,6). In contrast to pyrophosphate and ATP, the NMR for ADP gives no direct indication for a 2:1 cobalt to polyphosphate complex. The results are nevertheless consistent with the interpretation that a 2:1 complex is highly unstable toward hydrolysis, a view supported by the kinetic results.

The principal method used to follow the rate of hydrolysis was periodic quenching of reactant solutions with NaOH (1). This procedure halts the reaction and releases the various phosphate species into solution. The ratio of the integrals over the signals at +4 and +6 ppm (free AMP and P_i) and +14 ppm (complexed P_i) to the total integral gives the percentage of ADP hydrolyzed.

The rate of hydrolysis depends markedly on pH and on the cobalt to ADP stoichiometric ratio; it depends also on the nature of any pre-equilibration procedures. Variations in the ionic strength ($NaClO_4$) have a minor effect.

The influence of pH is illustrated by the following percentages of ADP hydrolyzed after 20 minutes at each pH for solutions which were 0.36 M in tn_2Co^{III}(aq) and 0.12 M in ADP (20°C; solutions first equilibrated in the 1:1 ratio for 30 minutes at pH 4.0 before adding the extra cobalt and raising the pH; Cl^- included as a counter ion but similar results obtained with ClO_4^-): pH 4.0, 10%; pH 5.5, 24%; pH 7.0, 43%; pH 8.5, 36%; pH 10.0, 6%. As with pyrophosphate and ATP, the maximum is at pH ∿7. The fall-off at higher pH is undoubtedly related to competing hydrolysis at the metal center.

Figure 2, which summarizes values of the pseudo first-order rate constant for ADP hydrolysis at pH 7.0 and 20°C, illustrates the influences of the stoichiometric cobalt to ADP ratio, the total stoichiometric concentration, and preformation of the 1:1 complex at pH 4.0. The small values of k_{obs} observed for the 1:1

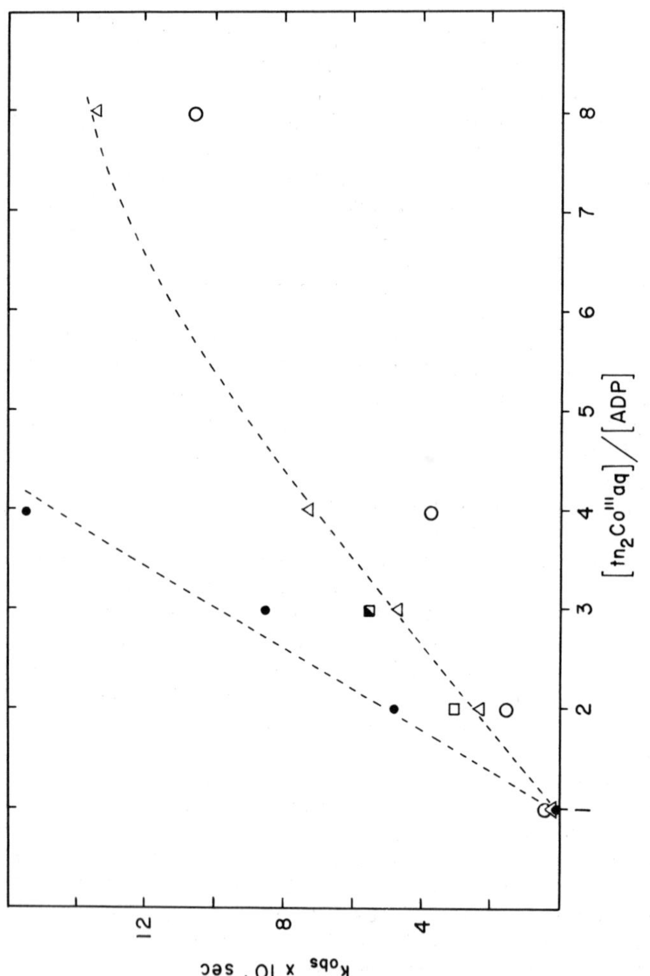

Figure 2. [31]*P NMR spectra at pH 7.0 and 20.0° C for 0.1M in ADP, 1:1 complex preformed* (●); *0.02M in ADP, 1:1 complex preformed* (△, *20.0° C and* □, *25.0° C*)*; 0.02M in ADP, 1:1 complex preformed with* $en_2Co(III)$ *(aq)* (◪)*; 0.02M in ADP, directly mixed (O).*

ratio demonstrate that the 1:1 complex is formed almost quantitatively and that it is of low reactivity. For each of the three series the values of k_{obs} are approximately proportional to the concentration of cobalt in excess of that required for stoichiometric formation of the 1:1 complex; this result is consistent with the view that the reactive species contains two cobalts per ADP. Comparison between the preforming and direct mixing experiments for 0.02 M ADP shows an advantage for preforming the 1:1 complex, and illustrates that in direct mixing equilibrium reactions leading to the 2:1 complex are not fully established (i.e., that formation of the 2:1 complex is, at least for this series, part of the rate-determining process). For the experiments with 0.1 M ADP preforming the 1:1 complex is necessary because of solubility limitations. The larger values of k_{obs} for given Co/ADP ratios is probably due to higher proportions of ADP being tied up as the 2:1 complex.

When the 1:1 complex $tn_2Co^{III}ADP$ was preformed and 1 equivalent of $en_2Co^{III}(aq)$ was added at pH 7.0 (20°C, 0.02 M ADP) no accelerated hydrolysis could be observed. Similar results were obtained on adding $trenCo^{III}(aq)$, Cu^{2+} or Mg^{2+}. However, when the 1:1 complex $en_2Co^{III}ADP$ was preformed and 2 equivalents of $tn_2Co^{III}(aq)$ were added at pH 7.0, the hydrolysis rate was significant and slightly higher than for the parallel experiment of adding 2 equivalents of $tn_2Co^{III}(aq)$ to preformed $tn_2Co^{III}ADP$ (1 eq tn_2Co^{III} + 2 eq $tn_2Co^{III}(aq)$, $k_{obs} = 4.8 \times 10^{-4}$ sec^{-1}; 1 eq en_2Co^{III} + 2 eq $tn_2Co^{III}(aq)$, $k_{obs} = 5.6 \times 10^{-4}$ sec^{-1}; 20°C 0.02 M ADP). These rates and those for related cobalt(III) promoted hydrolysis reactions (1,2,7) can be understood in terms of the substitution labilities of the cobalt centers. Clearly the cobalt in the six-membered ring can be inert, but if the added cobalt center is too inert the rates of complex formation and hence of catalyzed hydrolysis will be low.

The high reactivity of the 2:1 complex can be attributed to the availability on the second cobalt of a coordinated hydroxide which can attack a phosphorus center (preferably P_β). The other cobalt, in the six-membered ring, can assist the process by functioning as an electron withdrawing center. The acceleration for the hydrolysis of ADP is marked ($\sim 10^5$) and comparable to the effect we have previously seen for pyrophosphate and ATP.

Acknowledgment. Support provided by the National Science Foundation is gratefully acknowledged.

Literature Cited

1. Hübner, P. W. A.; Milburn, R. M. *Inorg. Chem.* 1980, 19, 1267.
2. Hediger, M.; Milburn, R. M. Abstracts of Papers, Second Chemical Congress of the North American Continent, Las Vegas, 1980, Inorg. 33; submitted for publication.
3. Cooperman, B. S. "Metal Ions in Biological Systems", Sigel, H., Ed.; Marcel Dekker: New York, 1976, Vol. 5; p. 79.
4. Anderson, B.; Milburn, R. M.; Harrowfield, J. M.; Robertson, G. B.; Sargeson, A. M. *J. Am. Chem. Soc.* 1977, 99, 2652.

5. Cornelius, R. D.; Hart, P. A.; Cleland, W. W. Inorg. Chem. 1977, 16, 2799.
6. Seel, F.; Bohnstedt, G. Z. Anorg. Allg. Chem. 1978, 441, 237.
7. Cornelius, R. D. Inorg. Chim. Acta 1980, 46, L109.

RECEIVED June 30, 1981.

PMR Measurements of Chair-Twist Conformational Equilibria for Diastereomeric P-Derivatives of Thymidine Cyclic 3′,5′-Monophosphate

Possible Implications for Naturally Occurring Cyclic Nucleotides

ALAN E. SOPCHIK, GURDIP S. BAJWA, KEITH A. NELSON, and WESLEY G. BENTRUDE

Department of Chemistry, University of Utah, Salt Lake City, UT 84112

The factors influencing the unusually exothermic hydrolysis of cAMP to 5'-AMP (1) have been considered worthy of considerable investigation. A recent estimate (2) is that 4-5 kcal/mol of strain is imparted to cAMP by the trans nature of the phosphate-ribose ring fusion. Also of interest in such molecules is an understanding of how such strain may affect the conformational properties of the phosphate ring including the energetics of the conversion of that ring from its normal chair conformation into the twist form.

We have examined by PMR the conformational properties of certain P-derivatized thymidine cyclic nucleotides, I, and have compared them to those for the model compounds, II (3, 4). The

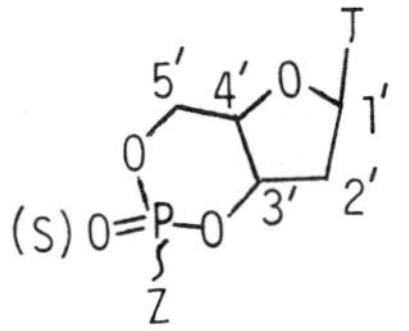

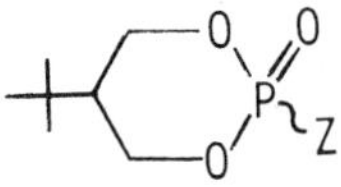

derivatives (III-VI) were prepared by methods we reported earlier (5-9) and separated into individual diastereomers of MPLC. Configurations at phosphorus were assigned on the basis of the well-known ^{31}P NMR chemical shift criterion (10) and confirmed by ^{13}C NMR (6). X-ray crystallography was applied to VIb (8).

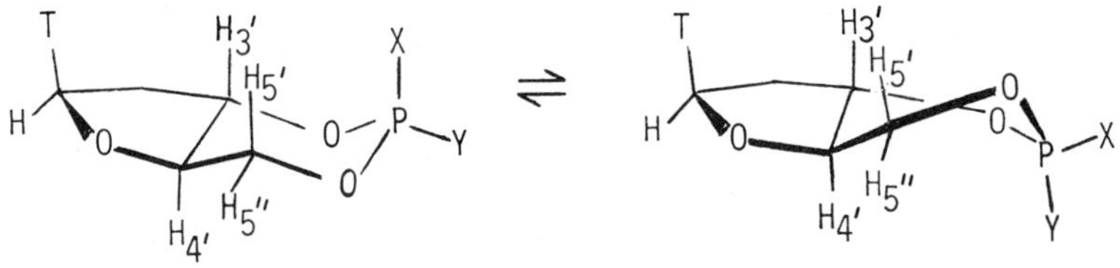

(1)

IIIa	X = MeO, Y = O	Va	X = Me_2N, Y = O
IIIb	X = O, Y = MeO	Vb	X = O, Y = Me_2N
IVa	X = MeO, Y = S	VIa	X = Me, Y = O
IVb	X = S, Y = MeO	VIb	X = O, Y = Me

0097–6156/81/0171–0217$05.00/0

Table I contains the proton coupling constants (300 MHz) for the individual diastereomers of I(III-VI). In the chair conformation, $J_{5'P}$ should be small (0.5-2.0 Hz) and $J_{5''P}$ large (20-22 Hz) as they are in various nucleoside cyclic 3',5'-monophosphates. (11) Ring reversal can give only <u>one</u> twist form of relatively low energy (because of the trans ring fusion (Dreiding models)) in which the values of $J_{5'P}$ and $J_{5''P}$ are essentially interchanged.

It is clear that IIIa and IVa are almost entirely in the chair conformation with MeO axial as it prefers to be in II (3). The same is true for Vb and VIb with bulky Me and Me_2N groups equatorial. However, Va, IIIb and IVb cannot be entirely in chair conformations since $J_{5'P}$ and $J_{5''P}$ are, respectively, too large and too small in each case. By using $J_{5'P}$ and $J_{5''P}$ values from Vb for those of the chair conformer of Va along with reasonable estimates of 19 Hz and 0.5 Hz for $J_{5'P}$ and $J_{5''P}$, respectively, in the twist structure, one can estimate that Va is about 65-75% in this twist conformation depending on solvent and temperature (7). The sterically large size of Me_2N (3, 12) results in destabilization of the chair conformation causing the twist structure to be highly populated.

If one uses a similar approach for IIIb, it is estimated that about 30-35% of IIIb is in the twist conformation with MeO pseudoaxial in accordance with its usual preference. The twist population for IVb is probably about the same.

Derivative VIa shows only a small increase in $J_{5'P}$ and decrease in $J_{5''P}$, and the sum of the couplings is somewhat lower than expected. Whether conformational averaging or only ring distortion occurs here is not certain.

The percentage of twist conformation populated by Va is similar to that for cis II with Z = Me_2N (<u>t</u>-Bu and Z cis) (4). We have argued elsewhere (7) that this means that unless the syn-axial repulsions are unusually high for chair-form Va, the VII → VIII chair-twist component of the conformational change depicted by eqn. 1 (X = Me_2N, Y = O) is probably only about 1 kcal/mol as it is for the model systems (4, 13).

VII ⇌ VIII

Further support for the idea that the accessibility of twist conformations is similar in the cyclic nucleotide derivatives to what it is for II comes from the results for phosphate IIIb. In similar solvents, the ΔG^{O}_{25} for the equilibrium IX ⇄ X is ∿1.5 kcal/mol in favor of X with MeO axial (14). In the case of IIIb, relief of steric interactions involving the small axial P=O is not a factor in the isomerization of eqn. 1. The 33% population of twist conformation is an indication that ΔG^{O}_{25} for the XI → XII

IX ⇌ X

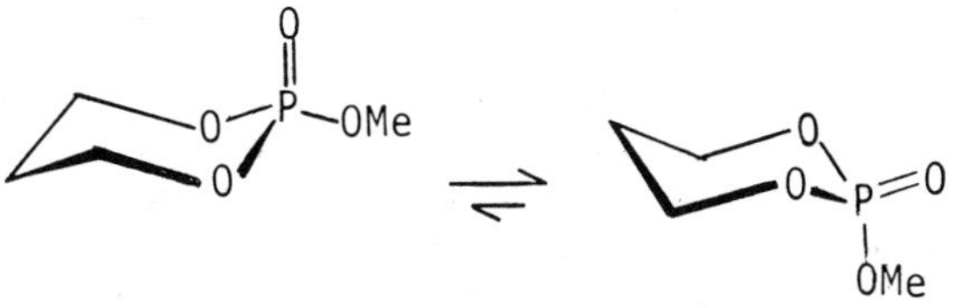

Table I. Coupling Constants (Hz) for III - IV at 300 MHz (26°)

Compounds	$J_{5'P}$	$J_{5''P}$	$J_{5'5''}$	$J_{3'P}$	$J_{4'5'}$	$J_{4'5''}$
IIIa[a]	0.9	21.5	-9.4	<0.5	10.4	4.7
IVa[a]	∿1.0	22.4	-9.3	1.6	10.7	5.1
Va[a]	12.2	8.4	-9.2	∿1.0	10.4	5.9
Va[b]	14.4	6.0	-9.1	<0.5	10.1	6.0
VIa[c]	2.5	17.2	-9.2	<0.5	9.9	5.5
IIIb[a]	7.0	14.8	-9.4	<0.5	10.4	5.5
IVb[a]	7.4	17.4	-9.6	2.2	10.3	5.4
Vb[a]	0.9	21.6	-9.3	1.5	10.8	4.8
VIb[c]	4.7	20.2	-9.4	<0.5	9.0	5.3

[a] acetone-d_6 [b] toluene-d_8 [c] DMSO-d_6

component of eqn. 1 is nearly offest by the ∿1.5 kcal/mol gain in energy accompanying the axial-equatorial exchange of MeO and P=O. Therefore, the XI → XII component must be only about 1.5 to 2.0 kcal/mol itself.

XI ⇌ XII

The fact that the trans fusion of the thymidine-based systems appears to have little effect on the chair ⇄ twist equilibrium suggests strongly that like the model systems, the cyclic nucleotide ΔG^{o}_{25} values for chair → twist interconversions like VII → VIII or XII → XIII are only 1 to 2 kcal/mol. It seems very likely that similar values for chair-twist isomerizations of the natural cyclic nucleotides such as cAMP and cGMP (Z = X = O or O) can be expected. (See e.g. XIII → XIV.)

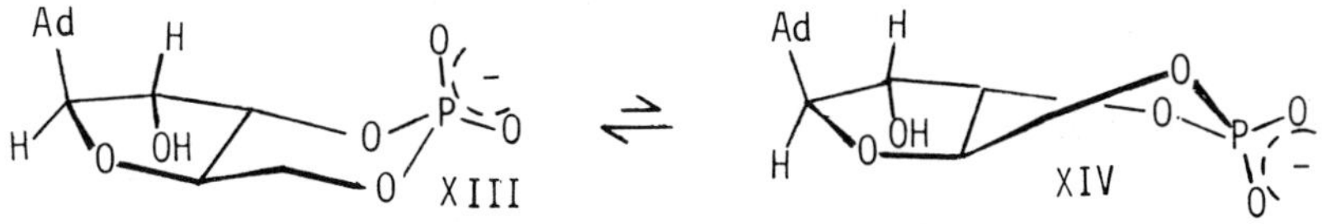

XIII ⇌ XIV

Large energy differences between conformers are difficult to overcome on complexation by an enzyme. However, the small differences our results indicate are likely involved with cAMP and cGMP enhance the possibility that the natural cyclic nucleotides could be enzyme-bound in a twist conformation. It must be noted clearly that there is no experimental evidence at present for such binding or any obvious chemical advantage to it. Our results, however, do emphasize the fact that such a possibility should not be overlooked in future interpretations of experimental findings.

It is also significant that the equilibrium of eqn. 1 for Va is strongly perturbed by changes in solvent polarity (including several solvents not shown in Table I). The twist form is *disfavored* as the dielectric constant is increased. The equilibrium for IIIb is only slightly affected and responds in the opposite direction to solvent polarity changes.

Acknowledgment: This work was supported by the National Cancer Institute of the Public Health Service. (Grant CA-11045.)

LITERATURE CITED

1. Gerlt, J.A.; Westheimer, F.H.; Sturtevant, J.H. *J. Biol. Chem.* 1975, *250*, 5059.
2. Gerlt, J.A.; Gutterson, N.I.; Data, P.; Belleau, B.; Penny, C.L. *J. Am. Chem. Soc.* 1980, *102*, 1655; Marsh, F.J.; Weiner, P.; Douglas, J.E.; Kollman, P.A.; Kenyon, G.L.; Gerlt, J.A. *ibid* 1980, *102*, 1660.
3. Bentrude, W.G.; Hargis, J.H. *Chem. Commun.* 1969, 1113.
4. Bentrude, W.G.; Tan, H.W. *J. Am. Chem. Soc.* 1973, *95*, 4666.
5. Bajwa, G.S.; Bentrude, W.G. *Tetrahedron Lett.* 1978, 421.
6. Bajwa, G.S.; Bentrude, W.G. *Tetrahedron Lett.* 1980, 4683.
7. Sopchik, A.E.; Bentrude, W.G. *Tetrahedron Lett.* 1980, 4679.
8. Newton, M.G.; Pantaleo, N.S.; Bajwa, G.S.; Bentrude, W.G. *Tetrahedron Lett.* 1977, 4457.
9. Sopchik, A.E.; Bentrude, W.G. *Tetrahedron Lett.* 1981, 307.
10. Data tabulated by Maryanoff, B.M.; Hutchins, R.O.; Maryanoff, C.A. *Top. in Phosphorus Chem.* 1979, *11*, 187.
11. Lee, C.H.; Sarma, R.H. *J. Am. Chem. Soc.* 1976, *98*, 3541; Blackburn, B.J.; Lapper, R.D.; Smith, I.C.P. *ibid.* 1973, *95*, 2873; Kainosho, M.; Ajisaka, K. *ibid.* 1975, *97*, 6839; Robins, M.J.; MacCoss, M.; Wilson, J.S. *ibid.* 1977, *99*, 4660.
12. Mosbo, J.A.; Verkade, J.G. *J. Am. Chem. Soc.* 1973, *95*, 4659; Majoral, J.P.; Bergounhou, C.; Navech, J. *Bull. Soc. Chim. Fr.* 1973, 3146.
13. Bentrude, W.G.; Yee, K.C. *J. Chem. Soc., Chem. Commun.* 1972, 169.
14. See ref. 7, p. 220.

RECEIVED July 7, 1981.

Phosphonate Inhibitors of Carboxypeptidase A

NEIL E. JACOBSEN and PAUL A. BARTLETT

Department of Chemistry, University of California, Berkeley, CA 94720

Stable compounds which resemble the transition-state structure of a substrate in an enzymatic reaction are expected to behave as potent reversible inhibitors (1). Based on the X-ray crystallographic structure of the active site of carboxypeptidase A (*CPA*) (2), a mechanism was proposed in which a water molecule adds directly to the scissile carbonyl group of the substrate to give the tetrahedral intermediate 1, which collapses to products (3). We proposed to mimic this tetrahedral intermediate, similar to the transition state, with the stable tetrahedral phosphonic acid derivatives 2, 3, 4, and 5.

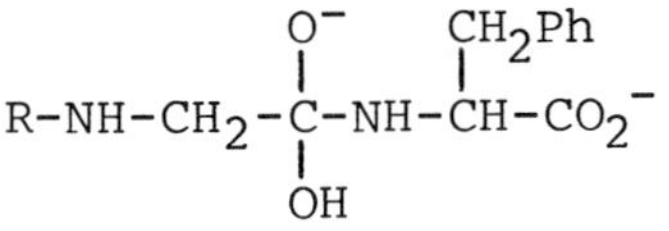

$$\text{R-NH-CH}_2\text{-C(O}^-\text{)(OH)-NH-CH(CH}_2\text{Ph)-CO}_2^-$$

1

$$\text{Cbz-NH-CH}_2\text{-P(=O)(X)-Y-CH(CH}_2\text{Ph)-CO}_2^-$$

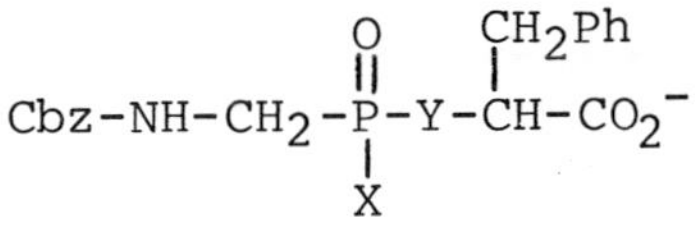

2: $X = O^-$, $Y = NH$ 4: $X = S^-$, $Y = NH$

3: $X = NH_2$, $Y = NH$ 5: $X = O^-$, $Y = O$

N-[[[(Benzyloxycarbonyl)amino]methyl]hydroxyphosphinyl]-L-phenylalanine, dilithium salt, 2 (4), was prepared in six steps from dimethyl phthalimidomethylphosphonate (5). The pure dilithium salt,

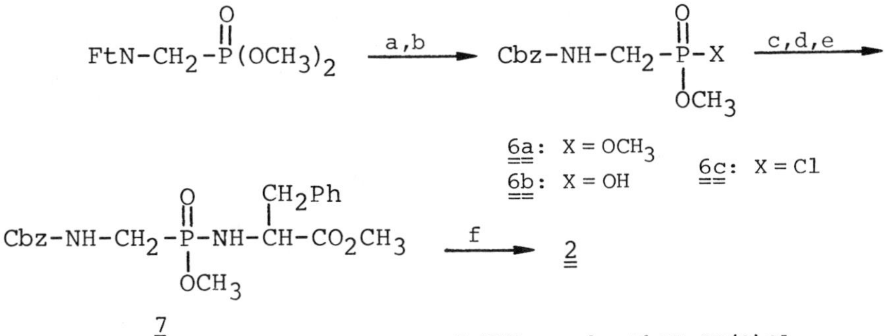

$$\text{FtN-CH}_2\text{-P(=O)(OCH}_3)_2 \xrightarrow{a,b} \text{Cbz-NH-CH}_2\text{-P(=O)(X)(OCH}_3) \xrightarrow{c,d,e}$$

6a: $X = OCH_3$

6b: $X = OH$

6c: $X = Cl$

$$\text{Cbz-NH-CH}_2\text{-P(=O)(OCH}_3\text{)-NH-CH(CH}_2\text{Ph)-CO}_2\text{CH}_3 \xrightarrow{f} \mathbf{2}$$

7

a: H_2NNH_2; b: $PhCH_2OC(O)Cl$;

c: 2N NaOH, H^+; d: $SOCl_2$; e: L-Phe•OCH_3, Et_3N; f: LiOH, H_2O, CH_3CN

0097–6156/81/0171–0221$05.00/0

obtained by anion-exchange chromatography, showed potent competitive inhibition of *CPA* with $K_i = 8.9 \times 10^{-8}$ M. Inhibition by 2 is rapid and reversible and added zinc chloride does not diminish the inhibitory activity. The hydrolysis of 2 is not catalyzed by *CPA*.

The preparation of N-[amino[[(benzyloxycarbonyl)amino]methyl]-phosphinyl]-L-phenylalanine, 3, was approached via the methyl ester 8a. The model diamides 8b, 8c, and 8d were prepared in the same manner. Treatment of 8a with NaOH gave only a small amount of 3, which decomposed during work-up and anion-exchange chromatography.

$R-CH_2-P(O)(OCH_3)-Cl \xrightarrow{a} R-CH_2-P(O)(OCH_3)-NHR' \xrightarrow{b} R-CH_2-P(O)(NH_2)-NHR'$

a: $R'NH_2$, Et_3N;
b: PCl_5, NH_3

8a: R = CbzNH, R' = $CH(CH_2Ph)CO_2CH_3$
8b: R = H, R' = CH_2Ph
8c: R = H, R' = $CH(CH_2Ph)CO_2CH_3$
8d: R = H, R' = $CH(CH_2Ph)CO_2CH_2Ph$

The model diamide 8b was unchanged under the same conditions, but 8c gave (by ^1H- and ^{31}P-NMR) a 1:1 mixture of the desired diamide 9 and the rearranged product 10. Under the reaction conditions, 9 slowly hydrolyzes to give the dianion 11. At lower pH in buffered solution, the decomposition of 9 to give 11 is accelerated, with $t_{1/2}$ less than 0.3 min at pH 7.5.

8c $\xrightarrow{a}$ 10: $CH_3-P(O)(O^-)-NH-CH(CH_2Ph)-CONH_2$ and 9

8c $\xrightarrow{a}$, 8d $\xrightarrow{b}$ 9: $CH_3-P(O)(NH_2)-NH-CH(CH_2Ph)-CO_2^-$

9 $\xrightarrow{c}$ 14: $CH_3-P(O)(OCH_3)-NH-CH(CH_2Ph)-CO_2^-$

9 $\xrightarrow{d}$ 11: $CH_3-P(O)(O^-)-NH-CH(CH_2Ph)-CO_2^-$

12: cyclic $CH_3-P(O)$–NH–$CH(CH_2Ph)$–C(=O)–HN– (ring)

a: OH^-, H_2O; b: H_2/Pd;
c: CH_3OH; d: H_2O

Thus it appears that the starting methyl ester 8c undergoes base-catalyzed cyclization to give the intermediate 12, which is opened by hydroxide attack at phosphorus to give 10. The desired diamide 9, formed either by direct hydrolysis of 8c or by opening of the intermediate 12 at the carbonyl carbon, loses ammonia with assistance from the free carboxylate to give 11.

When the reaction is carried out with catalytic methoxide in methanol, the presumed intermediate 12 is opened by methanol to give, quantitatively, the rearranged product 13. Hydrolysis of 13 with aqueous NaOH is extremely rapid and gives only the phenylalanine amide derivative 10.

$$\mathbf{8c} \xrightarrow[CH_3OH]{K_2CO_3} CH_3-\overset{O}{\overset{\|}{P}}(OCH_3)-NH-CH(CH_2Ph)-CONH_2 \xrightarrow[H_2O]{OH^-} \mathbf{10}$$

13

When the diamide 9 was formed directly by hydrogenolysis of the benzyl ester 8d, only the solvolysis products 11 or 14 were observed. Thus it appears that the free carboxylate of either 3 or 9, necessary for binding to *CPA*, participates in the solvolytic loss of ammonia and prevents the assessment of these compounds as inhibitors at neutral pH.

The model phosphonamidothioate 16, incorporating the essential features of the desired inhibitor 4, was prepared from methylphosphonothioic dichloride in two steps. The two diastereomers of the diester 15 were separated by high pressure liquid chromatography and individually deprotected with base to give pure 16a and 16b.

$$CH_3-\overset{S}{\overset{\|}{P}}-Cl_2 \xrightarrow{a,b} CH_3-\overset{S}{\overset{\|}{P}}(OCH_2CH_2CN)-NH-CH(CH_2Ph)-CO_2CH_3 \xrightarrow{c} CH_3-\overset{O}{\overset{\|}{P}}(S^-)-NH-CH(CH_2Ph)-CO_2^-$$

15a,b

16a: $K_i = 5.3\ \mu M$

16b: $K_i = 8.5\ \mu M$

a: $HOCH_2CH_2CN$, Et_3N ;

b: L-Phe•OCH_3, Et_3N ; c: NaOH

Initial enzymatic studies show that 16a and 16b bind to *CPA* somewhat less strongly than the corresponding phosphonamidate 11 (K_i of 11 = 1.2 μM).

The phosphonate 5, prepared from L-β-phenyllactic acid and the phosphonochloridate 6c, binds to *CPA* equally as tightly as does the more labile phosphonamidate 2.

We are currently pursuing the preparation and testing of the phosphonamidothioates 4 and the preparation of inhibitors analogous to 2-5 for the zinc peptidases thermolysin and collagenase.

$$\underset{\displaystyle HO-CH-CO_2H}{\overset{CH_2Ph}{|}} \xrightarrow{a,b} Cbz-NH-CH_2-\overset{O}{\overset{||}{\underset{\underset{OCH_3}{|}}{P}}}-O-\overset{CH_2Ph}{\overset{|}{CH}}-CO_2CH_3 \xrightarrow{c,d} \underline{\underline{5}}$$

a: CH_3I, K_2CO_3, 18-crown-6; b: 6c, Et_3N; c: t-$BuNH_2$, Δ (see ref. 6); d: NaOH, H_2O

LITERATURE CITED

1. Wolfenden, R. Ann. Rev. Biophys. Bioeng., 1976, 5, 271.
2. Lipscomb, W.N. Tetrahedron 1974, 30, 1725.
3. Breslow, R.; Wernick, D.L. Proc. Natl. Acad. Sci. U.S. 1977, 74, 1303.
4. Jacobsen, N.E.; Bartlett, P.A. J. Am. Chem. Soc. 1981, 103, 654.
5. Seyferth, D.; Marmor, R.; Hilberg, P. J. Org. Chem. 1971, 36, 1379.
6. Gray, M.; Smith, D. Tetrahedron Letters 1980, 859.

RECEIVED June 30, 1981.

46

"Illicit Transport" Systems for Organophosphorus Antimetabolites

MAQSOOD SHEIKH, BARRY GOTLINSKY, BURTON E. TROPP, and ROBERT ENGEL

The City University of New York, Queens College, Kissena Boulevard, Flushing, NY 11367

THOMAS PARKER

The Rockefeller University, 1230 York Avenue, New York, NY 10021

There have been synthesized in this laboratory over the past several years a variety of phosphonic acids which are nominally isosteric with natural monophosphate esters. Of this collection of compounds three are of particular interest for the present work. These are: (*S*)-3,4-dihydroxybutyl-1-phosphonic acid (A)(1), an isosteric analogue of *sn*-glycerol-3-phosphate; 4-hydroxy-3-oxobutyl-1-phosphonic acid (B)(2), an isosteric analogue of dihydroxyacetone phosphate; and 5-carboxy-4-hydroxy-4-methylpentyl-1-phosphonic acid (C)(3), an isosteric analogue of 5-phosphomevalonate.

A B C

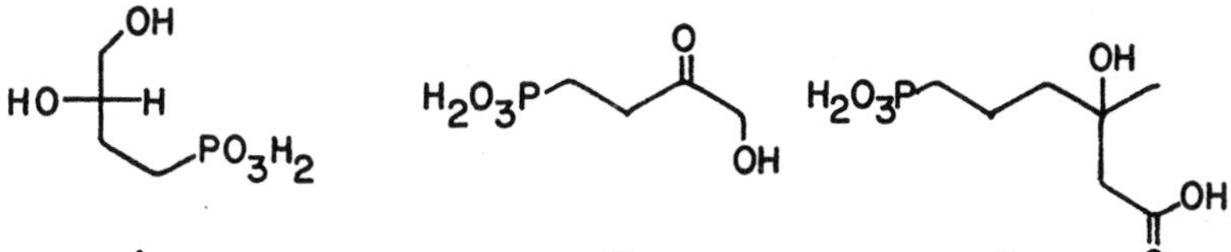

All of these have been found to exhibit *in vitro* enzymatic activity as inhibitors of normal biological processes. For example, (A) has been found to serve as a substitute for *sn*-glycerol-3-phosphate in reaction involving CDP-diglyceride: *sn*-glycerol-3-phosphate phosphatidyltransferase (4,5), (B) serves as substrate for L-glycerol-3-phosphate:NAD oxidoreductase (6), and (C) serves as a specific inhibitor of 5-phosphomevalonate kinase from rat liver (7). *In vivo* activity however has been found *only* for (A), as a growth inhibitor (1,8), when used with mutant strains of *Escherichia coli* or other bacteria which are capable of transporting the natural *sn*-glycerol-3-phosphate. If the mutant strain is deficient in the appropriate transport systems, no activity can be observed. For (B) *in vivo* bacterial activity *is not* observed, presumably as it is not transported; no transport route exists for the normal phosphate. Likewise, (C) is not active in whole liver cells, again presumably due to a lack of transport for it (and the natural phosphate).

0097–6156/81/0171–0225$05.00/0

The present work is directed toward the design and synthesis of specific chemical vectors which will facilitate transport of these potential chemotherapeutic agents into the appropriate cells or to particular sites in a complex organism. For this, two potential transport-facilitating "substituent" classes were considered.

The first of these involves peptides. The transport of tripeptide systems (and other peptides) into bacteria, into yeast, and into mammalian gut has been well documented during the past decade (9). For bacteria, an interesting characteristic of this transport is that aside from a requirement for the two peptide linkages and free termini of the tripeptide, there appears to be little other structural requirement for transport. That is, bulky and charged functions may be attached along the peptide chain at the "alpha-positions" of individual amino acid components, and the materials remain capable of being transported into the cells. Thus there would appear to be a significant potential for the use of such a tripeptide system for attachment to a potential metabolic regulator for "illicit transport" of it into cells bearing this transport characteristic.

We have for the present effort synthesized several conjugated regulator-peptide systems as illustrated (D)-(G). The fundamental tripeptide is constructed by standard methods bearing protective functions at the normal amino and carboxy termini, but bearing a free reactive site along the chain for coupling to the regulator. For the compounds (A) and (B), of interest in bacterial systems, such a coupling could easily be attained through a carboxyl function; thus an aspartate residue is

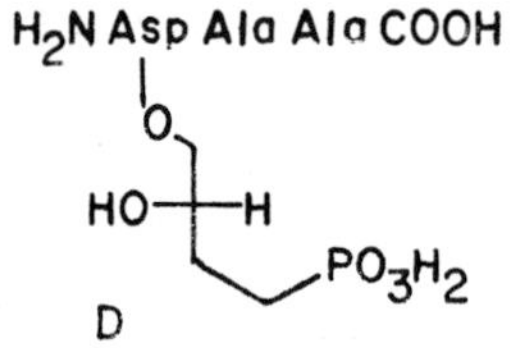

H_2N Asp Ala Ala COOH — O — HO–C–H — PO_3H_2

D

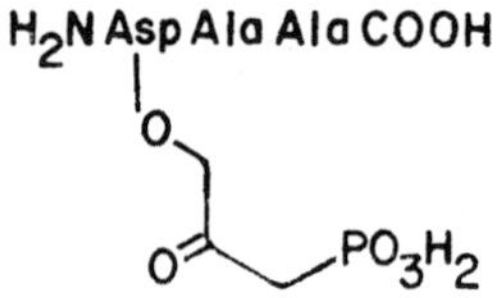

H_2N Asp Ala Ala COOH — O — O= — PO_3H_2

E

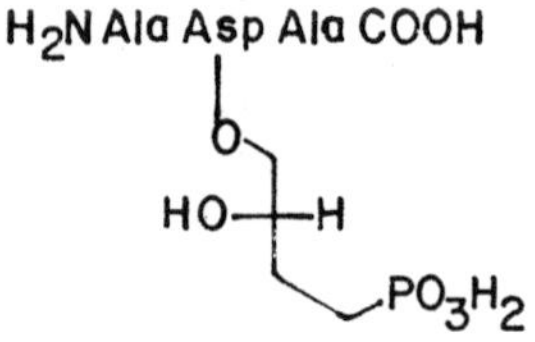

H_2N Ala Asp Ala COOH — O — HO–C–H — PO_3H_2

F

H_2N Ala Asp Ala COOH — O — O= — PO_3H_2

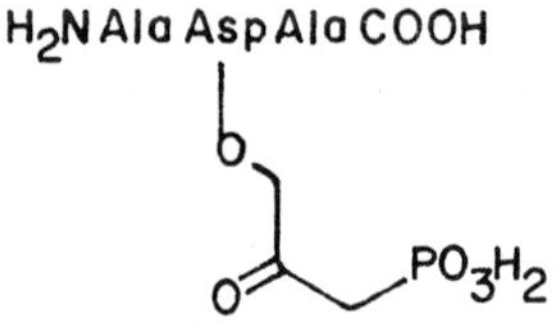

G

incorporated for utilization of its β-carboxylic acid site. Once the fundamental tripeptide is constructed, the potential regulator (as the phosphonic diester) is coupled through a free hydroxyl group, followed by complete deprotection. The peptide termini protecting groups are removed by standard methods and the phosphonic diester is cleaved selectively by treatment with trimethylsilyl iodide followed by water.

Two of these materials (E) and (F), have been investigated for inhibitory activity toward mutant strains of *Escherichia coli*. In treatment of *E. coli* strain 8, which transports glycerol-3-phosphate and (A), but *not* (B), with (E) at 1.5 mM concentration there is observed inhibition of growth approximating that produced by (A). Control experiments demonstrate this *not* to be the result of any species except the conjugated regulator-peptide. Likewise, when (F) was used at 0.8 mM concentration with *E. coli* strain 8 there was produced an inhibition approaching that produced by (A). More importantly, when *E. coli* strain M56, a glp T^- mutant, deficient in both glycerol-3-phosphate and hexose-6-phosphate transport systems and thereby not transporting (A), was treated with 0.8 mM concentration of (F) there resulted severe growth inhibition.

This inhibition of growth is presumably the result of transport of the conjugated species and the ultimate action of either the regulatory agent (A) or (B), liberated intracellularly by the action of a non-specific esterase. It is also possible that the conjugated species itself could produce the inhibition; the testing of this requires the use of (E) or (F) bearing a label in the regulator portion.

In regard to the regulation of cholesterol biosynthesis in liver, there has been observed (10) a transport system for taurocholic acid with isolated rat liver cells. There is also noted (11) evidence that this transport system is capable of serving for other bile acids. It was thereby postulated that, as the coupling of (C) to a reduced form of cholic acid using the carboxyl function of (C) would yield a species (H) of general structural similarity to taurocholic acid, the species might be capable of transport into intact liver cells, thereby delivering the regulator species.

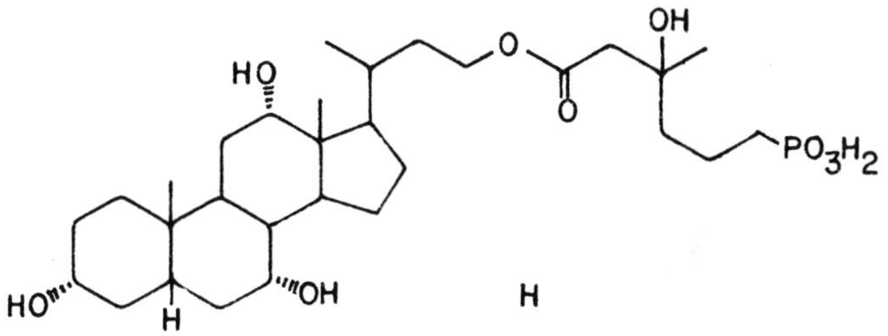

H

This synthesis has been accomplished by the lithium aluminum hydride reduction of cholic acid and the coupling of the resultant primary hydroxyl group to (C) by a standard method. Again, the coupling of (C) is performed using the phosphonic diester, and the final stage of the synthesis involves the trimethylsilyl iodide mediated cleavage of the ester functions, proceeding without damaging other structural features of the compound.

The material (H) has been tested with preparations of rat liver hepatocytes and has been found to enter them and to serve as an inhibitor of cholesterol biosynthesis therein. Using (H) bearing a specific radiocarbon label, studies have been performed in intact animals. The material has been injected IV and found to be cleared from the blood rapidly, the major portion being delivered to the liver. Small portions are found in bile and intestine, but not in other organs of the animals.

Literature Cited

1. Tang, K.-C.; Tropp, B.E.; Engel, R. Tetrahedron 1978, 34, 2873.
2. Goldstein, S.L.; Braksmayer, D.; Tropp, B.E.; Engel, R. J. Med. Chem. 1974, 17, 363.
3. Sarin, V.; Tropp, B.E.; Engel, R. Tetrahedron Lett. 1977, 351.
4. Cheng, P.-J.; Nunn, W.D.; Tyhach, R.J.; Goldstein, S.L.; Engel, R.; Tropp, B.E. J. Biol. Chem. 1975, 250, 1633.
5. Tyhach, R.J.; Engel, R.; Tropp, B.E. J. Biol. Chem. 1976, 251, 6717.
6. Cheng, P.-J.; Hickey, R.; Engel, R.; Tropp, B.E. Biochim. Biophys. Acta 1974, 341, 85.
7. Popjak, G.; Parker, T.S.; Sarin, V.; Tropp, B.E.; Engel, R. J. Am. Chem. Soc. 1978, 100, 8014.
8. Shopsis, C.S.; Engel, R.; Tropp, B.E. J. Bacteriol 1972, 112, 408.
9. Payne, J.W. Advances in Microbial Physiol. 1976, 13, 55.
10. Schwarz, L.R.; Burr, R.; Schwenk, M.; Pfaff, E.; Grein, H. Eur. J. Biochem. 1975, 55, 617.
11. Anwer, M.S.; Kroker, R.; Hegner, D. Biochem. Biophys. Res. Commun. 1976, 73, 63.

RECEIVED July 7, 1981.

ORGANIC SYNTHETIC METHODS BASED ON REAGENTS CONTAINING PHOSPHORUS

The Preparation of Phosphorus Esters and Thioesters from White Phosphorus

CHARLES BROWN, ROBERT F. HUDSON, and GARY A. WARTEW
Chemical Laboratory, University of Kent at Canterbury, Kent, CT2 7NH, England

HAROLD COATES
Albright & Wilson Limited, P.O. Box 3, Oldbury, Worcestershire, England

Many attempts have been made to synthesise organophosphorus compounds directly from the element with varying degrees of success.[1,2] In general mixtures of products are obtained, and the yields are frequently low and variable. The main difficulty is the insolubility of phosphorus in organic liquids and consequently the reactions are heterogeneous with the attendant problems of diffusion and local concentration differences.

Our approach was based on the following considerations.[3] 1) White phosphorus, a strained tricyclic system has a low nucleophilic reactivity but is highly electrophilic. 2) Attack by a nucleophile produces a reactive phosphide ion which is rapidly protonated in hydroxylic media. For high yields of a required product it is essential to trap this ion with a suitable electrophile. 3) Owing to the mutual reaction of nucleophile and electrophile the number of useful combinations is limited. In principle a combination of hard nucleophile and soft electrophile is preferred.

Phosphines and phosphides react rapidly with positive halogen and hence polyhalogen compounds, e.g. tetrachloromethane, are particularly suitable and have the advantage that the chlorophosphine formed is rapidly attacked by the nucleophile. In principle therefore the phosphorus atoms can be fully substituted by reactions of the following type,

$$RO^- + P_4 \longrightarrow RO\text{–}P_4^{(-)} \xrightarrow[ROH]{CCl_4} RO\text{–}P_4\text{–}Cl + CHCl_3 \xrightarrow{RO^-} RO\text{–}P_4\text{–}OR + Cl^-$$

0097–6156/81/0171–0231$05.00/0

The reactions were carried out in an excess of alcohol and tetrachloromethane, and the concentration of alkoxide varied, the white phosphorus being added, under nitrogen, in the form of a fine sand.[3] The progress of the reaction was followed by GLC after the products of reaction had been identified by ^{31}P NMR. In initial experiments, stoicheiometric quantities of phosphorus and alkoxide were used, according to the equation,

$$P_4 + 6NaOR + 6ROH + 6CCl_4 \longrightarrow 4P(OR)_3 + 6CHCl_3 + 6NaCl$$

Although quantitative yields of chloroform were found, low yields of trialkyl phosphite were produced and these decreased with time as the yields of trialkyl phosphate and dialkyl phosphate increased, e.g. in the reaction of sodium n-butoxide (0.06 mol) with phosphorus (0.009 mol) in 30 ml of methanol and 50 ml of n-butanol 47% of phosphite was produced in 3 h. together with 19% of phosphate and 7% of phosphonate. The product composition changed to 34%, 22% abd 21% respectively after 7 h.

The former is no doubt produced by the well known reaction[4]

$$(RO)_3P + CCl_4 \longrightarrow (RO)_3\overset{+}{P}\text{-}Cl \xrightarrow{ROH} (RO)_4\overset{+}{P}Cl^- \begin{cases} \longrightarrow (RO)_3P{=}O + RCl \\ \xrightarrow{RO^-} (RO)_3P{=}O + R_2O \end{cases}$$

De-alkylation by alkoxide ion increases the acidity of the medium and this accounts for the formation of dialkyl phosphonate in the later stages of the reaction.

For these reasons, higher concentrations of alkoxide were used, firstly to increase the initial rate and secondly to pre-erve alkalinity throughout the reaction. With two equivalents of alkoxide, high yields of trialkyl phosphite were obtained within 1-2 h. at 25°. Again, the yield decreased with time owing to the subsequent oxidation, e.g. 82% trimethylphosphite and 76% triethylphosphite after 1 h.

When the reaction was carried out in the probe of a ^{31}P NMR spectrometer, no evidence for the accumulation of intermediates was obtained (vide infra).

These results establish the conditions for the formation of trialkyl phosphites in high yields. However attempts to distil the ester from the reaction mixture were unsuccessful as co-distillation and some decomposition always occurred. Other workers who carried out similar experiments independently,[5] report a 50% yield of isolated trimethyl phosphite.

Attention was turned to the analogous reaction of thiols. Here the subsequent oxidation can be neglected, and the products were readily separated by distillation. Again, low yields of triester were obtained when equivalent quantities of thioalkoxide and phosphorus were used.[6] Evidence of incomplete conversion was obtained from the ^{31}P NMR spectra of the reaction mixtures using ethane thiol. Three absorptions in the ratio 2:1:1 were observed corresponding to triethyl phosphorotrithioite (δp 114.1), diethyl phosphorochloridodithioite (δp 184.5) and diethyl trichloromethyl phosphonodithioite (δp 121.6).

The two intermediates probably arise in the later stages of the reaction when the thioalkoxide concentration is low. The presence of these intermediates is evidence of the intermediate formation of the trichloromethyl anion, in reactions of the following kind.

$$(RS)_2P\text{-}P(SR)_2 \xrightarrow[CCl_4]{RS^-} (RS)_3P + (RS)_2PCl + CCl_3^- \longrightarrow (RS)_3P + (RS)_2P.CCl_3 + Cl^-$$

These reactions occur because the thioalkoxide is depleted by reaction with tetrachloromethane in the following side reactions,[7]

$$4NaSR + CCl_4 + RSH \longrightarrow CH(SR)_3 + RSSR + 4NaCl$$

The chloridodithioite does not react with neutral thiol, whereas the corresponding dialkyl phosphorochloridite reacts rapidly with ethanol. Consequently the latter reaction proceeds to completion even when the alkoxide has been neutralised.

With two equivalents of thioalkoxide, the triester only is produced and this can be distilled from the reaction mixture in high yield (e.g. 97% from ethanethiol and 82% from butanethiol).

The following scheme is suggested for the breakdown of the P_4 molecule and the formation of triester (X = O, S).

$$P_4 \longrightarrow \underset{\text{I}}{(RX)P\text{-}P_2\text{-}P(XR)\ \text{(bicyclic)}} \longrightarrow \underset{\text{II}}{cyclo\text{-}[P(XR)]_4}$$

$$\downarrow$$

$$4(RX)_3P \longleftarrow \underset{\text{IV}}{2(RX)_2P\text{—}P(XR)_2} \longleftarrow \underset{\text{III}}{(RX)_2P\text{—}P(XR)\text{—}P(XR)\text{—}P(XR)_2}$$

The absence of detectable reaction intermediates suggests that the initial heterogeneous reaction is rate determining. The bicyclic intermediate, I, gives the cyclotetraphosphite II with release of ring strain. Cyclic molecules of this kind have been isolated from reactions of elemental phosphorus, e.g. tetraalkyl cyclotetraphosphines from the combined action of Grignard reagents and n-alkyl bromides.[8]

The subsequent stages involving tetraphosphine, triphosphine and diphosphine derivatives proceed rapidly owing to the high reactivity of the P-P bond.

These reactions appear to be restricted to strongly basic nucleophiles, as we found no reaction with phenoxides and thiophenoxides. This lack of reactivity is attributed to the reversibility of the nucleophilic substitution, promoted by the increased leaving group ability of the nucleophile, e.g.

$$P_4 + PhO^- \rightleftharpoons PhO\text{-}P_4^{(-)}$$

LITERATURE CITED

1. Rahut, M.M. Topics in Phosphorus Chemistry Vol. 1, p.1, Interscience, N.Y., 1964
2. Maier, L. Fortsch. Chem. Forsch. 1971, 19, 1
3. Brown, C.; Hudson R.F.; Wartew, G.A.; Coates, H. Phosphorus and Sulphur 1979, 6, 481
4. Burn, A.J.; Cadogan, J.I.G. J. Chem. Soc. 1963, 5788
5. Lehmann, H.A.; Schadow, H.; Richter, H.; Kurze, R.; Oertel, M. Ger. Patent, 127,188, 1977
6. Brown, C.; Hudson, R.F.; Wartew, G.A. J.Chem. Soc. Perk. I., 1979, 1979
7, Backer, J.H.; Stedehonder, P.L. Rec. Trav. Chim. 1933, 52, 437
8. Cowley, A.H.; Punnell, R.P. Inorg. Chem. 1966, 5, 1463

RECEIVED June 30, 1981.

48

Thermal Rearrangement and Condensation of *O,O*-Dimethyl-*O*-phenylphosphorothionate

HERBERT TEICHMANN

Zentralinstitut für Organische Chemie der Akademie der Wissenschaften der DDR, DDR-1199 Berlin-Adlershof, GDR

GERHARD SCHRAMM

Orthopädische Klinik der Medizinischen Akademie, Erfurt, DDR-50 Erfurt, GDR

It is now more than three decades that dialkylarylphosphorothionates such as parathion and parathion-methyl took a dominant role among organophosphorus pesticides. Owing to occasional incidents in production and handling, various efforts have been made in the past to elucidate the thermal behaviour of such compounds (cf. (1, 2, 3) and references cited therein). However, our knowledge of the reactions involved is still far from being satisfactory. With the aim of a better understanding of the thermal rearrangement and decomposition processes in this class of commercially highly important substances, the thermolysis of $(MeO)_2(PhO)PS$ (I) as a model compound has been reexamined.

In the temperature range of 125-140°C two periods clearly can be distinguished. During the first one which consumes about 70% of the total reaction time, only 30% conversion of the thionate I takes place. So far the decrease of I follows an autocatalytic law and corresponds quite well to the increase of the isomeric S-methyl thiolate II. In the shorter and exothermic second period isomerization of I soon comes to completion. Here, however, the thiolo isomer II also rapidly decomposes after reaching a maximum concentration of about 65%. The final product, free from I and II, constitutes a water soluble, hygroscopic substance of unchanged elemental composition, exhibiting strong acidic reaction and containing $Me_3S^{\oplus}$ ions (about half of the total sulfur) and condensed phosphates. An identical product is obtained from pure II after a much shorter heating time.

The thiono-thiolo rearrangement I→II is catalyzed by II as well as by decomposition products of II: Me_2S induces an O-dealkylation/S-realkylation sequence (4), and $Me_3S^{\oplus}$ acts as an alkylating agent superior

0097–6156/81/0171–0235$05.00/0

to I or II; S-alkylation of I-analogues with $R_3O^{\oplus}$ instead of $R_3S^{\oplus}$ salts to form crystalline trialkoxy-alkylthiophosphonium salts and dealkylation of them by Me_2S to yield II-analogues has been demonstrated before (5, 6). Relative reaction times until total loss of I at 135°C are (hours, approximately): without catalyst 7; in the presence of 1 equivalent of II or of 0,06 equivalents of $Me_3S^{\oplus}$ $SbCl_6^{\ominus}$ 2; in the presence of 0,06 equivalents of Me_2S 1; removing Me_2S by passing through a stream of N_2 18. Further evidence for catalysis by II arose from crossing experiments (7).

Soon after total isomerization of I the decomposition of its isomer II also is completed. The conversion of II into the final mixture of ionic substances consists of two basic steps: dealkylation by Me_2S to form the S-methyl-O-phenylphosphorothiolate anion III (eq. 3), and nucleophilic attack at phosphorus by the anion III to give a S-methyl diphosphate which is O-dealkylated to IV by the leaving group $MeS^{\ominus}$ (eq. 1).

$$\underset{\text{III}}{{}^{\ominus}O\text{-}\overset{\overset{O}{\|}}{\underset{\underset{OPh}{|}}{P}}\text{-}SMe} \xrightarrow{+\ II} MeO\text{-}\overset{\overset{O}{\|}}{\underset{\underset{OPh}{|}}{P}}\text{-}O\text{-}\overset{\overset{O}{\|}}{\underset{\underset{OPh}{|}}{P}}\text{-}SMe \xrightarrow{+\ MeS^{\ominus}} \underset{\text{IV}}{{}^{\ominus}O\text{-}\overset{\overset{O}{\|}}{\underset{\underset{OPh}{|}}{P}}\text{-}O\text{-}\overset{\overset{O}{\|}}{\underset{\underset{OPh}{|}}{P}}\text{-}SMe} + Me_2S \quad (1)$$

Preparative application of this condensation principle has been shown (8) to yield, e.g., 83% diphenyldiphosphate from II and (MeO)(PhO)POOK after 5 hours reflux in Bu_2O. Analogous condensation of II with IV or even higher condensed phosphorothiolate anions would enable a stepwise formation of longer chains.

The Me_2S produced in every single condensation step must consume an equivalent amount of O-methylester functions because no sulfur is lost. Therefore, after isomerization of I, one half of II is required for sulfonium salt formation, and a reasonable stoichiometry of the II-conversion results from the sum of equations (2) and (3). Since by reaction (3) a large supply of anion III is offered, III will be the dominant nucleophile for condensation with II, and one may expect a large number of chains, i.e., a low average condensation degree b.

TLC on DEAE cellulose with 0.2 m HCl permits the ionic species to be separated and thus the course of the thermolysis to be followed. Besides an initial tiny uncertain spot (obviously $(MeO)(PhO)PO_2^{\ominus}$ according to R_F value), four different compounds emerge successively: III, after it IV, and finally the sulfur-

$$n\ MeO\text{-}\overset{O}{\overset{\|}{P}}(OPh)\text{-}SMe + a\ {}^{\ominus}O\text{-}\overset{O}{\overset{\|}{P}}(OPh)\text{-}SMe \rightarrow a\ {}^{\ominus}\left[O\text{-}\overset{O}{\overset{\|}{P}}(OPh)\right]_b O\text{-}\overset{O}{\overset{\|}{P}}(OPh)\text{-}SMe + n\ Me_2S \quad (2)$$

II   III

$$n\ MeO\text{-}\overset{O}{\overset{\|}{P}}(OPh)\text{-}SMe + n\ Me_2S \longrightarrow n\ {}^{\ominus}O\text{-}\overset{O}{\overset{\|}{P}}(OPh)\text{-}SMe + n\ Me_3S^{\oplus} \quad (3)$$

II   III

$$2n\ MeO\text{-}\overset{O}{\overset{\|}{P}}(OPh)\text{-}SMe \rightarrow a\ {}^{\ominus}\left[O\text{-}\overset{O}{\overset{\|}{P}}(OPh)\right]_b O\text{-}\overset{O}{\overset{\|}{P}}(OPh)\text{-}SMe + c\ {}^{\ominus}O\text{-}\overset{O}{\overset{\|}{P}}(OPh)\text{-}SMe + n\ Me_3S^{\oplus} \quad (4)$$

II   III

$$a \cdot b = a + c = n$$

free diphosphate V and triphosphate VI. By ^{31}P-NMR spectroscopy exactly the same components are identified. Phosphorus determinations in isolated spots of the final product are in line with values obtained from integration of NMR signals (see table).

Table: Thermolysis of I at 135°C, composition of final product

	% of total P from TLC	% of total P from P-NMR	$\delta\,^{31}P$ (ppm) 1)
III	15	14	16,29
IV	62	65	P^A 12,72(d);P^B-18,05(d) 2)
V	11	9	-18,26
VI	12	13	P^A-18,20(d);P^B-29,14(t) 3)

1) in CH_2Cl_2 2) P^A bond to SMe; J_{PP} 26,4 Hz
3) P^A terminal P; J_{PP} 17,7 Hz

Hydrolytic stability of the condensed species decreases in the series V > VI > IV; the latter could not be isolated but was identified unambiguously by its hydrolysis products (eq. 5, b = 1) in two dimen-

$${}^{\ominus}\left[O\text{-}\overset{O}{\overset{\|}{P}}(OPh)\right]_b O\text{-}\overset{O}{\overset{\|}{P}}(OPh)\text{-}SMe + H_2O \rightarrow {}^{\ominus}O\left[\overset{O}{\overset{\|}{P}}(OPh)\text{-}O\right]_b^{\ominus} + {}^{\ominus}O\text{-}\overset{O}{\overset{\|}{P}}(OPh)\text{-}SMe + 2\ H^{\oplus} \quad (5)$$

b = 2: V   III
b = 3: VI

sional TLC as well as in ^{31}P-NMR spectra. Higher condensed thiolo oligophosphates must be regarded to be extremely sensitive to moisture and to act as precursors of V and VI.

If the thermolysis temperature is allowed to rise to 180°C and above, evolution of Me_2S takes place with sulfur loss up to 60%. Simultaneously the content of V and VI and of other unidentified condensation products increases markedly at the expense of IV. Loss of $Me_3S^{\oplus}$ approximately parallels that of thiolo ester groups, however; no MeO groups are detectable in the H-NMR spectra. From these results one must conclude that the uncharged thiolo sulfur will prevail over the anionic oxygen functions as nucleophile toward the $Me_3S^{\oplus}$ ion (eq. 6); the resulting

$$\ominus\left[\text{O-}\overset{\overset{\text{O}}{\|}}{\underset{\underset{\text{OPh}}{|}}{\text{P}}}\text{-}\right]_b\text{O-}\overset{\overset{\text{O}}{\|}}{\underset{\underset{\text{OPh}}{|}}{\text{P}}}\text{-SMe} + Me_3S^{\oplus} \not\rightarrow \text{Me-}\left[\text{O-}\overset{\overset{\text{O}}{\|}}{\underset{\underset{\text{OPh}}{|}}{\text{P}}}\text{-}\right]_b\text{O-}\overset{\overset{\text{O}}{\|}}{\underset{\underset{\text{OPh}}{|}}{\text{P}}}\text{-SMe} + Me_2S \quad (6)$$

$$\downarrow$$

$$\ominus\left[\text{O-}\overset{\overset{\text{O}}{\|}}{\underset{\underset{\text{OPh}}{|}}{\text{P}}}\text{-}\right]_b\text{O-}\overset{\overset{\text{O}}{\|}}{\underset{\underset{\text{OPh}}{|}}{\text{P}}}\text{-}\overset{\oplus}{\text{S}}Me_2 + Me_2S \longrightarrow \left[\text{O-}\overset{\overset{\text{O}}{\|}}{\underset{\underset{\text{OPh}}{|}}{\text{P}}}\text{-}\right]_{b+1} + 2\ Me_2S$$

monomeric (b = 0) or oligomeric metaphosphate species open an alternative route for building up condensed phosphates even after total consumption of II.

Literature cited

1. Hilgetag, G.; Schramm, G.; Teichmann, H. J. Prakt. Chem. 1959, 8, 73.
2. Teichmann, H.; Lehmann, G. Sitzungsber. DAW Berlin Kl. Chem., Geol., Biol. 1962, No. 5.
3. Engel, R. R.; Liotta, D. J. Chem. Soc. C 1970, 523.
4. Hilgetag, G.; Teichmann,H. Angew. Chem. internat. Edit. 1965, 4, 914.
5. Teichmann, H.; Hilgetag, G. Chem. Ber. 1963, 96, 1454.
6. Schulze, J. Thesis, Humboldt University Berlin, 1971.
7. Teichmann, H. Angew. Chem. internat. Edit. 1965, 4, 993.
8. Hilgetag, G.; Krüger, M.; Teichmann, H. Z. Chem. 1965, 5, 180.

RECEIVED July 7, 1981.

49

Synthesis and Reactivity of (Silylamino)phosphines

ROBERT H. NEILSON, PATTY WISIAN-NEILSON, DAVID W. MORTON, and H. RANDY O'NEAL

Department of Chemistry, Texas Christian University, Fort Worth, TX 76129

Compounds containing the Si-N-P linkage combine the structural and stereochemical diversity of phosphorus with the reactivity of the silicon-nitrogen bond. Indeed, much of the derivative chemistry and synthetic potential of these compounds, especially the (silylamino)phosphines such as $(Me_3Si)_2NPMe_2$, is based on this difunctional character. We report here a general, "one-pot" synthesis of (silylamino)phosphines and describe their use in the preparation of several types of phosphorus-containing materials.

Synthesis of (Silylamino)phosphines. In a typical preparation, addition of one molar equivalent of PCl_3 to a stirred solution of $LiN(SiMe_3)_2$ in ether at $-78^\circ C$ followed by one or two equivalents of an alkyl Grignard or lithium reagent at $0^\circ C$ gives the corresponding mono- or dialkylphosphine. This proce-

$$(Me_3Si)_2NH \xrightarrow[\text{(2) } PCl_3]{\text{(1) } \underline{n}\text{-BuLi}} (Me_3Si)_2NPCl_2$$

$$(Me_3Si)_2NPCl_2 \xrightarrow[\text{or RLi}]{\text{RMgX}} (Me_3Si)_2N{-}P(R)Cl$$

$R = \underline{i}$-Pr, $\underline{t}$-Bu, CH_2SiMe_3

$$(Me_3Si)_2NPCl_2 \xrightarrow{\text{2 RMgX}} (Me_3Si)_2NPR_2$$

$R = Me, Et, CH_2SiMe_3$

0097–6156/81/0171–0239$05.00/0

dure routinely affords high yields (ca 75%) and large quantities (ca 100-200 g) of the phosphine product. Moreover, the method has been generalized to include the use of other silylamines, $PhPCl_2$ instead of PCl_3, and different organometallic reagents.

Synthesis of Polyphosphazenes. We are investigating a new, direct synthesis of phosphazene polymers which involves the thermally-induced elimination of silanes from suitably constructed N-silylphosphinimines. As a route to linear polyphosphazenes, this method

$$Me_3SiN{=}\overset{X}{\underset{R'}{P}}{-}R \longrightarrow Me_3SiX + \frac{1}{n}\left(N{=}\overset{R}{\underset{R'}{P}}\right)_n$$

offers the distinct advantage of incorporating the desired phosphorus substituents directly into an easily-prepared starting material, thereby eliminating the need for preparing the dihalo polymers $(X_2PN)_n$.

Many of the "suitably constructed" phosphinimines (1) are prepared via the bromination of (silylamino)-phosphines. The P-bromo compounds are easily converted to the corresponding dialkylamino or alkoxy derivatives.

$$(Me_3Si)_2NPRR' + Br_2 \xrightarrow{0^oC} Me_3SiBr + Me_3SiN{=}\overset{Br}{\underset{R'}{P}}{-}R$$

$R,R' = Me, Et, Ph, OCH_2CF_3$

We find that the nature of the leaving group (X) is important in determining both the relative stability of the starting material as well as the degree of oligomerization of the phosphazene products. For example, while the $P-NMe_2$ and P-OMe phosphinimines are stable to at least 250°C, the P-Br analogues decompose at lower temperatures (100-150°C) to give Me_3SiBr and cyclic phosphazenes $(R_2PN)_n$ where n = 3,4.

Most significantly, however, polymeric phosphazenes (2) are obtained almost exclusively when the leaving group is OCH_2CF_3. The products are the first fully characterized polyphosphazenes containing only direct P-C bonded substituents. These materials are soluble film-forming or elastomeric polymers with molecular weights in the 50-70,000 range. The

$$Me_3SiN{=}P(OCH_2CF_3)(R)(R') \xrightarrow[\text{2-5 days}]{180\text{-}190^{\circ}C} Me_3SiOCH_2CF_3 + \frac{1}{n}{+}N{=}P(R)(R'){+}_n$$

$R,R' = Me, Et, Ph$

complete characterization of these and related materials is under active investigation.

Nucleophilic Reactions of (Silylamino)phosphines. The reactions of (silylamino)phosphines with simple aldehydes and ketones proceed via nucleophilic attack by phosphorus followed by a [1,4] silyl migration from nitrogen to oxygen to yield new N-silylphosphinimines (3). With α,β-unsaturated carbonyl compounds, 1,4-addi-

$$(Me_3Si)_2NPMe_2 + R{-}C({=}O){-}R' \xrightarrow[0^{\circ}C]{CH_2Cl_2} Me_3SiN{=}PMe_2{-}C(R)(R'){-}OSiMe_3$$

tion occurs producing silyl enol ethers which, upon hydrolysis, yield γ-carbonyl phosphine oxides.

$$(Me_3Si)_2NPMe_2 + CH_2{=}CH{-}C({=}O){-}R \xrightarrow{0^{\circ}C}$$

$$Me_3SiN{=}PMe_2{-}CH_2CH{=}C(R)(OSiMe_3) \xrightarrow{H_2O} Me_2P({=}O)CH_2CH_2C({=}O)R$$

(Silylamino)phosphines also react with various halides including ethyl bromoacetate, allyl bromide, and chloroformates. The initially-formed phosphonium salts eliminate silyl halides to afford functionalized phosphinimines $Me_3SiN{=}P(R)Me_2$ where R = $-CH_2C(O)OEt$, $-CH_2CH{=}CH_2$, and $-C(O)OR'$.

Two-coordinate (Silylamino)phosphines. Certain chloro(silylamino)phosphines bearing a trimethylsilylmethyl group can be dehydrohalogenated to yield stable (methylene)phosphines. The very low field (δ 309.9) ^{31}P shift confirms the two-coordinate nature of this compound while the low field (δ 7.09) ^{1}H signal shows it to be the P=CH rather than the P=N isomer. Three modes of reactivity of this (methylene)phosphine have been observed: addition to the P=C bond, oxidative addition at phosphorus, and complexation with transition metal centers.

$$(Me_3Si)_2N\text{-}P(Cl)\text{-}CH_2SiMe_3 \xrightarrow[-HCl]{base} (Me_3Si)_2N\text{-}P{=}C(H)(SiMe_3)$$

(Silylamino)phosphines with P-H Bonds. Treatment of sterically crowded chloro(silylamino)phosphines with *i*-PrMgCl yields, unexpectedly, the P-H derivatives in a process which must involve hydride donation by the Grignard reagent. Typically, the P-H phosphines are formed along with varying amounts of the expected

$$(Me_3Si)_2N\text{-}P(R)(Cl) + (CH_3)_2CHMgCl \xrightarrow{-MgCl_2}$$

$$(Me_3Si)_2N\text{-}P(R)(H) + CH_3CH{=}CH_2$$

R = *i*-Pr, *t*-Bu, CH_2SiMe_3, $N(SiMe_3)_2$

alkylation product $(Me_3Si)_2NP$(*i*-Pr)R.

Acknowledgement. This research is generously supported by the U.S. Army Research Office, the Office of Naval Research, and the Robert A. Welch Foundation.

Literature Cited

1. Wisian-Neilson, P.; Neilson, R.H. *Inorg. Chem.* 1980, 19, 1875.
2. Wisian-Neilson, P.; Neilson, R.H. *J. Am. Chem. Soc.* 1980, 102, 2848.
3. Morton, D.W. *Ph.D. Dissertation*, Texas Christian University, 1981.

RECEIVED July 7, 1981.

Addition of Lithium Dialkylcuprates to α,β-Unsaturated Phosphoryl Compounds

Nucleophilic Properties of Adducts

R. BODALSKI

Institute of Organic Chemistry, Technical University, 90-924, Łódź, Poland

T. MICHALSKI, J. MONKIEWICZ, and K. M. PIETRUSIEWICZ

Centre of Molecular and Macromolecular Studies, Polish Academy of Sciences, 90-362, Łódź, Poland

The addition of lithium dialkylcuprates to α,β-unsaturated compounds is one of the effective and simple methods for carbon -carbon bond formation (*1*). In recent years significant synthetic importance has been gained by the reaction sequence in which organocopper adducts are utilized as nucleophiles (*2*, *3*, *4*). This sequence enables geminal functionalization of activated olefins particularly useful for the synthesis of natural products.

In contrast to the reactions of lithium dialkylcuprates with carbonyl and sulphonyl olefins, reactions with phosphoryl analogues have not been a subject of broad interest (*5*, *6*).

In this communication we wish to report results of our studies on the structure, some nucleophilic properties and selected synthetic applications of lithium dialkylcuprate adducts to various α,β-unsaturated organophosphorus compounds.

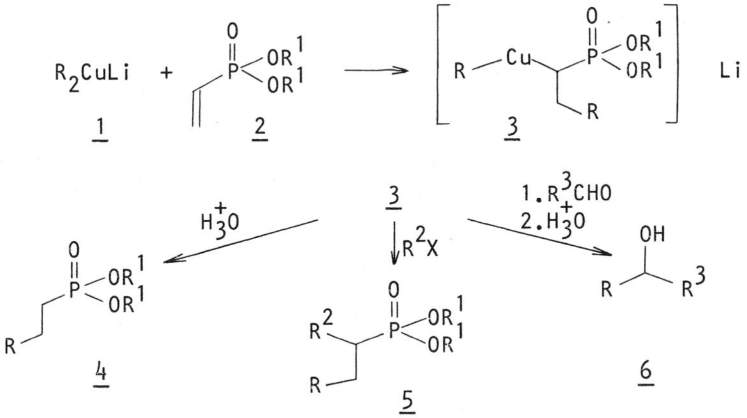

R_2CuLi (1) + $CH_2{=}CH{-}P(O)(OR^1)_2$ (2) → $[R{-}Cu{-}CH(CH_2R){-}P(O)(OR^1)_2]\,Li$ (3)

3 → (H_3O^+) → $RCH_2CH_2P(O)(OR^1)_2$ (4)

3 → (R^2X) → $RCH_2CH(R^2)P(O)(OR^1)_2$ (5)

3 → (1. R^3CHO; 2. H_3O^+) → $R{-}CH(OH){-}R^3$ (6)

R-Me, Bu; R^1-Et; R^2-allyl; R^3-pentyl, Ph

Scheme 1.

0097–6156/81/0171–0243$05.00/0

Reactions of vinylphosphonates 2 with an equimolar amount of lithium dialkylcuprates 1 result in the formation of complexes 3 containing two different organic ligands. These complexes react with electrophiles in various ways. In each of them a diverse ligand plays the role of a nucleophile. Hydrolysis of 3 or alkylation with alkyl halides affords the corresponding phosphonates 4 and 5 comprising extended saturated carbon chains bonded to phosphorus. However, in a number of reactions with aldehydes, the complexes 3 were found to undergo almost completely selective transformation into carbinols 6

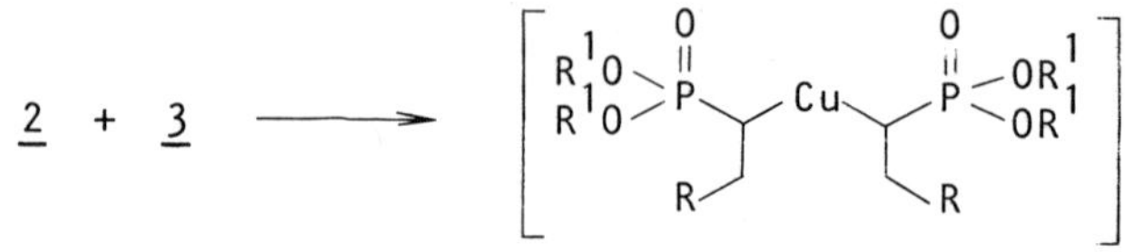

2 + 3 ⟶ [7]

R-Me, Bu; R^1-Et

Scheme 2.

An interesting alternative shows that the conversion of 3 into the secondary adducts 7 can be effected by treatment with additional equivalents of the cuprates 1. In such a conversion both alkyl groups of 1 are exploited in functionalization of the olefinic system. The nucleophilic properties of the new complexes appear to be remarkable. Although 7 are reactive towards water and alkyl halides giving 4 and 5 respectively, no reactions under standard conditions were observed with aldehydes and ketones.

The stereochemistry of the addition has been recognized on the basis of structural features and chemical behavior of the complexes 9 produced in selected reactions of 1 with 1-ethoxy-1-oxo-2-phospholene 8. Neutralization of these complexes gave exclusively cis-3-substituted phospholane oxides 10 while alkylation resulted in stereoselective formation of cis-3, cis-4-disubstituted phospholane oxides 11. The progress of the latter reaction was strongly dependent on the size of the alkyl groups involved.

The experimental data and well documented stereochemical observations concerning alkylation of chiral cuprates (*7*, *8*) allow us to propose the syn-3,4 addition mechanism for the reaction studied. The crucial step of this mechanism evidently involves pre-complexation of the cuprate 1 by the phosphoryl oxygen of 8.

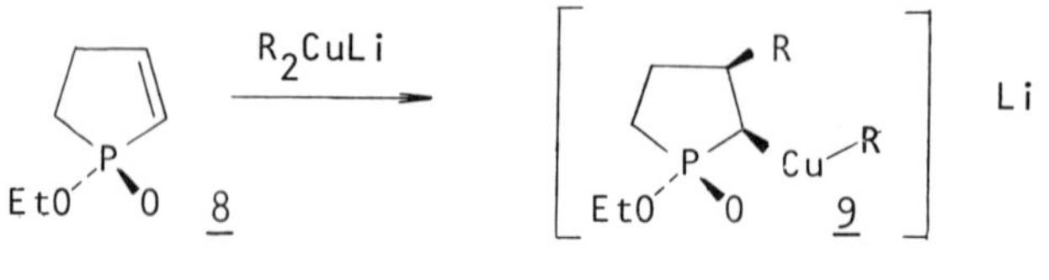

8 $\xrightarrow{R_2CuLi}$ [9] Li

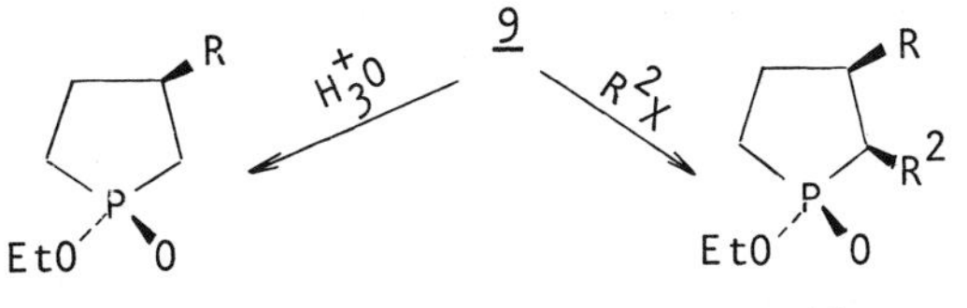

9

H_3^+O ; R^2X

10 ; 11

R-Me,Bu,octyl; R^2-Me, allyl, heptyl

Scheme 3.

A noteworthy synthetic application for the reaction of 1 with α,β-unsaturated phosphoryl compounds is represented by the addition involving hitherto unknown (-)-(S_P)-methylphenylvinylphosphine oxide 12. The resulting tertiary phosphine oxides 13 with saturated carbon chains and known stereochemistry at phosphorus constitute attractive starting materials for the preparation of optically pure phosphines. The organophosphorus substrate 12 was obtained by decarbomenthoxylation of the enantiomeric ester (-)-(S_P)-14 (*9*).

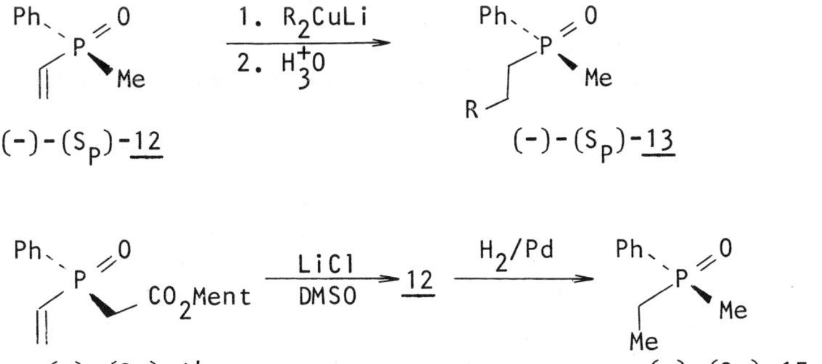

1. R_2CuLi 2. H_3^+O

(-)-(S_P)-12 → (-)-(S_P)-13

(-)-(S_P)-14 —LiCl, DMSO→ 12 —H_2/Pd→ (-)-(S_P)-15

R- Me, Bu

Scheme 4.

The configuration of 12 was unambiguously confirmed via chemical correlation with the phosphine oxide (-)-(S_P)-15 (*10*).

Significant synthetic advantage has been attained from the already mentioned sequence 8→9→11. This sequence, based on the appropriate lithium cuprate 16 and the alkyl halide 17, was successfully used for the preparation of diastereomeric 11-deoxy-9-prostanoids 18. The natural trans-configuration of the prostanoids was obtained by alkylation of the intermediate complex in the presence of hexamethylphosphorous triamide. It seems likely that the observed alteration of stereochemistry is closely connected with the known transformation of lithium cuprates into organolithium and organocopper species (_5_).

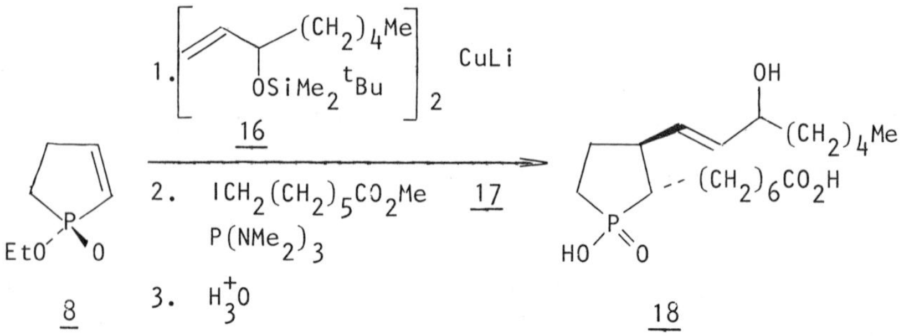

1. $\left[\ \text{CH}_2\text{=CH-CH(OSiMe}_2{}^t\text{Bu)-(CH}_2)_4\text{Me}\ \right]_2$ CuLi (16)
2. $ICH_2(CH_2)_5CO_2Me$, $P(NMe_2)_3$ (17)
3. $H_3\overset{+}{O}$

8 → 18

Scheme 5.

Literature Cited:

1. House, H.O. Acc.Chem.Res. 1976, 9, 59.
2. Stork, S.; Isobe, H. J.Am.Chem.Soc. 1975, 97, 6260.
3. Alvarez, F.S.; Wren, D.; Price, A. J.Am.Chem.Soc. 1972, 94, 7823.
4. Posner, G.M.; Stirling, J.J.; Whitten, C.E.; Lentz, C.M.; Brunelle, D.J. J.Am.Chem.Soc. 1974, 97, 107.
5. Berlan, J.; Battioni, J.P.; Koosba, K. Tetrahedron Lett. 1976, 3351.
6. Bodalski, R.; Michalski, T.; Monkiewicz, J. Phosphorus and Sulfur 1980, 9, 121.
7. Kropp, P.J. J.Am.Chem.Soc. 1969, 91, 5783.
8. Brown, W.C.; Kabalka, G.W. J.Am.Chem.Soc. 1970, 92, 712.
9. Bodalski, R.; Rutkowska-Olma, E.; Pietrusiewicz, K.M. Tetrahedron 1980, 36, 2353.
10. Korpiun, O.; Lewis, R.A.; Chickos, J.; Mislow, K. J.Am.Chem.Soc. 1968, 90, 4842.

RECEIVED June 30, 1981.

Phosphorylated Ketenes

O. I. KOLODYAZHNYI, V. I. YAKOVLEV, and V. P. KUKHAR

The Institute of Organic Chemistry, Academy of Sciences of the Ukrainian SSR, Kiev 252660, USSR

In our contribution we review methods of preparation and properties of phosphorylated ketenes unknown earlier. A high reactivity of phosphorylated ketenes opens synthetic ways to various organic and organophosphorus compounds.

The developed methods of synthesis of phosphorylated ketenes can be classified into two groups. The first one consists of the reactions which are traditionally used for the synthesis of organic ketenes (1,2). Phosphorylated ketenes are obtained by thermolysis of derivatives of phosphorylated carboxylic acids (1,2).

$R_2P(O)CHR'-CO_2H$

$\downarrow (CF_3CO)_2O$

$R_2P(O)CHPhCOCl$ $\quad$ $R_2P(O)CHR'CO_2COCF_3$ $\quad$ $R_2P(O)CHR'-COXEt$, $X = O,S$

Δ, $-HCl$ $\quad$ $\Delta \downarrow -CF_3COOH$ $\quad$ Δ, $-EtXH$

$R_2P(O)CR' = C = O$

$R = AlkO, Alk, Alk_2N$
$R' = Me, Pr, Ph, SMe, CO_2Me$

Acceptor substituents R' favor elimination of HCl, EtXH, CF_3COOH. The most convenient route for synthesis of ketenes is the decomposition of mixed anhydrides of trifluoroacetic and α-phosphorylated carboxylic acids. The bulky substituents at phosphorus facilitate the conversion of phosphorylated CH-acids into ketenes.

The second group of the reactions resulting in the formation of phosphorylated ketenes is based on the specific properties of trivalent phosphorus compounds (2-5). The Phosphines bearing an activated hydrogen atom in α-position readily react with carbon tetrahalides

0097-6156/81/0171-0247$05.00/0

to yield P-halogenylids. The P-halogenylides with carbalkoxyl group are unstable above 30°C and undergo the decomposition to give alkyl halides and ketenes.

$$R_2P\text{-}CHR'\text{-}COOMe \xrightarrow[-CHHlg_3]{CHlg_4} R_2\underset{\substack{|\\Hlg}}{P}{=}CR'\text{-}COOMe \xrightarrow[-MeHlg]{>30^\circ} R_2\underset{\substack{\|\\O}}{P}\text{-}CR'{=}C{=}O$$

Another method of preparation of α-phosphorylated cumulenes,including ketenes,is the reaction of P-halogenylides with carbon dioxide or isocyanates.

$$t\text{-}Bu_2P\text{-}CH_2R \xrightarrow[-CHHlg_3]{CHlg_4,\,-20^\circ C} t\text{-}Bu_2\underset{\substack{|\\Hlg}}{P}{=}CHR \xrightarrow{O=C=X}$$

$$\longrightarrow t\text{-}Bu_2\underset{\substack{|\\Hlg}}{\overset{+}{P}}\text{-}CHR\text{-}\underset{\substack{|\\O^{\ominus}}}{C}{=}X \longrightarrow t\text{-}Bu_2\underset{\substack{\|\\O}}{P}\text{-}CHR\text{-}\underset{\substack{|\\Cl}}{C}{=}X$$

$$\xrightarrow[-t\text{-}Bu_2P(CH_2R)Hlg_2]{t\text{-}Bu_2P(Hlg)=CHR} t\text{-}Bu_2\underset{\substack{\|\\O}}{P}\text{-}CR{=}C{=}X$$

X = O,N-R; R = H,Me,Ph.

The phosphorylated ketenes obviously represent the most stable group among ketenes. In reactions with nucleophils having hydrogen atoms they are more reactive than ordinary organic ketenes (diphenylketene).

$$Ph_2C{=}C{=}O \xrightarrow{p\text{-}MeOC_6H_4OH} Ph_2CH\text{-}COOAr \qquad \tau_{1/2} = 2500\text{min}$$

$$(EtO)_2\underset{\substack{\|\\O}}{P}\text{-}\underset{\substack{|\\Ph}}{C}{=}C{=}O \xrightarrow{p\text{-}MeOC_6H_4OH} (EtO)_2\underset{\substack{\|\\O}}{P}\text{-}CHPh\text{-}COOAr \qquad \tau_{1/2} = 1.5\text{min}$$

CCl_4 solutions,20°C.

Ketenes easily add water,alcohols, amines and thiols to give derivatives of phosphorylated carboxylic acids (4-7). Phosphorylated ketenes undergo reactions of cycloaddition to vinyl ethers,diazomethane,styrene, cyclopentadiene, etc. From phosphorylated ketenes other phosphorylated cumulenes can be obtained,for example allenes,isocyanates,ketenimines

$R_2P(O)$-CR'=C=O (reaction scheme):

- HA (A=OH, OR, SR, R_2N) → $R_2P(O)CHR'$-CO-A
- >C = C< → $R_2P(O)$-CR'-C=O (four-membered ring with C—C)
- CH_2N_2 → $R_2P(O)$-CR'-C=O / CH_2-CH_2 + $R_2P(O)$-CR'-CH_2 / CH_2-C=O
- Ph_3P=N-R'' → $R_2P(O)$-CR'=C=N-R''
- $Ph_3P=CR_2$' → $R_2P(O)$-CR'=C=CR_2''
- HN_3 → $R_2P(O)$-CHR'-CON_3 —Δ→ $R_2P(O)$CHR'N=C=O

Phosphorylated ketenes add halogens to give halides of phosphorylated halogenoacetic acids.

$$R_2P(O)\text{-}CR'{=}C{=}O \xrightarrow{Hlg_2} R_2P(O)\text{-}CR'Hlg\text{-}CO\text{-}Hlg$$

Phosphorylated ketenes react with aromatic ortho - hydroxy aldehydes to form coumarins by the Wittig - Horner - type reaction.

$R_2P(O)$-CR'=C=O + X-C_6H_3(OH)CHO → $R_2P(O)$-CHR-CO-O-C_6H_3(CHO)X —NaOH→ X-coumarin (3-R')

(X-C_6H_4(ONa)CHO → coumarin)

Labels: OH, CHO, ONa, CHO, X, R', O, OHC, X, NaOH

We have also developed convenient methods of preparation of the phosphorylated ketenimines. Like the phosphorylated ketenes the phosphorylated ketenimines are stable on storage and they readily react with variety of reagents having an active hydrogen atom, as shown in the Scheme.

$R_2P(O)$-CPh=C=NMe:

- HCl → $R_2P(O)$-CPh=C<(NHMe)(Cl)
- R'XH → $R_2P(O)$-CPh=C(XR')-NHMe
- H_2O → $R_2P(O)$-CHPh-CONHMe

Literature Cited

1. Kolodiazhnyi, O.I.; Zh.Obsch.Khim. 1979,49,716.
2. Kolodiazhnyi, O.I.; Zh.Obsch.Khim. 1980,50,1485.
3. Kolodiazhnyi, O.I.; Kukhar, V.P.; Zh.Org.Khim.1978, 14,1339.
4. Kolodiazhnyi, O.I.; Yakovlev, V.I.; Kukhar, V.P. Zh.Obsch.Khim.1979,49,2458.
5. Kolodiazhnyi, O.I. Tetrahedron Letters,in press.
6. Kolodiazhnyi, O.I.; Yakovlev, V.I. Zh.Obsch.Khim. 1980,50,55.
7. Kolodiazhnyi, O.I.; Yakovlev, V.I.; Kukhar, V.P. Zh.Obsch.Chim.1980,50,1418.

RECEIVED July 7, 1981.

Preparation and Properties of *N*-(Hydroxycarbonylmethyl)aminomethyl Alkyl- and Arylphosphinic Acids and Derivatives

LUDWIG MAIER

Ciba-Geigy Limited, Agricultural Division, CH-4002 Basel, Switzerland

The reaction of alkyl- and aryldichlorophosphines (or phosphonous acids) with N-benzylglycine and formaldehyde in acidic solution yields (N-benzyl-N-hydroxycarbonylmethyl-aminomethyl) alkyl- and -arylphosphinic acids of structure 1 (R = alkyl, $HOCH_2$) or 2 (R = CCl_3, CH_2Cl, C_6H_5), depending on the electronegativity of R:

$$[RPCl_2 + 2\ H_2O \rightarrow]\ R\text{-}P(=O)(H)OH + C_6H_5CH_2NHCH_2CO_2H + CH_2O \xrightarrow{H^+}$$

$$R\text{-}P(=O)(OH)\text{-}CH_2\overset{+}{N}H(CH_2C_6H_5)(CH_2CO_2H)\cdot Cl^- \quad (1) \qquad \text{or} \qquad R\text{-}P(=O)(O^-)\text{-}CH_2\overset{+}{N}H(CH_2C_6H_5)(CH_2CO_2H) \quad (2)$$

Debenzylation with hydrogen in the presence of Pd/C as a catalyst in acetic acid or alcohol/water produces N-hydroxycarbonylmethyl-aminomethyl)alkyl- and -arylphosphinic acids (3)

$$R\text{-}P(=O)(OH)\text{-}CH_2N(CH_2C_6H_5)(CH_2CO_2H) + H_2 \xrightarrow{Pd/C} R\text{-}P(=O)(OH)\text{-}CH_2NHCH_2CO_2H\ (3) + C_6H_5CH_3$$

In the case of the trichloromethyl derivative excess hydrogen must be avoided, otherwise dechlorination occurs and (N-hydroxycarbonylmethyl-aminomethyl)-dichloromethylphosphinic acid (4)

$$CCl_3\text{-}P(=O)(OH)\text{-}CH_2N(CH_2C_6H_5)(CH_2CO_2H) + 2\ H_2 \xrightarrow{Pd/C} CHCl_2\text{-}P(=O)(OH)\text{-}CH_2NHCH_2CO_2H\ (4)$$

0097-6156/81/0171-0251$05.00/0

is obtained. All phosphinic acid derivatives of structure 3 are obtained as crystalline, white solids of high decomposition points. Depending on the electronegativity of R in 3, the products crystallize either as semihydrochlorides or hydrochloride-free. The strong dependence of the ^{31}P-chemical shift on the pH indicates that all derivatives possess the betaine structure. Thus 3 (R = CH_3), dissolved in water, shows a ^{31}P-chemical shift of 30 ppm; at pH 1 (adjusted with HCl) apparently a hydrochloride is formed with ^{31}P 32.8 ppm, at pH 8 the monosodium salt (^{31}P 35.6 ppm) and at pH 10 the disodium salt (^{31}P 39.1 ppm) is produced.

An attempt to synthesize N-hydroxycarbonylmethyl-aminomethyl-phenylphosphinic acid (5) by oxidation of bis(N-hydroxycarbonyl-methyl)-aminomethyl-phenylphosphinic acid with oxygen in the presence of catalysts failed. Only decomposition products were

$$(HO_2CCH_2)_2NCH_2P(=O)(C_6H_5)OH + O_2 \xrightarrow{\text{cat.}} HO_2CCH_2NHCH_2P(=O)(C_6H_5)OH$$

(5)

obtained. On the other hand, this procedure was successfully used in the preparation of N-hydroxycarbonylmethyl-aminomethyl-phosphonic acid [1].

The phosphinic acids of structure 3 give crystalline mono-amine salts, e.g.

$$CH_3-P(=O)(OH)-CH_2NHCH_2CO_2H \cdot H_2NC_3H_7\text{-i}, \text{ m.p. } 203\text{-}204^{\circ}C \text{ (dec.)}$$

and form monoesters when treated with alcohol and hydrogen chloride, e.g. 6

$$R-P(=O)(OH)-CH_2NHCH_2CO_2H + R^1OH \xrightarrow{HCl} R-P(=O)(OH)-CH_2NHCH_2CO_2R^1 + H_2O$$

(6)

The diesters (7) are obtained directly by heating a mixture of O-alkylphosphonites and tris(N-ethoxycarbonylmethyl)hexa-hydrotriazine:

$$3\ R-P(=O)(H)OR^1 + \text{1,3,5-tris}(CH_2CO_2R^2)\text{-hexahydrotriazine} \longrightarrow 3\ R-P(=O)(OR^1)-CH_2NHCH_2CO_2R^2$$

(7)

However, interaction of O-ethyl-2-chloroethylphosphonite and tris(N-ethoxycarbonylmethyl)hexahydrotriazine produces the 1-ethoxycarbonylmethyl-1,3-azaphospholidine-3-ethoxy-3-oxide (8):

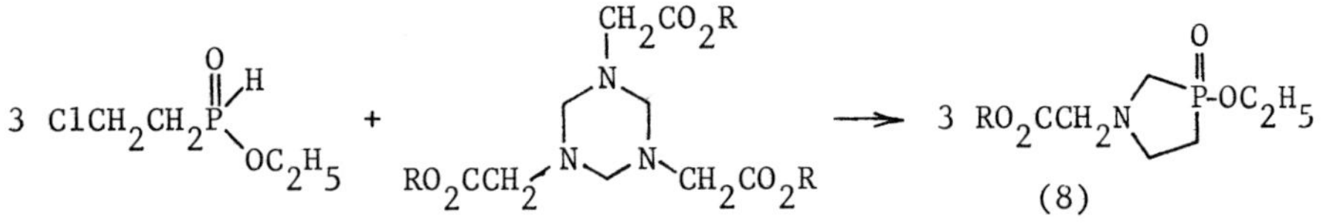

$$3\ ClCH_2CH_2P(O)(H)OC_2H_5 + (RO_2CCH_2N\text{-}CH_2)_3 \longrightarrow 3\ RO_2CCH_2N\langle CH_2CH_2P(O)(OC_2H_5)CH_2\rangle$$

(8)

Furthermore, cyanomethyldichlorophosphine (b.p.$_{15}$ 85-88°C, ^{31}P 159.47 ppm) and 2-chloroethyldichlorophosphine (b.p.$_{93}$ 98-102°, ^{31}P 182.8 ppm) when treated with benzylglycine and formaldehyde in acetic acidic solution give 9 and 10, respectively.

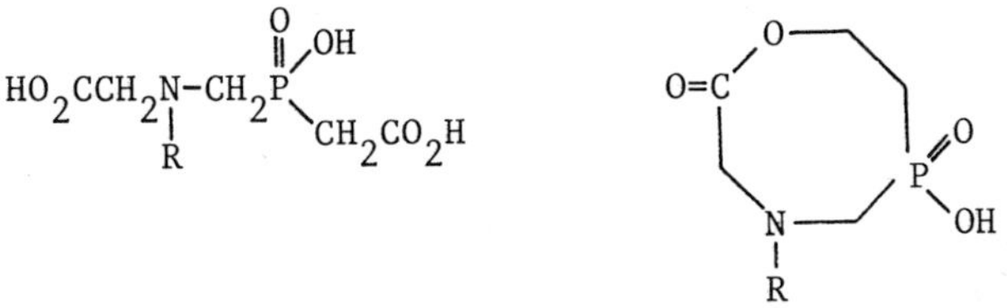

$$HO_2CCH_2N(R)\text{-}CH_2P(O)(OH)CH_2CO_2H$$

9, R = $CH_2C_6H_5$

11, R = H

10: eight-membered ring O=C–O–CH$_2$–CH$_2$–P(=O)(OH)–CH$_2$–N(R)–CH$_2$–

10, R = $CH_2C_6H_5$

12, R = H

These on debenzylation yield 11 (N-glycinomethyl-carboxymethyl-phosphinic acid) and 12 (1,4,6-oxazaphosphocane-2-oxo-6-hydroxy-6-oxide), respectively. The structure of the compounds has been confirmed by mass- and nmr-spectroscopic investigations.

Literature Cited

1. Monsanto Co., U.S. Patent 3 969 398 (1976)

RECEIVED June 30, 1981.

53

Some Aspects of Aminoalkylphosphonic Acids Synthesis by the Reductive Amination Approach

P. SAVIGNAC

Equipe CNRS-SNPE, 2-8 Rue Henry Dunant, 94320 Thiais, France

N. COLLIGNON

I.N.S.C.I.R., B.P. 08, 76130 Mont-Saint-Aignan, France

The discovery by Horiguchi and Kandatsu in 1960 of 2-aminoethylphosphonic acid (AEPA) represents the first example of the occurence of a covalent C-P bond in biological materials (1). Several laboratories attempted to elucidate the biosynthesis of the C-P bond. Horiguchi, who was studying the problem of AEPA induction, proposed two approaches (2) (Scheme 1). In the first, phosphonopyruvic acid (II) a substance produced by rearrangement of phosphoenolpyruvate (I) is readily decarboxylated to phosphonoacetaldehyde (III) and then via amination converted to AEPA. In the second, phosphonopyruvic acid (II) is at first transaminated to phosphonoalanine (IV) and then decarboxylated to AEPA. Recently Horiguchi has suggested that phosphonoalanine (IV) was deaminated in preference to decarboxylation (3).

I II III IV

$-CO_2$ $-CO_2$

AEPA $NH_2CH_2CH_2P(O)(OH)_2$

As seen in Scheme 1 a route similar to the biological pathway has now been explored by the independent synthesis of each precursors by chemical means. Work presented in this communication describes the production of synthetic 2-aminoethylphosphonic acids by the controlled reductive amination of 2-oxoalkylphosphonate diesters

(Scheme 1)

Briefly we review the chemical improvements which we achieved in the oxoalkylphosphonates field which represent the key compounds. Phosphonic aldehydes are obtained in adapting the Arbuzov procedure to β or γ haloketals (4). A modification of the phosphonylation conditions (t°, stoichiometry) followed by removal of the protecting group in dilute acid and then continuous extraction allows synthesis of suitably branched compounds on a large scale (5).

0097–6156/81/0171–0255$05.00/0

$$BrCH_2\overset{R^1}{\overset{|}{(CH)}}_nCH(OR)_2 + (RO)_3P \longrightarrow (RO)_2\underset{O}{\underset{\|}{P}}CH_2\overset{R^1}{\overset{|}{(CH)}}_nCH(OR)_2 \xrightarrow{H+} (RO)_2\underset{O}{\underset{\|}{P}}CH_2\overset{R^1}{\overset{|}{(CH)}}_nCHO$$

$n = 0,1$

$$(RO)_2\underset{O}{\underset{\|}{P}}CH_2\overset{R^1}{\overset{|}{(CH)}}_nCH(OR)_2 \xrightarrow[R^2\text{-}X]{\underline{n}\text{-BuLi}} (RO)_2\underset{O}{\underset{\|}{P}}\overset{R^2}{\overset{|}{C}}H\overset{R^1}{\overset{|}{(CH)}}_nCH(OR)_2 \xrightarrow{H+} (RO)_2\underset{O}{\underset{\|}{P}}\overset{R^2}{\overset{|}{C}}H\overset{R^1}{\overset{|}{(CH)}}_nCHO$$

Because of the Perkow reaction, the above route for the production of ketophosphonates and phosphonopyruvates was abandoned. By coupling α-copperalkylphosphonates with acyl chlorides, we have been able to produce in one step ketophosphonate correctly functionalized as well as phosphonopyruvates in good yield and high purity (6).

$$(RO)_2\underset{O}{\underset{\|}{P}}\overset{R^1}{\overset{|}{C}}H_2 \xrightarrow{\underline{n}\text{-BuLi}} (RO)_2\underset{O}{\underset{\|}{P}}\overset{R^1}{\overset{|}{C}}HLi \xrightarrow{CuI} (RO)_2\underset{O}{\underset{\|}{P}}\overset{R^1}{\overset{|}{C}}HCu \xrightarrow{R^2COCl} (RO)_2\underset{O}{\underset{\|}{P}}\overset{R^1}{\overset{|}{C}}H\underset{O}{\underset{\|}{C}}R^2$$

$$(RO)_2\underset{O}{\underset{\|}{P}}\overset{R^1}{\overset{|}{C}}HCu \xrightarrow{ClCOCOOR} (RO)_2\underset{O}{\underset{\|}{P}}\overset{R^1}{\overset{|}{C}}H\underset{O}{\underset{\|}{C}}COOR$$

R = Me,Et R^1 = H,Me,Et R^2 = alkyl, aryl...

Phosphonopyruvate systems were the first candidates submitted to reductive amination. Because of the presence of the ester group, the α-ketoester carbonyl is less reactive than traditional oxoalkylphosphonates, and yields of isolated amino-esters never exceeded 55 %. The reaction is run at pH 7 in ethanol and it is general for ammonia and primary amines. Steric hindrance represents the second limiting factor since α-substituted or thionophosphonopyruvates react sluggishly ; thus secondary amines cannot be introduced. When the reductive amination process is effected the exclusive by-product is the α-hydroxyester which arises by reduction of the carbonyl group (7-8).

$$(RO)_2\underset{X}{\underset{\|}{P}}\overset{R^1}{\overset{|}{C}}H\underset{O}{\underset{\|}{C}}COOR \xrightarrow[H_2NR^3]{NaBH_3CN,EtOH} (RO)_2\underset{X}{\underset{\|}{P}}\overset{R^1}{\overset{|}{C}}H\underset{NHR^3}{\underset{|}{C}}HCOOR \xrightarrow{H+} (HO)_2\underset{X}{\underset{\|}{P}}\overset{R^1}{\overset{|}{C}}H\underset{NHR^3}{\underset{|}{C}}HCOOH$$

X = O,S R^3 = H, Me, Et 25 - 70 %

The amino-esters were hydrolyzed in aqueous HCl to give phosphonoalanine as a large variety of derivatives.

Reductive amination was next carried out with the ketophosphonates. Our results indicate that the steric hindrance around the carbonyl is the limiting factor. However the yields can be increased by increasing the reaction time without any side reactions lowering the purity of the products. Ammonia, primary and secondary amines can be introduced ; each one reacts in the enaminophosphonate form. Using ketophosphonate bearing functional groups we have observed either a participation of the functional group (halogen , ester or unsaturated group) to the reaction or the complete conservation of the function (aromatic group). The reaction conduces after hydrolysis to acids containing an asymetric carbon.

$$(RO)_2P(O)CH(R^1)C(O)R^2 \xrightarrow[HNR^3R^4]{NaBH_3CN} (RO)_2P(O)CH(R^1)CH(NR^3R^4)R^2 \xrightarrow{H+} (HO)_2P(O)CH(R^1)CH(NR^3R^4)R^2$$

$R^3 = R^4 = H$, Me, Et 55 - 80 %

The third type of compound studied were phosphonic aldehydes which are more reactive. Steric factors are absent but we observe a difference in behavior among phosphonicaldehydes according to the length of the carbon chain. Phosphonoacetaldehyde reacts almost exclusively in the enaminophosphonate form which is less reactive than the iminophosphonate form observed in the case of homologous compounds. As for the aminating reagent we observe some differences. Ammonia always leads to a mixture of aminophosphonate (i) and aminodiphosphonate (ii) while primary and secondary amines lead specifically to monocondensed compounds (i) (9).

$$(RO)_2P(O)(CH(R^1))_nCHO \xrightarrow[HNR^3R^4]{NaBH_3CN} \begin{cases} \text{i: } (RO)_2P(O)(CH(R^1))_nCH_2NR^3R^4 \xrightarrow{H+} (HO)_2P(O)(CH(R^1))_nCH_2NR^3R^4 \\ \text{ii: } [(RO)_2P(O)(CH(R^1))_nCH_2]_2NH \xrightarrow{H+} [(HO)_2P(O)(CH(R^1))_nCH_2]_2NH \end{cases}$$

$n = 1,2$

The above route for the production of primary aminophosphonic acids was abandoned following their successful synthesis, in good yield and high purity, by the successive debenzylation (H_2/Pd/C)-hydrolysis (aqueous HCl) of benzylaminoalkylphosphonate diesters (10).

$$(RO)_2P(O)(CH(R^1))_nCHO \xrightarrow{H_2NCH_2\phi} (RO)_2P(O)(CH(R^1))_nCH{=}NCH_2\phi \longrightarrow (HO)_2P(O)(CH(R^1))_nCH_2NH_2$$

$n = 1,2$ 47 - 90 %

In addition since the reaction of 2-oxoethyl - phosphonate with amines showed a preferential formation of an enaminophosphonate, we used that structure for the production of specifically nitrogen substituted compounds by the following sequence of reactions(11).

$$(RO)_2\underset{\underset{O}{\parallel}}{P}CH_2CHO \xrightarrow{H_2NCH_2\phi} (RO)_2\underset{\underset{O}{\parallel}}{P}CH{=}CH\overset{H}{\overset{|}{N}}CH_2\phi \xrightarrow[R^2-X]{NaH} (RO)_2\underset{\underset{O}{\parallel}}{P}CH{=}CH\overset{R^2}{\overset{|}{N}}CH_2\phi$$

$$\xrightarrow{H_2/Pd/C,\ H+} (HO)_2\underset{\underset{O}{\parallel}}{P}CH_2CH_2NHR^2$$

R^2 = Me, Et, Pr 35 - 55 %

All the aminophosphonic acids prepared were obtained in good yield and purified by the use of ion exchange resins (Amberlite IRA 410, OH^- form).

Literature Cited

1. Horiguchi, M ; Kandatsu, M. Agr. Chem. Soc. Japan. 1960,24,565
2. Horiguchi, M ; Kittredge, J.S ; Roberts, E. Biochim .Biophys Acta 1968,165,164
3. Horigane, A ; Horiguchi, M ; Matsumoto, T ; ibid. 1979,572,385
4. Razumov, A.I ; Liorber, B.G ; Moskva, V.V ; Sokolov, M.P Russ. Chem. Rev. 1973,42,538
5. Varlet, J.M ; Fabre, G ; Sauveur, F ; Collignon, N ; Savignac, P Tetrahedron (in preparation)
6. Mathey, F ; Savignac, P ; Tetrahedron 1978,34,649
7. Borch, R.F ; Bernstein, M.D ; Durst H.D.J. Am. Chem. Soc 1971, 93, 2897
8. Varlet, J.M ; Collignon, N ; Savignac, P ; Can . J . Chem 1979, 57, 3216
9. Isbell, A.F ; Englert, L.F ; Rosemberg, H ; J. Org. Chem 1969, 34, 755
10. Szczepaniak, W ; Kuczynski, K ; Phosphorus and Sulfur 1979,7, 333
11. Nagata, W ; Hayase, Y ; J. Chem. Soc (c), 1969, 460

RECEIVED July 7, 1981.

PHOSPHORUS HETEROCYCLES

Recent Results on Open-Chain and Cyclic Phosphanes and Organylphosphanes

MARIANNE BAUDLER

Institut für Anorganische Chemie der Universität Köln, Greinstrasse 6, D-5000 Köln 41, FRG

Compounds containing a skeleton of cumulated P-P bonds have been rare until recently, as they are in general highly reactive. This means that they can easily be oxidized and have a strong tendency to disproportionate. Nevertheless, considerable progress has been achieved in this field lately. Thereby a significant analogy of phosphorus- to carbon-chemistry became apparent with respect to structural features.

Phosphorus Hydrides

Since 1965 we have found an unexpected number of binary phosphorus hydrogen compounds in addition to the well-known hydrides PH_3 and P_2H_4 (Table I). These phosphanes have been detected in the hydrolysis products of calcium phosphide or in the thermolysis products of P_2H_4 (1, 2) by mass spectroscopy. Only the compounds P_3H_5, P_4H_6, P_5H_5, and P_7H_3 could be isolated in pure form so far, whereas the other phosphanes have been obtained only as mixtures. As the detailed structures are mostly unknown, a ^{31}P-NMR spectroscopic investigation was initiated. Beginning with tetraphosphane(6), the structural situation of the open-chain phosphanes becomes more and more complex due to the existence of constitutional and configurational isomers. The low temperature $^{31}P\{^1H\}$-NMR spectrum of P_4H_6 could be simulated very satisfactorily by the superposition of two AA'BB'-spin systems for the d,l- and meso-isomer of n-P_4H_6 and one AB_3 spin system for i-P_4H_6 with a branched P-skeleton (3). In a similar way the four NMR spectroscopically distinguishable isomers of pentaphosphane(7) have been identified. The branched i-P_5H_7 shows the highest relative frequency.

The chemistry of the higher homologues of PH_3 was until the present almost unknown. We have therefore investigated first the metallation of diphosphane (5). The reaction of P_2H_4 with n-BuLi or $LiPH_2$ yields the polyphosphide $Li_3P_7 \cdot 3$ solvents as the final product. A significant feature of the tricyclic P_7^{3-} ion is its fluctional behavior which is analogous to that in bullvalene, as found by studying the temperature dependence of the ^{31}P-NMR spec-

0097–6156/81/0171–0261$05.00/0

Table I. Phosphorus Hydrides (Phosphanes)

	P_nH_{n+2}	P_nH_n	P_nH_{n-2}	P_nH_{n-4}	P_nH_{n-6}	P_nH_{n-8}	P_nH_{n-10}	P_nH_{n-12}	P_nH_{n-14}
P_3	P_3H_5	P_3H_3							
P_4	P_4H_6	P_4H_4	P_4H_2						
P_5	P_5H_7	P_5H_5	P_5H_3						
P_6	P_6H_8	P_6H_6	P_6H_4	P_6H_2					
P_7	P_7H_9	P_7H_7	P_7H_5	P_7H_3					
P_8	P_8H_{10}	P_8H_8	P_8H_6	P_8H_4	P_8H_2				
P_9	P_9H_{11}	P_9H_9	P_9H_7	P_9H_5	P_9H_3				
P_{10}		$P_{10}H_{10}$	$P_{10}H_8$	$P_{10}H_6$	$P_{10}H_4$	$P_{10}H_2$			
P_{11}					$P_{11}H_5$	$P_{11}H_3$			
P_{12}					$P_{12}H_6$	$P_{12}H_4$	$P_{12}H_2$		
P_{13}						$P_{13}H_5$			
P_{14}						$P_{14}H_6$	$P_{14}H_4$	$P_{14}H_2$	
P_{15}							$P_{15}H_5$		
P_{16}							$P_{16}H_6$	$P_{16}H_4$	$P_{16}H_2$
P_{17}								$P_{17}H_5$	
P_{18}								$P_{18}H_6$	$P_{18}H_4$

trum of Li_3P_7. Above room temperature all P-atoms become equivalent, so that a total of 1680 tautomeric forms exist (5). The synthesis of the pure hydride P_7H_3 could be achieved via trimethylsilylation of Li_3P_7 and subsequent solvolysis with MeOH. Through the reaction with methyl bromide, the first polycyclic organylphosphane, P_7Me_3, has been obtained (5).

Three-Membered Phosphorus Ring Compounds

A short time ago numerous triorganyl-cyclotriphosphanes have become accessible as pure substances on various synthetic routes (2). Moreover, three-membered phosphorus heterocycles with carbon, silicon, germanium, arsenic, boron, and sulfur as hetero ring atoms could be synthesized (2). Recently, we have found that such compounds with elements of the fifth period are also stable enough for existence. Through the reaction of $K(Bu^t)P\text{-}P(Bu^t)K$ with Bu^tSbCl_2 or $Bu^t{}_2SnCl_2$ at -78^oC, the diphosphastibirane $(PBu^t)_2SbBu^t$ (6) and the diphosphastannirane $(PBu^t)_2SnBu^t{}_2$ (7), respectively, have been obtained. Whereas the antimony heterocycle could not be worked up because of its instability, the tin compound could be isolated in its pure form in 52% yield. It forms light-yellow crystals, is not inflammable in contact with air and can be stored at -30^oC for several weeks without decomposition.

The thiadiphosphirane $(PBu^t)_2S$ was already obtained previously via ring contraction of phosphorus sulfur heterocycles of larger ring sizes (2). A considerably better preparative approach, however, is the cyclization of $Cl(Bu^t)P\text{-}P(Bu^t)Cl$ with $(Me_3Sn)_2S$ (7). Using $(Me_3Sn)_2Se$ as a reactant in the [2+1]-cyclocondensation, the first selenadiphosphirane $(PBu^t)_2Se$ has been obtained. Like the sulfur compound, this phosphorus selenium three-membered heterocycle can be isolated in pure form by high vacuum fractional distillation (7). It is a yellow liquid of extremely bad odor that can be stored in a refrigerator for weeks. The structure can immediately be deduced from its high field ^{31}P-NMR chemical shift and is unambiguously confirmed by all other methods of characterization.

Polycyclic Organylphosphanes

Whereas the dehalogenation of $RPCl_2$ yields monocyclic organylphosphanes $(PR)_n$, the synthesis of polycyclic organylphosphanes P_nR_m could be achieved by reacting $RPCl_2$ with magnesium in the presence of PCl_3 or of white phosphorus (8). However, a mixture of various compounds is always formed that must be separated by chromatography. Our results to date are summarized in Table II. With the exception of the iso-propyl derivatives, all compounds have been isolated pure in a preparative scale. The structures of the various polycyclic organylphosphanes have been elucidated by ^{31}P-, ^{13}C-, and ^1H-NMR spectroscopy. In addition, an X-ray analysis was done in some cases by Prof. Tebbe in Köln. The various compounds

Table II. Polycyclic Organylphosphanes

P_nR_m / R	P_6R_4	P_7R_3	P_7R_5	P_8R_6	P_9R_5
CH_3					
C_2H_5					
$C_3H_7^i$					
$C_4H_9^t$		—	—		—

of the series P_8R_6 demonstrate the fact that organylphosphanes of the same molecular formula may display entirely different structures, depending on the substituent R.

The tetra-tert-butylhexaphosphane, $P_6Bu^t_4$, has the structure of a bicyclo[3.1.0]hexaphosphane, in which the three-membered ring and the nearly planar five-membered ring are approximately perpendicular to each other. This structure can be deduced from the tricyclic P_7-skeleton by removing one of the three monoatomic bridges. In the same way, the pentaorganyl-nonaphosphane structure, that has been found to be analogous to that of the hydrocarbon noradamantane, is formed by substituting a five-membered ring for the three-membered ring of the tricyclic P_7-species. And finally, the pentaorganyl-heptaphosphanes with a structure analogous to bicyclic norbornane, merely differ from the tricyclic P_7-skeleton by opening one bond of the three-membered ring.

As obviously small substituents do not influence the structures of the polycyclic phosphane skeletons, the substitution of methyl groups by hydrogen atoms should not cause any change. Hence, the formulae for the methyl compounds simultaneously represent the hitherto unknown structures of the phosphorus hydrides P_7H_3, P_7H_5, P_8H_6, and P_9H_5 (1, 2). Small differences between the hydrogen and methyl substituted phosphanes should exist only with respect to the relative amounts of the various configurational isomers.

Acknowledgement - This work was supported by the Deutsche Forschungsgemeinschaft and the Fonds der Chemischen Industrie.

Literature Cited

1. Baudler, M.; Ständeke, H.; Borgardt, M.; Strabel, H.; Dobbers, J. Naturwissenschaften 1966, 53, 106.
2. Baudler, M. Pure & Appl. Chem. 1980, 53, 755.
3. Baudler, M.; Krause, U. M.; Kloth, B.; Skrodzki, H. D. Z. Anorg. Allg. Chem., submitted for publication.
4. Baudler, M.; Krause, U. M. Z. Anorg. Allg. Chem., submitted for publibation.
5. Baudler, M.; Ternberger, H.; Faber, W.; Hahn, J. Z. Naturforsch. 1979, B 34, 1690.
6. Baudler, M.; Klautke, S. Z. Anorg. Allg. Chem., in press.
7. Baudler, M.; Suchomel, H. Z. Anorg. Allg. Chem., in press.
8. Baudler, M.; Aktalay, Y.; Hahn, J. Angew. Chem., submitted for publication.

RECEIVED June 30, 1981.

Formation of Phosphorus Oxoacids with P-P-P-P-P-P and P-P-P-P-P Frameworks and Related Compounds

TOSHIO NAKASHIMA[1], HIROHIKO WAKI, and SHIGERU OHASHI

Department of Chemistry, Faculty of Science, Kyushu University 33, Hakozaki, Higashiku, Fukuoka, 812 Japan

The following ten kinds of phosphorus oxoacids with P-P bonds are known at present (1, 2): P^{IV}-P^{IV}, P^{IV}-P^{IV}-O-P^{IV}-P^{IV}, (-P^{IV}-P^{IV}-O-$)_2$, P^{IV}-P^{IV}-O-P^{III}, P^{IV}-P^{IV}-O-P^{V}, P^{III}-O-P^{IV}-P^{IV}-O-P^{III}, P^{II}-P^{II}, P^{II}-P^{IV}, P^{IV}-P^{III}-P^{IV} and (-P^{III}-$)_6$. Among these compounds we reported the formation of P^{III}-O-P^{IV}-P^{IV}-O-P^{III} by the substitution reaction between P^{IV}-P^{IV} and P^{III}-O-P^{III} at the ICPC., 1979 (2).

The last compound in the above series, (-P^{III}-$)_6$, was first prepared by Blaser and Worms in 1959 (3). This compound has a six-membered P-P bond ring structure as shown in Formula 1. It has been known that this ring acid is decomposed by hydrolysis into a complicated mixture of various phosphorus oxoacids and finally converted into a mixture of monomeric phosphorus oxoacids; phosphinic acid (hypophosphorous acid, P^{I}), phosphonic acid (phosphorous acid, P^{III}) and orthophosphoric acid (P^{V}). Therefore, there is a possibility that some new phosphorus oxoacids with P-P bonds may be found in the decomposition products by partial hydrolysis of (-P^{III}-$)_6$ ring acid. This paper describes the separation of new phosphorus oxoacids with P-P-P-P-P-P and P-P-P-P-P frameworks and related compounds from the decomposition products of (-P^{III}-$)_6$ ring acid by ion-exchange chromatography and their identification by the measurements of their UV absorption spectra and by the charge determination of anionic species of these compounds. Decomposition products by oxidation of (-P^{III}-$)_6$ ring acid were also examined.

Formula 1

Experimental

The mixed potassium sodium salt of (-P^{III}-$)_6$ ring acid was

[1]Current address: Department of Inorganic and Analytical Chemistry, Hebrew University of Jerusalem, Jerusalem, Israel.

0097-6156/81/0171-0267$05.00/0

prepared by the Blaser and Worms' procedure (3), i.e., the oxidation of red phosphorus with bromine and the separation of the desired compound from the other products.

The potassium sodium salt of $(-P^{III}-)_6$ ring acid was hydrolyzed in dilute hydrochloric acid at 21°C for 40-60 min. In some cases the salt of $(-P^{III}-)_6$ ring acid in potassium hydrogen carbonate solution was oxidized with hydrogen peroxide at 11-20°C for 20-25 h.

The decomposition products were separated by eluting with potassium chloride solution on a column packed with anion-exchange resin, Dowex 1x4 in chloride form. During the elution the concentration of potassium chloride in the eluent was increased exponentially. The phosphorus contents in effluent fractions were determined colorimetrically with a molybdenum(V)-molybdenum(VI) reagent (4).

For UV spectral measurements a Hitachi recording spectrophotometer ESP-3T was employed with a 5-cm quartz cell.

Charges of anionic species of phosphorus oxoacids were determined by an ion-exchange equilibrium method. For this purpose distribution ratios, D, of phosphorus oxoacid between an anion-exchange resin Dowex 1x4 phase and an aqueous solution phase containing tetramethylammonium chloride as a supporting electrolyte were obtained from the absorbancies of phosphorus oxoacid in the aqueous solution phase before and after equilibration. D was defined as the ratio of the concentration of phosphorus in the resin phase to the concentration of phosphorus in the solution phase.

Results and Discussion (5, 6)

A complicated mixture produced by hydrolysis of $(-P^{III}-)_6$ ring acid was separated by ion-exchange chromatography. The elution peaks due to Compounds 1 and 2 were always observed in the latest part of an elution pattern. Compounds 1 and 2 have characteristic UV absorption at 272 and 255 nm, respectively,and were presumed from their elution positions to be phosphorus oxoacids having a rather long P-P chain.

D values of Compounds 1 and 2 in an aqueous solution of tetramethylammonium chloride of various concentrations at pH about 7 were measured. The plots of log D values against log $[Cl^-]$ values showed a linear relationship between these two variables. The slopes of these straight lines indicated that anion charges of Compounds 1 and 2 at neutral pH are -6.0 and -4.8, respectively. A method for this analysis was discussed in reference (5).

The results mentioned above suggest that Compound 1 is produced immediately after the hydrolytic ring opening of the parent compound, $(-P^{III}-)_6$.

$$(-P^{III}-)_6 \xrightarrow{H_2O} P^{II}-P^{III}-P^{III}-P^{III}-P^{III}-P^{IV} \quad (P_{6d})$$

All the protons of P_{6d} except one of two protons at the end P^{IV}-unit are considered from the dissociation behavior of other known phosphorus oxoacids to be completely dissociated at neutral pH (1). Therefore, P_{6d} may have a charge of about -6 under the present experimental conditions.

Further hydrolytic decomposition of $(-P^{III}-)_6$ ring acid may produce the following three kinds of phosphorus oxoacids with a P-P-P-P-P framework (P_{5d}).

$$(-P^{III}-)_6 \xrightarrow{2H_2O} \begin{cases} P^{IV}-P^{III}-P^{III}-P^{III}-P^{IV} + P^{I} & (P_{5d}-1) \\ P^{II}-P^{III}-P^{III}-P^{III}-P^{IV} + P^{III} & (P_{5d}-2) \\ P^{II}-P^{III}-P^{III}-P^{III}-P^{II} + P^{V} & (P_{5d}-3) \end{cases}$$

At neutral pH, P_{5d}-1, P_{5d}-2 and P_{5d}-3 may have a charge of -5 ~ -6, -5 and -5, respectively. The result of the charge determination mentioned above suggests that Compound 2 is any one or a mixture of these three compounds. An anion charge of a copper complex of Compound 2 was determined by the anion-exchange equilibrium method to be -3.2. This means that P_{5d}-3 is the most probable compound or the most predominant compound in a mixture of these three species.

Since the number of phosphorus atoms in a molecule of Compound 1 or 2 is considered to be six or five, the molar absorptivity, ε, of each can be calculated. The curves of log ε vs. wavelength for Compounds 1 (P_{6d}) and 2 (P_{5d}) and the parent $(-P^{III}-)_6$ ring acid were obtained (5).

A comparison of the position of the longest wavelength peaks (or shoulders) for these new longer chain species (P_{6d} and P_{5d}) and the known shorter chain species, $P^{IV}-P^{IV}$ (P_{2d}) and $P^{IV}-P^{III}-P^{IV}$ (P_{3d}) is very interesting. It was revealed that there is a linear relationship between the wavenumber of the above mentioned peak or shoulder and the number of phosphorus atoms in a molecule. A species which has a shoulder at the position corresponding to P_{4d} was also found in a mixture of hydrolytic products of $(-P^{III}-)_6$ ring acid, though it could not always be detected in the chromatographic separation. This behavior mentioned above may be analogous to that observed in polysulfide series (7). This spectral relation also supports the structures of P_{6d} and P_{5d}.

It was very difficult to obtain the new compound of P_{6d} or P_{5d} as a solid salt, because a very small amount of P_{6d} or P_{5d} was present in very dilute aqueous solution. However, the preparation of a small volume of P_{6d} solution of relatively high concentration was successfully achieved by the following procedure; the enrichment of phosphorus oxoacid species into a certain amount of an anion-exchange resin and the subsequent elution with a calculated volume of potassium chloride solution of a proper concentration. By adding small volumes of magnesium chloride solution and methyl alcohol to the resulting P_{6d} solution, 4 mg of magnesium salt of P_{6d} was obtained.

Decomposition products of $(-P^{III}-)_6$ ring acid with hydrogen peroxide in potassium hydrogen carbonate solution were also examined in a similar way. Compound 3 having a characteristic absorption spectrum with a peak at 268 nm was found in the decomposition products. At pH 3.2, the spectrum of Compound 3 gradually changed into a spectrum of P_{5d}. A considerable amount of diphosphate, P^V--O-P^V, was always found in the same decomposition products. This suggests that the oxidation of phosphorus oxoacid with a framework composed of only P-P bonds leads to the formation of a compound having a P-O-P bond.

In order to calculate the average oxidation number of phosphorus atoms in Compound 3, the final hydrolysis products of Compound 3, i.e., P^I, P^{III} and P^V, were separated and determined by ion-exchange chromatography. The average oxidation number obtained by this analysis was +3.43. The facts mentioned above may suggest that Compound 3 has a structure of Formula 2. The average oxidation number of phosphorus atoms in Formula 2 is calculated to be +3.40. Since Compound 3 is rather unstable even at neutral pH, some of the techniques such as the anion-exchange equiliblium method for charge determination, etc.,cannot be used for further investigation.

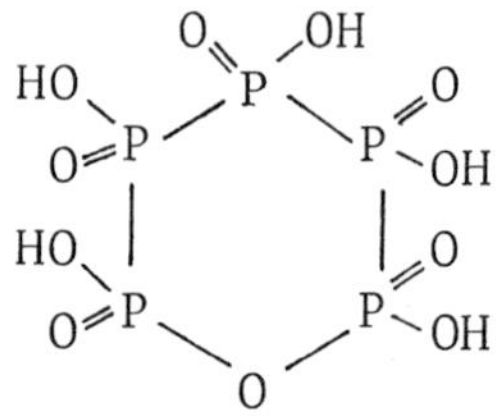

Formula 2

Literature Cited

1. Grayson, M. ; Griffith, E. J., Ed. ; "Topics in Phosphorus Chemistry", Vol. 1, Interscience Publishers, New York 1964; pp. 113-187.
2. Yoza, N.; Yoshidome, H., Ohashi, S. J. Chromatogr. 1978, 150, 393.
3. Blaser, B.; Worms, K. H. Z. Anorg. Allgem. Chem. 1959, 300, 237.
4. Yoza, N.; Ishibashi, K.; Ohashi, S. J. Chromatogr. 1977, 134, 497.
5. Nakashima, T.; Waki, H., Ohashi, S. : J. Inorg. Nucl. Chem. 1977, 39, 1751.
6. Nakashima, T., Nakamura, S.; Sutoh, H.; Ohashi, S. J. Inorg. Nucl. Chem. 1981, 43, 612.
7. Fehér, F., Münzner, H. Chem. Ber. 1963, 96, 1131.

RECEIVED July 7, 1981.

NMR Characterization of Homologous Cyclic Phosphoramides

JACK E. RICHMAN,[1] ROBERT B. FLAY, and O. D. GUPTA

Department of Chemistry, University of Idaho, Moscow, ID 83843

In 1974 a DuPont patent[2] disclosed the synthesis of cyclen phosphine oxide, 2,[3] by hydrolysis of cyclen fluorophosphorane, 1.

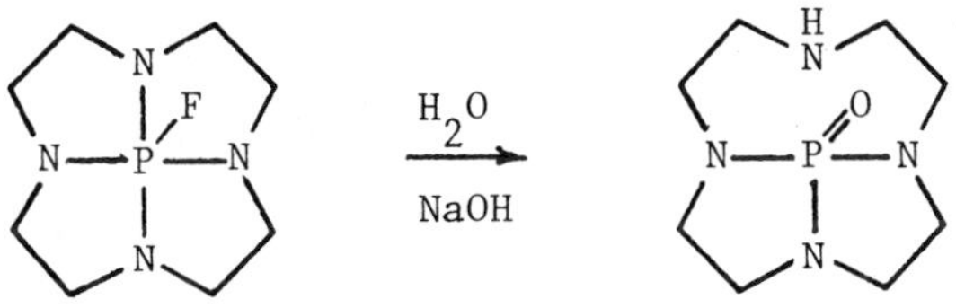

$$\mathbf{1} \xrightarrow[\text{NaOH}]{H_2O} \mathbf{2}$$

In this paper we wish to report the synthesis of a series of compounds (structure 4) that are higher homologues of 2. Oxides 4 have been synthesized by hydrolysis of the corresponding ionic chloride salts, 3, which we previously reported.[4]

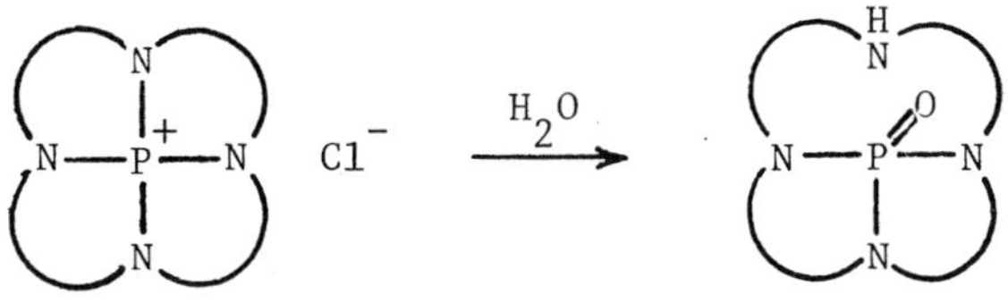

$$\mathbf{3}\ Cl^- \xrightarrow{H_2O} \mathbf{4}$$

(Curved lines in structures 3 and 4 represent ethylene or trimethylene bridges.)

Oxides 4 in each case are volatile compounds that have an intense molecular ion in their mass spectra. ^{31}P NMR studies have been useful in characterizing these compounds as phosphine oxides. Physical and NMR properties of structures 4 are collected in the Table.

0097–6156/81/0171–0271$05.00/0

Table: Physical and Spectroscopic Data for **4**.

structure[a]	Peripheral Ring Size	b.p./pressure (°C/Torr)	m.p. (°C)	^{31}P Chemical Shift[b] 5,5	5,6	6,6
2,2,2,2 (**2**)	12	150-160/0.1	215-17[c]	[d]		
2,2,2,3 (**8**)	13	120/0.03	37.5-40	+40.5	+23.8[e]	
2,2,3,3 (**7**)	14	140-150/0.05	132-136	[f]	24.5	15.2
2,3,2,3 (**5**)	14	125-135/0.05	144-147		25.8	
2,3,3,3 (**9**)	15	140/0.07	100-103		25.9	12.2
3,3,3,3 (**6**)	16	145/0.07	136-139			11.7

[a] Numbers represent bridge sequence in structure **4**.

[b] ^{31}P chemical shifts in $CDCl_3$ are positive in downfield direction from the reference, external 85% H_3PO_4. Numbers represent the smaller rings sizes within which phosphorus is fused.

[c] Cyclen phosphine oxide is extremely hygroscopic. The hydrate melts ca. 180° with partial decomposition.

[d] See footnote 3.

[e] Chemical shifts obtained at -52°, see text.

[f] A minor peak at +40 ppm my represent a small amount of the 5,5-fused isomer in addition to the 5,6- and 6,6-fused isomers.

Two of the oxides, **5** and **6**, shown in the Table can exist in only one structural form.

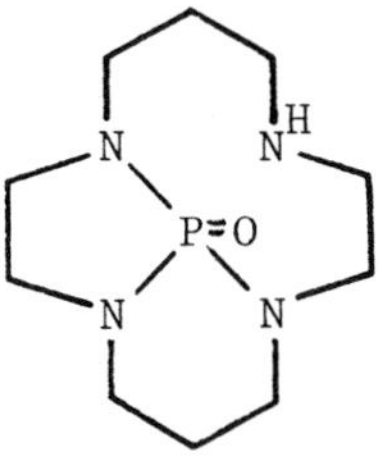

5 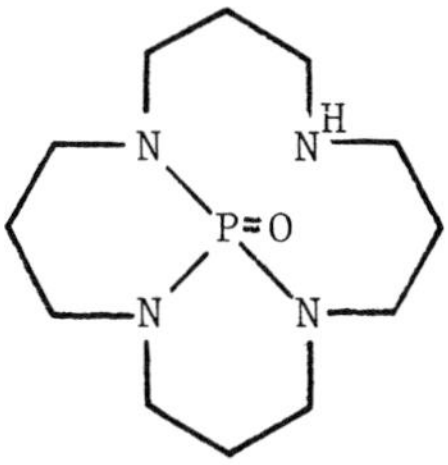 **6**

Both **5** and **6** show only one phosphorus NMR signal at +25.8 and +11.7 ppm, respectively, similar to the shift of hexamethylphosphoric triamide, δ +23.4.[5] The higher field chemical shift of **6** compared to **5** is consistent with the effect of bond angle changes at phosphorus which for **5** and **6** are located at bridgehead positions of fused 5,6- and 6,6-membered rings.[6] The other members of the series of oxides shown in the Table exhibit structural isomers and in each case show two or more ^{31}P NMR signals. The assignments of chemical shifts in the Table to the various isomeric forms are analogous to the shifts of **5** and **6** and are consistent with the expected effects of angle changes at phosphorus fused in rings of different sizes. The chemical shifts change about 15 ppm to higher field on expansion from fused 5,5- to 5,6- and 5,6- to 6,6-membered rings. This is comparable to the trends observed previously for analogous polycyclic compounds containing tetravalent[4] and pentavalent[7] phosphorus atoms.

The assignments for the 2,2,3,3 isomer, **7**, are shown below.

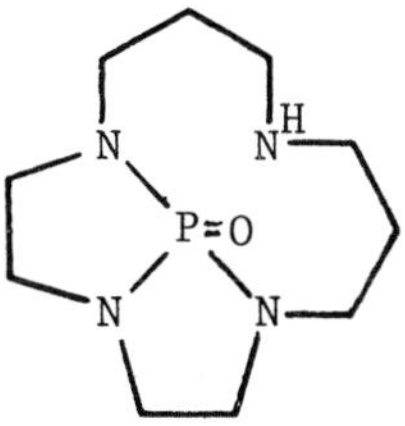

	7a	**7b**	**7c**
$\delta\ ^{31}P$:	+40	+25.4	+15.2
Integral:	trace	2	1

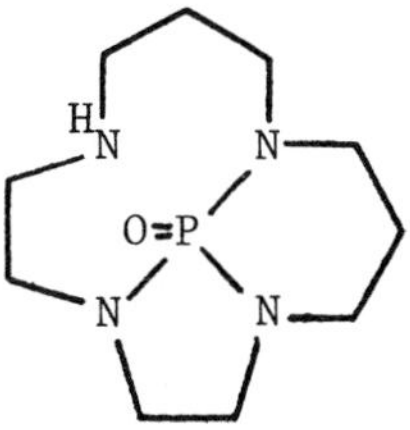

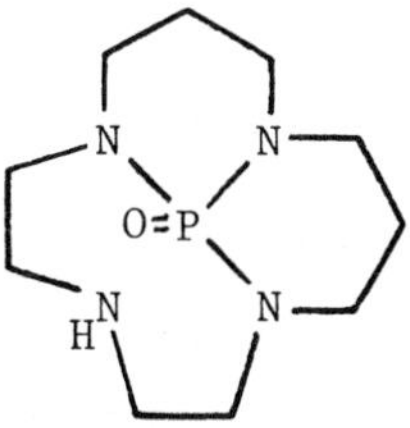

The statistical distribution of isomers 7a, b and c would be 1:2:1. The near absence of isomer 7a indicates unfavorable thermodynamic stability for the phosphorus fused in two five membered rings compared to 5,6- and 6,6-fusion. Greater stability for the larger ring fusion is also exhibited for compounds 8 and 9 which show low field to high field peaks in the approximate ratios 1:8 and 1:2 respectively (statistical abundances are 1:1).

In addition to these phenomena, the 13-membered compound 8 also shows a temperature dependent NMR effect in the ^{31}P spectra. The ^{31}P values shown in the Table were observed at -52°. At higher temperatures the two phosphorus signals broaden and move closer together. This behavior suggests rapid tautomerism of the two isomeric forms of this compound. The high temperature limit where the two signals for 8a and 8b coalesce was not reached in chloroform-d.

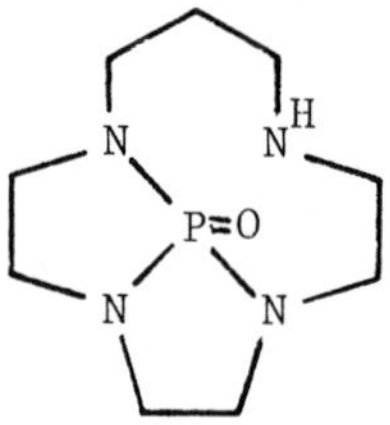

8a 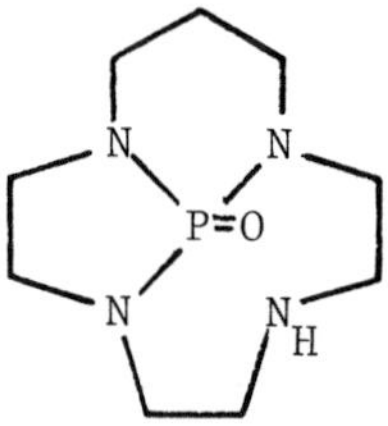 8b

Further studies of tautomerism and dimerization of cyclen phosphine oxide are forthcoming.[3]

Notes and References

(1) Present address: 3M Center, St. Paul, MN 55144.

(2) Richman, J. E., U.S. Patent 3 987 128, 1976.

(3) Unusual structural properties indicated in a detailed NMR investigation of 2 will be reported in a subsequent communication by J. E. Richman and J. J. Kubale.

(4) Richman, J. E.; Gupta, O. D.; Flay, R. B. J. Am. Chem. Soc. 1981, 103, 1291-1292.

(5) Ramirez, F.; Smith, C. P. Tetrahedron Lett. 1966, 3651.

(6) A similar relationship between phosphorus chemical shifts and O-P-O bond angles in phosphate esters has been reported by Gorenstein, D. G. J. Am. Chem. Soc. 1975, 97, 898-900.

(7) Richman, J. E.; Flay, R. B. J. Am. Chem. Soc. 1981, 103, in press.

RECEIVED July 7, 1981.

57

Synthesis and Chemical Behavior of Some Bicyclophosphanes

C. BONNINGUE, O. DIALLO, D. HOUALLA, A. KLAEBE, and R. WOLF

Equipe de Recherche N° 926. Associée au CNRS Université Paul Sabatier, 31062 Toulouse Cédex, France

Although numerous acyclic amidophosphites of type 1, 2 and 3 have been described very early, until recently no bicyclic compound like 4 was known.

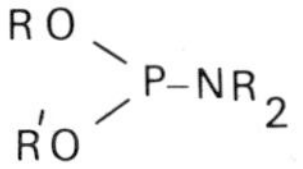

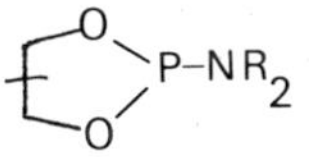

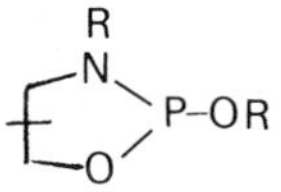

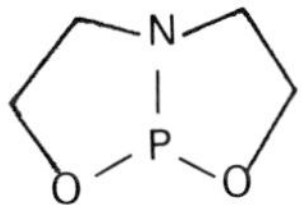

In a note published in 1977 (1), we have described the 2,8-dioxa-5-aza-1-phospha^{III}bicyclo[3.3.0]octanes 4 and 5. In spite of their structural simplicity, they were, to our knowledge, the first phosphorus derivatives of this kind ever described, although the arsenic homologue of 4 has been known since 1970 (2).

Such compounds were not obtained sooner due to two difficulties we met during our approach. (i) First the compounds tend to oligomerize quite easily, however the presence of substituents slows down the phenomenon. (ii) The second difficulty comes from the fact that these compounds can add any molecule with a mobile hydrogen (ROH, R_2NH etc...) leading mainly to phosphoranes with a P-H bond 6 (X = RO-, R_2N-).

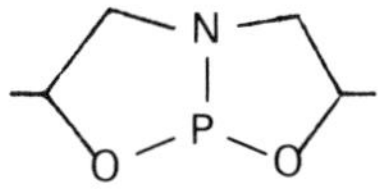

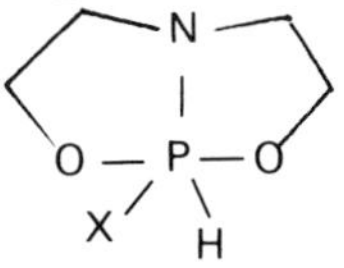

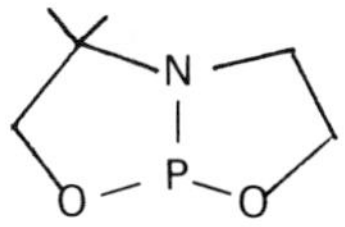

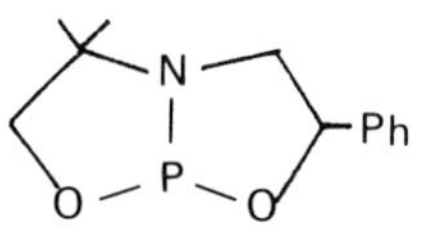

These two observations show that double cyclisation confers on the bicyclic phosphanes particular properties and we were prompted to continue our research in this field.

We have described recentlyfour new bicyclic phosphanes 7, 8 9 (3) and 10 (4) while three interesting articles have been devoted to 10 (5), 4 (6) and 11 (7).

0097-6156/81/0171-0275$05.00/0

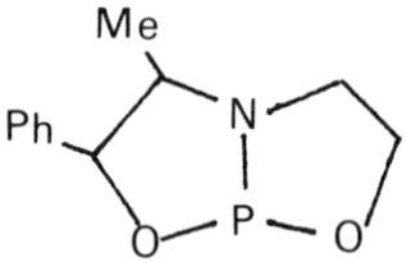

Synthesis : Simple in its principle with the corresponding aminodiol acting either on $(Me_2N)_3P$ or on PCl_3 in the presence of a base, the synthesis is rather delicate owing to the secondary products which are formed. The difficulties encountered in the preparation of 7 have been described (3).

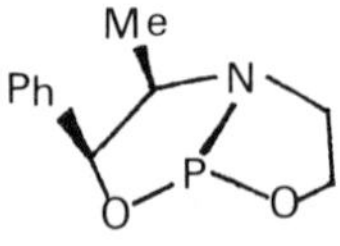

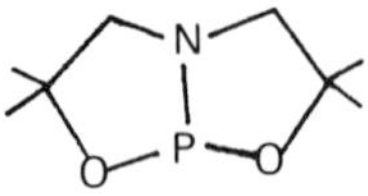

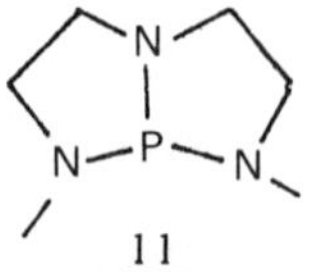

Stereochemistry : The presence in derivatives 5, 8 and 9 of a chiral phosphorus atom and of one or two chiral carbon atoms leads to interesting stereochemical problems. Neglecting the highly constrained conformation in which the phosphorus and nitrogen lone pairs are trans, the number of expected diastereoisomers is equal to 3 for 5, 2 for 8 and 9. In the case of 5, only two diastereoisomers were obtained in a 60/40 ratio. The two diastereoisomers of 8 were obtained in a 85/15 ratio while the synthesis of 9, using an aminodiol with a nor-ephedrine branch, leads to a single diastereoisomer. Starting with the ligand prepared with + nor-ephedrine we isolated 9* in pure enantiomeric form (α^{17}_{546} = +86.8 c = 2.10^{-2} M in benzene).

Conformation : The skeleton of these molecules appears as a two-pitched roof, the lone pairs of phosphorus and nitrogen being located on the roof.

The study of the conformation of the five membered rings was performed in the classical way by detailed analysis of the proton NMR parameters. The conclusions are : the five membered rings have a blocked conformation. This is to be expected as cyclisations are accompanied by strong constraints. The most likely blocked conformation taken on by the rings is an envelope form, the carbon atoms linked to oxygen being at the end of the folds. These folds can be oriented in respect to the pseudo π position of the lone pairs of phosphorus and nitrogen. The detailed examination of the proton NMR parameters leads us to conclude that quite generally the phosphanes studied have an endo-endo conformation 12

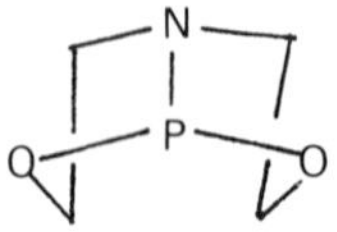

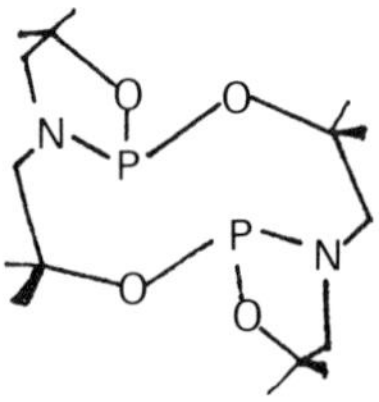

Oligomerization : The marked tendency of these bicyclophosphanes towards oligomerization is an obstacle to their purification. In specific cases, 7 and 10, this difficulty became an advantage. It appeared that the oligomerization was a step by step process and we were able to isolate pure dimers noted $[7]_{Dim.}$ and $[10]_{Dim}$. Our results show the generality of the macrocyclisation process, considered as a set of equilibria between n-mers, (n = 1,2,3...) described first (8) and developed by J.B. Robert et al. (9). Our examples are original under several points : (i) the dimers are readily available, (ii) we are not obliged to sulfurize the tricovalent phosphorus atom in order to isolate them, (iii) $[7]_{Dim}$ and $[10]_{Dim}$ appear in N.M.R. (^{1}H and ^{31}P) as a unique diastereoisomer, among the six which are possible. An X-Ray determination has confirmed this statement for $[10]_{Dim}$. It also shows that the dimerization takes place exclusively by P-O bond rupture (10). The cyclodecane central part of the molecule has a boat-chair-boat conformation and the relative orientation of the two phosphorus lone pairs is trans.

Chemical Properties : The reactivity of bicycloamidophosphites has been explored by means of ^{31}P N.M.R. spectroscopy, using mainly the racemic mixture 9 as a representative derivative of this class of compounds. As we have already mentioned these compounds add easily X-H molecules to give the corresponding hydrido phosphoranes 6. With alcohols ROH (R = Me, Et, n.Pr) the reaction takes place immediately and the potential tautomers 9a, 9b and 9c do not appear. With t-BuOH the addition is slowed down and

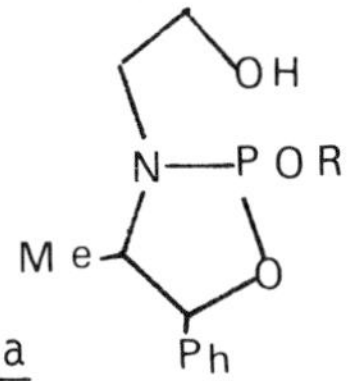

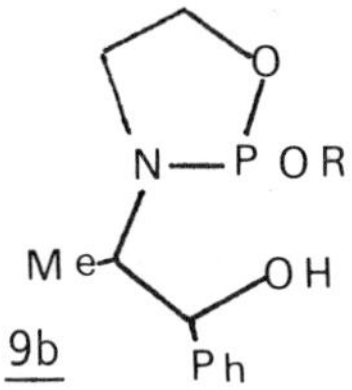

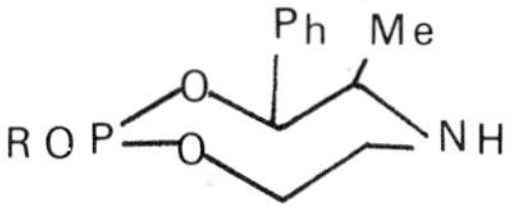

the formation of the phosphorane has been followed kinetically. The reaction is of second order and presents a significant isotopic effect using t-BuOD. Several phenols have given similar results, the oxidative addition beeing also sensitive to steric hindrance. These results enable us to understand more complicated reactions as with pyrocatechol. With benzoic acid, the addition compound 6 (X = $C_6H_5-\underset{\|}{\overset{}{C}}-O-$ O) : $\delta\ ^{31}P = -43.3$ $^{1}J_{P-H} = 820$ Hz

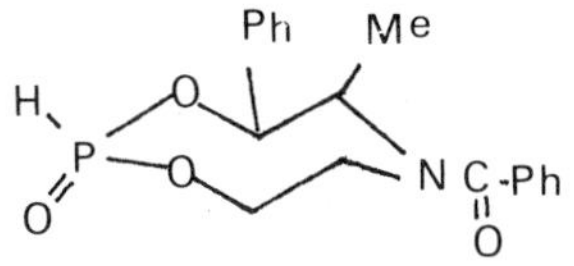

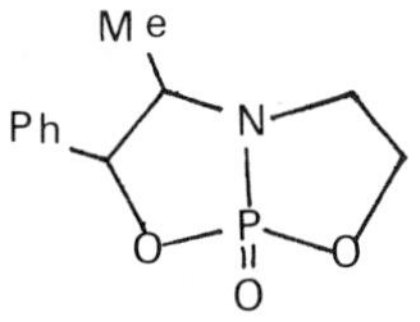

is observed as an intermediate and the final product is the eight membered ring 13 which has been characterized, isolated and analysed : $\delta\ ^{31}P = +5.7$, $^{1}J_{P-H} = 733$ Hz.

With amines, the variety of results is larger. There is no reaction with pyrrole, diphenylamine or N-methylaniline, but with aniline, the equilibrium phosphane + amine $\rightleftharpoons$ aminophosphorane is shifted to the right side (>65%). With propylamine and methylamine, only the phosphorane is present.

The oxidation of 9 occurs with pyridine oxide, dimethylsulfoxide and N_2O_4. The expected final product 14 has been detected (δ ^{31}P = +18) isolated and characterized. The detailed oxidation process is not simple and intermediates are detected before the accomplishment of the oxidation. Compound 14 also collapses into several tetracoordinated phosphorus entities.

Conclusion : As expected, the double cyclisation of bicyclophosphanes induces a strong ring constraint. The marked tendency of these compounds to oligomerization by rupture of the P-O bond is certainly favoured because it minimizes this tension. The driving force of the addition reaction of X-H molecules is the great stability of the bicyclophosphorane structure. Depending upon the nature of X, the P^{III} tautomer forms of the last compounds could be observed or not.

REFERENCES

1- D. Houalla, F.H. Osman, M. Sanchez and R. Wolf, Tetrahedron Letters 1977, 3041.

2- K. Sommer, W. Lauer and M. Becke-Goering, Z. Anorg. Chem. 1970, 379, 48.

3- C. Bonningue, D. Houalla, M. Sanchez, R. Wolf and F.H. Osman, J.C.S. Perkin II 1981, (1), 19.

4- C. Bonningue, Thèse de l'Université Paul Sabatier. Toulouse Juin 1980.

5- D. Grec, L.G. Hubert-Pfalzgraff, J.G. Riess and A. Grand, J. Amer. Chem. Soc. 1980, 102, (23), 7133.

6- D.B. Denney, D.Z. Denney, P.J. Hammond, Chialang Huang and Kuo-Shu Tseng, J. Amer. Chem. Soc. 1980, 102, (15), 5073.

7- D.B. Denney, D.Z. Denney, D.M. Gavrilovic, P.J. Hammond, Chialang Huang and Kuo-Shu Tseng, J. Amer. Chem. Soc. 1980, 102, (23), 7072.

8- J.P. Albrand, J.P. Dutasta and J.B. Robert, J. Amer. Chem. Soc. 1974, 96, 4584.

9- (a) J.B. Robert and H. Weichmann, J. org. Chem. 1978, 43, (15), 3031.
(b) J. Martin and J.B. Robert, Nouveau J. Chimie, 1980, 4, (8-9) 515.

10- J. Jaud, J. Galy, D. Houalla, C. Bonningue and R. Wolf, to be published.

RECEIVED July 7, 1981.

Reactions of 2,4-Bis(4-methoxyphenyl)-1,3,2,4-dithiadiphosphetane 2,4-Disulfide

SVEN-OLOV LAWESSON

Department of Organic Chemistry, Chemical Institute, University of Aarhus, DK-8000 Aaarhus C, Denmark

The reaction of anisole with P_4S_{10} produces in high yields the title compound, LR, (1) (now commercially available)

$$CH_3O-C_6H_4-P(=S)(\mu\text{-}S)_2P(=S)-C_6H_4-OCH_3 \quad \text{(LR)}$$

In the solid phase LR is in the E-form, but in solution the ^{31}P NMR spectrum shows about 10 absorptions showing that different species are present and that precautions should be exercised when considering mechanisms. We now want to report on LR as a thiation reagent and in synthesis of P-heterocycles. It should be noted that LR also reacts with nucleophiles and is a deoxygenation reagent.

LR is superiour to most of the thiation agents hitherto known as it is

- easily prepared from inexpensive materials
- stable at room temperature and easy to handle
- reacting at relatively low temperature (-20-140°C)
- forming a stable product after the $>C=O \rightarrow >C=S$ transformation.

It has thus been found that LR smoothly transforms

- ketones into thioketones
- amides into thioamides
- lactams into thiolactams
- N',N'-disubstituted hydrazides into thiohydrazides
- esters into O-substituted thioesters
- S-substituted thioesters into dithioesters
- lactones into thionolactones
- thiololactones into dithiolactones

0097-6156/81/0171-0279$05.00/0

- pyrazolidinediones into pyrazoline-5-thione disulfides
- enaminones into enamino-thiones
- peptides into thiopeptides

(see Scheme 1, reactions 1-5 and Scheme 2, reaction 10)

Scheme 1.

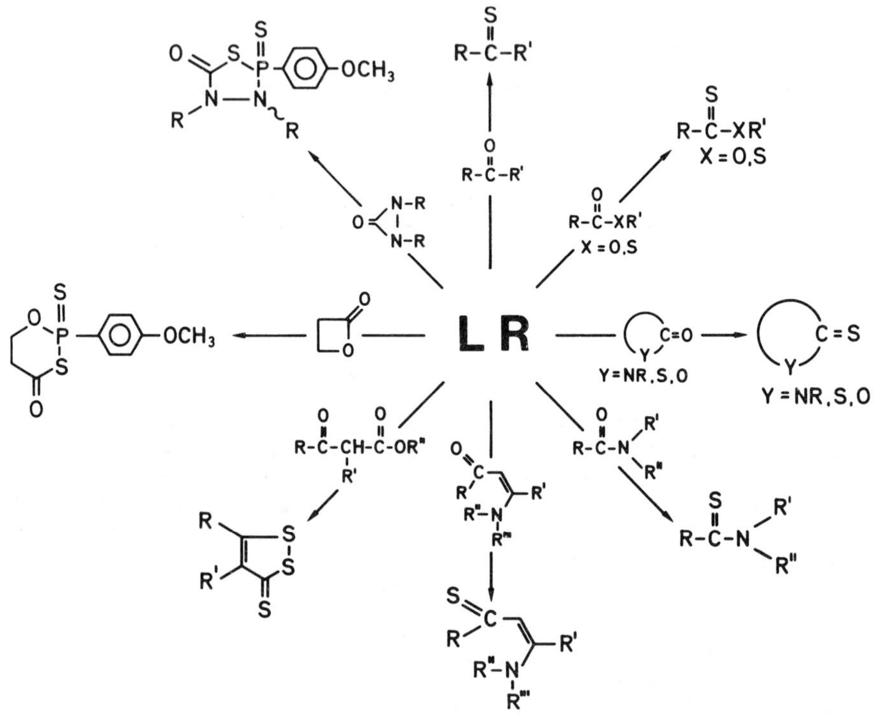

It is assumed that LR exists in solution as the dimer (E and/or Z) or as monomer (tricoordinated pentavalent species (2))

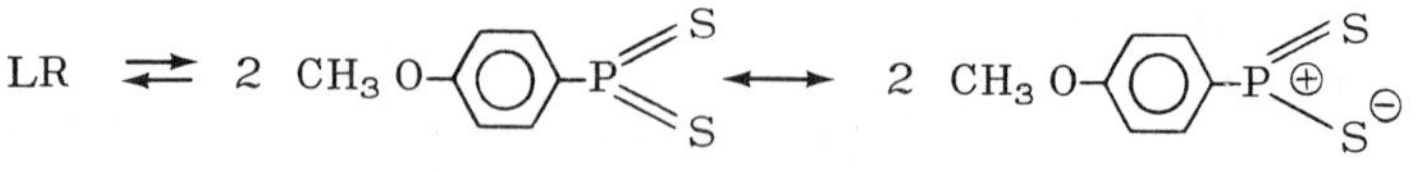

$$LR \rightleftarrows 2\ CH_3O\text{-}C_6H_4\text{-}P(=S)_2 \longleftrightarrow 2\ CH_3O\text{-}C_6H_4\text{-}P^{\oplus}(=S)S^{\ominus}$$

When reacting LR with different bifunctional substrates (Schemes 1 and 2: reactions 7-9, 11-16) different types of P-heterocycles are formed. Also dispiro-trithiaphosphetanes are formed from cycloalkanones.

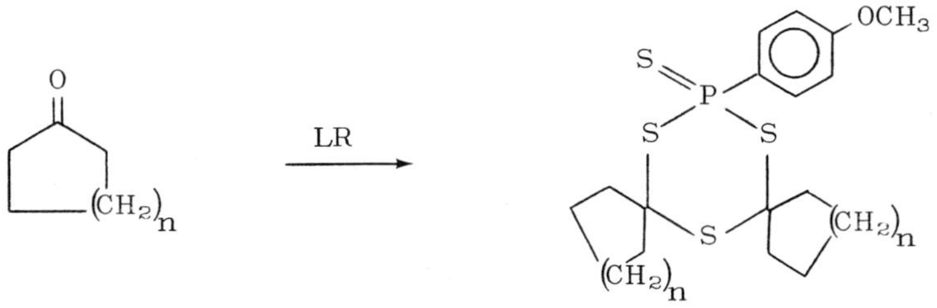

Scheme 2.

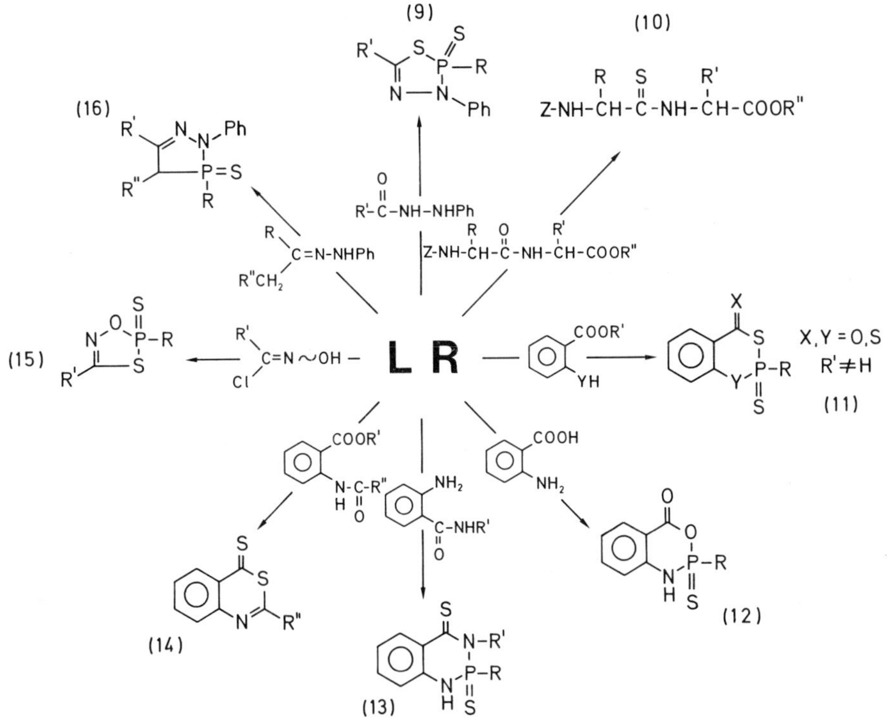

1. Thomsen, I., Clausen, K., Scheibye, S., Lawesson, S.-O. Org. Synth. (submitted).
2. Bertrand, G., Majoral, P., Baceiredo, A. Tetrahedron Letters 1980, 5051.

RECEIVED June 30, 1981.

Small Rings with Tervalent Phosphorus

EKKEHARD FLUCK and HORST RICHTER

Institute of Inorganic Chemistry, University of Stuttgart, Pfaffenwaldring 55, D-7000 Stuttgart 80, FRG

In recent years diaza-λ^3,λ^3-diphosphetidines (I) have found increasing interest

$$\text{R-P}\overset{\text{N(R')}}{\underset{\text{N(R')}}{\diamond}}\text{P-R} \qquad \text{I}$$

They can occur as cis and/or trans isomers (1).

While R is always a dialkylamino, a bis(trimethylsilyl)amino group or a halogen atom, R' varies more. As to the compounds which are described in literature R' is very often a phenyl group (2), in other cases a t-butyl (3) or trimethylsilyl (4) and in very few cases an acyl group (5).

We are able to synthesize the diazadiphosphetidine (II) in which R' is a phosphoryl group. By reaction of amidophosphoric diethyl ester with tris(diethylamino)phosphane in the absence of a solvent, II can be obtained in 80% yield according to eq. (1) and (2):

$$(C_2H_5O)_2\overset{O}{\overset{\|}{P}}\text{-NH}_2 + P[N(C_2H_5)_2]_3 \rightarrow \underset{\text{III}}{(C_2H_5O)_2\overset{O}{\overset{\|}{P}}\text{-}\underset{H}{\underset{|}{N}}\text{-P}[N(C_2H_5)_2]_2} + HN(C_2H_5)_2 \qquad (1)$$

0097–6156/81/0171–0283$05.00/0

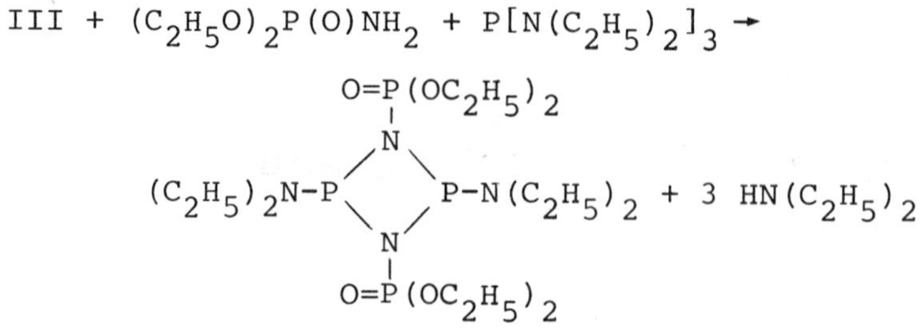

$$\mathrm{III} + (C_2H_5O)_2P(O)NH_2 + P[N(C_2H_5)_2]_3 \rightarrow \tag{2}$$

$$\begin{array}{c} O{=}P(OC_2H_5)_2 \\ | \\ N \\ (C_2H_5)_2N{-}P \diagup \diagdown P{-}N(C_2H_5)_2 \\ \diagdown \diagup \\ N \\ | \\ O{=}P(OC_2H_5)_2 \\ \mathrm{II} \end{array} + 3\ HN(C_2H_5)_2$$

II forms colorless crystals which are extremely sensitive towards oxidation and hydrolysis. The chemical shift of the ring phosphorus atoms (δ=+169.7 ppm) indicated, that the trans isomer is formed. This assumption was verified by X-ray structural analysis (molecular symmetry c_i).

If the reaction is carried out in the presence of a solvent such as toluene the intermediate product III does not split off diethylamine. By proton migration IV is formed as the only product of reaction:

$$\underset{\mathrm{III}}{(C_2H_5O)_2\overset{O}{\overset{\|}{P}}{-}\underset{H}{\underset{|}{N}}{-}P[N(C_2H_5)_2]_2} \rightarrow \underset{\mathrm{IV}}{(C_2H_5O)_2\overset{O}{\overset{\|}{P}}{-}N{=}\underset{H}{\underset{|}{P}}[N(C_2H_5)_2]_2}$$

The nmr coupling constants in IV are $^1J(PH)$ = 580 Hz, $^2J(PP)$ = 46.7 Hz.

Reaction of amidophosphoric diphenylester with tris(diethylamino)phosphane yields only 10% of the diazadiphosphetidine, even in the absence of a solvent. The main product is compound V.

$$(C_6H_5O)_2\overset{O}{\overset{\|}{P}}{-}N{=}\underset{H}{\underset{|}{P}}[N(C_2H_5)_2]_2 \qquad \mathrm{V}$$

By oxidation of II with mercury(II)-oxide the diaza-λ^3,λ^5-diphosphetidine VI was obtained, colorless crystals of m.p. 62 ^oC. The same product is obtained when pure oxygen is used for oxidation though it is always accompanied by impurities of the corresponding diaza-λ^5,λ^5-diphosphetidine VII.

With elemental sulfur II can be oxidized to give mono- (VIII) and disulfide (IX). Both form colorless crystals having melting points of 56-57 ^oC and 135 ^oC, respectively.

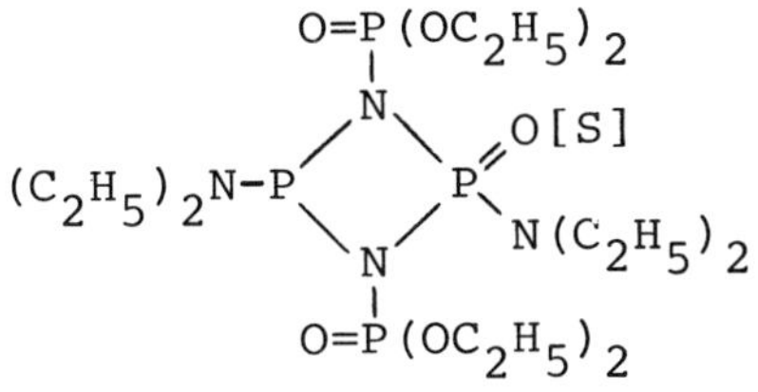

VI [VIII]

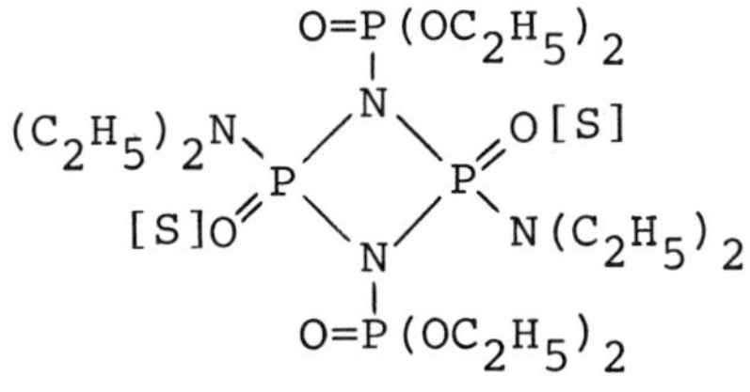

VII [IX]

Reaction of phenylphosphonic dianilide with tris-(diethylamino)phosphane yields colorless crystals of compound X (m.p. 156-157 °C). Reaction of N,N′-diphenylsulfamide with the same reagent leads to compound XI, which is the first member of the class of thiadiazaphosphetidines containing tervalent phosphorus. The colorless crystals melt at 94-95 °C. With elemental sulfur the corresponding sulfide is obtained.

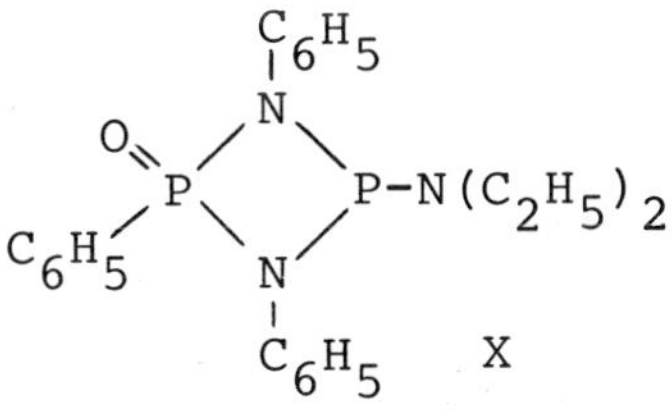

X

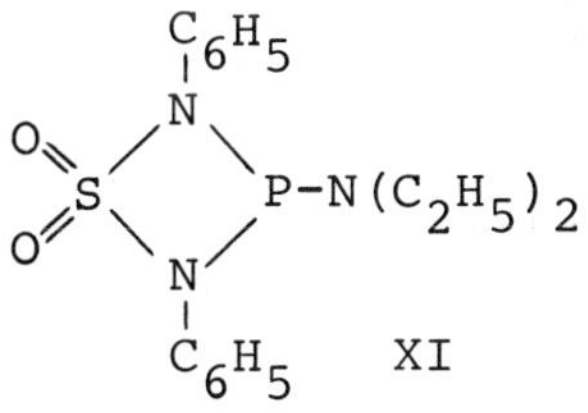

XI

In compound X the diethylamino group can be exchanged for chlorine when it is reacted with PCl_3. A mixture of isomers of XIIa and XIIb was observed.

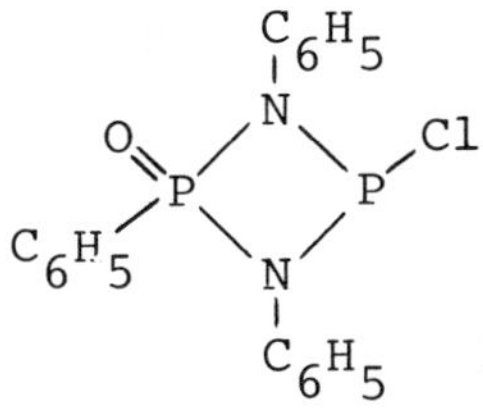

XIIa

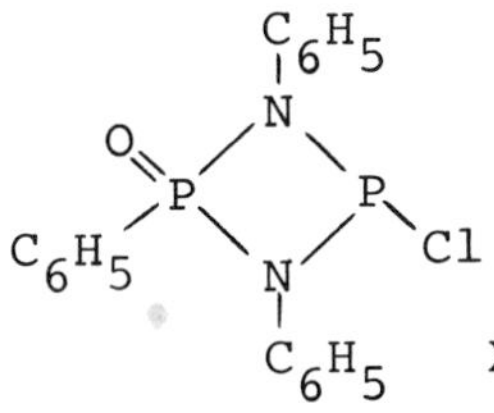

XIIb

In a similar way the diethylamino group in the reaction product of N,N′-dimethylurea and tris(diethylamino)phosphane, compound XIII (6), was reacted with PCl_3. The product,however, was compound XIV.

$$\underset{\text{XIII}}{O{=}C\langle N(CH_3)\rangle_2 P{-}N(C_2H_5)_2} + 2\ PCl_3 \rightarrow \underset{\text{XIV}}{O{=}C[N(CH_3){-}PCl_2]_2} + PCl_2[N(C_2H_5)_2]$$

Literature Cited

1. see e.g. Pohl, S. Z. Naturforsch. 1979, 34b, 256. - Schwarz, W.; Hess, H.; Zeiss, W. Z. Naturforsch. 1978, 33b, 723. - Keat, R.; Keith, A. N.; Macphee, A.; Muir, K. W.; Thompson, D. G. J. Chem. Soc. Chem. Commun. 1978, 372.
2. see e.g. Grimmel, H. W.; Guenther, A.; Morgan, J.F. J. Am. Chem. Soc. 1946, 68, 539. - Zeiss, W.; Weis, J. Z. Naturforsch. 1977, 32b, 485. - Fluck, E.; Wachtler, D. Liebigs Ann. Chem. 1979, 1125.
3. Scherer, O. J.; Klusmann, P. Angew. Chem. 1969, 81, 743.
4. Niecke, E.; Flick, W. J. Organometal. Chem. 1976, 104, C23. - Zeiss, W.; Feldt, Ch.; Weis, J.; Dunkel, G. Chem. Ber. 1978, 111, 1180.
5. Devillers, J.; Willson, M.; Burgada, R. Bull. Soc. Chim. France 1968, 4670. - Bowden, F. L.; Dronsfield, A. T.; Haszeldine, R. N.; Taylor, D. R. J. Chem. Soc. Perkin I, 1973, 516.
6. Bermann, M.; Van Wazer, J. R. J. Chem. Soc. Dalton 1973, 813.

RECEIVED June 30, 1981.

60

The Unexpected Formation of 1,2-Oxaphosphol-3-ene 2-Oxides in the Reaction of Diacetone Alcohol with Phosphonous Dihalides

KURT MOEDRITZER and RAYMOND E. MILLER

Monsanto Agricultural Products Company, Research Department, 800 North Lindbergh Boulevard, St. Louis, MO 63166

In exploring various synthetic methods for the preparation of 1,2-oxaphospholene 2-oxides we investigated the reaction of diacetone alcohol (4-hydroxy-4-methyl-2-pentanone) with methyl and phenylphosphonous dichloride, which was reported (1) to proceed as shown in eq (1) to give 2,3,3,5-tetramethyl-1,2-oxaphosphol-4-ene 2-oxide or the corresponding 2-phenyl derivative,

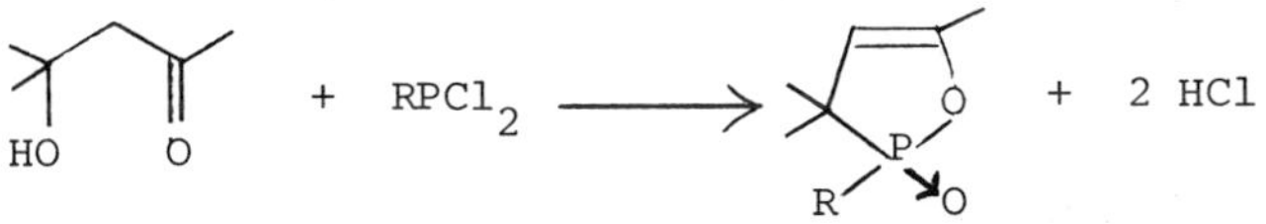

HO O + $RPCl_2$ ⟶ (R, P, O, O) + 2 HCl (1)

Ia: $R=CH_3$
IIa: $R=C_6H_5$

respectively. In the course of these studies we found, however, that the distilled product samples of the reaction of eq (1) displayed two ^{31}P nmr signals. For $R=CH_3$ (bp 120-144°/0.2 mm, 61% yield) the two signals (in the approximate intensity ratio 8:1) were at +77.0 and +61.8 ppm, respectively, with the chemical shifts measured versus 85% H_3PO_4 and with downfield shifts being positive. For $R=C_6H_5$ (bp 160°/0.1 mm, 80% yield) the two signals were at +65.9 and +52.3 ppm, respectively, also in the approximate intensity ratio of 8:1.

By slow spinning band distillation in each case ($R=CH_3$, C_6H_5) the two compounds corresponding to the two ^{31}P nmr signals were separated with the major product in both instances having the downfield ^{31}P nmr chemical shift ($R=CH_3$, δ ^{31}P + 75.1, 36% yield; $R=C_6H_5$, δ ^{31}P + 65.9, 58% yield). These compounds were identified spectroscopically (^1H and ^{13}C nmr, MS, IR) and by microanalysis as the expected 1,2-oxaphosphol-4-ene derivatives of the reaction of eq (1).

The minor product, having in each instance the more upfield ^{31}P nmr chemical shift ($R=CH_3$, δ ^{31}P + 59.4, 10% yield); $R=C_6H_5$, δ ^{31}P + 52.3, 5% yield), was identified spectroscopically (^1H and ^{13}C nmr, MS, IR) and by microanalysis as the isomeric 1,2-oxaphosphol-3-ene derivative of the structure shown below.

0097-6156/81/0171-0287$05.00/0

Ib: $R=CH_3$
IIb: $R=C_6H_5$

Key features of the proton nmr spectra of Ib and IIb are the following. The olefinic proton appears as doublet of a poorly resolved quartet, Ib, δ 6.57, $^3J_{HP}$ = 38 Hz; IIb, δ 6.53, $^3J_{HP}$ = 38 Hz. The protons of the two geminal methyl groups on the carbon next to O in the ring are not coupled to phosphorus, however, by virtue of the chirality at phosphorus appear as two singlets, Ib, δ 1.42 and 1.50; IIb, δ 1.48 and 1.57. The protons of the methyl group attached to the olefinic carbon atom appear as a doublet of doublets due to coupling to phosphorus and the olefinic proton, Ib, δ 1.97, $^3J_{HP}$ = 13, $^4J_{HH}$ = 1.7 Hz; IIb, δ 1.83, $^3J_{HP}$ = 13, $^4J_{HH}$ = 2 Hz. Further support for the structure of Ib and IIb is derived from ^{13}C nmr spectra shown in Table I, with the assignments confirmed by off-resonance decoupling.

The isolation of a solid intermediate under certain reaction conditions from the reaction of eq (1) for $R=CH_3$ and its identification as 3-chloro-2,3,5,5-tetramethyl-1,2-oxaphospholane 2-oxide, shown as compound D in Scheme I, (mp 97°, δ ^{31}P +65.7; ^1H nmr,

Table I. ^{13}C Nmr Chemical Shifts (in ppm) and ^{13}C-^{31}P Coupling Constants (in parentheses in Hz)[a]

	Ib	IIb	D
C-1	129.31 (103.0)	--[b]	64.34 (85.5)
C-2	148.46 (18.4)	148.80 (19.1)	51.65 (7.4)
C-3	87.20 (3.7)	87.71 (0.0)	85.13 (0.0)
C-4	14.71 (11.5)	11.66 (14.7)	29.15 (4.4)
C-5	28.77 (0.0)	29.06 (0.0)	31.93 (0.0)
C-6	27.71 (2.9)	27.54 (3.0)	25.55 (0.0)
C-7	15.84 (94.9)	--[c]	10.69 (101.5)

a) Spectra were obtained at 25.05 MHz on a JOEL FX-100 FT NMR Spectrometer; peaks are referenced versus TMS with downfield shifts being positive.
b) Only the upfield peak of the doublet for C-1 is seen at 127.38 ppm, the downfield peak overlaps with the aromatic carbons.[c]
c) Resonances for the aromatic carbon atoms not shown here.

SCHEME I

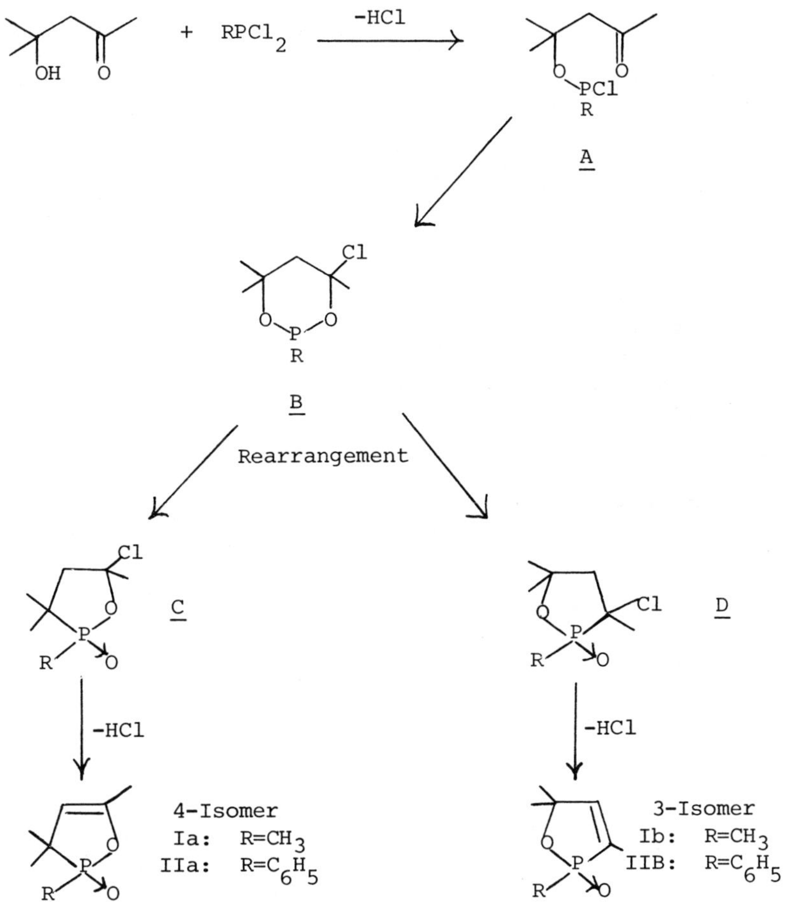

OH O
+ $RPCl_2$
-HCl
O O
PCl
R
A
Cl
O P O
R
B
Rearrangement
Cl
O
P
R O
C
O
P
Cl
R O
D
-HCl
-HCl
O
P
R O
4-Isomer
Ia: $R=CH_3$
IIa: $R=C_6H_5$
O
P
R O
3-Isomer
Ib: $R=CH_3$
IIB: $R=C_6H_5$

geminal CH_3 groups at δ 1.53 and 1.55; CH_3 on ring carbon atom next to P δ 1.81, d ($^3J_{HP}$ = 14.9 Hz); CH_3 on P, δ 1.86, d ($^2J_{HP}$ = 13 Hz); CH_2 is represented by an ABX (X=P) spin coupling pattern with δ_A 2.54, δ_B 2.47, $^2J_{AB}$ = 15.1, $^3J_{AX}$ = 10.5 and $^3J_{BX}$ = 10.5 Hz; ^{13}C nmr data in Table I) gives support to a suggested mechanism for the unexpected formation of the isomeric 1,2-oxaphosphol-3-ene 2-oxides as summarized in Scheme I.

In the mechanism of Scheme I it appears reasonable to assume, in agreement with earlier studies (1), that solvolysis is the first step of the reaction of eq (1). In a second step, in analogy to the well documented (2) reaction of P^{III}-Cl compounds with carbonyl groups, A may cyclize to B, with B having the capability of rearranging by two pathways, resulting in C and/or D (with D having been isolated for R=CH_3) which upon dehydrochlorination give the 4-isomer or the 3-isomer, respectively.

Literature Cited

1. Arbuzov, B. A.; Rizpolozhenskii, N. I.; Vizel, A. O.; Ivanovskaya, K. M.; Mukhametov, F. S.; Gol'dfarb, E. I. Izvest. Akad. Nauk SSSR, Ser. Khim. (Engl. Transl.), 1972, 1827.
2. Fild, M.; Schmutzler, R.; Peake, S. C.; "Organic Phosphorus Compounds"; Eds. Kosolapoff, G. M.; Maier, L.; Wiley-Interscience, New York, N.Y.; 1972; Vol. 4, p 169.

RECEIVED June 30, 1981.

Dihydrophenophosphazines via the Interaction of Diarylamines and Phosphorus Trichloride: Applications and Limitations

HAROLD S. FREEMAN and LEON D. FREEDMAN

Department of Chemistry, North Carolina State University, Raleigh, NC 27650

In 1971 it was shown (1) that the interaction of diphenylamine and phosphorus trichloride at 210° (followed by treatment of the reaction mixture with water) yields not only the previously described phosphine oxide 1 but also the spirophosphonium chloride 2.

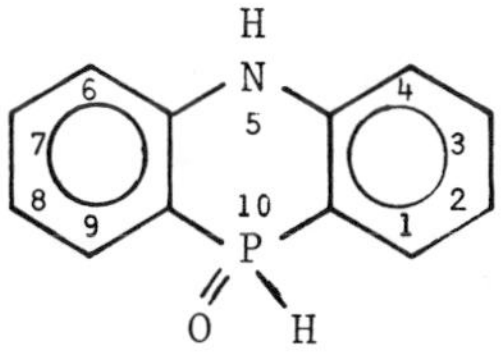

1

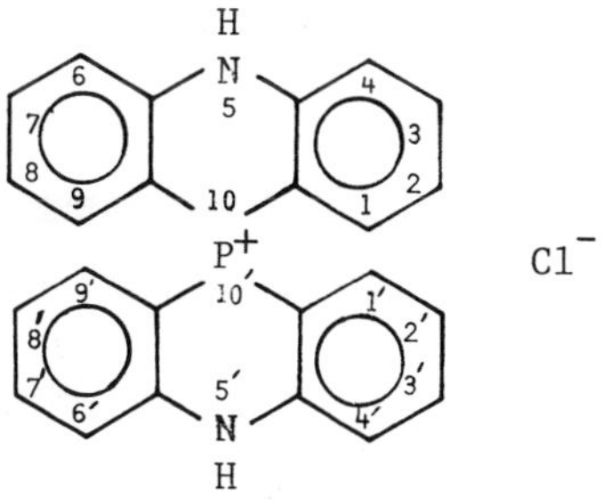

2

The mechanism of formation of the latter compound (a derivative of P^V) from phosphorus trichloride has not been elucidated, but the following mechanism suggested (2) for the formation of 1 seems reasonable:

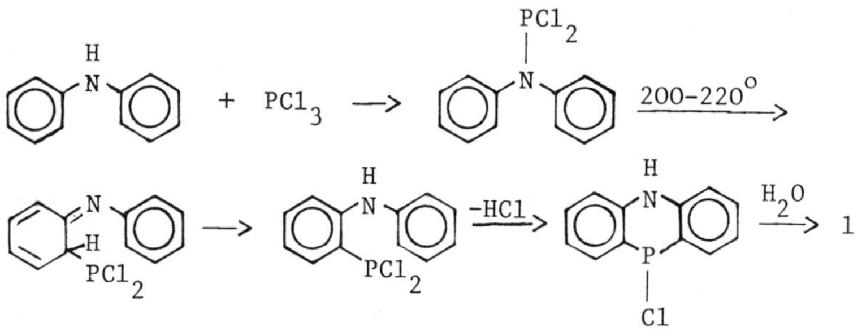

$$Ph_2NH + PCl_3 \longrightarrow Ph_2N\text{-}PCl_2 \xrightarrow{200-220^\circ} \cdots \xrightarrow{-HCl} \cdots \xrightarrow{H_2O} 1$$

0097-6156/81/0171-0291$05.00/0

Studies (2,3) have also been reported of the reaction of phosphorus trichloride with diarylamines containing *p*-methyl or *p*-chloro substituents. In every case, the expected ring-substituted derivatives of 1 and 2 were obtained after the reaction mixture was treated with water. The interaction of *N*-phenyl-*o*-toluidine and phosphorus trichloride at 200°C also gave a reaction mixture from which the expected phosphine oxide was isolated (3). None of the corresponding spirophosphonium chloride, however, could be obtained. The failure to isolate this substance can not be explained simply by the presence of an ortho substituent in the diarylamine, since it had been previously found that a 34% yield of a spirophosphonium chloride can be obtained via the interaction of *N*-phenyl-1-naphthylamine and phosphorus trichloride (2). No dihydrophenophosphazine derivatives at all were obtained by the interaction of di-*o*-tolylamine and phosphorus trichloride at 200°C (3). It was possible, however, to obtain a small yield of the expected phosphine oxide (isolated as the phosphinic acid) by converting di-*o*-tolylamine to the corresponding phosphoramidous dichloride, $(o\text{-}MeC_6H_4)_2NPCl_2$, dehydrohalogenating the latter substance, and treating the reaction mixture with water.

The present paper is concerned with the reaction between phosphorus trichloride and the *meta*-substituted diarylamines listed in Table I. The

TABLE I

DIARYLAMINES STUDIED

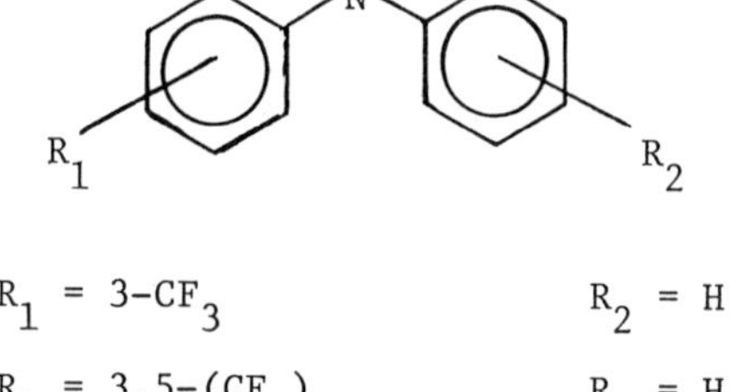

3a	R_1 = 3-CF_3	R_2 = H
3b	R_1 = 3,5-$(CF_3)_2$	R_2 = H
3c	R_1 = 3,5-Me_2	R_2 = H
3d	R_1 = 3-Me	R_2 = 3-CF_3
3e	R_1 = 3-CF_3	R_2 = 3-CF_3

substituents include both the activating, *ortho*, *para*-directing methyl group and the deactivating, *meta*-directing trifluoromethyl group. Each of the amines was heated with phosphorus trichloride at 220-250°C for about 17 h, and the reaction mixtures were then treated with water. The first three amines (3a-3c) yielded modest amounts of phosphine oxides but no spirophosphonium chlorides, while the other two amines gave no organophosphorus compounds at all. A phosphine oxide was obtained from the fourth amine (3d), however, via dehydrohalogenation of the phosphoramidous dichloride at 220°C. The last amine (3e) could also be converted to a phosphoramidous dichloride, but the latter substance could not be dehydrohalogenated to a dihydrophenophosphazine derivative even at 255°C.

An ^{1}H NMR study of the phosphine oxides obtained from amines 3a and 3d indicated that each oxide consists of a single isomer. Of the two possible isomeric products from 3a, only 3-trifluoromethyl-5,10-dihydrophenophosphazine 10-oxide (4a) was actually isolated. The ^{1}H NMR spectrum of this substance in deuterated DMSO showed a single peak for the N-H group and two signals assigned to H_a and H_b (J_{P-H} for each signal was about 13 Hz). Similarly, an analysis of the ^{1}H NMR spectrum of the phosphine oxide obtained from amine 3d indicated that only 3-methyl-7-trifluoromethyl-5,10-dihydrophenophosphazine 10-oxide (4b) was present; no evidence for the formation of any of the three other possible isomers was obtained.

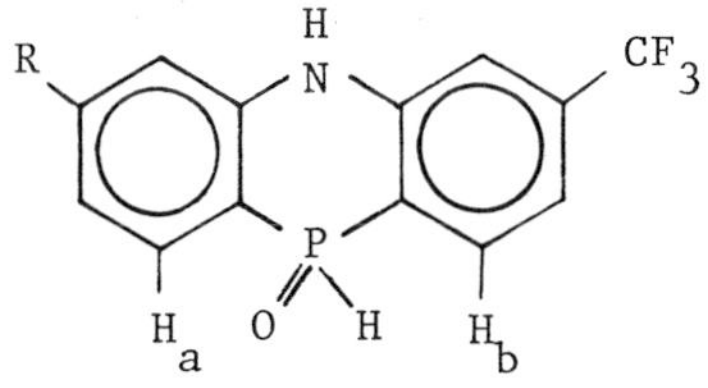

4a R = H

4b R = Me

Steric factors probably play a key role in determining the regiospecificity of the reactions leading to the phosphine oxides 4a and 4b. An intermediate involved in both cases is probably of type 5, in which ring closure preferentially occurs *para* rather than *ortho* to the bulky CF_3 group. The preferential formation of 5b from the corresponding phosphoramidous dichloride is undoubtedly associated with the fact that

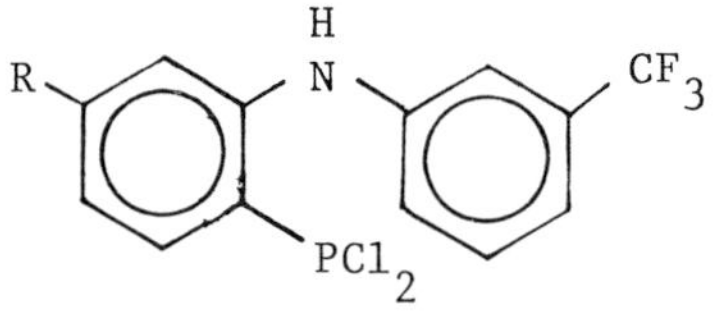

5a R = H

5b R = Me

the 6-position (*para* to the methyl group) is less hindered than the 2-position (which is *ortho* to both the methyl group and the $NHC_6H_4CF_3$ group).

The results obtained in this investigation suggest that dihydrophenophosphazine synthesis from a diarylamine requires that at least one ring of the amine must be free of a *meta*-directing group. If this requirement is fulfilled, the PCl_2 group bonded to the nitrogen of the phosphoramidous dichloride can migrate to an *ortho* position of that ring, and cyclodehydrohalogenation will subsequently occur. The regiospecificity noted in the two cases where more than one isomeric product seemed possible indicates that steric factors must be important in determining both the site to which the PCl_2 group of the phosphoramidous dichloride migrates and the site at which ring closure occurs.

Literature Cited

1. Jenkins, R.N.; Freedman, L.D.; Bordner, J. Chem. Commun. 1971, 1213.
2. Jenkins, R.N.; Freedman, L.D. J. Org. Chem. 1975, 40, 766.
3. Butler, J.R.; Freeman, H.S.; Freedman, L.D. Phosphorus Sulfur 1981, 9, 269.

RECEIVED July 7, 1981.

PHOSPHAZENES

Contributions to the Chemistry of *N*-Phosphoryl Phosphazenes

L. RIESEL, E. HERRMANN, A. PFÜTZNER, J. STEINBACH, and B. THOMAS

Departments of Chemistry, Humboldt University of Berlin and Technical University Dresden, GDR

N-phosphorylated phosphazenes are characterized by P=N-PO-sequences. N-dialkoxyphosphoryl trialkoxyphosphazenes ("pentaesters"), $(R'O)_2PO-N=P(OR)_3$, especially those with different alkyl groups are biologically active compounds. Tetraesters of imidodiphosphoric acid, $(RO)_2PO-NH-PO(OR)_2$, formally derived from the former by replacing one alkyl group by a hydroxyl group, are known as indeed good chelate ligands (1). Complex compounds with ligands of this kind are used for separating metal ions by extraction methods.

Esters of N-phosphoryl phosphazenes are usually formed by the Staudinger reaction which requires the handling of the extremely toxic phosphoric acid ester azides (eq. 1). For developing new synthetic

$$(R'O)_2P(O)N_3 + P(OR)_3 \longrightarrow (R'O)_2P(O)-N=P(OR)_3 + N_2 \qquad (1)$$

approaches to obtain N-phosphoryl phosphazenes we thoroughly studied both the reaction of $Cl_3P=N-POCl_2$ with O-nucleophiles and reactions for P=N-P-bridge formation which avoid handling phosphoryl azides.

Solvolysis of $Cl_3P=N-POCl_2$. P_2NOCl_5 reacts with alcohols in the molar ratio 1:1, forming alkoxydichlorophosphazenes (eq. 2; n=1).

$$Cl_2OP-N=PCl_3 + n\ ROH \longrightarrow Cl_2OP-N=PCl_{3-n}(OR)_n + n\ HCl \qquad (2)$$

With excessive alcohol a further substitution occurs at the same phosphorus atom, though due to by-reactions to a considerably smaller extent. For instance in the reaction with ethanol $(EtO)Cl_2P=N-POCl_2$ is formed in 97 % yield and $(EtO)_2ClP=N-POCl_2$ in 75 % yield but $(EtO)_3P=N-POCl_2$ only to an extent of less than 15 %, even when using a great excess of ethanol. These by-reactions are alkylation reactions (eq. 3), phosphazene-phosphazane rearrangements (eq. 4) and olefin eliminations (eq. 5). In all cases imidodiphosphoryl compounds are formed.

0097-6156/81/0171-0297$05.00/0

$$(RO)X_2P{=}N{-}P(O)X_2 + Nuc^- \longrightarrow Nuc{-}R + [X_2P(O){\cdots}N{\cdots}P(O)X_2]^- \quad (3)$$

$$(RO)X_2P{=}N{-}P(O)X_2 \xrightarrow{\sim R} X_2P(O){-}NR{-}P(O)X_2 \quad (4)$$

$$(RO)X_2P{=}N{-}P(O)X_2 \longrightarrow X_2P(O){-}NH{-}P(O)X_2 + \text{olefin} \quad (5)$$

X denotes Cl or OR.

The exclusive substitution at the phosphazene side of the molecule as well as these three types of by-reactions can be understood by taking into consideration the following mesomeric form of N-phosphoryl phosphazenes: $(RO)X_2P^{\oplus}{-}N{=}P({-}O^{\ominus})X_2$. The electrophilic nature of the phosphazene phosphorus atom determines the substitution order and effects a strong tendency for forming a double bond between phosphorus and oxygen connected with a partial transfer of the electrophilic properties to the alkyl group. Therefore alkoxyphosphazenes act as fairly good alkylation agents for nucleophiles such as HCl, RO^- and $PhNMe_2$ forming RCl, R_2O and $PhNRMe_2^+$, respectively.

In the absence of external nucleophiles self-alkylation of the molecule is possible, in which the electrophilic alkyl group will attack both the phosphoryl oxygen atom and the nitrogen atom. As only in the latter case a substance without alkylating properties is formed self-alkylation finally results in the rearrangement of O-alkyl into N-alkyl compounds. N-alkylation takes place particularly readily in the case of monoalkoxyphosphazenes, $(RO)Cl_2P{=}N{-}POCl_2$, with R = Me, Et and $CH_2C_6H_5$ (2). Monoalkoxyphosphazenes having different alkyl groups (R = C_3H_7, iso-C_3H_7, C_4H_9, C_5H_{11}, C_6H_{13}) form the N-alkyl rearrangement products only at higher temperatures (about 75°C) and to an extremely small extent (about 1 %). "Pentaesters" can be rearranged into N-alkyl derivatives at about 150°C (eq. 4; X = OR; R = Et, Bu, Hex). This rearrangement is catalysed by alkyl iodides.

After all this reaction can also be regarded as a chlorine-oxygen exchange. Such exchange generally seems to occur in reactions of N-phosphoryl trichlorophosphazenes with O-nucleophiles. This is demonstrated by the reactions of P_2NOCl_5 with dimethylformamide and dimethylsulfoxide (eq. 6, 7) as well as by the fact that all attempts to synthesize N-dialkoxyphosphoryl trichlorophosphazenes, $(RO)_2P(O){-}N{=}PCl_3$, have failed so far. Both in the reaction of $(EtO)_2PONCl_2$ with PCl_3 (eq. 8) and in the reaction of $(EtO)_2P(O)NHSiMe_3$ with PCl_5 (eq. 9) isomeric diethoxyphosphazene $(EtO)_2ClP{=}N{-}POCl_2$ is obtained instead of trichlorophosphazene as should be expected. Only N-diphen-

$$Cl_3P{=}N{-}POCl_2 + Me_2NCHO \longrightarrow [Me_2N{=}CClH]^+[Cl_2PO{-}N{-}POCl_2]^- \quad (6)$$

$$Cl_3P{=}N{-}POCl_2 + (CH_3)_2SO \longrightarrow CH_3{-}S{-}CH_2Cl + (Cl_2PO)_2NH \quad (7)$$

$$(EtO)_2P(O)NCl_2 + PCl_3 \longrightarrow (EtO)_2ClP{=}N{-}POCl_2 + Cl_2 \quad (8)$$

$$(EtO)_2PONHSiMe_3 + PCl_5 \xrightarrow[-HCl]{} (EtO)_2ClP{=}N{-}POCl_2 + Me_3SiCl \quad (9)$$

oxyphosphoryl trichlorophosphazene, $(PhO)_2P(O){-}N{=}PCl_3$, (3) and probably compounds having different aryl groups seem to be stable.

Due to the by-reactions (eq. 3 – 5) alcoholysis of $Cl_3P{=}N{-}POCl_2$ does not yield the desirable "pentaesters" $(RO)_3P{=}N{-}PO(OR)_2$. Depending on molar ratio, temperature and time the reaction mixtures contain mono-, di- and trialkoxyphosphazenes, tetrachloride and several esterchlorides of imidodiphosphoric acid. In the presence of a great excess of alcohol the final products are exclusively tetraesters of imidodiphosphoric acid, $(RO)_2PO{-}NH{-}PO(OR)_2$. Indeed these can be converted into "pentaesters" by reacting with diazoalkanes (eq. 10) (2), which, however, is not a convenient way for synthesizing "pentaesters".

$$(RO)_2PO{-}NH{-}PO(OR)_2 + R'N_2 \longrightarrow (RO)_3P{=}N{-}PO(OR)_2 + N_2 \qquad (10)$$

Using alcoholates instead of alcohols "pentaesters" (4,5) and trialkoxyphosphazenes, $(RO)_3P{=}N{-}POCl_2$, can be obtained in good yields. However, this method does not allow to synthesize pure "pentaesters" with different alkyl groups.

Formation of P=N-P-Bridges. A simple reliable synthesis of N-phosphorylated phosphazenes which avoid the handling of the dangerous phosphoryl azides consists in a direct reaction of di- and trialkylphosphites and carbon tetrachloride with sodium azide in a single step procedure (eq. 11). The phosphoryl azides formed intermediately instantly react with the trialkylphosphites present. Therefore, their

$$(RO)_3P + (R'O)_2P(O)H + CCl_4 + NaN_3 \longrightarrow (RO)_3P{=}N{-}PO(OR')_2 + CHCl_3 + NaCl \qquad (11)$$

concentration is kept small all the time. In contrast to the classic Atherton-Todd reaction addition of amines is not necessary. In the absence of trialkylphosphite no reaction occurs; the formation of phosphoryl azides does not occur either. We used phosphites of aliphatic and aromatic alcohols and synthesized about 40 different "pentaesters" with yields up to 95 %. As by-products $Cl_3C{-}P(O)(OR)_2$ and small amounts of rearrangement products, $(RO)_2P(O){-}NR{-}P(O)(OR')_2$, (<1 %) are obtained. Only in reactions with trimethylphosphite the formation of rearrangement products is considerable (up to 30 %).

NMR Spectroscopic Investigations. The ^{31}P NMR data of the alcoholysis products of $Cl_3P{=}N{-}POCl_2$ and $(Cl_2PO)_2NH$ are listed in table I. Under certain conditions each of these compounds can be found in the reaction mixtures of P_2NOCl_5 with either ethanol or ethylate.

What is very surprising is that some "pentaesters" (e.g. R = Et, Bu, Oct) and probably also some "tetraesters", $(RO)_3P{=}N{-}PO(OR)Cl$, show only one singlet in ^{31}P NMR spectra at room temperature. However, a detailed investigation (6) revealed that these singlets will split into two doublets both at higher and lower temperatures. The occurrence of the singlet at room temperature is caused by the different temperature dependence of the chemical shifts of the two phosphorus nuclei contained in the molecule. In the case of R = Et

Table I. NMR data of $Cl_{3-n}(RO)_nP_A{=}N{-}P_BOCl_{2-m}(OR)_m$ and $Cl_{2-k}(RO)_kP_C(O){-}NH{-}P_D(O)Cl_{2-l}(OR)_l$ (R = Et).

n	m	δ_A/ppm	δ_B/ppm	$^2J_{PP}$/Hz	k	l	δ_C/ppm	δ_D/ppm	$^2J_{PP}$/Hz
0	0	-1	-13	16-23	0	0	0		
1	0	+1	-11	36-42	1	0	-4	- 8	47-48
2	0	-1	- 9	45-50	2	0	-5	-11	42-45
3	0	-4	- 8	51-54	2	1	-3	+ 4	18-21
3	1	-3.7			2	2	-1		6.8
3	2	-3.2		69-70					

the change of chemical shifts caused by rising temperature amounts to -0.524 Hz/K for the phosphazene phosphorus atom and -0.079 Hz/K for the phosphoryl phosphorus atom. This behaviour of "pentaesters" demonstrates that the chemical equivalence of phosphorus nuclei in the NMR spectra need not necessarily correspond to a structural equivalence of the phosphorus atoms in the molecule.

$(PhO)_2P(O){-}N{=}PCl_3$ also shows only one singlet at -11.9 ppm but in this case it is most interesting that even the variation of temperature does not bring about two signals. Only by using shift reagents (e.g. $Pr(fod)_3$) the structural non-equivalence of the phosphorus atoms is indicated by the occurrence of two doublets with a P-P coupling constant of 47 Hz.

An entirely different example for substances with two different phosphorus atoms in the molecule showing only a single NMR signal are O-silylated N-phosphoryl phosphazenes, $X_2P(OSiMe_3){=}N{-}POX_2$ (X = NMe_2, OEt) (7,8). However, in the case of these substances the occurrence of a singlet is caused by a fast, reversible exchange of the silyl group between the two oxygen atoms in the molecule. By lowering temperature migration of the silyl group will be slowed down, with two doublets appearing.

Literature Cited

1 Herrmann, E.; Hoang ba Nang; Dreyer, R. Z. Chem. 1979, 19, 187.
2 Riesel, L.; Willfahrt, M.; Grosse, W.; Kindscherowsky, P.; Khodak, A. A.; Gilyarov, V. A.; Kabachnik, M. I. Z. anorg. allg. Chem. 1977, 435, 61.
3 Kirsanov, A. V.; Zhmurova, I. N. Zh. cbshch. Khim. 1958, 28, 2478.
4 Kireev, V. V.; Kolesnikov, G. S.; Titov, S. S. Zh. obshch. Khim. 1970, 40, 2015.
5 Khodak, A. A.; Gilyarov, V. A.; Shcherbina, T. I.; Kabachnik, M. I. Izv. Akad. Nauk SSSR, Ser. Khim. 1979, 1884.
6 Riesel, L.; Steinbach, J.; Thomas, B. Z. anorg. allg. Chem. 1979, 451, 5.
7 Riesel, L.; Clausnitzer, A.; Ruby, C. Z. anorg. allg. Chem. 1977, 433, 200.
8 Kabachnik, M. I.; Zaslavskaya, N. N.; Cilyarov, V. A.; Petrovskij, P. V.; Svoren, V. A. Dokl. Akad. Nauk SSSR 1976, 228, 849.

RECEIVED July 7, 1981.

Conjugation in Phosphazenes: Pyrrylphosphazenes and Phosphazenyl Carbanions

K. D. GALLICANO, R. T. OAKLEY[1], R. D. SHARMA[2], and N. L. PADDOCK

Department of Chemistry, University of British Columbia, Vancouver, British Columbia, Canada

The bonding system in phosphazenes is closely analogous to that in phosphoryl compounds, in which conjugative P-C interactions require electron release from the substituent. Although there is only weak conjugation between phenyl and phosphoryl groups (1), the electronic spectra of tris(pyrryl)phosphine oxides show a strong low frequency band attributable to pπ-dπ bonding between the pyrryl and phosphoryl groups (2); its nature, for acylpyrroles, has been confirmed theoretically (3).

In the expectation that phosphazenyl and phosphoryl groups would behave similarly, we have prepared the 1-methyl-2-(fluorophosphazenyl)pyrroles $N_nP_nF_{2n-1}\cdot C_4H_3NMe$ (n = 3-6; 2,a-d) and have measured their electronic spectra. Figure 1 illustrates possible modes of conjugation to phosphoryl and N_3P_3 rings. Like the pyrrylphosphine oxides, the fluorophosphazenylpyrroles show a band characteristic of pπ-dπ conjugative charge transfer, and an interaction of this type is equally well established for the two series. The numerical results are shown in Table I.

Table I: Electronic spectra of fluorophosphazenylpyrroles.[a]

Cpd.	λ(ε)	λ(ε)	Cpd.	λ(ε)	λ(ε)
1[b]	213(6.7)	-	2c(n=5)	235(11.5)	220(9.9)
2a(n=3)	237(10.9)	220(10.1)	2d(n=6)	240(11.8)	220(10.1)
2b(n=4)	240(12.4)	220(10.4)	3[c]	243(12.9)	218(8.4)

[a] λ(nm) and ε($\times 10^{-3}$) at maxima. [b] 1-methylpyrrole. [c] tris(1-methylpyrrol-2-yl)phosphine oxide (2).

[1] Current address: Department of Chemistry, University of Calgary, Alberta, Canada.
[2] Current address: Department of Chemistry, Simon Fraser University, British Columbia, Canada.

0097-6156/81/0171-0301$05.00/0

The wavelength of the charge transfer band increases in the series phosphazenyl < phosphoryl < formyl (289.5 nm)(3), so that the degree of conjugation increases in that order, the first two being comparable. The slight alternation in the phosphazenyl series, if real, is consistent with the transfer of charge to an antibonding π-level of the phosphazene ring. Ground state stabilisation of 2b is demonstrated by its molecular structure (4) (Figure 2), the P-C bond being 0.06 Å shorter than in phenylphosphazenes (5).

Conjugative interactions have important preparative consequences (6). Primarily, deprotonation of an alkylphosphazene is promoted by conjugation in the carbanion so formed, and a number of new phosphazene derivatives have been made by the reaction of the lithio-derivatives with electrophiles; some are illustrated in Figure 3. The reactions of the monocarbanion formed from gem-$N_3P_3Ph_4Me_2$ are the simplest, and compounds of Type 4 have been obtained in high yield. Rings can be joined, either by the use of a difunctional halide, or through a coupling reaction, involving $CuCl/O_2$(7), to give 5. The tricarbanion formed from $N_3P_3Me_6$ gives the expected derivatives $N_3P_3Me_3(CH_2R)_3$; both *cis-cis-cis* and *cis-trans-trans* stereo-isomers of the tribromo-derivative have been isolated.

Tetrasubsubstituted derivatives 6 are easily obtained from $N_4P_4Me_8$, and π-electron calculations show that antipodal di-substitution is to be expected for electropositive substituents, as found for R = Me_3Si, Me (7). Monocarbanions are not obtained, principally because the intra-ring exothermic electrostatic interactions are reduced less by charge transfer from the second ylidic group than the first; if the added base is strong enough to remove the first proton, it will also remove the second.

Normally electronegative substituents facilitate full substitution, but if the withdrawal is strong enough, charge transfer may remove all charge from the ylidic groups of, e.g., $N_4P_4Me_4(CH_2R)_2(CH_2^-)_2$, so that it is no longer a nucleophile. This evidently occurs in the reaction of $N_4P_4Me_4(CH_2Li)_4$ with ethyl benzoate, which gives a high yield of the vicinal di-derivative (Figure 3, 8); the reaction cannot be made to go further. The effect of the substituent can be modelled as a perturbation of the local Coulomb parameter α_p, and Figure 4 shows that, as this phosphorus atom is made more electronegative, the favoured mode of substitution changes from antipodal to vicinal, and the ylidic charge decreases, becoming zero near the point where the vicinal isomer is the more stable. The dicarbanion is then no longer a nucleophile, and the reaction stops. This is evidently what happens in the reaction of $N_4P_4Me_4(CH_2^-)_4$ with ethyl benzoate.

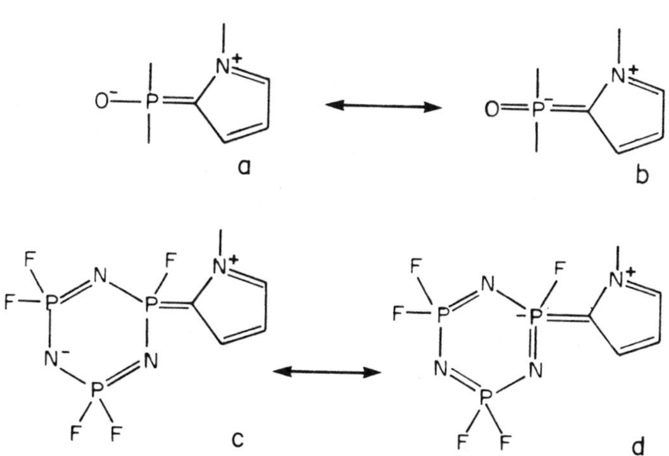

Figure 1. Some canonical forms involving conjugation in 1-methylpyrrol-2-yl derivatives. Valence shell expansion is implied in b *and* d.

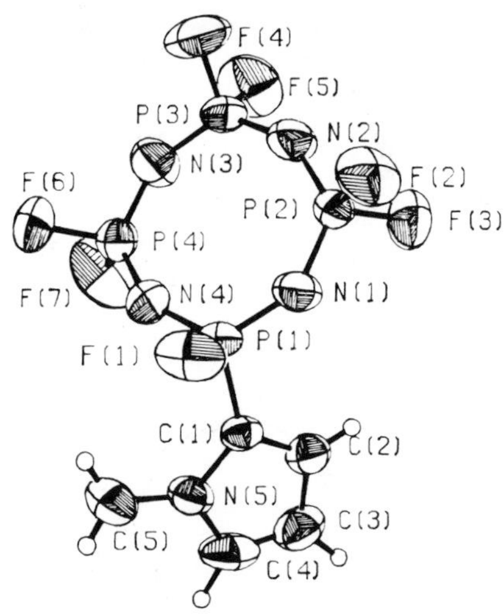

Canadian Journal of Chemistry

Figure 2. The molecular structure of 2b (4). P(1)C(1) = 1.765(5) Å, P(1)F(1) = 1.561(4) Å; mean length of other P–F bonds = 1.536 Å. Bonds from P(1) to N(1), N(4), lengthened by 0.027(12) Å.

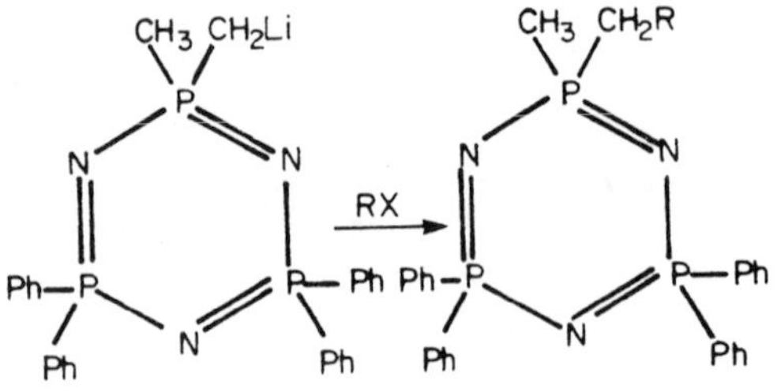

Canadian Journal of Chemistry

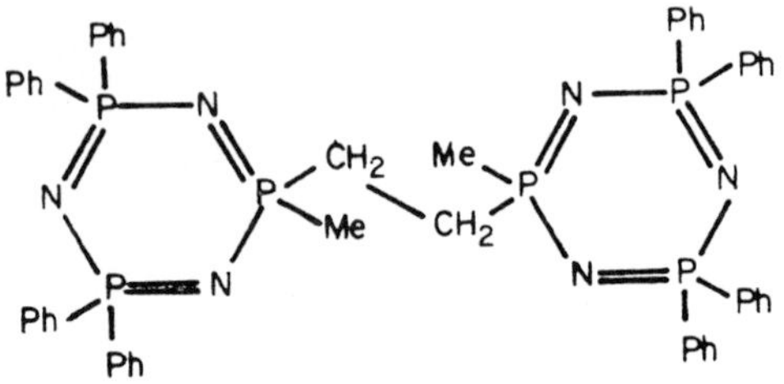

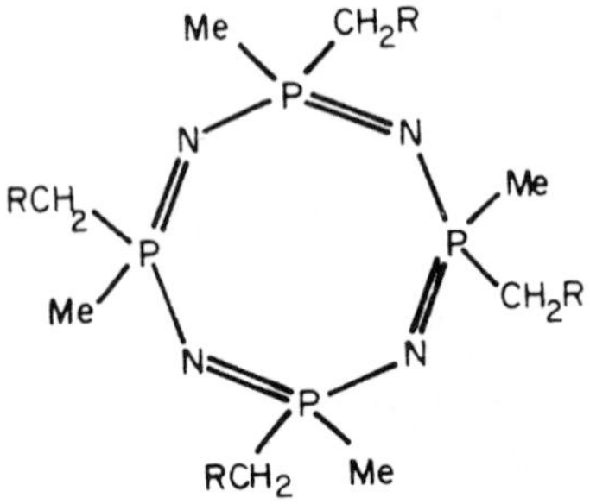

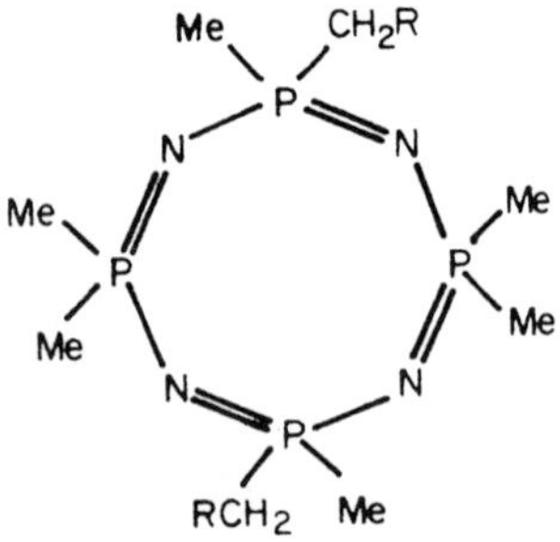

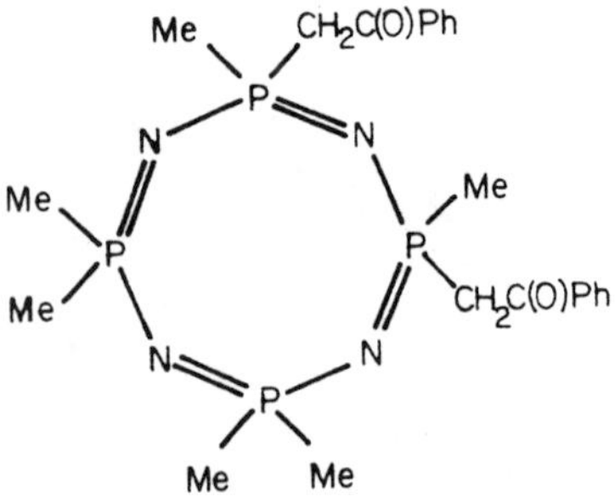

Figure 3. Products of carbanion reactions (6): 4, R = Me_3Sn, Br, PhC(0), COOH; 6, R = Me, Me_3Si, Me_3Ge, Me_3Sn, Br, I, Me_2As; and 7, R = Me_3Si, Me_3Ge, Me_3Sn, Me.

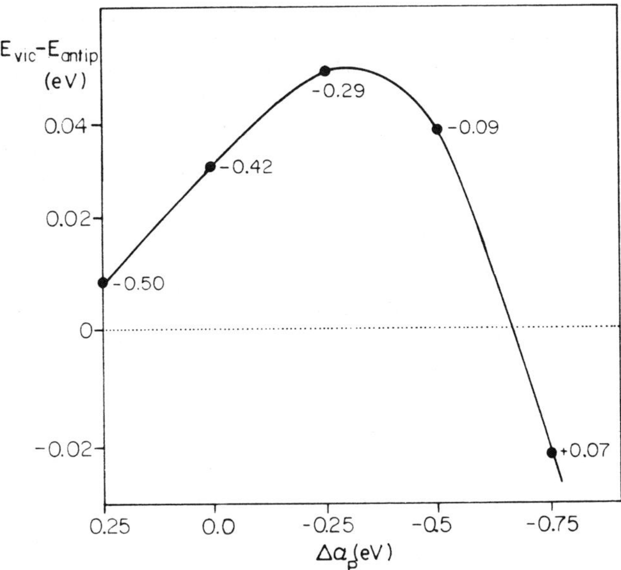

Figure 4. π-Electron energy of vicinal relative to antipodal $N_4P_4Me_4(CH_2R)_2(CH_2^-)_2$, as a function of the perturbation $\Delta\alpha_p$ induced by R(6). The residual π-charge on the ylidic carbon atoms is marked on the curve.

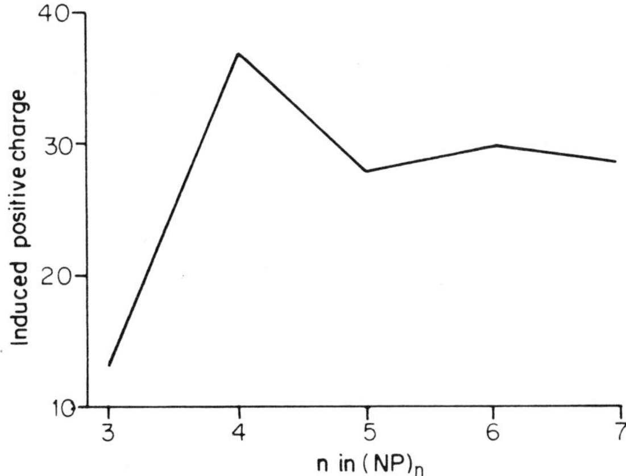

Figure 5. Positive charge (×10^4) induced at P(2) by a perturbation of −0.5 eV at P(1) (8).

As an interesting contrast, the benzoylation of $N_3P_3Me_6$ goes to completion, a fact which has a related explanation. The effect of an electronegative perturbation at phosphorus is to concentrate electron density locally, at the expense of the next nearest atoms. In a homomorphic system, the induced electron density is an oscillating function of ring size (Figure 5)(8). The positive charge induced at P(2) by an electronegative substitution at P(1) is greatest for the 8-membered and least for the 6-membered ring, and the secondary withdrawal from the exo-carbon atoms must be greater for the tetramer. If the same parameters are used as for the tetramer, the exo-charge on a disubstituted trimer is -0.2 when it is 0.0 in the tetramer. This is much reduced from the value of -0.54 found for the unperturbed molecule, but it still corresponds to pronounced nucleophilic character, and demonstrates one of the differences between $N_3P_3Me_6$ and $N_4P_4Me_8$ arising from π-interactions.

Literature Cited.

1. Jaffé, H.H.; Freedman, L.D. J. Am. Chem. Soc. 1952, 74, 1069, 2930.
2. Griffin, C.E.; Peller, R.A.; Martin, K.R.; Peters, J.A. J. Org. Chem. 1965, 30, 97.
3. Matsuo, T.; Shosenji, H. Bull. Chem. Soc. Japan. 1972, 45, 1349.
4. Sharma, R.D.; Rettig, S.J.; Paddock, N.L.; Trotter, J. Can. J. Chem. 1981, in press.
5. Ahmed, F.R.; Singh, P.; Barnes, W.H. Acta Cryst. 1968, B25, 316.
6. Gallicano, K.D.; Oakley, R.T.; Sharma, R.D.; Paddock, N.L. Can. J. Chem. 1981, in press.
7. Kauffmann, T.; Kuhlmann, D.; Sahm, W.; Schrecken, H. Angew. Chem. Int. Ed. Engl. 1968, 7, 541.
8. Craig, D.P.; Paddock, N.L. in "Non-Benzenoid Aromatics", ed. Snyder, J.P., Academic Press, New York, 1971, p. 273.

RECEIVED June 30, 1981.

Structure, Conformation, and Basicity in Cyclophosphazenes and Related Compounds

R. A. SHAW and S. N. NABI

Department of Chemistry, Birkbeck College, University of London, Malet Street, London, WC1E 7HX, England

The results so far achieved by means of basicity studies on cyclophosphazenes can be summarised as follows (1): (i) establishment of a basicity series in homogeneously substituted compounds, e.g., for the six-membered ring system, $N_3P_3X_6$, X = NHAlk > Alk > Ph ~ NHPh > OAlk > SAlk > SPh > OPh > OCH_2CF_3 > Cl. A similar, slightly less comprehensive, series was established for the eight-membered ring system, $N_4P_4X_8$; (ii) proof that in all of the above systems protonation occurred at a ring nitrogen atom; (iii) distinction between positional (but not geometric) isomers in partially substituted derivatives, $N_3P_3X_nX'_{6-n}$. If $N_3P_3X_6$ is more basic than $N_3P_3X'_6$, then the geminal isomers $N_3P_3X_nX'_{6-n}$ (n = 3,4) are more basic than the nongeminal ones; (iv) evaluation of substituent constants α_X, $\alpha_{X'}$, γ_X and $\gamma_{X'}$ for $N_3P_3X_nX'_{6-n}$; these represent the basicity contributions to a ring nitrogen atom by substituents X and X' on a phosphorus atom α or γ to this ring nitrogen atom.

Selected Substituent Constants (in pK'_a units).

X	α_X	γ_X
NMe_2	5.6	2.8
Ph	4.2	2.3
OMe	3.6	1.8
SEt	3.6	1.8
OPh	3.1	1.3

(v) observation that with phosphazenylcyclophosphazenes, e.g., $N_3P_3X_4X'(NPPh_3)$ two sets of differing, but self-consistent, basicity data are obtained. These were related to the conformations of the $NPPh_3$ group relative to the local NPN ring segment and an explanation in terms of endo or exo protonation was tentatively offered. The conformational changes have been amply demonstrated. The endo versus exo-protonation is still a hypothesis; (vi) the triphenylphosphazenyl group, Ph_3PN, was shown to be the most electron-releasing substituent hitherto observed in basicity studies on cyclophosphazenes.

0097–6156/81/0171–0307$05.00/0

The present paper reports progress in four areas of structure, conformation and basicity: (1) the effect of the conformation of amino groups on the values of the substituent constants; (2) further studies of phosphazenylcyclophosphazenes; (3) the effect of replacing chlorine by fluorine in structurally related phosphazenes; (4) the relative basicities of ring segments P-N-P, S-N-S, C-N-C and As-N-As, and the preferred sites of protonation in mixed ring systems containing nitrogen and phosphorus and/or sulphur, carbon, arsenic.

(1) In earlier studies the substituent constants α_X and γ_X were derived by averaging the data from a number of compounds. Conformational effects were not considered. The results obtained from basicity and X-ray crystallographic studies on phosphazenylcyclophosphazenes (v), together with an increasing number of crystallographic studies on other compounds, where it was shown that in many cases the conformations of the substituent X varied considerably for the environments $\equiv$PClX (nongeminal) and $\equiv PX_2$ (geminal) focussed attention on the effect of conformation on the values of the substituent constants. Whilst those averaged values had given fair to excellent agreement between calculated and observed values for most compounds investigated, there remained some obvious discrepancies. Perhaps the most widely investigated group in these studies is the dimethylamino group. For this, the substituent constants α_{NMe_2} and γ_{NMe_2} were found to be 5.6 and 2.8 respectively. However, on passing from structure (I) to structure (II) (asterisks indicate points of protonation), the value deduced for the additional α_{NMe_2} substituent is only 4.4. Similarly from averaged results α_{pip} = 5.6, γ_{pip} = 2.8, α_{pyr} = 5.9 and γ_{pyr} = 3.0, (pip = piperidino, pyr - pyrrolidino), but for the additional group added on passing from structure (I) to (II), the values are lower, *viz.* α_{pip} = 4.5 and α_{pyr} = 4.9. It is known from numerous crystallographic investigations that NMe_2 groups in the moieties $\equiv$PCl(NMe_2) and $\equiv P(NMe_2)_2$ adopt type I and type III conformations respectively. Other secondary amines no doubt behave similarly, and it can be deduced that in conformation (III) the contribution to basicity is less than in (I).

(I) (II)

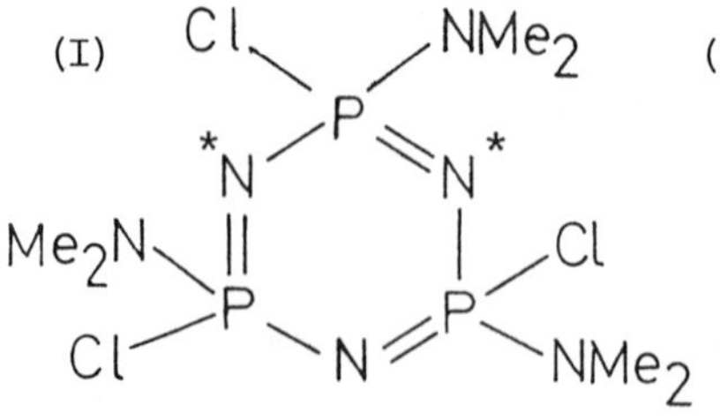

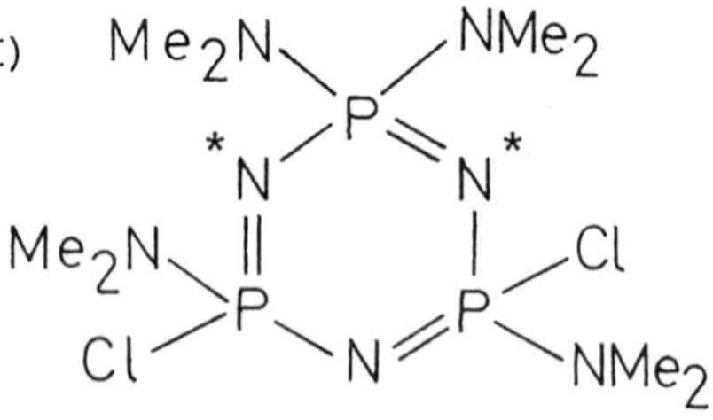

(2) Further crystallographic and basicity data have been obtained for phosphazenylcyclophosphazenes. Both $NPPh_3$ groups in g.-$N_3P_3Cl_4(NPPh_3)_2$ have been shown by X-ray crystallography to have type II conformations (2). This is in line with evidence from ^4J(PP) coupling constants and disproves the deduction made

earlier from basicity data. It thus leaves open the question whether so far unexplained differences in basicity should be attributed to different sites of protonation (exo versus endo) or to different conformations of substituents contributing differently to the pK'$_a$ values measured [see (3) below].
ng.-*trans*-$N_3P_3F_3(NPMe_3)_3$ has been shown to have two of its substituents in type I, the third in type II conformation (3). $NPMe_3$ may have different substituent constants in different environments, but these are always equal to or greater than that of $NPPh_3$.
(3) A great deal of basicity data is available on substituted chlorophosphazenes. None has hitherto been published on fluorophosphazenes. A priori, one surmises a greater electron-withdrawing inductive effect for fluorine, which could be compensated for by the greater potential of fluorine to back-donate by a mesomeric effect. Data is now compared on structurally related chloro- and fluoro-derivatives, e.g.,

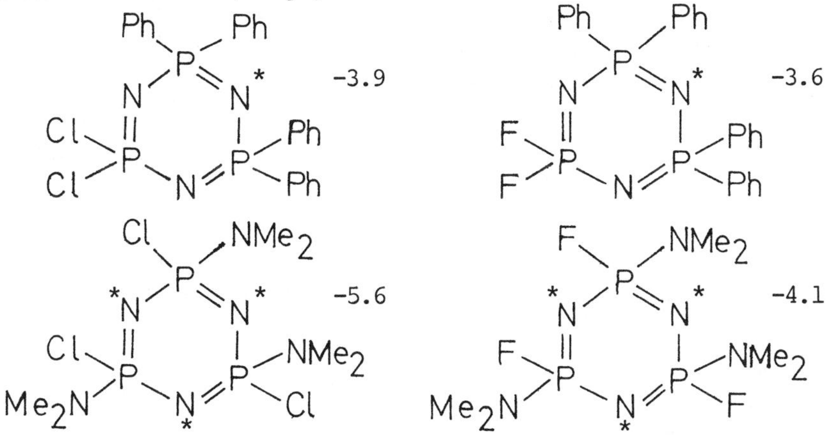

-3.9 -3.6

-5.6 -4.1

Basicity values for other structurally related compounds are: ng.-$N_3P_3X_4(NPPh_3)_2$ X = Cl or F +0.2; ng.-$N_3P_3X_4(NMe_2)(NPPh_3)$ X = Cl, -4.8; X = F, -4.2; ng.-$N_3P_3X_2(NMe_2)_4$ X = Cl, -1.4; X = F, +0.6. It can be seen that, in some cases, the fluoro compounds have approximately the same basicity as their chloro analogues, whilst in other cases, they are markedly more basic. On the as yet limited data available, no significant conclusions can be drawn. Related fluoro- and chloro-compounds have the same basicity, except that when NMe_2 substituents are present, they seem to increase the basicity of the fluoro-analogue by about 0.5pK'$_a$ units per NMe_2 group.
(4) A considerable amount of basicity data has been published on cyclophosphazenes, i.e., systems containing only P-N-P segments. A small number of measurements have been published on mixed P-N-S ring systems, (4), e.g., $(NPR_2)_2(NSOR)$ and $(NPR_2)(NSOR)_2$. These and other data show that protonation of ring segments is preferred in the order P-N-P > P-N-S > S-N-S. The values of substituent constants on phosphorus and sulphur appear to be the same.

Some basicity data on 1,3,5-triazines, $(NCR)_3$, and mixed P-N-C systems, $(NPR_2)_2(NCR)$, is now available. Our measurements indicate that in the mixed ring systems protonation is preferred at P-N-C rather than at P-N-P segments, in line with n.m.r. studies by Schmidpeter and Ebeling (5). Basicity data on this mixed ring system and on 1,3,5-triazines has been evaluated. α-Values of substituent constants appear to be similar for phosphorus and carbon, but γ-values seem to be higher for the latter than the former.

Using the basicity data available, we can deduce that the electron-withdrawing power of the groupings discussed is SOCl > PCl_2 > CCl. Hence the probable order of basicities of the parent chlorides is (with tentative *calculated* values where available, in brackets): $(NCCl)_3$, (-12.4); $(NCCl)_2(NPCl_2)$; $(NCCl)(NPCl_2)_2$, (-13 to -17); $(NPCl_2)_3$, (-20.4); $(NPCl_2)_2(SOCl)$, (-24.5); $(NPCl_2)(NSOCl)_2$; $(NSOCl)_3$.

Finally, basicity values for two compounds containing the As-N-As segments, $N_3As_3Ph_6$ (+6.0) and $N_4As_4Ph_8$ (+8.6) can now be compared with their phosphorus analogues, $N_3P_3Ph_6$ (+1.5) and $N_4P_4Ph_8$ (+2.2). The much greater basicity of the arsenic analogues is noteworthy. The basicity order of the segments As-N-As > P-N-P > C-N-C > S-N-S can now be tentatively established for those compounds where only one hetero element apart from nitrogen is present in the ring system.

Literature cited

1. Shaw, Robert A. Z. Naturforsch. 1976, 31b, 641-667
2. Krishnaiah, M.; Ramamurthy, L.; Ramabrahman, P.; Manohar, H. submitted for publication.
3. Bullen, G.J.; Tam, K.O. personal communication.
4. Faucher, Jean-Paul; van de Grampel, Johan C.; Labarre, Jean-Francois; Nabi, Syed Nurun; de Ruiter, Barteld; Shaw, Robert A. J.Chem. Research, 1977, (S) 112-113; (M) 1257-1294.
5. Schmidpeter, A.; Ebeling, J. Chem. Ber. 1968, 101, 3883-3901.

RECEIVED July 7, 1981.

Phosphazene Rings and High Polymers Linked to Transition Metals or Biologically Active Organic Species

H. R. ALLCOCK

Department of Chemistry, The Pennsylvania State University, University Park, PA 16802

Macromolecules that contain phosphorus as a skeletal atom have been studied in detail for many years. Yet, compared to the vast array of known carbon-backbone polymers, macromolecules based on phosphorus have occupied only a small and very specialized niche. We have been systematically exploring the prospect that a broad new class of high polymers, the poly(organophosphazenes) (III), can be synthesized in which phosphorus rather than carbon plays a key role in the skeletal chain (1-3).

$$(NPCl_2)_3 \text{ (I)} \xrightarrow{250°C} \left[N = P(Cl)_2 \right]_n \text{ (II)} \xrightarrow{\text{nucleophile}} \left[N = P(X)_2 \right]_n \text{ (III)}$$

($n \simeq 15{,}000$, X = OR, NR_2, alkyl, or aryl)

For most organic substituent groups, X, polymers of type III are hydrolytically stable and offer unusual combinations of physical and chemical properties not found in biological- or petrochemical-based macromolecules.

Two key principles have played a pivotal role in our exploration and development in this field. First, unlike most macromolecules, nearly all poly(organophosphazenes) are prepared by a substitutive technique, in which a broad range of different substituent groups are introduced via a reactive polymeric intermediate (II). Second, because substitution reactions play such an important role in the chemistry of these

0097-6156/81/0171-0311$05.00/0

polymers, and because the reactions of macromolecules are usually complex, we have made extensive use of species such as I as small molecule models for the reactions of II. Thus, the reactions of cyclic trimers and tetramers have been investigated in tandem with the reactions of the high polymers (4).

In this paper we consider two specific challenges. First, how might transition metals be linked to phosphazene high polymers? Such systems are of interest as immobilized catalysts or materials with unusual electrical properties. Second, how can bioactive agents be attached to polyphosphazenes to prepare, for example, targeted, slow release chemotherapeutic agents? An important link in this process would be the use of a carrier polymer that could biodegrade to harmless small molecules.

Five different approaches have been developed for the linkage of transition metals to cyclic or high polymeric phosphazenes. The first three make use of organic side groups as coordination ligands, the fourth utilizes the coordination power of the backbone nitrogen atoms, and the fifth involves the synthesis of derivatives in which the side group is itself an organometallic unit linked to the skeleton through phosphorus-metal bonds. These possibilities are illustrated in structures IV-VIII.

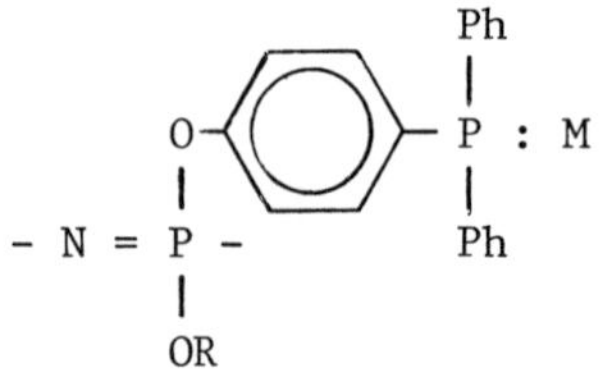

```
                      Ph
                      |
      O—C6H4—P : M
      |               |
- N = P -             Ph
      |
      OR
```

IV

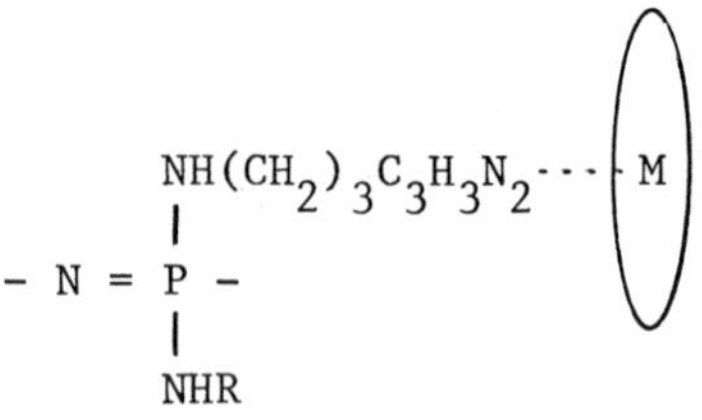

```
      NH(CH2)3C3H3N2···M
      |
- N = P -
      |
      NHR
```

V

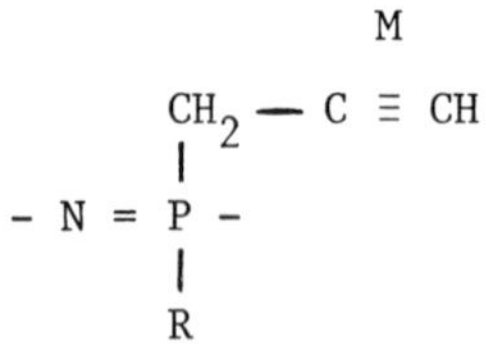

```
                 M
      CH2 — C ≡ CH
      |
- N = P -
      |
      R
```

VI

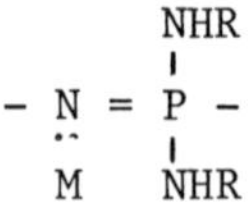

```
          NHR
          |
- N   =   P -
  ..      |
  M       NHR
```

VII

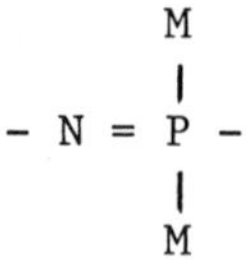

```
      M
      |
- N = P -
      |
      M
```

VIII

Species IV were prepared from p-bromophenoxy-substituted phosphazene trimers and high polymers, by metal-halogen exchange to yield the p-lithio-derivative, followed by reaction with diphenylchlorophosphine. Both cyclic trimers and high polymers containing the pendent phosphine reacted with $H_2Os(CO)_{10}$,

$MnCp(CO)_3$, AuCl, CuI, or $Rh_2Cl_2(CO)_4$ to yield the appropriate phosphazene-transition metal complex. The skeletal nitrogen atoms did not interfere with the process. On the other hand, it was shown earlier that $PtCl_2$ residues are bound strongly to the skeletal nitrogen atoms of $[NP(CH_3)_2]_4$, $[NP(NHCH_3)_2]_4$, and $[NP(NHCH_3)_2]_n$ (VII) (5). The latter compound is a prospective polymer-bound antitumor agent. The pendent imidazolyl derivative (V) coordinated strongly to heme or hemin in aqueous media (6) to yield products that are of interest both as heme-protein models and as redox systems. Our recent discovery of a facile route to the formation of propynyl phosphazenes via organocopper intermediates (7) has allowed the synthesis of pi-coordination derivatives of structure VI. The complex formed with $Co_2(CO)_8$ is especially stable.

The linkage of transition metals to skeletal phosphorus in a phosphazene has presented an unresolved challenge for many years. We have recently succeeded in the preparation of iron and ruthenium phosphazenes by the reaction shown below (8).

$$N_3P_3F_6 \xrightarrow[-NaF]{NaFeCp(CO)_2} \text{IX} \xrightarrow[-CO]{h\nu} \text{X}$$

IX

X

An important feature of this approach is the isolation of species IX (as a yellow, crystalline solid) and, following mild photolysis, the red, metal-metal bonded product, IX. X-Ray crystal structures have been obtained for both compounds.

The linkage of steroid molecules to both cyclic and high polymeric phosphazenes has been studied (9). Steroids with hydroxyl groups at the 3-position can be converted to their sodium salts by treatment with sodium hydride. If the A-ring is aromatic, linkage of the steroid to the phosphazene skeleton occurs in a manner reminiscent of the behavior of simple aryloxides. However, if the A-ring is alicyclic, complex reactions occur, including dehydration of the A-ring by the phosphazene.

Two types of side group structures attached to a phosphazene ring or chain induce hydrolytic breakdown -- amino acid ester and imidazole residues. The former decompose to alcohol, amino acid, phosphate, and ammonia (10). Hexakis(imidazolyl)cyclotriphosphazene hydrolyzes rapidly in the pH range 6.5 to 7.8 by a mechanism that involves autocatalysis by the free imidazole liberated. Thus, either type of side group could facilitate biodegradation of chemotherapeutic carrier macromolecules.

Acknowledgments

The following coworkers have contributed to this work: T. L. Evans, K. Lavin, N. M. Tollefson, L. J. Wagner, P. P. Greigger, J. P. O'Brien, R. W. Allen, J. L. Schmutz, T. J. Fuller, K. Matsumura, K. M. Smeltz, D. P. Mack, P. J. Harris, and R. A. Nissan. Financial support from the Army Research Office, the Office of Naval Research, and the National Institutes of Health is gratefully acknowledged.

References

1. Allcock, H. R., Kugel, R. L., Valan, K. J. Inorg. Chem. 1966, 5, 1709.
2. Allcock, H. R., Kugel, R. L. Inorg. Chem. 1966, 5, 1716.
3. Allcock, H. R. Makromol. Chem. 1981, Suppl. 4, 3.
4. Allcock, H. R. Accounts Chem. Res. 1979, 12, 351.
5. Allcock, H. R., Allen, R. W., O'Brien, J. P. J. Am. Chem. Soc. 1977, 99, 3984.
6. Allcock, H. R., Greigger, P. P., Gardner, J. E., Schmutz, J. L. J. Am. Chem. Soc. 1979, 101, 606.
7. Allcock, H. R., Harris, P. J., Nissan, R. A. J. Am. Chem. Soc. 1981, 103, 2256.
8. Allcock, H. R., Greigger, P. P., Wagner, L. J., Bernheim, M. Y. Inorg. Chem. 1981, 20, 716.
9. Allcock, H. R., Fuller, T. J. Macromolecules, 1980, 13, 1338.
10. Allcock, H. R., Fuller, T. J. J. Am. Chem. Soc. 1981, 103, 2250.

RECEIVED July 7, 1981.

Polymerization of Hexachlorocyclotriphosphazene

JOHN W. FIELDHOUSE and DANIEL F. GRAVES

The Firestone Tire and Rubber Company, Central Research Laboratories, 1200 Firestone Parkway, Akron, OH 44317

Hexachlorocyclotriphosphazene (or simply trimer) can be thermally polymerized in bulk or solution at 200-270°C producing hydrolytically unstable polydichlorophosphazene (or simply chloropolymer). Allcock (1,2) obtained chloropolymer soluble in organic solvents by limiting the conversion of trimer to chloropolymer. This discovery permitted the chloropolymer to be converted to hydrolytically stable polyphosphazenes by chlorine substitution with an appropriate nucleophile.

Polymerization catalysts such as sulfur (3), water (4,5,6), oxygenated organics (7-10) and silica from the surface of glass (11,12) have been used to promote the polymerization. Many of these catalysts promote the formation of crosslinked chloropolymer at conversions above 50 percent, thus rendering them unsuitable for reaction with nucleophiles.

We have discovered that boron halides or boron halide•triarylphosphate complexes polymerize trimer at 160-250°C to soluble chloropolymer in yields up to 100 percent. The trimer used in this study was purified by sublimation at 130-140°C and 20-30 mm Hg vacuum. Sublimation under these conditions allows entrapped hydrogen chloride to escape. Trimer (30g) and catalysts were placed in a 35 ml pyrex tube (previously washed with 20% aqueous NaOH, water and then heated at 350°C/24 hours) and sealed under vacuum prior to polymerization.

Using a boron trichloride to trimer molar ratio of 1:15, almost 100% conversion to chloropolymer could be obtained in 16 hours at 180°C or in 2 hours at 250°C (D.S.V.=0.30 dl/g at 1.00% in cyclohexane). At 200°C using a molar ratio of 1:1280 a higher molecular weight chloropolymer was obtained (D.S.V.=1.17 dl/g at 1.00% in cyclohexane). Boron tribromide and

0097-6156/81/0171-0315$05.00/0

phenylboron dichloride gave results comparable to boron trichloride, although the former is faster and the latter is slower than boron trichloride.

Boron halides form complexes with phosphorus oxyhalides (13) and triarylphosphates (14,15). Compared to boron trichloride, boron trichloride•phosphorus oxychloride at comparable conditions reduced the rate of polymerization about two fold, whereas the molecular weight, as measured by dilute solution viscosity remained unchanged. Use of boron trichloride•triphenyl phosphate (recrystallized from a 1:1 by weight carbon tetrachloride solution) 1 gave high molecular weight chloropolymer (2.5 dl/g, 1% in chloroform) at a catalyst to trimer molar ratio of 1:5600 and low molecular weight chloropolymer (0.03 dl/g, 1% in cyclohexane) at a catalyst to trimer molar ratio of 1:38.

A comparison of catalytic activity of the triphenyl phosphate complexes of boron trifluoride, boron trichloride and boron tribromide showed that comparable rates of polymerization were obtained using the chloride or bromide complexes. The fluoride complex gave about half the rate of the bromide or chloride; comparable molecular weights were obtained in all three cases.

The molecular weight of the chloropolymer made via 1 is relatively insensitive to the time and temperature of the polymerization. This is shown in Tables I and II.

Table I. Polymerization Of 30g Trimer Using 0.11 mmol 1 At 220°C

Hours	% Conversion	DSV (dl/g)
4.5	28	0.92
10.0	53	0.97
18.0	55	0.93
24.0	67	0.95
40.0	83	0.90

Table II. Polymerization Of 30g Trimer Using 0.22 mmol 1

Hours	T°C	% Conversion	(η)
90	160	30	0.60
64	180	64	0.52
16	220	80	0.45
7	250	89	0.63

Attempted homo- or co-polymerization of octachlorocyclotetraphosphazene (tetramer) with trimer at 220°C gave no conversion of tetramer into chloropolymer. At a molar ratio of 1:1 trimer to tetramer, there was no change in chloropolymer molecular weight, but a reduction in the rate of polymerization (86% conversion of 100% trimer vs. 20% conversion using 1:1 trimer-tetramer).

Complex 1 is reported (14) to be thermally stable at 200°C but decomposes rather than dissociates at 300°C. This was confirmed in our laboratory via thermogravimetric analysis when only 10% weight loss occurred below 175°C but a rapid loss occurred between 200-250°C. ^{31}P NMR (reference is 85% H_3PO_4) was used to confirm that thermal rearrangement of 1 (-27.8ppm) occurred to produce 2 (-5.5ppm) and 3 (1.8 ppm) in a ratio of 1.74:1 respectively. A mass spectrograph showed 2 and 3 to be diphenyl chlorophosphate and phenyl dichlorophosphate respectively, which were shown to not initiate polymerization. The remaining fragments from this rearrangement may be phenyl dichloroborinate 4 and diphenyl chloroboronate 5 as shown in Scheme I.

Scheme I

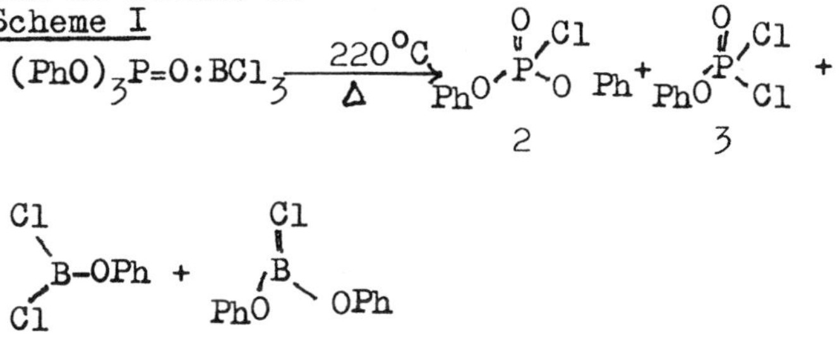

$$(PhO)_3P{=}O{:}BCl_3 \xrightarrow[\Delta]{220^\circ C} \underset{2}{PhO{-}P(=O)(Cl){-}O\,Ph} + \underset{3}{PhO{-}P(=O)(Cl){-}Cl} +$$

$$\underset{4(?)}{Cl_2B{-}OPh} + \underset{5(?)}{Cl{-}B(OPh)_2}$$

An equilibrium mixture of 4 and 5 was prepared by the ligand exchange of boron trichloride and triphenyl borate (16) and found to be an effective catalyst. This suggests our effective catalyst may be produced by the *in situ* rearrangement of 1 to give 4 and/or 5 which then initiates the polymerization.

The ^{31}P NMR spectra of a polymerization of a 1:1 molar ratio of trimer and 1 is shown in Figure 1. A small amount of 1 remains (-27.4ppm) along with unpolymerized trimer (19.77ppm) and tetramer (-6.7ppm). The sharp singlet at -18.3ppm represents internal $-PCl_2-$ units. The sharp singlet at -4.87 represents diphenyl chlorophosphate 3. Based on model linear phosphorus compounds (17,18), the shoulder of peaks at -14 to -18ppm may represent $-PCl_2-$ units adjacent to a $Cl_3P{=}N-$ group while the $Cl_3P{=}N\!=$ group might be at 3.08ppm. Additional studies are needed to conclusively elucidate the mechanism of initiation.

The hydrolysis of trimer produces hydrogen chloride and 2,2-dihydroxy-4,4,6,6-tetrachlorocyclotriphosphazene 6 which may be present in impure trimer. A ^{31}P NMR of 6, prepared according to Stokes (19), is shown in Figure 2. Since 1 is sensitive to

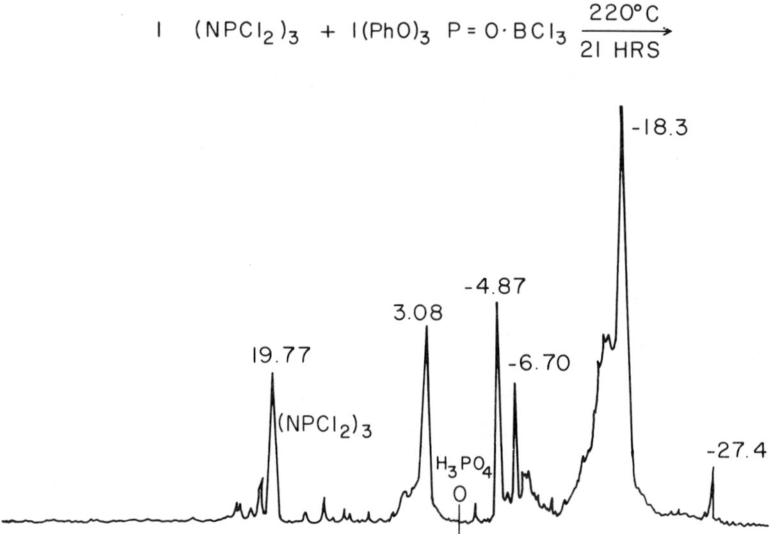

Figure 1. The ^{31}P NMR ($CDCl_3$) of the polymerization of 1 mol trimer with 1 mol boron trichloride·triphenyl phosphate after 21 h at 220° C.

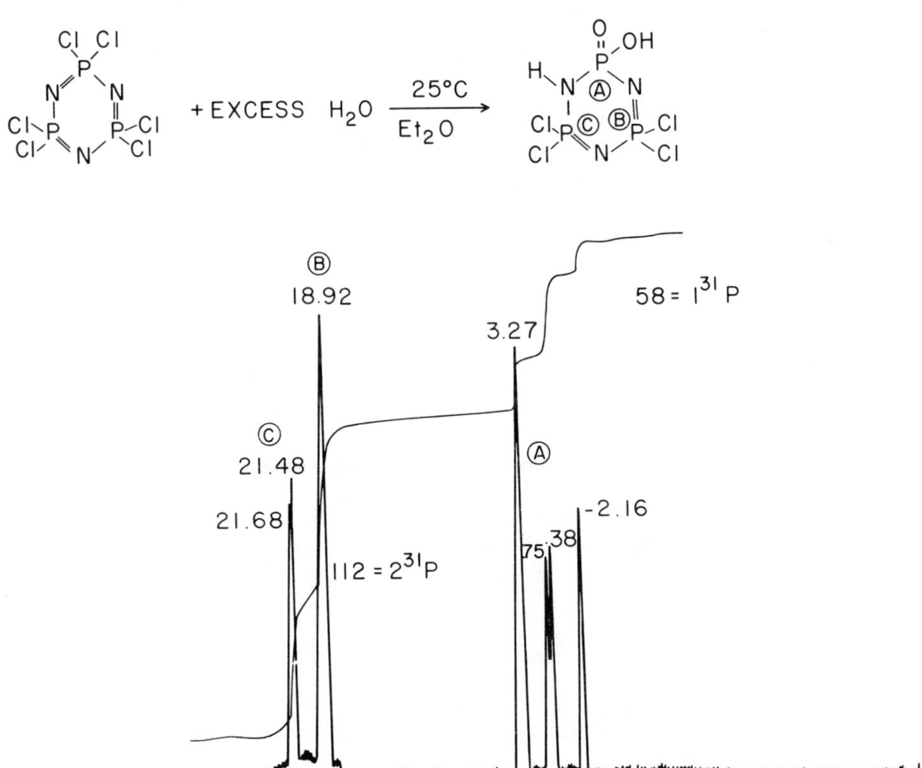

Figure 2. The ³¹P NMR (THF) of 2,2-dihydroxy-4,4,6,6-tetrachlorocyclotriphosphazene.

most protic materials, we expected 6 to hydrolyze 1. Reaction of 2 moles 1 with 1 mole of 6 gave a total conversion of 6 to trimer, along with by product triphenyl phosphate. A similar reaction occurred between boron trichloride and 6, thus showing the great propensity for boron trichloride to 'rechlorinate' hydrolyzed cyclophosphazenes.

Acknowledgement. The authors wish to thank Dr. Tom Antkowiak for his guidance in this work and The Firestone Tire and Rubber Company for permission to publish.

Literature Cited

1. Allcock, H.R.; Kugel, R.L. J. Am. Chem. Soc., 1965, 87, 4216-7.
2. Allcock, H.R.; Kugel, R.L.; Valan, K.J. Inorg. Chem. 1966, 5, 1709-15.
3. MacCallum, J.R.; Tanner, J. J. Pol. Sci 1969, 7, 743-747.
4. Allcock, H.R.; Gardner, J.E.; Smeltz, K.M. Macromolecules 1975, 8 (1), 36-42.
5. Korsak, V.V.; Vinogradova, S.V.; Tur, D.R.; Kasarova, N.N.; Komarova, L.I.; Gilman, L.M. Acta. Polymerica 1979, 30 (5), 245-8.
6. U. S. 4,137,330.
7. Konecny, J.O.; Douglas, C.M. J. Pol. Sci. 1959, 36, 195-203.
8. Konecny, J.O.; Douglas, C.M.; Gray, M.J. J. Pol. Sci. 1960, 42, 383-90.
9. Gimblett, F.G.R. Polymer 1960, 1, 418-24.
10. MacCallum, J.R.; Werninck, A. J. Pol. Sci. 1967, A-1 5, 3061-70.
11. Gimblett, F.G.R. Plast. Inst. Trans. 1960, 28, 65-73.
12. Emsley, J.; Udy, P.B. Polymer 1972, 13, 593-4.
13. Peach, M.E.; Waddington, T.C. J. Chem. Soc. 1962, 3450-3.
14. Frazer, M.J.; Gerrard, W.; Patel, J.K. J. Chem. Soc. 1960, 726-730.
15. Allcock, H.R.; Levin, M.; Fieldhouse, J.W. Acta. Crysta. 1981.
16. Colclough, T.; Gerrard, W.; Lappert, M.F. J. Chem. Soc. 1955, 907-11.
17. Fluck, E. Z. Anorg. Chem. 1962, 315, 181.
18. Becke-Goehring, M.; Fluck, E. Angew. Chem. (Int. Ed.) 1962, 1, 281.
19. Stokes, H.N. Amer. Chem. J. 1985, 17, 275-290.

RECEIVED June 30, 1981.

Alkenylfluorocyclotriphosphazenes

CHRISTOPHER W. ALLEN, RANDALL P. BRIGHT, and KOLIKKARA RAMACHANDRAN

Department of Chemistry, University of Vermont, Burlington, VT 05405

The use of organometallic reagents such as organolithium, Grignard and organocopper (1) species in substitution reactions with halocyclophosphazenes has led to the synthesis of a large number of organophosphazene derivatives (2). The vast majority of these compounds have been the aryl and alkyl derivatives. Recently, we have initiated an investigation of the synthesis, characterization and reactions of alkenylphosphazenes (3). Our interest in these materials is three-fold: comparison of the stereochemistry of substitution reactions with that observed for other organometallic reagents (2), elucidation of the electronic interaction between the phosphazene and unsaturated organic moieties (4), and synthetic transformations of the organofunctional exocyclic group (5).

We have previously reported the preparation of propenyl phosphazenes *via* the reactions of propenyl lithium reagents with hexafluorocyclotriphosphazene, $N_3P_3F_6$ (3). These materials undergo a broad variety of reactions ranging from simple hydrogenation (2) to copolymerization with styrene (6). Given the potential

$$N_3P_3F_6 + LiC(CH_3) = CH_2 \rightarrow N_3P_3F_5C(CH_3){=}CH_2$$

$$N_3P_3F_5C(CH_3){=}CH_2 + H_2 \xrightarrow{Pd/C} N_3P_3F_5CH(CH_3)_2$$

$$N_3P_3F_5C(CH_3){=}CH_2 + C_6H_5CH = CH_2 \rightarrow [(\underset{\displaystyle C_6H_5}{\underset{|}{C}}HCH_2)_x(\underset{\displaystyle N_3P_3F_5}{\underset{|}{\overset{\displaystyle CH_3}{\overset{|}{C}}}}CH_2)_y]_n$$

technological utility of these copolymers as flame retardant materials (6), we have now reinvestigated the reaction of 2-propenyl lithium with $N_3P_3F_6$. In addition to the expected propenylphosphazene, we obtain an oily high-boiling by-product. The ratio of propenylphosphazene to by-product is strongly dependent on the source of lithium used to prepare the organolithium reagent. The average molecular weight of a typical by-product

0097-6156/81/0171-0321$05.00/0

is around 1000 and the mass spectrum of a carefully fractionated material shows only one $N_3P_3F_5$ unit. We propose that anionic attack by the organolithium reagent on the olefinic center in the propenylphosphazene initiates the formation of the observed by-products.

$$N_3P_3F_5C(CH_3)=CH_2 + LiC(CH_3)=CH_2 \rightarrow N_3P_3F_5-\underset{Li}{\overset{CH_3}{\overset{|}{\underset{|}{C}}}}-CH_2-\overset{CH_3}{\overset{|}{C}}=CH_2$$

Our experience with propenylphosphazenes suggested that one could avoid some of the complications in the synthesis of alkenylphosphazenes by dealing with alkenylphosphazenes containing electron donating functions on the olefin thus counteracting the electron withdrawing effect of the phosphazene. We tested this hypothesis by metalating ethylvinyl ether with tert butyl lithium and allowing the organo-lithium reagent to react with $N_3P_3F_6$. Using this reaction, we were able to prepare both the mono and disubstituted derivatives, $N_3P_3F_5C(OC_2H_5)=CH_2$ and $N_3P_3F_4[C(OC_2H_5)=CH_2]_2$, without any of the by-product observed in the case of the propenyl lithium reaction. Similar results were obtained starting with methylvinyl ether. Attempts to achieve trisubstitution led to a complex series of reactions involving

$$N_3P_3F_6 + nLiC(OR)=CH_2 \rightarrow N_3P_3F_{6-n}[C(OR)=CH_2]_n$$
$$R=C_2H_5, CH_3;\ n=1,2$$

such processes as cleavage of the ethoxy group from the ethoxyvinylphosphazene. The disubstituted material was shown to have the geminal configuration (2,2-$N_3P_3F_4[C(OC_2H_5)=CH_2]_2$) by ^1H, ^{13}C, ^{19}F and ^{31}P nmr spectroscopy. The monosubstituted materials show long-range fluorine coupling with exocyclic group. These interactions are not observed in the disubstituted derivatives. The ^{31}P and ^{19}F nmr spectra clearly show, by the phosphorus-fluorine coupling patterns, the presence of $\equiv PF_2$ and $\equiv PR_2$ centers and the absence of $\equiv PFR$ centers thus confirming the geminal nature of the disubstituted materials.

The chemical studies discussed above strongly support our hypothesis that the polarity of the organic function in alkenylphosphazenes can be significantly altered by the nature of the olefin substituents. We sought alternative methods to evaluate these effects and found that ^{13}C nmr spectroscopy is a very useful tool in this regard. Generally, the β carbon chemical shift in substituted olefins is sensitive to electronic effects at the α position (7). Thus on going from ethylvinyl ether to $N_3P_3F_5C(OC_2H_5)=CH_2$, the β carbon atom experiences a 15 ppm downfield shift and the propenyl derivative, $N_3P_3F_5C(CH_3)=CH_2$, is 32 ppm downfield from the ethoxyvinyl derivative. Similar observations hold for mixed phenyl/alkoxyvinyl and dimethylamino/alkoxyvinyl fluorophosphazenes. This later observation clearly shows the large difference in charge distribution in these two olefins. Examination of the ^{31}P data also has uncovered some interesting trends, for example, there is a significant difference in ^{31}P data chemical shifts when an ethoxyvinyl group is replaced by a phenyl group ($\Delta\delta$ = 6 ppm). The effect of the various vinyl substituents on the ^{31}P shifts is also significant.

Having demonstrated that the ethoxyvinylphosphazene can be converted to the disubstituted derivative, we decided to explore a variety of reactions involving the ethoxyvinylphosphazene. Since much of our interest in alkenylphosphazenes is related to their incorporation into traditional vinyl polymers (5,6), we were pleased to observe facile copolymerization of $N_3P_3F_5C(OC_2H_5){=}CH_2$ with styrene. We also have prepared some mixed substituent derivatives. The reaction of phenyl lithium with

$$N_3P_3F_5C(OC_2H_5){=}CH_2 + C_6H_5CH{=}CH_2 \xrightarrow{R\cdot} [(\underset{C_6H_5}{\underset{|}{C}}HCH_2)_x(\overset{OC_2H_5}{\overset{|}{\underset{P_3N_3F_5}{\underset{|}{C}}}}CH_2)_y]_n$$

$N_3P_3F_5C(OC_2H_5){=}CH_2$ leads to the geminally substituted material 2,2-$N_3P_3F_4(C_6H_5)C(OC_2H_5){=}CH_2$; the same product is obtained from the reaction of ethoxyvinyl lithium with phenylpentafluorocyclotriphosphazene. However, the reaction of two molecules of phenyl lithium with $N_3P_3F_6$ is known to follow a predominantly non-geminal pathway (8).

$$N_3P_3F_5C(OC_2H_5){=}CH_2 \xrightarrow{C_6H_5Li} \text{2,2-}N_3P_3F_4(C_6H_5)C(OC_2H_5){=}CH_2$$

$$N_3P_3F_5C_6H_5 \xrightarrow{LiC(OC_2H_5)=CH_2} \text{2,2-}N_3P_3F_4(C_6H_5)C(OC_2H_5){=}CH_2$$

In another series of experiments, we found that the reaction of $N_3P_3F_5C(OC_2H_5){=}CH_2$ with dimethylamine leads to the non-geminal isomers, 2,4-$N_3P_3F_4[N(CH_3)_2]C(OC_2H_5){=}CH_2$; the same product is obtained from the reaction of ethoxyvinyl lithium with dimethylaminopentafluorocyclotriphosphazene.

$$N_3P_3F_5C(OC_2H_5){=}CH_2 \xrightarrow{(CH_3)_2NH} \text{2,4-}N_3P_3F_4N(CH_3)_2\overset{OC_2H_5}{\overset{|}{C}}{=}CH_2$$

$$N_3P_3F_5N(CH_3)_2 \xrightarrow{LiC(OC_2H_5)=CH_2} \text{2,4-}N_3P_3F_4N(CH_3)_2\overset{OC_2H_5}{\overset{|}{C}}{=}CH_2$$

The structures of all the mixed substituent derivatives were unambiguously established by ^{31}P and ^{19}F nmr spectroscopy.

The observation of the variation of the stereochemical course of the reaction of ethoxyvinyl lithium with various monosubstituted pentafluorocyclotriphosphazenes with the nature of the phosphazene substituent clearly demonstrates that there is a directive effect based on the ring substituent which is operative in these reactions. These findings are at variance with current thinking which stresses the control of the incoming reagent on the stereochemistry of the substitution reaction (9). We propose the following model for cases where substituent control of directive effects occurs. We have previously shown that exocyclic electron release in organofluorophosphazenes is through the σ system (4). The net result of this effect is that ring nitrogen atom lone pair electron density is preferentially transferred to the $\equiv PF_2$ center, thus leaving the $\equiv$PFR center as the site of nucleophilic attack. When a π donating substituent, such as dimethylamine, is on the ring, electron release is into the phosphorus atom system at the substituted phosphorus center. This mechanism reduces the formal positive charge at the substituted phosphorus atom and hence leaves the $\equiv PF_2$ center as the site of substitution.

Acknowledgement

This work was supported, in part, by the Office of Naval Research.

Literature Cited

1. Allcock, H.R.; Harris, P.J. J. Am. Chem. Soc. 1979, 101, 6221-9.
2. Allen, C.W. Ind. Eng. Chem. Prod. Res. Dev. 1981, 20, 77-79.
3. Dupont, J.G.; Allen, C.W. Inorg. Chem. 1978, 17, 3093-6.
4. Allen, C.W.; Green, J.C. Inorg. Chem. 1980, 19, 1719-22.
5. Allen, C.W.; Dupont, J.G. Ind. Eng. Chem. Prod. Res. Dev. 1979, 18, 80-1.
6. Dupont, J.G.; Allen, C.W. Macromolecules 1979, 12, 169-72.
7. Strothers, J.B. "Carbon-13 NMR Spectroscopy"; Academic Press: New York, N.Y., 1972; p. 184.
8. Allen, C.W.; Moeller, T. Inorg. Chem. 1978, 7, 2177-83.
9. Shaw, R.A. Z. Naturforsch. 1976, 31b, 641-67.

RECEIVED June 30, 1981.

The Reactions of Halophosphazenes with Organometallic Reagents

PAUL J. HARRIS
Department of Chemistry, Virginia Polytechnic Institute and State University, Blacksburg, VA 24061

HARRY R. ALLCOCK
Department of Chemistry, Pennsylvania State University, University Park, PA 16802

The interactions of cyclic halophosphazenes with organometallic reagents are some of the most complex and least understood reactions in Main Group chemistry.[1-4] Until recently it was thought that the initial products formed during these reactions were acyclic or "ring opened" species.[3,4] However, we have undertaken a detailed investigation of the reactions of both Grignard reagents and organocopper reagents with hexachlorocyclotriphosphazene (I) and find that the products from these reactions contain intact phosphazene rings. The results of these studies are presented in this paper.

The reactions of hexachlorocyclotriphosphazene (I) with a variety of Grignard reagents (RMgCl where R = Me, Et, _n_-Pr, _n_-Bu, _i_-Pr, _t_-Bu, Ph) were investigated. It was found that these reactions did not lead to initial clevage of the phosphazene ring, but led exclusively to the formation of two well defined types of products, both of which contained intact phosphazene rings as shown in Scheme I.

These products were either mono-or di-substituted phosphazene trimers of type II and III,[5,8] or compounds IV in which two cyclic phosphazene rings were linked together through a direct P-P bond. The ratios of the two types of products formed in the different reactions studied were found to be markedly dependent on both the reaction temperature and the Grignard reagent employed. Thus, while methyl- or phenyl-magnesium chloride led almost exclusively to formation of the respective dimer (IV, R = Me, Ph), independant of reaction temperature, _i_-propyl magnesium chloride gave the mono-substituted compound (II, R = _i_-Pr) as the only product. Reactions involving _n_-butyl magnesium chloride were found to be extremely temperature dependent; at -10°C dimer formation (IV, R = _n_-Bu) was preferred, while at reflux temperatures (~66°C) the mono-and di-alkyl products predominated (II and III, R = _n_-Bu).

More detailed studies of the reactions indicated that two competing mechanisms were in operation. Compounds of type II and subsequently III resulted from a simple nucleophilic attack by

0097-6156/81/0171-0325$05.00/0

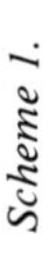

Scheme 1.

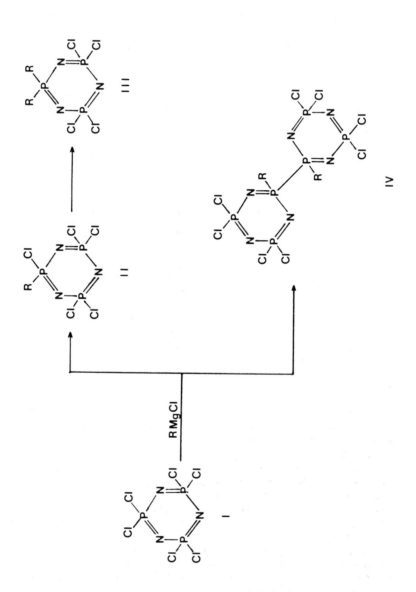

the Grignard reagent at phosphorus, followed by elimination of a chloride ion. Compounds of type IV were found to result from a metal-halogen exchange type reaction. The alkyl chloride generated by this reaction between the Grignard reagent and the phosphazene was detected in some cases. The initial reaction product, a metallophosphazene, was too reactive to isolate and attacked another molecule of $[NPCl_2]_3$ to generate the bicyclic compound IV.

Thus it appears that the more nucleophilic Grignard reagents, or higher reaction temperatures, favor the formation of compounds II and III, while weaker nucleophiles undergo the metal-halogen exchange reaction, a reaction that is also preferred at low temperatures.

An investigation into the reactions of organocopper reagents with hexachlorocyclotriphosphazene (I) was also undertaken.[6,9] The organocopper reagent was generated in situ by the reaction between a Grignard reagent and the complex $[\underline{n}\text{-}Bu_3PCuI]_4$, which was used as a soluble form of copper[I]. These reactions were found to proceed exclusively by the metal-halogen exchange pathway.[9] In these cases however, the copper was able to stabilize the reduced phosphazene ring and thus dimers of type IV were not observed. The initial products formed from these reactions were found to be metallophosphazenes of type V. These intermediates have proved to be extremely useful in the synthesis of a variety of phosphazene compounds as shown in Scheme II. Thus reaction with alkyl halides such as allyl bromide or propargyl bromide allow for the introduction of olefinic[7,8] or acetylenic[10] side groups onto the phosphazene ring VI, while alcohol leads to the formation of hydrido-phosphazene complexes[6,9] VII. The hydrogen in these compounds can be replaced with halogen[5] to yield the first series of iodophosphazene compounds VIII.

Thus, from the series of reactions presented in this paper it appears that a major reaction pathway for chlorocyclophosphazenes with certain organometallic reagents, involves a metal-halogen exchange process. Subsequent chemistry however appears to be extremely dependent on the metal involved in the organometallic reagent.

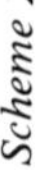

Scheme 2.

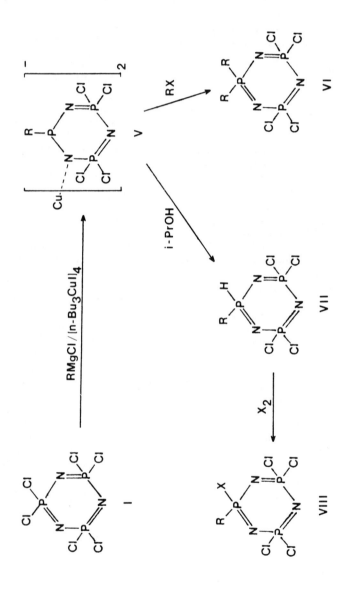

Acknowledgement

We thank the Office of Naval Research for the support of this work.

Literature Cited

1. DuPont, J. G.; Allen, C. W. Inorg. Chem. 1978, 17, 3093.
2. Ranganathan, T. N.; Todd, S. M.; Paddock, N. L. Inorg. Chem. 1973, 12, 316.
3. Biddlestone, M.; Shaw, R. A. J. Chem. Soc. (A), 1971, 2715.
4. Biddlestone, M.; Shaw, R. A. Phosphorus, 1973, 3, 95.
5. Allcock, H. R.; Harris, P. J. Inorg. Chem. in Press.
6. Harris, P. J.; Allcock, H. R. J. Am. Chem. Soc. 1968, 100, 6512.
7. Harris, P. J.; Allcock, H. R. J. Chem. Soc. Chem. Comm. 1979, 714.
8. Allcock, H. R.; Harris, P. J., Connolly, M. S. Inorg. Chem. 1981, 20, 11.
9. Allcock, H. R.; Harris, P. J. J. Am. Chem. Soc. 1979, 101, 6221.
10. Allcock, H. R., Harris, P. J., Nissan, R. A. J. Am. Chem. Soc. in Press.

RECEIVED June 30, 1981.

NEW ORGANOPHOSPHORUS COMPOUNDS OF COMMERCIAL INTEREST

1,2-Bis(dichlorophosphino)alkanes

E. H. UHING and A. D. F. TOY

Eastern Research Center, Stauffer Chemical Company, Dobbs Ferry, NY 10522

This paper presents a new method for preparing 1,2-bis-(dichlorophosphino)alkanes based on the reaction of unsaturated hydrocarbons with PCl_3 and elemental phosphorus at 180-250°C under autogenous pressure[1]. The basic reaction with ethylene is shown in eq 1.

$$6CH_2{=}CH_2 + 8PCl_3 + P_4 \longrightarrow \underset{\underline{1}}{6Cl_2PCH_2CH_2PCl_2} \qquad (1)$$

A prior method for making 1 involved the reaction of 1,2-bis-(diphenylphosphino)ethane with PCl_3 at 280°C using $AlCl_3$ as a catalyst[2].

To show that eq 1 could be feasible we hypothesized the reaction occurring via eq 2 and 3. The addition of PCl_3 across

$$6CH_2{=}CH_2 + 6PCl_3 \longrightarrow \underset{\underline{2}}{6ClCH_2CH_2PCl_2} \qquad (2)$$

$$6(\underline{2}) + 2PCl_3 + P_4 \longrightarrow 6(\underline{1}) \qquad (3)$$

a double bond using uv or other free radical initiators to form 2-chloroalkylphosphonous dichlorides is well known[3]. The thermal initiation of the reaction shown in eq 2 has not been reported. When excess PCl_3 and ethylene (eq 2) are reacted at 200-250°C under pressure, our study showed only trace amounts of 2 being formed. At 250°C, with an excess of ethylene, a black solid formed. We believe the low yield of 2 might be due to an unfavorable equilibrium or decomposition reaction. Trapping of 2 by some further reaction as shown in eq 3 could allow eq 2 to proceed. The reaction of alkyl chlorides with PCl_3 and elemental phosphorus at 200-300°C to yield alkylphosphonous dichlorides has been reported[4]. Therefore, if 2 reacts like an alkyl chloride then eq 3 could form 1.

One of the proposed mechanisms for the uv initiated addition of PCl_3 to isobutene involves a cyclic or bridged free radical

0097-6156/81/0171-0333$05.00/0

intermediate to account for the absence of cross-transfer products[5]. Analogously, the formation of a transitory cyclic addition complex between PCl_3 and ethylene as shown in Scheme I could also be an alternate reaction pathway for the formation of 1.

Scheme I

$$CH_2{=}CH_2 + PCl_3 \longrightarrow \left[\begin{array}{c} CH_2 — CH_2 \\ \diagdown \; {+} \; \diagup \quad Cl^- \\ PCl_2 \end{array}\right]$$

3

$$\underline{3} + P_4 \longrightarrow \left[\begin{array}{c} CH_2 - CH_2 \\ \diagdown \; \diagup \\ P \\ | \\ Cl \end{array}\right] + PCl_3$$

4

$$\underline{4} + PCl_3 \longrightarrow \underline{1}$$

A cyclic phosphirane related to 4 has been reported[6].

The practical scope of our new method for preparing 1,2-bis-(dichlorophosphino)alkanes using the ratio of alkene to elemental phosphorus shown in eq 1 and a 50-150% excess of PCl_3 at 200-250°C for 4-6 hours under autogenous pressure is shown in Table I.

Table I Yields of $Cl_2PCH_2CH(R)PCl_2$

1-Alkene Used	R		% Yield
ethylene	H	(1)	70
propylene	CH_3	(5)	66
1-butene	CH_2CH_3	(6)	47
1-pentene	$(CH_2)_2CH_3$	(7)	41
1-octene	$(CH_2)_5CH_3$	(8)	20

Table I shows a sharp reduction in yields as the molecular weight of the 1-alkene increases. For alkenes above 1-propene, I_2 was added as a catalyst even though its beneficial effect was not clearly established. All the above reactions were run in a 316 Stainless Steel autoclave rated at 5000 psig. A typical charge for making 1 in a 300 mL autoclave is 1.0m PCl_3, 0.45m ethylene and 0.3 gram-atom of white phosphorus. The maximum pressure generated at 200°C is 2000 psig. All products were isolated by vacuum distillation.

In Table II we show the ^{31}P-NMR of these bis compounds. The obvious point of interest for the ^{31}P-NMR spectra shown in Table II is that the 1 and 5 both are singlets at about the same frequency. Since 5 is not a symmetrical molecule, we ran the

Table II ^{31}P-NMR Spectra of Compounds in Table I

Compound	δ (+ppm from H_3PO_4)	Jp-p(Hz)
1	190.6	
5	190.7	
6	194.8(d);191.4(d)	10
7	194.5(d);191.6(d)	10
8	194.5(d);191.3(d)	10

^{13}C-NMR spectra to be sure that a 1,3-bis product did not form. The ^{13}C-NMR spectra of 1 and 5 are shown in Table III. Compound 1 has the spectrum of a symmetrical 1,2-bis molecule.

Table III ^{13}C-NMR Spectra of 1 and 5

Compound	δ (+ ppm from $(CH_3)_4Si$)	^{1}Jc-p	^{2}Jc-p	^{3}Jc-p (Hz)
1	36.1(d of d)	50	10	
5	45.5(d of d)	53	13	
	40.5(d of d)	50	11	
	16.0(t)			13.5

The ^{13}C-NMR spectrum of 5 shows the presence of a methyl group that is coupled to two magnetically equivalent phosphorus atoms. To confirm that 5 has the expected 1,2-bis structure, we hydrolyzed and oxidized it to the bis-phosphonic acid. The ^{31}P-NMR spectra of this acid shows two doublets consistent for a 1,2-bis structure. The ^{31}P-NMR spectra reported for the 1,3-bis-(phosphonic acid) propane is a singlet[2].

Several internal olefins were also tried in the reaction shown in eq 1. When the reaction of both cis-or trans-2-butene were run at 230°C, they gave a low (10%) yield of the 2,3-bis (dichlorophosphino)butane as evidenced by a single ^{31}P-NMR peak at 193.6 ppm and a symmetrical ^{13}C-NMR spectra. When the reaction temperature is raised to 250°C, both cis-and trans-2-butene yield a compound which has the same ^{31}P and ^{13}C-NMR spectra as 6, indicating a 1,2-bis(dichlorophosphino)butane product. Along with this rather complex isomer product change, there is an increase in the amount of butylphosphonous dichloride formed.

A branched alkene, isobutylene, gave a 23% yield of 1,2-bis-(dichlorophosphino)isobutane at a reaction temperature of 210°C along with a trace of tert-butylphosphonous dichloride and two isobutenylphosphonous dichlorides. Cyclohexene failed to produce even a trace of bis product under a variety of reaction conditions. The only product isolated was cyclohexylphosphonous dichloride which is one of the products previously reported for the thermal addition of PCl_3 to cyclohexene[7].

Since the 1,2-bis(dichlorophosphino)alkanes made by this new process are reactive intermediates they have a variety of potential uses. Compound 1 has been converted to the 1,2-bis (dimethoxyphosphino)ethane and used for making metal carbonyl complexes[8]. There also is a report on the conversion to 1,2-bis-(dimethylphosphino)ethane and 1,2-bis(diethylphosphino)ethane[9]. The tetra-sodium salt of ethylenediphosphinetetraacetic acid has been made using intermediate 1[10]. The reactions with phenols and cyclic aliphatic alcohols also have been reported[11].

LITERATURE CITED

1. Toy, A.D.F.; Uhing, E.H. (to Stauffer Chemical Co.) U.S. 3,976,690 (1976).
2. Sommer, K. Z. Anorg. Allg. Chem. 1970, 376, 37.
3. Fild, M.; Schmutzler, R. "Organic Phosphorus Compounds"; Kosolapoff, G. M.: Maier, L. Ed.; Wiley-Interscience, NY, 1972, Vol. 4; Chapter 8.
4. Bliznyuk, N.K.; Kvasha, Z.N.; Kolomiets, A.F. Zh. Obshch. Khim. 1967, 37, 890.
5. Little, J. R.; Hartman, P.F. J. Am. Chem. Soc. 1966, 88, 96.
6. Wagner, R. I.; Freeman, L. D.; Goldwhite, H., Rowsell, D. G. J. Am. Chem. Soc. 1967, 89, 1102.
7. Wunder, K.; Drawe, H.; Henglein, A. Z. Naturforschg. 1964, 19b, 999.
8. King, R. B.; Rhee, W. M. Inorg. Chem. 1978, 17, 2961.
9. Burt, R. J.; Chatt, J.; Hussain, W.; Leigh, G. J. J. Organomet. Chem. 1979, 182, 203.
10. Podlahova, J.; Podlaha, J. Collect. Czech. Chem. Commun. 1980, 45, 2049.
11. Uhing, E. H. (to Stauffer Chem. Co.) U.S. 4,263,230 (1981).

RECEIVED July 7, 1981.

Products of Peracid Oxidation of *S*-Alkyl Phosphorothiolate Pesticides

YOFFI SEGALL[1] and JOHN E. CASIDA

Pesticide Chemistry and Toxicology Laboratory, Department of Entomological Sciences, University of California, Berkeley, CA 94720

Biooxidation is an essential activation process for some organothiophosphorus neurotoxicants (1). m-Chloroperoxybenzoic acid (MCPBA) has been used to mimic some of these reactions but without identifying the products derived from phosphorothiolates (2,3). We observed that S-alkyl phosphorothiolates react with MCPBA to form a new and unexpected class of phosphinyloxysulfonates via a novel rearrangement process (Eq. 1).

1 phosphorothiolate —MCPBA→ [2 phosphorothiolate S-oxide] → [phosphoranoxide] →

Eq. 1

[3 phosphinyloxysulfenate] —MCPBA→ [4 phosphinyloxysulfinate] —MCPBA→ 5 phosphinyloxysulfonate

Peracid oxidation of S-alkyl phosphorothiolates (1) appears to proceed by initial formation of S-oxides (2) which undergo spontaneous and very rapid rearrangement, via phosphoranoxide intermediates, to the corresponding sulfenate esters (3) that are

[1]Current address: Israel Institute for Biological Research, Ness-Ziona, P.O.B. 19, Israel.

0097–6156/81/0171–0337$05.00/0

further oxidized to the oxysulfinate (4) and oxysulfonate esters (5) (Eq. 1). With equimolar MCPBA very little starting material undergoes reaction because most of the oxidant is used up in conversion of 3 and 4 to 5. As a result the most extensively oxidized product (5) is strongly favored.

These generalizations are based on studies with four S-alkyl phosphorothiolate pesticides (6a, 7a, 7b, 8) and related compounds as follows:

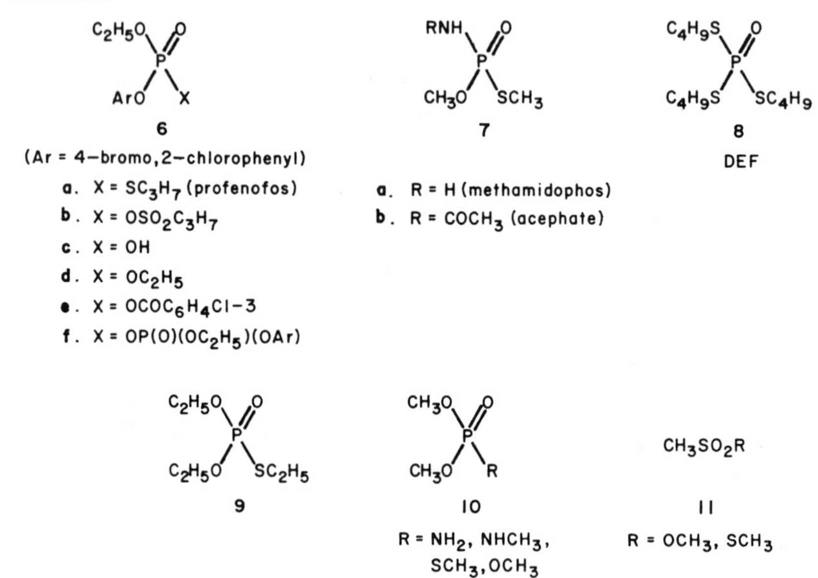

The oxidation rate with excess peracid decreases in the order 8 > 7a > 7b >> 6a; the first three compounds react readily below -30°C whereas 6a requires > 10°C for significant reaction in 24 hr. The nitrogen free electrons of 7a and 7b may facilitate their oxidation by increasing the polarizability of the sulfur.

Profenofos (6a) and its derivatives were selected for detailed examination. Sulfonate ester 6b is obtained pure in 86% yield on reacting 6a with five equivalents of MCPBA in ethanol-free chloroform at 25°C for 8 hr followed by rapidly extracting the solution with aqueous $NaHSO_3$ and $NaHCO_3$ at 0°C. Similar treatment at 25°C for 5 min results in complete hydrolysis of 6b to 6c and propylsulfonic acid. NMR studies (Table I) suggest conversion of the phosphorothiolate to a phosphate. Thus, on going from 6a to 6b there is a significant high field shift in the ^{31}P NMR signals and downfield shifts in the ^{1}H and ^{13}C signals of the α-methylene bonded to sulfur. In addition, neither the proton nor the carbon of the S-methylene in 6b is coupled with the phosphorus. Three minor phosphorus-containing products formed on oxidation of 6a are: the acid 6c (δ ^{31}P -6.82 ppm in $CDCl_3$); the 3-chlorobenzoyl

TABLE I. NMR Spectral Data for 6a and 6b ($CDCl_3$)

Chemical shifts denoted as follows: ^1H for diastereotopic protons of the methylene directly bonded to sulfur (downfield of TMS); ^{13}C for carbon directly bonded to sulfur (TMS); ^{31}P are negative when upfield of 1% trimethyl phosphate in C_6D_6 (δ ^{31}P = 0).

	6a		6b	
Nucleus	δ ppm	$J_{^{31}P\text{-nucleus}}$	δ ppm	$J_{^{31}P\text{-nucleus}}$
^1H	2.91	9.3 Hz	3.49	none
^{13}C	33.44	9.0 Hz	55.18	none
^{31}P	+22.91	-	-21.77	-

ester 6e (δ ^{31}P -16.51); the diastereomeric pyrophosphate 6f (two lines centered at δ ^{31}P -22.74, $\Delta\delta$ 0.03 ppm). Comparable oxidation of 9 (δ ^{31}P +31.28 in acetone, referenced to 1% trimethyl phosphate in $CDCl_3$) yields the analogous diethyl phosphoric acid (δ ^{31}P +1.80), diethyl 3-chlorobenzoyloxyphosphate (δ ^{31}P -10.30) and tetraethyl pyrophosphate (one line at δ ^{31}P -10.78).

Mixed anhydride 6b is a sulfonylating rather than a phosphorylating agent. Thus, hydrolysis with $H_2{}^{18}O$ gives acid 6c and propylsulfonic acid in which the ^{18}O isotope is incorporated only in the sulfonic acid (Eq. 2). Compound 6b reacts with either alcohols (methanol, ethanol, sec-butanol) or L-cysteine to yield acid 6c and with triethylamine to give the anion of 6c, probably via propylsulfene (4) (Eq. 2).

$$(C_2H_5O)(ArO)P(O)OSO_2C_3H_7 \ (\mathbf{6b}) \xrightarrow{H_2{}^{18}O} (C_2H_5O)(ArO)P(O)OH \ (\mathbf{6c}) + H^{18}OSO_2C_3H_7$$

$$\mathbf{6b} \xrightarrow[-(C_2H_5)_3\overset{+}{N}H]{(C_2H_5)_3N} \left[(C_2H_5O)(ArO)P(O)O{-}SO_2{-}\bar{C}HCH_2CH_3\right] \longrightarrow (C_2H_5O)(ArO)P(O)O^- + O_2S{=}CHCH_2CH_3 \xrightarrow{H_2O} HOSO_2C_3H_7 \quad \text{Eq. 2}$$

Phosphoramidothiolates 7a and 7b with MCPBA yield thermally-unstable products in contrast to the more stable 6b obtained on oxidation of 6a. Neutral aqueous hydrolysis of the reaction products of 7a then derivatization with diazomethane yields many compounds, including 10 and 11 identified by MS and NMR. These products indicate involvement of both the $-SCH_3$ and $-NH_2$ groups in the oxidation and decomposition processes. Phosphorotrithiolate 8 (δ ^{31}P +66.40 ppm in $CDCl_3$) reacts almost instantly with MCPBA (3-4 equivalents) at -30°C to give rearranged products (δ ^{31}P +0.50, -7.60 and -17.00) that do not contain any direct P-S bond as their proton coupled and decoupled ^{31}P NMR signals are identical to each other.

The biological properties of a phosphorothiolate S-oxide presumably depend in part on the relative rate at which it reacts as a phosphorylating agent as opposed to the rate it rearranges to the phosphinyloxysulfenate (Eq. 3). A simple experiment

phosphorothiolate —mfo [O] / *activation*→ [phosphorothiolate S-oxide] —hydrolase(s)→ phosphorylated hydrolase(s)

[phosphorothiolate S-oxide] ↓ *deactivation*

phosphinyloxy-sulfenate —H_2O→ hydrolysis products

Eq. 3

illustrates this point. On oxidation of 6a in ethanol the only phosphorus product isolated is the diethyl ester 6d undoubtedly obtained from phosphorylation of ethanol by the S-oxide intermediate; phosphorothiolate 6a by itself does not react with ethanol and phosphinyloxysulfonate 6b gives the acid 6c. Thus, if the S-oxide activation product is formed on mixed-function oxidase (mfo) activation within the cell it may immediately phosphorylate sensitive sites such as hydrolases including acetylcholinesterase. However, when the S-oxide is formed in an environment where rearrangement occurs faster than phosphorylation the overall result is the deactivation process of hydrolysis. This hypothesis warrants careful consideration in evaluating the toxicology and metabolism of S-alkyl phosphorothiolate pesticides.

Acknowledgment

Supported in part by Grant 5 P01 ES00049 from the National Institutes of Health.

Literature Cited

1. Eto, M. "Organophosphorus Pesticides: Organic and Biological Chemistry," CRC Press, Cleveland, 1974.
2. Bellet, E. M.; Casida, J. E. J. Agric. Food Chem. 1974, 22, 207.
3. Eto, M.; Okabe, S.; Ozoe, Y.; Maekawa, K. Pestic. Biochem. Physiol. 1977, 7, 367.
4. Michalski, J.; Radziejewski, C.; Skrzypczyński, Z. J. Chem. Soc. Chem. Comm. 1976, 762.

RECEIVED July 7, 1981.

Research on Organophosphorus Insecticides, Synthesis of *O*-Alkyl *O*-Substituted Phenyl Alkylphosphonothioates

WU KIUN-HOUO, SUN YUNG-MIN, and WANG SING-MIN

Department of Chemistry, Fudan University, Shanghai, China

In order to obtain the organo-phosphorus insecticides needed for agriculture in China, we have prepared more than one hundred organo-phosphorus compounds. After screening for the efficiency of insecticides, we found that o-1,3,-dichlorophenyl o-alkyl ethylphosphonothioates showed high activities against a wide variety of insects as an efficient contact insecticide. We have synthesized fifteen compounds of the homologs by the following reactions:

$$(C_2H_5)_4Pb + 3PCl_3 \rightarrow 3C_2H_5PCl_2 + C_2H_5Cl + PbCl_2 \quad (1)$$

$$C_2H_5PCl_2 + S \rightarrow C_2H_5PSCl_2 \quad (2)$$

$$C_2H_5PSCl_2 + ROH + (C_2H_5)_3N \rightarrow (C_2H_5)(RO)P(=S)Cl + (C_2H_5)_3N\cdot HCl \quad (3)$$

$$(C_2H_5)(RO)P(=S)Cl + HO\text{-}C_6H_4\text{-}X \xrightarrow[\text{Acetone}]{Na_2CO_3} (C_2H_5)(RO)P(=S)O\text{-}C_6H_4\text{-}X \quad (4A)$$

$$(C_2H_5)(RO)P(=S)Cl + NaO\text{-}C_6H_4\text{-}X \xrightarrow[\text{Butanone}]{} (C_2H_5)(RO)P(=S)O\text{-}C_6H_4\text{-}X \quad (4B)$$

For the last step of the preparation we used method 4A and 4B, and found that method 4B is better.

For the synthesis of the alkyldichlorothiophosphine, we also used the aluminum chloride complex method:

0097-6156/81/0171-0341$05.00/0

$$C_2H_5Cl + PCl_3 + AlCl_3 \rightarrow C_2H_5PCl_4{\cdot}AlCl_3 \xrightarrow{H_2O} C_2H_5POCl_2$$

$$C_2H_5POCl_2 + P_2S_5 \rightarrow C_2H_5PSCl_2 + P_2O_5$$

The molecular structural formulas, boiling points, melting points and the yields are listed in Table I.

Table I $(C_2H_5)(RO)P(=S)(OR')$

	Structural Formulas R	R'	Boiling Point & Melting Point °C/mmHg	Method	Yield (%)
1	CH_3	o,p-$Cl_2C_6H_3$	bp 128-129°/0.9	4B	70.4
2	CH_3	p-ClC_6H_4	bp 118-119°/0.9	4B	61.3
3	C_2H_5	o,p,$Cl_2C_6H_3$	bp 130-131°/0.5	4A	80.0
4	n-C_3H_7	o,p-$Cl_2C_6H_3$	bp 152-153°/0.8	4A	56.9
5	n-C_3H_7	p-ClC_6H_4	bp 128°/0.1	4A	78.6
6	C_2H_5	p-ClC_6H_4	bp 124-125°/0.6	4A	62.5
7	CH_3	p-t-BuC_6H_4	bp 130-131°/0.3	4B	66.1
8	C_2H_5	p-t-BuC_6H_4	bp 134.5°/0.7	4B	78.2
9	n-C_3H_7	p-t-BuC_6H_4	bp 135-136°/0.2	4B	77.2
10	CH_3	o,o,p-$Cl_3C_6H_2$	mp 49°	4B	92.3
11	C_2H_5	o,o,p,$Cl_3C_6H_2$	mp 43°	4B	90.0
12	n-C_3H_7	o,o,p,$Cl_3C_6H_2$	mp 48°	4B	77.1
13	CH_3	Cl_5C_6	mp 66-67°	4A	58.6
14	C_2H_5	Cl_5C_6	mp 79°	4A	58.8
15	n-C_3H_7	Cl_5C_6	mp 64-65°	4A	44.3

TABLE II. Toxicity Test for Drosophila Melanogaster

Sample Number		Numbers of insects for test (1)	(2)	(3)	(4)	Mortality % (17 hours) (1)	(2)	(3)	(4)
1	1	46	34	32	-	100	100	100	-
	2	-	-	31	33	-	-	100	100
2	1	33	37	31	-	100	100	100	-
	2	-	-	31	30	-	-	100	3.3
3	1	36	34	42	-	100	100	100	-
	2	-	-	33	36	-	-	100	100
4	1	45	37	38	-	100	100	100	-
	2	-	-	29	30	-	-	100	33.3
5	1	33	38	41	-	100	100	100	-
	2	-	-	35	28	-	-	100	14.3
6	1	42	47	36	-	100	100	100	-
	2	-	-	30	31	-	-	100	83.9
7	1	43	52	30	-	100	100	100	-
	2	-	-	25	24	-	-	73.3	25.0
8	1	34	30	32	-	100	100	62.5	-
	2								
9	1	38	28	40	-	100	100	55.0	-
	2								
10	1	38	28	-	-	100	100	-	-
	2	-	28	22	21	-	100	100	85.7
11	1	37	35	-	-	100	100	-	-
	2	-	32	33	22	-	100	81.8	40.9
12	1	25	12	-	-	73.5	34.2	-	-
	2								
13	1	38	42	43	-	7.9	0	0	-
	2								
14	1	42	35	44	-	0	0	0	-
	2								
15	1	43	34	41	-	0	0	0	-
	2								
Standard Sample Rogor	1	71	60	59	-	100	100	3.4	-

Notes A. (1) = 100 ppm (2) = 10 ppm
(3) = 1 ppm (4) = 0.1 ppm

B. The test solution made of 10% absolute alcoholic insecticide, was diluted gradually with water.

C. Determination by method of agar-agar poison diet.

*Rogor is O,O-dimethyl-S-(N-methyl carbomoyl methyl) dithiophosphate.

Table III. Results for Second Screening Test

Sample Number	Emulsion Concentration	Mortality of Insects %*		
		Myzus Persicae	Tetranychus Urticae	Agrotis Ipsilon
1	47.1%	55.1	10.0	77.8
3	50.4%	80.0	50.0	77.8
6	48.9%	65.7	0	70.0
10	18.9%	78.3	18.8	0
Rogor	25.0%	99.4	100	0

*Notes: The emulsion concentration in practice was diluted with 2000 volumes of solution.

Table IV. Test for First Generation of Tryporza Incertulas, First Insta Lava (Outdoors in Pot).

Insecticide Emulsion Concentran.	Quantity of Insecticide Jin/Mou*	No. of Rice Plants	No. of Insects Per Pot	Dead Heart %	Mortality %
No. 1 (47.1%)	1	58	60	1.7	100.0
No. 3 (50.4%)	1	49	60	2.0	98.7
No. 5 (48.1%)	1	26	60	30.7	95.5
No. 10 (18.9%)	3	28	60	32.1	89.5
Sumithion (50 % Sample)	1	45	60	0	100.0

*Notes: Jin; Chinese measure of weight, 1 Jin = 500 gm.
Mou; Chinese measure of area, 1 Mou = 6.67 acres

Literature Cited

1. Hoffman, F.W.; Moore, T.R. J. Am. Chem. Soc. 1958, 80, 1151.
2. Fukuto, T.R. J. Am. Chem. Soc. 1959, 81, 372.
3. Schegk, E.; Schrader, G. Ger. 1, 078, 124, Chem. Abstr. 1963, 58, 1492.

RECEIVED June 30, 1981.

72

Introduction of Phosphorus into the Polyethyleneterephthalate Molecule

G. BORISOV, K. TROEV, and A. GROZEVA

Central Laboratory for Polymers, Bulgarian Academy of Sciences, Sofia 1113, Bulgaria

The introduction of phosphorus into the polyethyleneterephthalate molecule is expected to improve not only the latter's resistance to combustion but also to affect the transesterification and polycondensation stages as well as the side reactions taking place during its synthesis.

The influence of various phosphorus-containing modifiers, i.e. diethyl phosphite (I), the sodium salt of diethyl phosphite (II) and the disodium salt of 1,2-dicarbomethoxyethylphosphonic acid (III), on the transesterification, polycondensation and side reactions were examined.

$C_2H_5O-P(=O)(H)-OC_2H_5$ (I) $C_2H_5O-P(ONa)-OC_2H_5$ (II) $CH_3OC(=O)-CH(P(=O)(ONa)_2)-CH_2-C(=O)OCH_3$ (III)

Dimethyl terephthalate was transesterified with ethylene glycol in the presence of the modifiers taken at different concentrations. The alcohol distillate (Table 1) obtained from conducting the process with diethyl phosphite as modifier revealed the increased presence of water, acetaldehyde and acetal as compared with the alcohol distillate from the transesterification of dimethyl terephthalate with ethylene glycol (1,2). These observations were in support of the conclusion that diethyl phosphite is unsuitable as a modifier for polyethyleneterephthalate.

Table I. Yields of by-products

By-products, %	Modifier								
	Diethyl-phosphite			Sodium salt of diethyl-phosphite			Disodium 1,2-dicarbomethoxyethyl-phosphonate		
	Phosphorus content, %								
	0.0	0.6	0.8	0.5	1.0	1.5	0.5	1.0	1.5
Water	1.02	3.60	4.80	2.50	3.84	4.30	2.37	3.42	3.85
2-methyl-dioxolane	0.13	0.15	0.16	0.04	0.04	0.04	0.02	traces	"
Acetaldehyd	0.01	0.12	0.13	traces	"	"	"	0.02	traces
Acetal	0.08	0.11	0.15	traces	"	"	"	"	"

0097–6156/81/0171–0345$05.00/0

The studies carried out on the alcohol distillate obtained from the co-transesterification of dimethyl terephthalate with ethylene glycol in the presence of the sodium salt of diethyl phosphite or the di-sodium salt of 1,2-dicarbomethoxyethylphosphonic acid in various concentrations showed (Table I) that the side reactions were markedly suppressed.

The hydroxyl value, methoxycarbonyl group content, acid value, melting point, content of diethylene glycol and of the obtained pre-condensates were determined in order to obtain more detailed information on the transesterification stage in the presence of the sodium salt of diethyl phosphite or di-sodium salt of 1,2-dicarbomethoxyethylphosphonic acid.

The results obtained (Table II) indicate that the precondensates modified with sodium salt of diethyl phosphite leads to a resin of following structure (3):

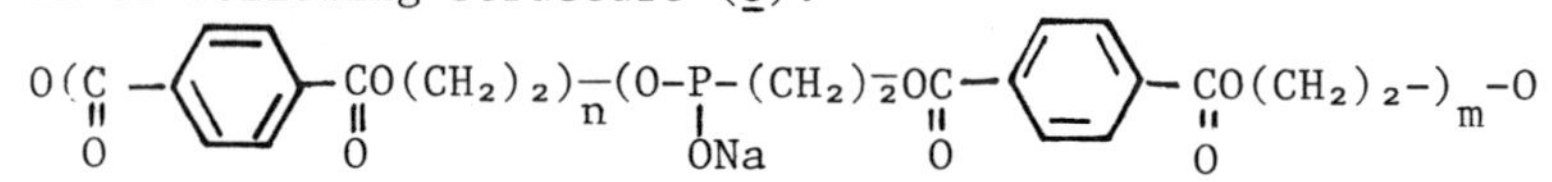

$$O(\underset{\|}{\overset{}{C}}{\text{-}}C_6H_4{\text{-}}\underset{\|}{C}O(CH_2)_2)_n{\text{-}}(O{\text{-}}\underset{|}{P}{\text{-}}(CH_2)_2O\underset{\|}{C}{\text{-}}C_6H_4{\text{-}}\underset{\|}{C}O(CH_2)_2{\text{-}})_m{\text{-}}O$$

(C=O groups; P bears ONa)

With the di-sodium salt of 1,2-dicarbomethoxyethylphosphonic acid a resin of the following structure (4) was obtained

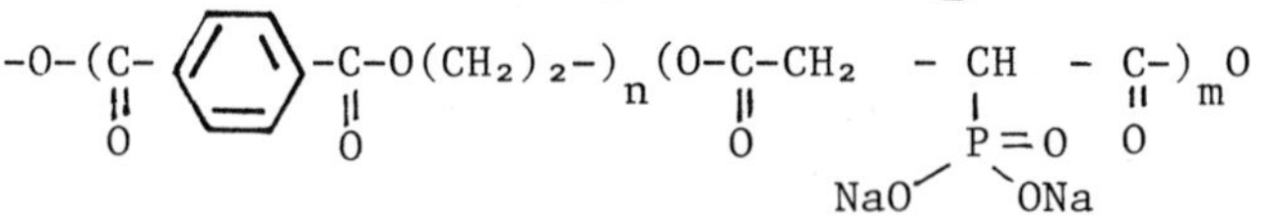

$$-O-(\underset{\|}{C}-C_6H_4-\underset{\|}{C}-O(CH_2)_2-)_n(O-\underset{\|}{C}-CH_2-\underset{|}{CH}-\underset{\|}{C}-)_m O$$

(C=O groups; CH bears P(=O)(ONa)(ONa))

The diethylene glycol and carboxyl group contents of the resin and its dispersity were used as a measure of the degree of side reactions taking place at the polycondensation stage.

With the sodium salt of diethyl phosphite as modifier the diethylene glycol content is within 0.13 to 2.6% against 1.5 to 2.0% for the un-modified resin.

Table II. Analyses of pre-condensates modified with sodium salt of diethyl phosphite or disodium 1,2-dicarbomethoxyethylphosphonate

Characteristics	Modifier: Sodium salt of diethyl phosphite — Phosphorus Content, %					Disodium 1,2-dicarbomethxyethylphosphonate — Phosphorus Content, %			
	0.0	0.5	1.0	1.5	1.9	0.4	0.7	1.2	1.5
OH value, mgKOH/g	440	436	403	398	379	186	268	310	378
CH_3O,%	2.0	2.5	4.4	7.0	9.9	1.7	2.6	3.3	3.2
Acid value, mgKOH/g	0.8	2.1	2.1	2.2	3.1	3.2	3.0	3.0	3.2
DEG, %	1.4	1.1	1.2	1.5	1.4	1.2	0.8	0.3	0.2
Melting point,°C	160	146	132	123	116	167	165	155	155

The carboxyl groups content is increased. The differential distribution curve (Fig.1) indicates a good polydispersity comparable with the one exhibited by the un-modified one.

Table III. Characteristics of polyethyleneterephthalate modified with sodium salt of diethyl phosphite or disodium 1,2-dicarbomethoxyethylphosphonate

Characteristics	Modifier								
	Sodium salt of diethyl phosphite					Disodium 1,2-dicarbomethoxy-ethylphosphonate			
	Phosphorus content, %								
	0.0	0.32	0.80	1.05	1.47	0.45	0.8	1.35	1.62
η	1.33	1.23	1.42	1.18	1.28	1.34	1.33	1.32	1.28
Melting point, °C	259	259	243	252	248	259	258	258	255
COOH.10^6 mgeq/g	67	99	120	117	132	44	40	38	34
DEG,%	1.4	0.13	2.5	2.6	2.5	0.48	0.55	0.5	0.70

DEG-diethylene glycol; η - relative viscosity;

With the di-sodium salt of 1,2-dicarbomethoxyethylphosphonic acid as modifier the diethylene glycol content and the carboxyl group content is very low (Table III). These results indicate that the used modifier has also thermostabilizing properties. The integral and differential curves (Fig.2) of molecular mass distribution show that the polydispersity of modified resin is comparable to that of the un-modified.

The thermogravimetric curves obtained indicate that the modified and un-modified resins begin to decompose at 300°C.

The combustion tests of resin modified with the sodium salt of diethyl phosphite samples indicate (Table IV) that with the increase in the phosphorus content the samples cease to burn and the overall time of combustion lengthens.

Table IV. Data from Combustion Tests

Modifier	P, %	Duration of combustion after ignition, sec.	Time required for complete combustion, sec.	Weight of residue, %	Oxygen Index
Without modifier	0.0	181	181	5.2	19
Diethyl phosphite sodium	0.32	extinguished	188	8.0	19.8
	0.80	"	220	21.0	21.6
	1.05	"	265	30.0	21.8
	1.47	"	275	29.9	22.4
Disodium 1,2-dicarbomethoxyethylphosphonate	0.45	"	202	25.0	21.4
	0.80	"	229	50.0	22.8
	1.35	"	227	52.0	23.6
	1.62	"	240	50.0	24.0

The residue after combustion increases considerably from 5.2% for the un-modified resin to 30% for the one containing 1.05% of phosphorus. The oxygen index changes from 19 for the un-modified resin to 22.4 for the one containing 1.47% of phosphorus.

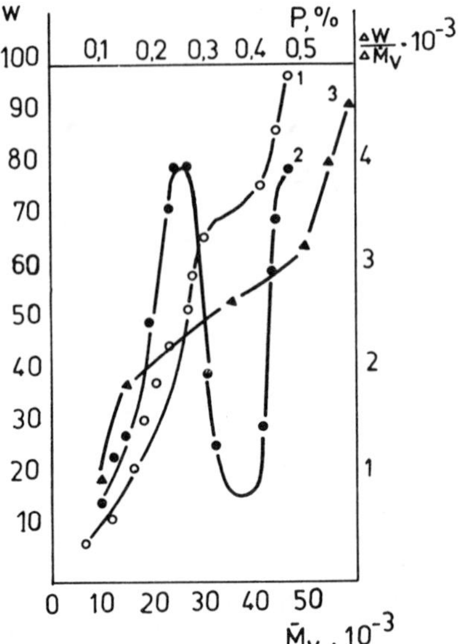

Figure 1. Molecular mass distribution curves from phosphorus-containing polyethyleneterephthalate sample with 0.74% phosphorus modified with sodium salt of diethyl phosphite. Key: 1, integral curve; 2, differential curve; and 3, integral curve of phosphorus distribution.

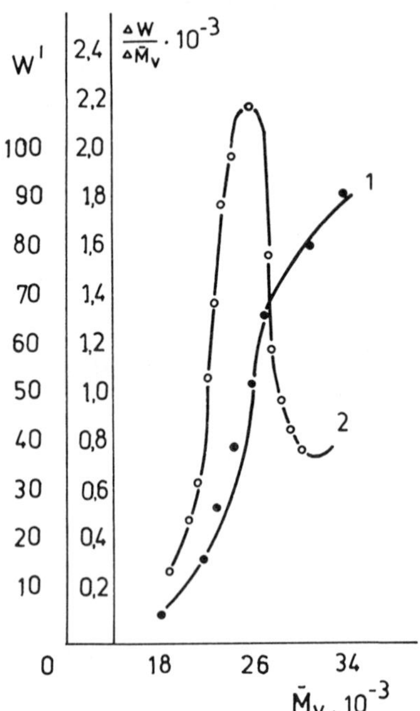

Figure 2. Molecular mass distribution curves from phosphorus-containing polyethyleneterephthalate sample with 0.80% phosphorus modified with disodium salt of 1,2-dicarbomethoxyethylphosphonic acid. Key: 1, integral curve; and 2, differential curve.

The samples modified with the di-sodium salt of 1,2-dicarbomethoxyethylphosphonic acid showed (Table IV) the weight of residue reaching 52%.

LITERATURE CITED

1. Troev, K.; Grozeva, At.; Borisov, G. European Polymer Journal 1979, 15, 437.
2. Troev, K.; Grozeva, At.; Borisov, G. European Polymer Journal 1981, 17, 27.
3. Troev, K.; Grozeva, At.; Borisov, G. European Polymer Journal 1979, 15, 1143.
4. Troev, K.; Grozeva, At.; Borisov, G. European Polymer Journal 1981, 17, 31.

RECEIVED July 7, 1981.

73

Selected Novel Trivalent Organophosphorus Processing Stabilizers for Polyolefins

J. D. SPIVACK, A. PATEL, and L. P. STEINHUEBEL

CIBA-GEIGY Corporation, Ardsley, NY 10502

Processing stabilizers are a special class of antioxidants used to inhibit polymer degradation during the processing steps subsequent to polymerization such as extrusion, injection moulding, spinning, etc. These steps are carried out at relatively high temperatures (220-320^0) in the presence of some oxygen.

Attempts to counteract degradation by the use of 2,6-di-tert.-butyl-4-methylphenol (BHT) and various organophosphorus compounds, such as tris-nonyl-phenylphosphite (7) and 3,9-dioctadecoxy-2,4,8,10,3,9-tetraoxadiphospha(5,5)-spiroundecane (8), have been only partially successful. BHT is volatile at high temperatures and contributes to discoloration during processing. 7 and 8 undergo hydrolysis even during ambient storage conditions leading to extrusion and spinning problems, corrosion of equipment and contamination of extrudate with hydrolysis and corrosion products, etc.

A generally accepted (1) mechanism for the oxidation of polyolefins in the presence of antioxidants of the hydrogen donor type, AH, involves the conversion to hydroperoxides of polymer radicals, R•, and polymer peroxy radicals, ROO•, formed during the initiation and propagation steps. Propagation is promoted by homolysis of ROOH to RO• and •OH. Chain transfer will take place to only an insignificant degree if A• is a stable free radical.

Equations 1 and 3 below show how trivalent phosphorus compounds can inhibit oxidation by converting hydroperoxides to non-radical products. Equation (2) illustrates how the phosphorus derivatives may also act as chain transfer agents with peroxy radicals.

ROO• + [2,6-di-tert-butylphenol with OH, R'] ⟶ [phenoxy radical O•, R'] + [cyclohexadienone O, R' OOR] + ROOH (1)

Similarly with RO• and R•

0097–6156/81/0171–0351$05.00/0

$$ROO\cdot + PR''_3 \longrightarrow [ROO\dot{P}R''_3] \longrightarrow RO\cdot + O{=}PR''_3 \quad (2)$$

However

$$ROOH + PR''_3 \longrightarrow \begin{matrix} RO \\ \\ HO \end{matrix} PR''_3 \longrightarrow ROH + O''PR''_3 \quad (3)$$

It is thus apparent that hydrogen donors AH and hydroperoxide decomposers, such as PR''_3, can act synergistically to inhibit radical initiated polymer chain oxidations.

The following six hindered phosphonite and phosphite esters were studied as processing stabilizers and compared with 7 and 8.

(A) O,O'-Biphenylene Phosphonites and Phosphites

1, R = C_6H_5— (ref. 2)

2, R = CH_3O— (ref. 3)

(B) Unsymmetrical Di-Hindered Phenylphosphonites

3, R = CH_3— (ref. 4)

4, R = $(CH_3)_3C$— (ref. 4)

(C) Symmetrical Di-Hindered Phenylphosphonites

5, R = CH_3— (ref. 4)

6, R = $(CH_3)_3C$— (ref. 4)

$[C_9H_{19}-C_6H_4-O]_3P$ 7

$n\text{-}C_{18}H_{37}O\text{-}P(O\text{-}CH_2)_2C(CH_2\text{-}O)_2P\text{-}O\text{-}n\text{-}C_{18}H_{37}$ 8

Oxidation during processing of polypropylene is principally accompanied by chain scission made evident by a reduction in melt viscosity. The oxidation during processing of polyethylene on the other hand is accompanied mainly by crosslinking. The following two tests are, therefore, used:

(1) Polypropylene - Melt flow rate after the first, third and fifth extrusion, at 500°F to determine the degree of chain scission (5). Y.I. index is also determined (6).

(2) High Density Polyethylene - [Both regular and high molecular weight (HMW)]. The time to increase in torque at 220°C. by the Brabender Plasticorder was determined as a measure of crosslinking. Yellowness Index after the Brabender test at 220°C. and after the first extrusion at 260°C. were also performed.

Moisture-Pickup and Hydrolysis

The moisture pickup and hydrolysis tests were carried out at room temperature and 80% R.H. to simulate storage. The results are shown in the following table.

Moisture Pickup and Hydrolysis at Room Temperature, 80% R.H.

Time - 1200 Hours

Compound	Water Pickup %	Hydrolysis %	Notes
1	0.5	ca. 0	(2)
2	0.1	ca. 0	(2)
3	0.1	ca. 0	(2)
4	0.5	ca. 0	(2)
5	0.0	0	(2)
6	0.0	0	(2)
7	18.9 (1)	ca. 100	(1,3)
8	34.0 (1)	ca. 100	(1,3)

Notes: (1) at 400 hrs.

(2) Little or no change by TLC and IR before and after exposure.

(3) TLC shows disappearance of starting compounds. IR shows strong hydroxyl band after exposure.

Summary of Processing Characteristics of Selected Organophosphorus Compounds

Polypropylene (Profax 6801, Hercules Inc.) Multiple Extrusion At 500°F (260°C)

On the basis of melt flow rate measurements after 1, 3 and 5 extrusions, compounds 1, 2 and 3 when used alone at 0.1% in the polypropylene base resin provide superior stabilization to BHT, 7 and 8. Under these conditions compound 2 provides the lowest Yellowness Index (Y.I.) color over other compounds evaluated.

Under the same test conditions on the basis of melt flow rate, compounds 1, 2, 3 and 4 provide significantly better performance than BHT, 7 and 8 when used in conjunction with phenolic AO-1 (neopentanetetrayl-tetrakis[3,5-di-tert.-butyl]-4-hydroxyhydrocinnamate). Compounds 2, 3 and 4 also develop less color than BHT and 8. Compounds 6 and 7 develop the most color in the presence of AO-1.

At 550°F (288°C.)

All six candidate compounds provide better process stability than BHT when evaluated alone after 5 extrusions. In conjunction with phenolic AO-1, all six compounds exhibit superior synergism and provide better process stability and color than (7) and (8) during multiple extrusion.

High Density Polyethylene (USI HDPE)LR 739X)) Color Improvement at 260°C.

When used in HDPE as color improvers for phenolic AO-1, compounds 1, 2 and 3 provide significantly improved color than obtainable with TNPP while being equivalent in resistance to discoloration to (7).

High Molecular Weight - High Density Polyethylene (Lupolen-BASF)

Employing the Brabender plasticorder at 220°C, 50 R.P.M. with the ram open compounds 1, 2, 3 and 4 prevent crosslinking about three times longer than BHT and AO-1 alone.

Conclusions

A number of ortho hindered alkyl-substituted phenyl phosphites and phosphonites were found to be effective process stabilizers for polypropylene and high density polyethylene combining more effective stabilization activity at high temperatures with good storage stability at relatively elevated humidity and ambient temperature, as well as resistance to discoloration.

REFERENCES CITED

(1) For Example, Chapter 12, J. A. Howard in Free Radicals Vol. II, Jay K. Kochi, Ed., John Wiley, N.Y. (1973).

(2) J. D. Spivack, U.S. 4,143,028 (March 6, 1979).

(3) J. D. Spivack, U.S. 4,196,117 (April 1, 1980).

(4) J. D. Spivack, U.S. 4,233,207 (November 11, 1980).

(5) ASTM Method 1238 Condition L.

(6) ASTM Method D1925-63T.

RECEIVED June 30, 1981.

Oligomeric Phosphorus Esters with Flame Retardant Utility

EDWARD D. WEIL, RALPH B. FEARING, and FRED JAFFE

Stauffer Chemical Company, Eastern Research Center, Dobbs Ferry, NY 10522

The term oligomer refers to a polymer-like material having only a few repeating units. The oligomeric phosphorus esters which are the subject of the present paper are generally viscous liquids having an average of two or more phosphate and/or phosphonate ester units per molecule.

Pioneering work on phosphorus ester oligomers has been done by Monsanto in the U.S. (1), by Hoechst in Germany (2), and in the Soviet Union. These studies involved synthesis and flame retardant applications. The polycondensation of 2-chloroethyl phosphates as a route to oligomeric phosphorus esters (Equation 1) was first reported by Korshak *et al.* (3). This Russian publication describes the polycondensation of tris(2-chloroethyl) phosphate at 240-280° under non-catalytic conditions. An acidic dark product was obtained. Besides the desired transalkylation, side reactions yielding acetaldehyde, vinyl chloride, phosphorus acids, and pyrophosphates were described by other Soviet researchers (4). Such a multitude of concurrent reactions is undesirable if this process is to be controllable and useful for flame retardant manufacture.

Polycondensation of 2-Chloroalkyl Phosphates and Phosphonates

Each step of the polycondensation can be described by the following general reaction:

$$(1)\quad 2RR'P(O)OCH_2CH_2Cl \rightarrow RR'P(O)OCH_2CH_2OP(O)RR' + ClCH_2CH_2Cl$$

Polycondensation reactions of 2-chloroalkyl phosphates or phosphonates to obtain products having a controllable degree of condensation and low acid or latent acid contents were accomplished in our laboratory using catalysts such as quaternary ammonium salts, amines, amides, sodium carbonate, or lithium chloride (5). Reduction of the temperature diminished the

0097–6156/81/0171–0355$05.00/0

undesirable side reactions. Further effort was necessary to eliminate latent acid products if the products were to be useful in, for example, urethane foams. Treating the crude product with ethylene oxide afforded a means for converting phosphorus acid groups to hydroxyethyl esters. Anhydride groups could also be removed by reaction with epoxides. Particularly persistent, however, were five-membered cyclic phosphate ester groups, which in the case of the polycondensation product of tris(2-chloroethyl) phosphate can occur as chain ends or as small molecules. These five-membered phosphate structures show a large ^{31}P shift (17-18 ppm) downfield from the acyclic phosphates.

Five-membered cyclic phosphate and phosphonate esters have rates of hydrolysis orders of magnitude greater than those of the corresponding acyclic esters (6), therefore such five-membered cyclic esters are undesirable components if the oligomers are used as flame retardants.

The presence of these cyclic esters in the crude polycondensation reaction product was found to be unavoidable; indeed some evidence was developed that the polycondensation at least in part proceeds *via* these cyclic esters. Considerable effort was expended to find means for eliminating these cyclic five-membered esters from our polycondensation products. The cyclic esters can be eliminated by either inducing them to polymerize by use of Lewis acid catalysts such as stannous octoate, or by subjecting them to ring opening by means of an alcohol or water (7).

Copolycondensation

A further variation on the transalkylation reaction described above is the cocondensation of different phosphorus esters (8); as shown in the following equation:

$$RP(O)(OCH_2CH_2Cl)_2 + CH_3P(O)(OCH_3)_2 \xrightarrow{-2CH_3Cl} \left[-OP(O)(CH_3)-OCH_2CH_2-OP(O)(R)-OCH_2CH_2- \right]_x$$

One embodiment of this general reaction led to a product which was commercially produced for several years by Stauffer as Fyrol® 76 (9), a copolycondensation product of dimethyl methylphosphonate with bis(2-chloroethyl) vinylphosphonate. The features of Fyrol 76 were high phosphorus content (20%), water solubility, and ability to be polymerized by means of a radical initiator to a crosslinked polymer. A related polycondensation product was developed from tris(2-chloroethyl) phosphate and dimethyl methylphosphonate. By control of the reagents and procedure used for neutralization, these oligomeric products were produced with primary alcohol functional groups (7).

Oligomers Via Metaphosphate(Phosphonate) Intermediates

The dissolution of P_4O_{10} in a phosphate or phosphonate ester brings about a reorganization reaction in which a metaphosphate or metaphosphonate/phosphate is formed (10). In the reaction of P_4O_{10} with dimethyl methylphosphonate, the initially-formed product mixture at 60-110° undergoes a structural reorganization when the mixture is held at reaction temperature. An increase in O,O-dimethylphosphoric anhydride end groups (δ -11.4, -11.6, -11.9) and P-methyl(meta)phosphonic "middle" groups (δ14.2, center of unresolved triplet and/or doublet of doublets) takes place at the expense of a decrease in O,P-dimethylphosphonic anhydride end groups (doublet at δ 25.7) and O-methyl (meta)-phosphoric anhydride middle groups (incompletely resolved triplet and/or doublet of doublets centered at -27.1). Branched phosphoric anhydride (δ-43) disappears early in the heating process.

The intermediate metaphosphate/phosphonate can then be made to react with ethylene oxide to effect the insertion of ethyleneoxy units into the P-O-P linkages. If a limited amount of water or an alcohol is added to the meta intermediate, the resultant oligomer can be produced with a controlled hydroxyl functionality (11, 12). An example of a functional oligomer made this way is the following:

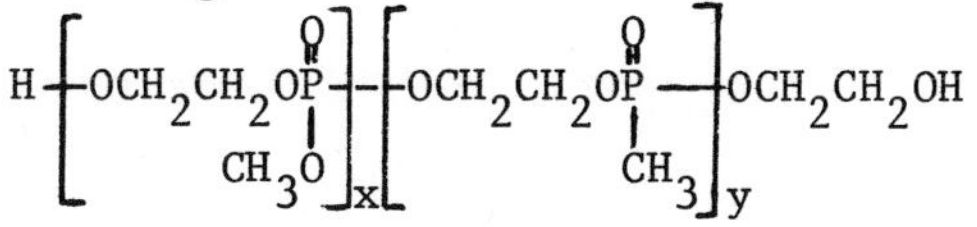

$$H\left[OCH_2CH_2O\overset{O}{\underset{CH_3O}{P}}\right]_x\left[OCH_2CH_2O\overset{O}{\underset{CH_3}{P}}\right]_y OCH_2CH_2OH$$

There is some evidence that the reaction of ethylene oxide with the metaphosphonate/phosphate may actually form some cyclic five-membered esters initially, as shown by ^{31}P signals at 18.4, 17.4 (phosphates) and 49 (phosphonate) which then are converted to acyclic esters. In another example of this route to oligomeric phosphorus esters, P_4O_{10} is reacted with tris(1,3-dichloroisopropyl) phosphate to prepare a metaphosphate which is ethoxylated with ethylene oxide to produce a substantially hydroxy-free phosphate oligomer (12).

Applications as Flame Retardants

Several commercial products have resulted from our phosphorus oligomer research. Fyrol 99, a 2-chloroethyl ethylene phosphate oligomer, has been successfully used as a flame retardant additive in rebonded urethane foam, in thermoset resins, in intumescent coatings, adhesives, paper air filters (13), and related uses. This product is less volatile and has a higher flame retardant efficacy than the parent compound tris(2-chloroethyl) phosphate. A related product was developed especially for use in flexible polyurethane foams. A vinylphosphonate/methylphosphonate oligomer whose chemistry was described above (Fyrol 76) was found effective for meeting the Federal flammability standard for children's cotton sleepwear (14).

A related methylphosphate/phosphonate oligomer has primary alcohol end groups, and can coreact with amino resins to form a water-resistant flame retardant resin finish on paper or on textile substrates. The application of this oligomer with a coreactant methylolmelamine on cotton upholstery fabric can enable furniture covered with this fabric to pass the Consumer Product Safety Commission's proposed cigarette ignition test.

LITERATURE CITED

1. Birum, G. H. (to Monsanto Co.), U.S. Pats. 3,014,956 (1961), 3,042,701 (1962).
2. Wortmann, J., Dany, F. J., and Kandler, J. (to Hoechst A.-G.), U.S. Pat. 3,850,859 (1974).
3. Korshak, V. V., Gribova, I. A., and Shitikov, V. K., Proc. Acad. Sci. USSR, Div. Chem. Sci., 196-201 (1958).
4. Kafengauz, I. M., Samigulin, F. K., Kafengauz, A. P., Polyakova, T. A., Tsarfin, Ya. A., and Gefter, E. L., Soviet Plastics, 73-75 (Apr. 1967).
5. Weil, E. D. (to Stauffer Chemical Co.), U.S. Pats. 3,896,187 (1975), 4,005,034 (1977), and 4,013,814 (1977).
6. Westheimer, F. H., Acc. Chem. Res. 1, 70-77 (1968).
7. Weil, E. D. (to Stauffer Chemical Co.), U.S. Pat. 3,891,727 (1975); Shim, K. S. and Walsh, E. N. (to Stauffer Chemical Co.), 3,959,414 (1976), 3,959,416 (1976); Walsh, E. N. (to Stauffer Chemical Co.), 4,012,463 (1977).
8. Weil, E. D. (to Stauffer Chemical Co.), U.S. Pats. 4,086,303 (1978), 4,152,371 (1979), 4,202,842 (1980), 4,225,522 (1980).
9. Weil, E. D. (to Stauffer Chemical Co.), U.S. Pats. 3,822,327 (1974), 3,855,359 (1974), 4,017,257 (1977), 4,067,927 (1978).
10. Schep, R. A., Coetzee, J. H. J., and Norval, S., J. S. Afr. Chem. Inst. 27, 63-69 (1974); Inorg. Chem. 12, 2711-13 (1973).
11. Fearing, R. B. (to Stauffer Chemical Co.), U.S. Pat. 4,199,534 (1980).
12. Hardy, T. A. and Jaffe, F. (to Stauffer Chemical Co.), S. Afr. Pat. 79/1120 (Mar. 26, 1980).
13. Weil, E. D., Leitner, G. J., and Kearnan, J. E., "Phosphorus Flame Retardants for Resin-Treated Paper", in Proc. TAPPI Paper Synthetics Conf., Orlando, FL, Sept. 27, 1978.
14. Eisenberg, B. J. and Weil, E. D., Textile Chemist and Colorist 6 (8), 180-2 (1974); Bruce, J., Am. Dyestuffs Rep. 62 (10), 68-70 (1973).

RECEIVED June 30, 1981.

INORGANIC PHOSPHATES

Crystalline Calcium Polyphosphate Fibers

E. J. GRIFFITH

Monsanto Company, St. Louis, MO 63166

Asbestos

Two molecular types of silicates are referred to as asbestos. Chrysotile is a magnesium silicate built upon a layered structure of silicate rings and $Mg(OH)_2$. The layered structure causes the sheets to roll into cylinders approximately 200Å in diameter. Amphibole asbestos may contain a variety of cations but is built upon a double chain silicate structure. The chrysotile asbestos is always found as an asbestiform crystal while the amphiboles may be either acicular or asbestiform.

The diseases attributed to asbestos are a result of the fiber morphology and stability of the fibers rather than any specific chemical reactions between the asbestos and a host organism. It is probable that any refractory substance of similar morphology should stimulate similar diseases.

Ingleman and Malgren (1) first demonstrated that long chain polyphosphates were enzymatically degraded by Aspergillius Niger in a manner similar to the enzymatic hydrolytic degradation of adenosine triphosphate in energy transfers of biological systems. Phosphatase chemistry has been the subject of numerous research efforts and the concepts are well established.

The utility of asbestos is a result of a number of extraordinary properties exhibited by the minerals. They are nonflammable, temperature resistant fibers composed of fibrals about 200Å in diameter, exhibit tensile strengths up to $1 \cdot 10^6$ psi, and moduli as great as $25 \cdot 10^6$ psi. The fibers are particularly resistant to attack by biological organisms and corrosive environments.

Phosphate Fibers

The chemistries of phosphates and silicates are similar, but the morphology of the crystals of the sparingly soluble phosphates are unsuited for fiber applications. Amorphous phosphate glasses can be easily spun into fibers in a process similar to the manufacture of fiberglass. Unfortunately, amorphous phosphates lack both strength and hydrolytic stability.

0097–6156/81/0171–0361$05.00/0

Calcium polyphosphates of a metaphosphate composition $[Ca(PO_3)_2]_n$ have properties and molecular structures which should yield a substance which is refractory in most environments but is rapidly degraded in systems containing active enzymes. None of the known systems of calcium phosphates yield fibrous crystals, although very short acicular crystals have been grown.(2)

When calcium polyphosphates are grown from a sodium or potassium ultraphosphate melt, a fibrous crystal with diameters as low as 3μ and lengths as long as 3cm were grown. The ultraphosphate matrix serves two functions: 1) it is a growing medium which allows the growth of long, very slender crystals; and 2) it controls the release of CaO to the crystallizing polyphosphate anions permitting the polyphosphate chains to grow to very high molecular weights.

When crystallization is completed the fibrous phosphate can be extracted from the ultraphosphate matrix by leaching the system with hot water.

Phase Chemistry

The walls for three of the sides of the three-dimensional triangular phase diagram have been published, but the internal details of the diagram are but poorly understood. It can be stated that fibrous calcium polyphosphates can be grown throughout the ultraphosphate diagram dominated by $[Ca(PO_3)_2]_n$ crystals. An approximate phase diagram is shown in Figure 1.

Property of Fibers

The calcium polyphosphate fibers are very very slowly dissolved in water. Even in boiling 0.1N HCl the fibers are resistant to degradation, but the fibers are not as resistant to boiling 0.1N NaOH solutions. See Figure 2.

The fibers are highly crystalline and single crystal tensile strength measurements range from $2 \cdot 10^6$ psi to $1 \cdot 10^6$ psi, while the modulus of elasticity ranges from $10 \cdot 10^6$ psi to $26 \cdot 10^6$ psi. These values are comparable to the published values for chrysotile asbestos.

Chrysotile asbestos is decomposed at temperatures below 500°C. Calcium polyphosphate fibers do not melt at temperatures below 970°C, but the β-phase phosphate is converted to α-phase phosphate at 940°C.

Animal Studies

All animal studies made with calcium phosphate fibers have shown that the fibers are degraded by the enzymes in living systems. To date, the phosphates have been tested as implants in the pleural or peritoneal cavities of rats. As was to be expected

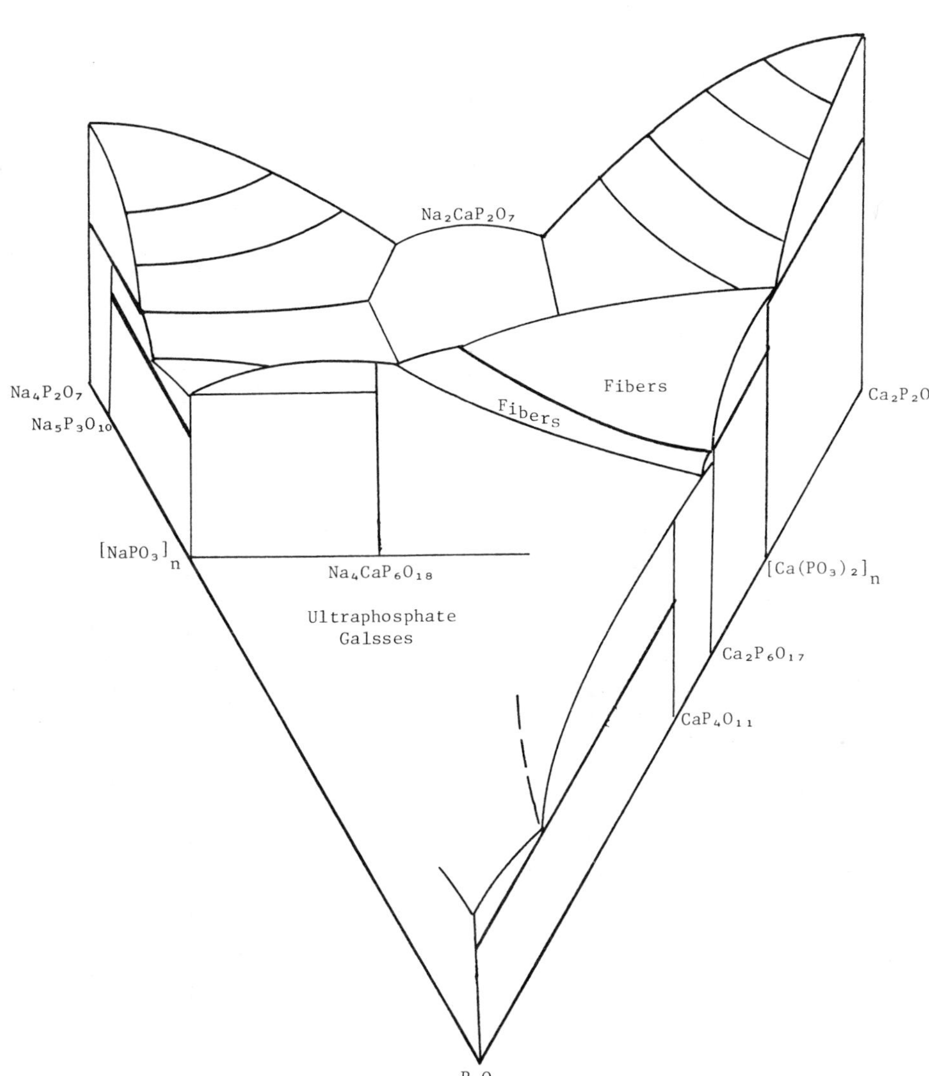

Figure 1. A schematic phase diagram for CaO–Na_2O–P_2O_5 system based on published diagrams. [Published diagrams: Na_2O–P_2O_5, CaO–P_2O_5, (4); and Na_2O–CaO–P_2O_5, (5).]

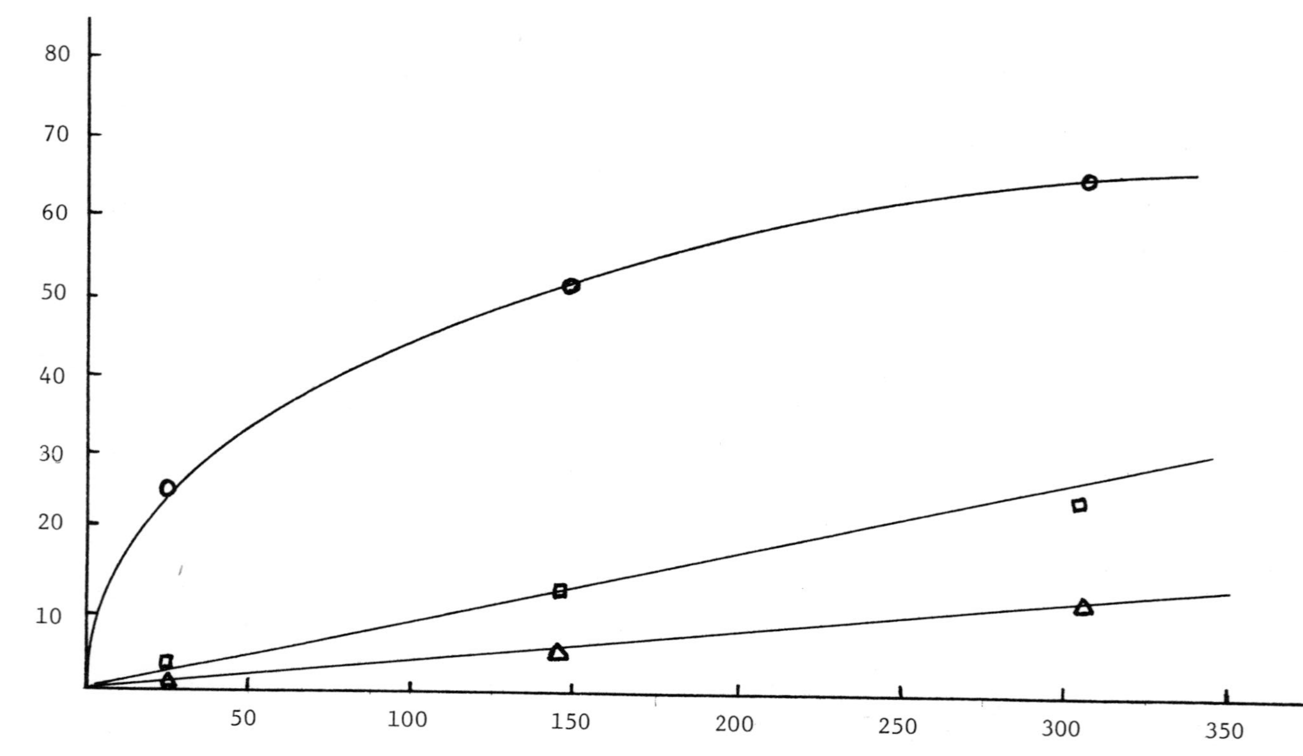

Figure 2. The corrosion of calcium phosphate fibers at boiling temperatures. Key: △, water; □, 0.1N HNO_3; and ○, 0.1N NaOH).

with implants of phosphates, the fibers and their degradation products are nutrients for all biological systems and no adverse results have been obtained.

Conclusion

It is concluded that calcium polyphosphate fibers are viable candidates as safe replacements for asbestos in many of its applications, particularly in those applications where human exposure is considered hazardous.

Literature Cited

1. Ingelman, B.; Malgren, H. Acta. Chem. Scand. 1947, 1, 422; 1948, 2, 365; 3, 157.
2. Abe, Y. Nature 1979, 282, 55.
3. Morey, G. W. and Ingerson, E. Am. J. Sci., 1944, 242, 4
4. Hill, W. L., Foust, G. T., and Reynolds, D. S. Am. J. Sci. 1944, 242, 547.--
5. Gremier, J. C., Martin, C., and Durif, A. Bull. Soc. Fr. Mineral. Cristallogr. 1970, 93, 52.

RECEIVED June 30, 1981.

Fluorination of Phosphoapatites

Possible Alterations of Their Structure

G. MONTEL, G. BONEL, J. C. HEUGHEBAERT, M. VIGNOLES, and M. HAMAD

Laboratoire de Physico-Chimie des Solides et des Hautes Températures, INP, ERA N° 263, CNRS, 38, Rue des 36 Ponts. 31400 Toulouse, France

G. BACQUET

Laboratoire de Physique des Solides, LA CNRS, Université Paul Sabatier, 118, Route de Narbonne. 31062 Toulouse Cédex, France

As soon as chemists become interested in natural apatites, the problem with fluorine appears. When phosphate ores are treated the fluorine must be removed. On the other hand it is sometimes necessary to fluorinate calcium phosphates in the mineral part of calcified tissues. Obviously, in industry and medecine it is necessary to know the mechanism of phosphate fluorination as well as the structure and the properties of products obtained.

The structure of fluorapatite $Ca_{10}(PO_4)_6F_2$ was described in the 30's owing to the existence of well-defined single crystals. It is crystallized in the hexagonal system (space group $P\ 6_{3/m}$) and it is characterized by the presence of channels, crossing the crystal from one end to the other. The fluoride ions (two for each unit-cell) are localized along the axis of the channel.

The methods of synthesis of fluorapatite have been widely discussed (1). It is for example possible to obtain fluorapatite by substituting the hydroxyl ion for the fluoride ion, either in aqueous solution at room temperature, or through a solid state reaction at 800°C. It can also be prepared by the action of β-tricalcium phosphate on calcium fluoride at about 800°C. Its solubility and thermal stability have already been established. While much is known about fluorapatite, many questions still exist concerning the mechanism of their formation, their composition and the structure of some of them. Two of these problems are dealt with here. First, we discuss the formation mechanism of fluorapatite by a solid state reaction between calcium fluoride and apatitic tricalcium phosphate. Then we present the preparation and the structure of a carbonated apatite rich in fluoride ions.

Mechanism of formation of fluorapatite from apatitic tricalcium phosphate and calcium fluoride in the solid state.

We studied (2) the mechanism of formation of fluorapatite from apatitic tricalcium phosphate and calcium fluoride in the solid state. Overall, the chemical reaction may be written :

$$Ca_9(HPO_4)(PO_4)_5(OH) + CaF_2 \longrightarrow Ca_{10}(PO_4)_6F_2 + H_2O$$

0097-6156/81/0171-0367$05.00/0

As the reaction occurs entirely in the apatitic phase, it can be neither an exchange reaction, nor an addition reaction. Our observations on this reaction point out that the mechanism is very complex. The following reactions occur successively with overlapping temperature :

(a) From 110 to 550°C, the dehydration of HPO_4^{2-} ions of apatite is observed giving rise to $P_2O_7^{4-}$ ions in solid solution in apatite.
(b) From 400 to 760°C, an exchange reaction between the OH^- ions of apatite and the F^- ions of calcium fluoride occurs. This phenomenon leads to an increase in the cristallinity, without any variation in the value of the unit-cell parameter a.
(c) From about 600°C to 900°C, the remaining calcium fluoride adds to with the fluorinated apatite. This reaction is accompanied by a decrease of the value of the unit-cell parameter a. A calcium deficient fluorapatite is then observed.
(d) From 600 to 1000°C, the $P_2O_7^{4-}$ ions in the apatite lattice react with the calcium oxide previously formed during the exchange reaction, leading to the fluorapatite.

However, from 200° to 500°C, a small quantity of $P_2O_7^{4-}$ ions react with the fluoride ions producing an unidentified fluorinated phosphate species. The fluorapatite obtained by this method, though very close to the ideal fluorapatite, is not exactly stoichiometric.

Apatites with large amounts of fluoride ions : the B-type carbonated fluorapatite.

LEHR and Mc CLELLAN (3) demonstrated numerous natural apatites and a correlation between the amount of fluoride ions and that of the carbonate ions. This led them to propose that PO_4^{3-} ions can be replaced by CO_3^{2-} ions associated with F^- ions. Such a hypothesis could explain the abnormally high amount of fluoride in some FRANCOLITES. However this type of substitution was not proved by the authors. We studied some synthetic apatites where fluoride and carbonate ions were simultaneously introduced. Samples of B-type carbonated fluorapatite (CO_3^{2-} substituting PO_4^{3-}) were obtained as a powder from an aqueous medium rich in fluoride ions and also an aqueous medium poor in fluoride ions.

Chemical analysis of the products prepared in the medium rich in fluoride ions showed the presence of more than two fluoride ions per unit cell (Table I) and a correlation between the amounts of carbonate and fluoride ions.

Table I

Sample	1	2	3	4	5	6
$n_{Ca^{2+}}$	9,74	9,72	9,78	9,72	9,84	9,87
nPO_4^{3-}	4,03	4,20	4,75	4,87	5,30	5,54
nCO_3^{2-}	1,97	1,80	1,25	1,13	0,70	0,46
n_{F^-}	3,46	3,30	2,60	2,59	2,36	2,20

These results agree with the hypothesis of LEHR and Mc CLELLAN that carbonate ions associated with F^- ions substitutes for PO_4^{3-}. Such apatites might have the formula :

$$Ca_{10}(PO_4)_{6-x}\ (CO_3F)_x\ F_2$$

However a lack of calcium is observed and the correlation between the amounts of fluoride ions and carbonate ions is not rigorous. This is due to the existence of a second type of substitution as originally proposed by LABARTHE (4-5). According to this author, the substitution of a PO_4^{3-} ions by a CO_3^{2-} ion is accompanied by the formation of vacancies in an oxygen site and also in the Ca^{2+} and F^- sites of the channel which are the closest to the missing oxygen. The B-type carbonated fluorapatite can be described by the formula :

$$Ca_{10-x+u}\ \square_{x-u}\ (PO_4)_{6-x}(CO_3,\square)_{x-u}(CO_3F)_u\ F_{2-x+u}\ \square_{x-u}$$

Both types of charge compensation are considered present in each case, but the value of u is low when the apatites are prepared in an aqueous medium poor in F^- (in which the second type of lattice substitution is dominant) (Table II) and higher for the apatites prepared in an aqueous medium rich in F^- ions (in which the first type of substitution dominates) (Table I)

Table II

Sample	1	2	3
$n_{Ca^{2+}}$	9,61	9,46	9,76
$n_{PO_4^{3-}}$	4,90	5,05	5,65
$n_{CO_3^{2-}}$	1,10	0,95	0,35
n_{F^-}	2,36	1,87	1,87

The X-band E.S.R. results (6) obtained with the carbonated apatites support the proposed model. Indeed, in X-irradiated samples we have observed the resonance of a defect in which an electron is trapped by an oxygen vacancy. The number of such defects is greater in the compounds prepared from an aqueous medium poor in F^- ions than in the apatites prepared from an aqueous medium rich in fluoride ions. Moreover there exists a close correlation between the spectral intensity and the number of oxygen vacancies calculated from the model we propose.

The properties of this materials will lead to a better insight into phosphates of biological interest.

Literature cited

1. WALLAEYS R. Ann. Chim., Paris, 12ème série, 1952, 7, 808-848.
2. HAMAD M. Thèse, Institut National Polytechnique de Toulouse, 1980.
3. LEHR J.R. ; Mc CLELLAN G.H. Colloque International sur les Phosphates Minéraux Solides. Toulouse 16-20 mai 1967. 2, 29-44.
4. LABARTHE J.C. Thèse, Université Paul Sabatier, Toulouse, 1972.
5. LABARTHE J.C. ; BONEL G. ; MONTEL G. Ann. Chim., Paris, 1973, 8, 289-301.
6. BACQUET G. ; VO QUANG TRUONG ; BONEL G. ; VIGNOLES M. J. Solid State Chem., 1980, 33, 189-195.

RECEIVED June 30, 1981.

Photo- and Thermo-Coloring of Reduced Phosphate Glasses

Y. ABE and R. EBISAWA
Nagoya Institute of Technology, Inorganic Materials, Showa-ku, Nagoya 466 Japan

D. E. CLARK and L. L. HENCH
University of Florida, Gainesville, FL 32611

Various reduced phosphate glasses show striking phenomenon. In the as-cast glasses which were made with a higher cooling rate from the melts, phosphorus may be atomically dispersed or it may form very fine particles; this leads to transparent and colorless glasses when the glasses contain no coloring agents such as transition metal ions. However, phosphorus associates into colloidal red phosphorus (∿50 nm) to give a reddish color to the glasses, when the glasses are subjected to a heat-treatment at moderately high temperatures (e.g., 400°-600°C) (1-3). This phenomenon is known as "striking." The colored glasses are bleached to be almost transparent and colorless again, when they are subjected to a heat-treatment above 600°C and subsequent quenching. These bleached glasses are referred to as PTC-RP glasses (3), since they were found to be Photo-and Thermo-Colorizable Reduced Phosphate glasses. They are easily red-colored either by sunlight (2) or UV light at room temperature or by heating above 200°C (3). Thermo-coloring of reduced glasses having composition of $3K_2O \cdot 12B_2O_3 \cdot 69P_2O_5$ and $6K_2O \cdot 22Al_2O_3 \cdot 72P_2O_5$, both of which have glass transition temperature >600°C have been reported (3).

In the present paper, only a simple composition, as an example of PTC-RP glass, is discussed. The dependence of sensitivity on the irradiation light wavelength is described in this paper.

Sample Preparation

The following three kinds of glasses were prepared.

(a) Glass reduced by silicon (CP-RS): A mixture of di-hydrogen calcium phosphate, $Ca(H_2PO_4)_2 \cdot H_2O$ and metallic silicon powder were melted in an alumina crucible at 1200°-1250°C for 1 hour and poured on a cold graphite plate. The silicon was 0.15 mole per mole of P_2O_5. (Formula: $CaO \cdot P_2O_5 \cdot 0.15SiO_2$)

(b) Glass reduced by ammonium salt (CP-RN): A mixture of

0097-6156/81/0171-0371$05.00/0

$Ca(H_2PO_4)_2$ H_2O, $CaCO_3$, and $NH_4H_2PO_4$ were used as raw materials. 20% of P_2O_5 of the glass was supplied by using $NH_4H_2PO_4$. Glasses were prepared in a similar way to CP-RS. (Formula: CaO P_2O_5)
(c) Glass not reduced (CP-OS): Glasses were prepared in the same way as CP-RS except that SiO_2 was substituted for Si in the batch. (Formula: CaO P_2O_5 $0.15SiO_2$)
All the glasses were annealed and all reagents were of reagent grade.

Striking (Coloring by Colloid Formation)

Reduced phosphate glasses of which as-cast glasses are transparent and colorless become tinged with red to yellow depending on the reheating time and temperature. This "striking phenomenon" is due to the formation of colloidal red phosphorus (1).

The specimens of CP-RS and CP-RN, both of which were prepared under a reducing condition, exhibit striking phenomenon but CP-OS does not. An example of the optical transmission curves of CP-RN glass specimen heated at 580°C is given in Figure 1, which shows similar changes to that for CP-RS previously reported (2), namely, that the absorption edge at shorter wavelength sides shifts to longer wavelength with increasing time of reheating. Thus, the specimens become tinged with red. The coloring is saturated for 30-50 hours at 580°C.

Photo-Coloring (Photochromism)

The bleached specimens (PTC-RP glasses) were prepared by heating the saturated-red colored specimens at ∿600°C for several minutes and subsequent quenching. Figure 2 gives an example of the transmission curves in coloring of the bleached CP-RN glass specimen exposed to solar rays. The bleached specimen of CP-RS (2) exhibited trends similar to that of CP-RN. Figure 3 shows the dependence of coloring sensitivity bleached CP-RS glass on the wavelength of the irradiating light, where Δe_{40} in the ordinate gives a parameter of coloring sensitivity. This parameter is the difference between the photon energy (ev) corresponding to wavelength at 40% of the transmission curve of a bleached glass before and after the irradiation. As is shown, it was found to be most sensitively colored by irradiating with light of ∿350 nm but it is not colored by a wavelength longer than ∿500 nm.

As an example of change in coloring of a PTC-RP specimen of CP-RS glass by the irradiation with a given wavelength (253.7 nm), a plot of $\lambda_{\frac{1}{2}}\cdot\tau$ is used as a measure of the absorption edge, indicating a wavelength corresponding to half of the maximum transmission of the spectrum (approximately a wavelength at a transmission of 40%).

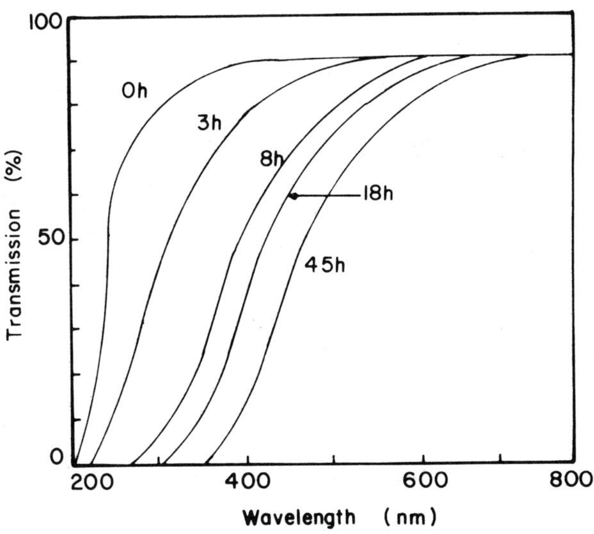

Figure 1. Transmission curves of CP–RN glass heated at 580° C (striking; 1 mm thick specimen).

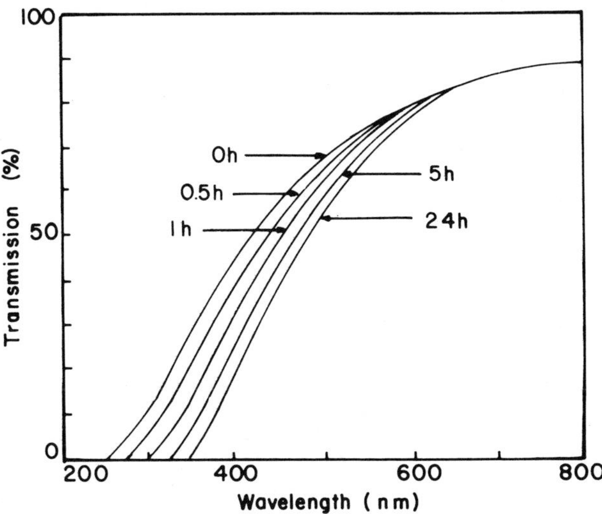

Figure 2. Transmission curves of a bleached CP–RN glass exposed to solar rays (photo-coloring; 1 mm thick specimen).

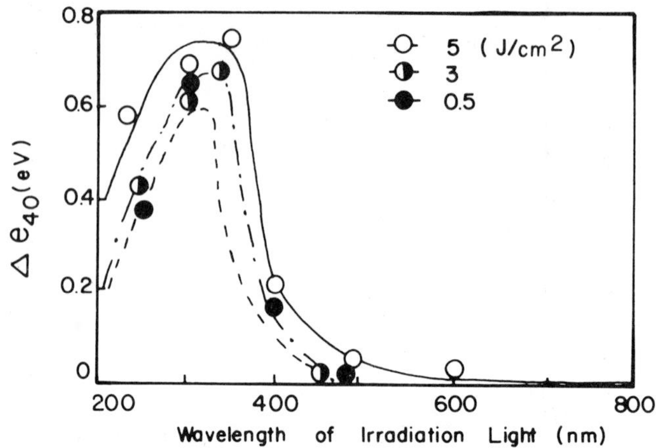

Figure 3. Dependence of coloring sensitivity of a bleached CP–RS glass on the wavelength of irradiation light (1 mm thick specimen).

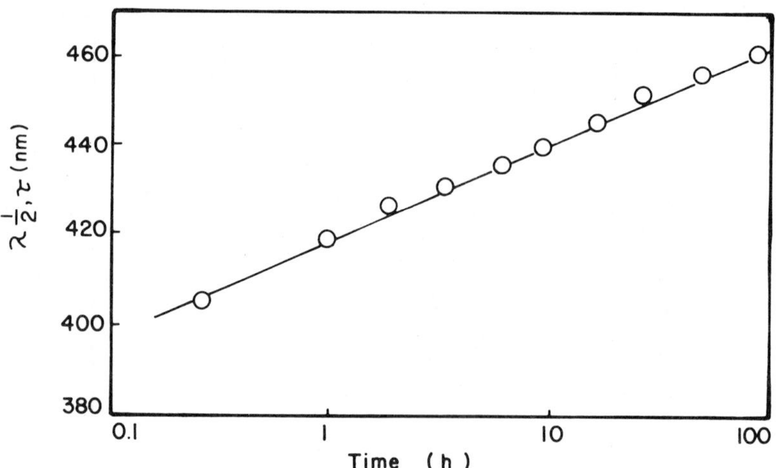

Figure 4. Coloring of a bleached CP–RS specimen by irradiation light (253.7 nm, 750 μW cm^{-2}, 1 mm thick specimen).

Thermo-Coloring (Thermochromism)

As previously reported on some glasses of different composition (3), PTC-RP glass is colored also by heating above 200°C. For convenience, we define a coloring paramter α in eq. [1].

$$\alpha = (\lambda_{\frac{1}{2}\cdot\tau} - \lambda_{\frac{1}{2}\cdot i}) / (\lambda_{\frac{1}{2}\cdot s} - \lambda_{\frac{1}{2}\cdot i}) \quad [1]$$

where, $\lambda_{\frac{1}{2}\cdot\tau}$ is $\lambda_{\frac{1}{2}}$ after heating for time τ,

$\lambda_{\frac{1}{2}\cdot i}$ is $\lambda_{\frac{1}{2}}$ before heating (initial specimen),

$\lambda_{\frac{1}{2}\cdot s}$ is $\lambda_{\frac{1}{2}}$ of a well colored (saturated) specimen.

An apparent activation energy for the thermo-coloring was estimated to be ∿35 Kcal/mol, according to the method described previously (3). This value is almost the same as that of the other glasses reported by Abe et al. (3).

Determination of Elemental Phosphorus Formed in the Reduced Glasses

The permanganometric method developed by Venugopalan (4) was modified in our laboratories. It was found that the reduced glass contained ∿0.2 wt % of elemental phosphorus.

Mechanism of Coloring and Bleaching

The coloring and bleaching phenomena are caused by changes in molecular configuration of colloidal phosphorus formed in the glasses. The P_4 molecule (liquid or white phosphorus) is colorless. It polymerizes thermally or photochemically to the

····–P(P|P)P–···· configuration which gives a reddish color; it dissociates into P_4 molecule again when melted.

Literature Cited:

1. A. Naruse, Y. Abe, J. Ceram. Soc. Japan, 1965, 73, 253-58.
2. Y. Abe, R. Ebisawa, A. Naruse, J. Am. Ceram. Soc., 1976, 59 453-54.
3. Y. Abe, K. Kawashima, S. Suzuki, J. Am. Ceram. Soc., 64, 206-09 (1981).
4. M. Venugopalan, K. U. Matha, Z. Anal. Chem., 1956, 151, 262-68.
   M. Venugopalan, K. J. George, Bull. Chem. Soc. Japan, 1957, 30, 51-53.

RECEIVED June 30, 1981.

A Gel Chromatographic Study on the Interactions of Long-Chain Polyphosphate Anions with Magnesium Ions

TOHRU MIYAJIMA, TOSHIMITSU ONAKA, and SHIGERU OHASHI

Department of Chemistry, Faculty of Science, Kyushu, University 33, Hakozaki, Higashiku, Fukuoka, 812 Japan

In spite of much information available for the interactions of various metal ions with small oxoanions of phosphorus, relatively little information has been obtained for the complex formation of long-chain polyphosphate ion. This may be due to the fact that the conventional methods useful for the study of the complex formation of a relatively small ligand are not always applicable to the polyanion complex formation system. Since a gel chromatographic method based on the same principle as the equilibrium dialysis method has been proved to be applicable in the field of inorganic complex chemistry (1), this method has been applied to the study of the binding of long-chain polyphosphate ions to magnesium ion.

A long-chain polyphosphate ion is composed mainly of middle PO_3-units and two end PO_3-units. A middle and an end unit have a formal charge of minus one and two, respectively. Therefore, the affinity of an end unit for a metal ion is much greater than that of a middle unit. When the chain length of polyphosphate is sufficiently long, the contribution of end units to the metal binding is small enough to be neglected. The complexation ability of middle units in a long-chain polyphosphate is expected to be different from that of middle units in a cyclic phosphate.

This work was undertaken in order to evaluate the binding of middle units of long-chain polyphosphate to magnesium ion from a view point of the mass action law.

A sample solution containing metal ion, M, high-molecular-weight ligand, L and their complexes is applied to a gel column conditioned with an eluent containing M of a specified concentration, $[M]_0$, and is eluted with the same eluent. The initial free metal concentration in the sample solution usually differs from $[M]_0$. During the elution, however, the sample ligand zone (zone α) is equilibrated with the metal solution of $[M]_0$ to reach a steady state, in which a free metal concentration in zone α, $[M]_\alpha$, is equal to $[M]_0$. In zone α the distribution of the chemical species is characterized by the free metal concentration, $[M]_0$. The ratio of the amount of bound metal to the total amount of the ligand, $\bar{n}$, can usually be expressed as the following formation function.

0097-6156/81/0171-0377$05.00/0

$$\bar{n} = \frac{\sum_{1}^{n} i\beta_i [M]_0^i}{1 + \sum_{1}^{n} \beta_i [M]_0^i} \qquad (1)$$

where β_i is the overall stability constant of an M_iL complex.

In this work a mixture of long-chain polyphosphates having various degrees of polymerization was used as the ligand. Therefore, $\bar{n}$ defined in eqn(1) cannot be calculated. Instead of $\bar{n}$, the average number of bound magnesium per one PO_3 unit, $\bar{m}$, was calculated. A value of $\bar{m}$ can be obtained by dividing the amount of bound magnesium, N_{Mg}, by the amount of total PO_3 units, N_P, each in zone α. A Perkin Elmer Model 403 atomic absorption spectrophotometer was used as a detector for magnesium (2).

$$\bar{m} = \frac{N_{Mg}}{N_P} \qquad (2)$$

It has been pointed out that the high and variable electric field at the surface of a charged polymer makes the quantitative description of its equilibria very complecated (3). In this work the following assumptions were made in order to calculate the stability constants of the complexes of middle PO_3 units with magnesium ion.

1) Concentrations are used in place of activities.

2) A set of adjacent PO_3 units, the number of which is n*, binds one magnesium ion to form a one-to-one complex, $Mg(PO_3)_{n*}$.

$$Mg + (PO_3)_{n*} \rightleftharpoons Mg(PO_3)_{n*} \qquad (3)$$

3) The binding ability of each magnesium ion to one site is not affected by the complex formation at other sites of the same chain.

The amount of bound magnesium, N_{Mg}, corresponds to the peak area of zone α. Since the amount of total ligand can be obtained by dividing N_P by n*, the amount of free ligand in zone α is $(N_P/n^* - N_{Mg})$. The stability constant of the complex can be defined as forlows.

$$\beta_1(n^*) = \frac{N_{Mg}}{[Mg]_0(N_P/n^* - N_{Mg})} \qquad (4)$$

Using eqn(2), eqn(4) can be expressed as

$$\beta_1(n^*) = \frac{\bar{m}}{[Mg]_0(1/n^* - \bar{m})} \tag{5}$$

With an appropriate n* value, a constant value of $\beta_1(n^*)$ may be obtained within a specific $[Mg]_0$ range. The pair of n* and $\beta_1(n^*)$ which accounts for the experimental plots indicates that n* PO_3 units bind one magnesium ion to form a specific complex in the $[Mg]_0$ range.

In order to estimate proper n* and $\beta_1(n^*)$ values which satisfy the experimental data, a curve fitting method was employed.

$$\bar{m} = \frac{\beta_1(n^*)[Mg]_0}{1 + \beta_1(n^*)[Mg]_0} \cdot \frac{1}{n^*} \tag{6}$$

$\bar{m}$ values were obtained for various $[Mg]_0$ values in 0.1 M tetramethylammonium chloride solution at 25°C. The calculated curves for n* = 3 through 7 were examined. When n* is assumed to be 4, 5 and 6, the calculated curves showed a good fit for the experimental plots, even though they are not satisfactory in the whole $[Mg]_0$ range. Since the calculated curve for n* = 3 or 7 greatly deviates from the plots, the formation of $Mg(PO_3)_{n^*=3}$ or $Mg(PO_3)_{n^*=7}$ type complexes in this $[Mg]_0$ range can be excluded. The stability constants ($\beta_1(n^*)$) thus obtained are listed in Table 1. It is noteworthy that the formation of $Mg(PO_3)_{n^*=4}$ type complex was clearly observed when $[Mg]_0$ was between 10^{-5} and 10^{-4} mol dm^{-3}.

In order to compare the complexation ability of middle PO_3 units of long-chain polyphosphate with those of relatively small polyphosphates, stability constants of the magnesium complexes of diphosphate, triphosphate, tetraphosphate, tetrametaphosphate, and hexametaphosphate were evaluated by the gel chromatographic method under the same experimental conditions. The stability constants of magnesium complexes with linear phosphates ($\beta_1(nl)$) and those with cyclic phosphates ($\beta_1(nc)$) are also tabulated in Table 1. By comparing the stability constants of linear phosphate complexes one another, it can be concluded that the addition of another PO_3 unit to a ligand does not necessarily contribute to the binding of the first magnesium ion, when the degree of polymerization of the ligand is more than 3. It is worthwhile to compare $\beta_1(n^*)$ with $\beta_1(nc)$, because a cyclic phosphate ion is composed of only middle PO_3 units. By comparison of $\beta_1(4^*)$ with $\beta_1(4c)$ and $\beta_1(6^*)$ with $\beta_1(6c)$, it can be concluded that $\beta_1(n^*)$ is always greater than $\beta_1(nc)$. This indicates that the complexation ability of middle PO_3 units of a long-chain polyphosphate ion is greater than that of the corresponding cyclic phosphate ion. This may be attributed to the flexibility of long-chain polyphosphate ion. It can be seen that the difference between $\beta_1(n^*)$ and $\beta_1(nc)$ decreases with an increase in n. The flexibility of cyclic phosphate ion may increase with n.

Table 1. Stability constants of magnesium complexes for different types of phosphates.

n, n* [a]	$\log(\beta_1(nl)/mol^{-1}dm^3)$ [b]	$\log(\beta_1(nc)/mol^{-1}dm^3)$ [b]	$\log(\beta_1(n^*)/mol^{-1}dm^3)$
2	5.57	-	-
3	6.53	1.80 [c]	-
4	6.53	3.40	5.60
5	-	-	6.03
6	-	5.50	6.22

a) the number of PO_3 units which constitutes the ligand.
b) I = 0.1, Supporting electrolyte; $(CH_3)_4NCl$, Temp.; 25°C.
c) obtained with ion-selective electrode (4).

Literature Cited

1. Miyajima, T; Ohashi, S. Bull. Chem. Soc. Jpn., 1978, 51, 2543.
2. Yoza, N; Ohashi, S. Anal. Lett., 1973, 6, 595.
3. Marinsky, J. A. Coord. Chem. 1976, 19, 125.
4. Kalliney, S. Y., "Topics in Phosphorus Chemistry", Vol. 7, Griffith E. J. and Grayson M., Eds., Interscience, New York, 1967, p 294.

RECEIVED June 30, 1981.

COMPOUNDS WITH MONOCOORDINATED AND DICOORDINATED PHOSPHORUS

Phosphaalkenes, $R_2C=PR'$, and Phosphaalkynes, $RC\equiv P$

H. W. KROTO and J. F. NIXON

School of Molecular Sciences, University of Sussex, Brighton, Sussex, England

The chemistry of trivalent phosphorus compounds in which phosphorus is one or two coordinate is rapidly developing. These systems contain sp or sp^2 hybridized phosphorus and multiple π-bonds between P and other elements; and until 1964 it was believed that bond formation involving pπ-pπ overlap was unfavourable. Subsequently certain -P=C systems resulted from the use of charged (1, 2, 3) and or delocalized systems (4) but it is only in the past few years that successful syntheses of phospha-alkenes, $R_2C=PR'$, and phospha-alkynes, $RC\equiv P$, have been reported. These novel compounds are the subject of this paper.

Pyrolysis techniques developed by Kroto et al. to produce sulphur and selenium doubly bonded species in a low pressure flow system where the lifetimes are of the order of seconds, enabled species such as H_2CS (5), F_2CS (6), CH_3CHS (7), CH_3CHSe (8), CH_2CCS (9) and HBS (10) to be identified by microwave and/or P.E. spectroscopy.

An extension of this approach led to the study of some simple molecules in which trivalent phosphorus is bonded to carbon by a double or a triple bond. These spectroscopic experiments have not only yielded conclusive identification but a wealth of molecular information such as geometric and electronic structural data. More importantly, the >C=P- and $-C\equiv P$ moieties could be considered as viable functional groups with interesting and well defined chemical properties.

In 1961 Gier (11) reported the synthesis of $HC\equiv P$ via a carbon arc discharge of PH_3. The compound which polymerised above -120° was long regarded as a chemical curiosity but in 1968 Kroto, Nixon et al. (12) reported the facile synthesis of $FC\equiv P$ by dehydrofluorination of CF_3PH_2.

$$CF_3PH_2 \longrightarrow CF_2{=}PH \longrightarrow FC\equiv P$$

NMR, microwave and He(I) photo-electron spectroscopic studies on this compound have been reported (12, 13).

0097-6156/81/0171-0383$05.00/0

A whole new class of compounds of the type RC≡P now exists, some of which are listed below (14-21).

HC≡P   FC≡P CF_3C≡P CH_3C≡P

NCC≡P   PhC≡P CH_2=CHC≡P   CH≡CC≡P

e.g. $RCH_2PCl_2 \longrightarrow$ RC≡P

$R_fCF_2PH_2 \longrightarrow R_fC$≡P

$HCP \xrightarrow{CNN_3}$ NCC≡P

Very recently (February 1981) a significant advance has been made with the syntheses of Me_3SiC≡P (½ life 50 min) (Appel *et al.*) (22) and the stable liquid t-BuC≡P (Becker *et al.*) (23) by the novel route shown below.

$(R_3Si)_3P \longrightarrow R_3Si{-}P{=}C(tBu){-}OSiR_3 \longrightarrow$ tBuC≡P

Our recent syntheses of CH_2=CH-C≡P and HC≡C-C≡P using the HX elimination approach has indicated that many more RC≡P compounds will be soon available. Our structural data indicate that the C≡P bond length 1.545 Å is relatively insensitive to the nature of the R substituent, whereas other properties, e.g. NMR parameters and I.P. data, are significantly affected by R.

The P-C distances in simple phosphines, phospha-alkenes and phospha-alkynes are summarised in Figure 1 and the results contrasted with C-C, C=C and C≡C bond lengths when a simple relationship is seen to exist. Additionally the data for the phosphabenzene systems shows excellent agreement with expectation.

Compounds of the type R_2C=PX were first postulated by Haszeldine and coworkers (24-27) as reaction intermediates in the reaction of certain perfluoroalkyl phosphines with bases. These species were first isolated and characterised by spectroscopic techniques by Kroto and Nixon and coworkers (28). Microwave and/or NMR data on CH_2=PH, CH_2=PCl and CF_2=PH were obtained on samples generated by the pyrolysis of suitable precursors, but subsequently a variety of alternative synthetic routes have been developed to obtain the following phospha-alkene molecules (28-34):

CH_2=PH, CH_2=PCl, CH_2=PBr, CH_2=PF,

CF_2=PH, CF_2=PCF_3, PhRC=PCl, etc.

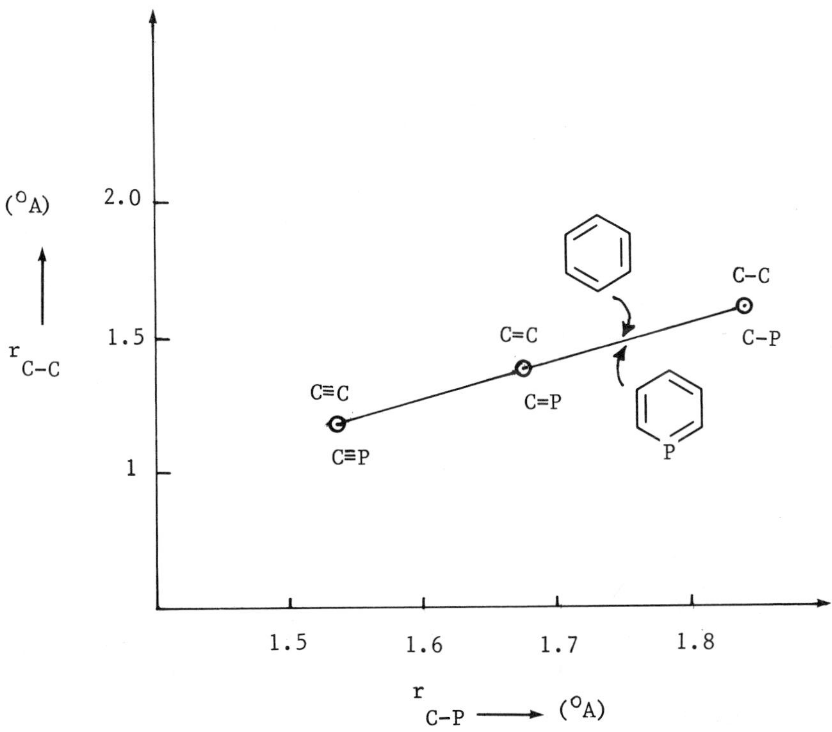

Figure 1. Plot of r_{C-P} *vs.* r_{C-C}. *Gives covalent radius of P: Sp³, 1.07 °A; Sp², 1.00; Sp, 0.94.*

$R_3SiCR'R''PX_2 \longrightarrow R_3SiX + CR'R''{=}PX$

$CF_3PH_2 \longrightarrow CF_2{=}PH$

Thermal stabilization of phospha-alkenes can be influenced by the attachment of a heteroatom to the P=C bond as evidenced by the elegant work of Becker (35-40, 43), Appel (41) and Issleib (42).

e.g. $RP(SiMe_3)_2 + tBuCOCl \longrightarrow R{-}P{=}C(OSiMe_3)(tBu)$

$PhP(SiMe_3)_2 + R'N{=}C(Cl)(R) \longrightarrow (R')(Me_3Si)N{-}C(R){=}P{-}Ph$

In 1978 Bickelhaupt et al. (44) reported the first stable phospha-alkene with carbon substituents only, using the following synthetic approach.

$RPCl_2 \longrightarrow RPClCHPh_2 \longrightarrow RP{=}CPh_2$ (R = mesityl)

P=C(Ph)(Ph) [mesityl]

(L)

Kroto, Nixon et al. have reported several examples of complexes of the phospha-alkene, L, $RP{=}CPh_2$ (R = mesityl) (45). In principle phospha-alkenes can coordinate to transition metals in any of the three modes shown but so far evidence only for type (a) has been obtained.

(a) R'—P(→M)=C(R)—R

(b) R'—P(:)=C(R)—R, η² to M

(c) R'—P(→M)=C(R)—R, η² to M

(a) (b) (c)

Complexes obtained are: cis-$M(CO)_4L_2$ (M = Cr,Mo,W); $M(CO)_5L$ (M = W); $RhCl(PPh_3)_2L$; $RhCl(CO)L_2$; $Rh(C_9H_7)L_2$; PtX_2L_2 (X = Cl,I,Me); $PtCl_2(PEt_3)L$. The latter is the subject of an X-ray structural analysis (46). These results are important in relation to other recent developments in the coordination

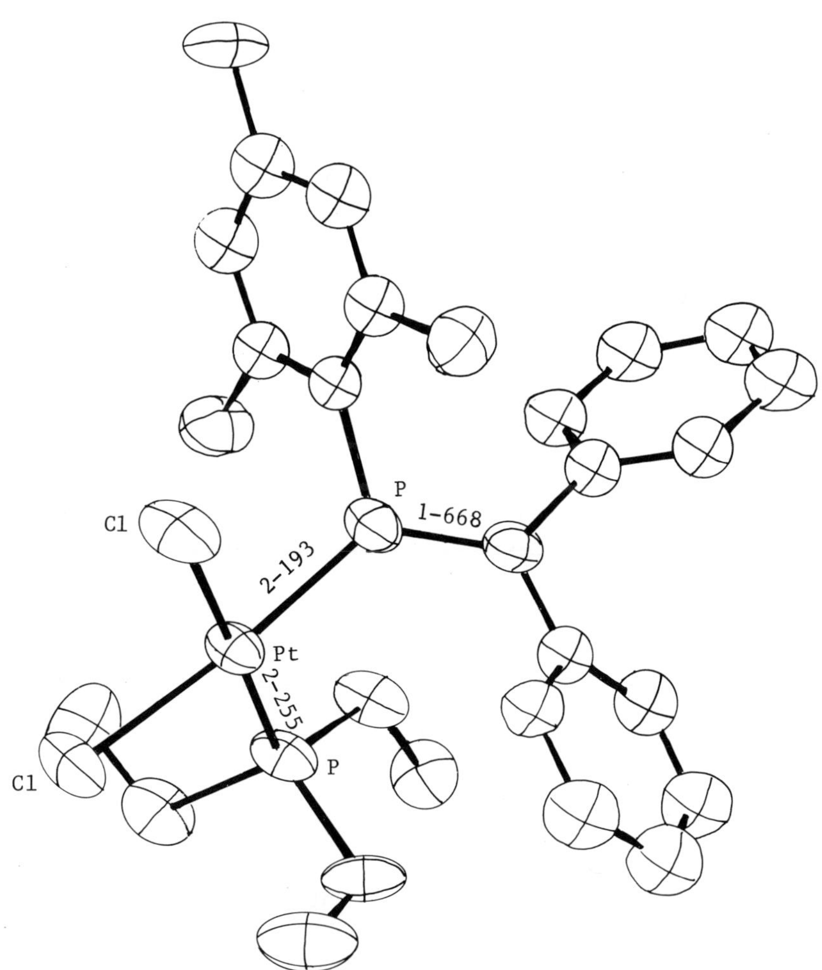

Figure 2. Structure of cis-*$PtCl_2(PEt_3)(P(mes)=CPh_2)$.*

chemistry of RPNR' and RPO (47) where phosphorus is the coordinating site. Of particular interest is the recently determined structure of the complex cis-$PtCl_2(PEt_3)$ (P(mes) = CPh_2) shown in Figure 2. The previously postulated (46) mode of coordination of the phospha-alkene to the metal atom using the lone pair on phosphorus has now been confirmed. The phosphorus-carbon double bond is 1.668 Å in excellent agreement with our previously published data from microwave spectroscopy on simpler uncoordinated phospha-alkenes (it should be noted that the structure of free P(mes)=CPh_2 is not known). The angles subtended at phosphorus namely <CPPt are 120.2° and 127.8° respectively while the CPC angle is 112°.

The bond length between platinum and the phospha-alkene is only 2.193 Å which is much shorter than the Pt-PEt_3 bond (2.255Å) and is only slightly longer than the shortest known Pt-P distance (to PF_3 2.141 Å) in cis-$PtCl_2(PF_3)(PEt_3)$. Clearly the s-character of the phospha-alkene lone pair is significantly enhanced. These results and a discussion of their relationship with Pt-Cl distances and $^1J_{PtP}$ will be the subject of future papers.

Acknowledgment

It is a pleasure to acknowledge the support of S.R.C. and the experimental skill of the coworkers mentioned in individual references.

Literature Cited

1. Dimroth, K.; Hoffman, P. Angew. Chem. Intl. Edn. 1964, 3, 384.
2. Allman, R. Angew. Chem. Intl. Edn. 1965, 4, 150.
3. Dimroth, K. Topics in Current Chem. 1973, 38.
4. Märkl, G. Angew. Chem. Intl. Edn. 1966, 5, 846.
5. Kroto, H.W.; Suffolk, R.J. Chem. Phys. Letters 1972, 15, 545.
6. Careless, A.J.; Kroto, H.W.; Landsberg, B.U. Chem. Phys. 1973, 1, 371.
7. Kroto, H.W.; Landsberg, B.M. J. Mol. Spect. 1976, 62, 346.
8. Hutchinson, M.W.; Kroto, H.W. J. Mol. Spect. 1978, 76, 347.
9. Georgiou, K.; Kroto, H.W.; Landsberg, B.M. Chem. Comm. 1974, 739.
10. Kroto, H.W.; Suffolk, R.J.; Westwood, N.P.C. Chem. Phys. Letters 1973, 22, 495.
11. Gier, T.E. J. Amer. Chem. Soc. 1961, 83, 1769.
12. Kroto, H.W.; Nixon, J.F.; Simmons, N.P.C.; Westwood, N.P.C. J. Amer. Chem. Soc. 1978, 100, 446.
13. Kroto, H.W.; Nixon, J.F.; Simmons, N.P.C. J. Mol. Spect. 1980, 82, 185.
14. Hopkinson, M.J.; Kroto, H.W.; Nixon, J.F.; Simmons, N.P.C. Chem. Phys. Letters 1976, 42, 460.

15. Westwood, N.P.C.; Kroto, H.W.; Nixon, J.F.; Simmons, N.P.C. J. Chem. Soc. (Dalton) 1979, 1405.
16. Kroto, H.W.; Nixon, J.F.; Simmons, N.P.C. J. Mol. Spect. 1979, 77, 270.
17. Cooper, T.; Kroto, H.W.; Nixon, J.F.; Ohashi, O. J. Chem. Soc. Chem. Comm. 1980, 333.
18. Ohno, K.; Kroto, H.W.; Nixon, J.F. J. Mol. Spect. Submitted for publication.
19. Kroto, H.W.: Ohno, K.; Nixon, J.F. J. Mol. Spect. Submitted for publication.
20. Burckett-St. Laurent, J.C.T.R.; Kroto, H.W.; Nixon, J.F.; Ohno, K. J. Mol. Spect. Submitted for publication.
21. King, M.A.; Klapstein, D.; Kroto, H.W.; Maier, J.P.; Marthaler, O.; Nixon, J.F. J. Mol. Spect. Manuscript in preparation.
22. Appel, R. Personal communication.
23. Becker, G.; Gresser, G.; Uhl, W. Z. Naturforsch 1981, 361, 16.
24. Burch, G.M.; Goldwhite, H.; Haszeldine, R.N. J. Chem. Soc. 1964, 572.
25. Goldwhite, H.; Haszeldine, R.N.; Isles, B.R.; Rowsell, D.G. J. Chem. Soc. 1965, 6879.
26. Green, M.; Haszeldine, R.N.; Isles, B.R.; Rowsell, D.G. J. Chem. Soc. 1965, 6875.
27. Haszeldine, R.N.; Taylor, D.R.; White, E.W. J. Fluorine Chem. 1977, 10, 27.
28. Hopkinson, M.J.; Kroto, H.W.; Nixon, J.F.; Simmons, N.P.C. J. Chem. Soc. Chem. Comm. 1976, 513.
29. Hosseine, H.E.; Kroto, H.W.; Nixon, J.F.; Morton, J.R.; Preston, K.F. J. Chem. Soc. Chem. Comm. 1979, 653.
30. Hosseini, H.E.; Kroto, H.W.; Nixon, J.F.; Ohashi, O. J. Organometal. Chem. 1979 (C1) 181.
31. Kroto, H.W.; Nixon, J.F.; Ohno, K.; Simmons, N.P.C. J. Chem. Soc. Chem. Comm. 1980, 709.
32. Kroto, H.W.; Nixon, J.F.; Ohno, K. J. Mol. Spect. Submitted for publication.
33. Kroto, H.W.; Nixon, J.F.; Taylor, M.J.; Burckett-St. Laurent, J.C.T.R.; Walton, D.R.M. Papers in preparation.
34. Appel, R.; Westerhaus, A. Angew. Chem. Int^{l} Ed^{n} 1980, 19, 556.
35. Becker, G. Z. Anorg. Chem. 1976, 423, 242.
36. Becker, G.; Mundt, O.; Rössler, M.; Schneider, E. Z. Anorg. Chem. 1978, 443, 42.
37. Becker, G.; Rössler, M.; Schneider, E. Z. Anorg. Chem. 1978, 439, 121.
38. Becker, G.; Mundt, O. Z. Anorg. Chem. 1978, 443, 53.
39. Becker, G. Z. Anorg. Chem. 1977, 430, 66.
40. Becker, G.; Beck, H.P. Z. Anorg. Chem. 1977, 430, 77.
41. Appel, R.; Barth, V. Angew. Chem. Int^{l} Ed^{n} 1979, 18, 469.

42. Issleib, K.; Schmidt, H.; Meyer, H. J. Organometal. Chem. 1978, 160, 47.
43. Becker, G.; Rössler, M.; Uhl, W. Z. Anorg. Chem. 1981, 473, 7.
44. Klebach, T.C.; Lourens, R.; Bickelhaupt, F. J. Amer. Chem. Soc. 1978, 100, 4886.
45. Hosseini, H.E.; Kroto, H.W.; Nixon, J.F.; Maah, M.J.; Taylor, M.J. J. Chem. Soc. Chem. Comm. 1981, 199.
46. Frew, A. and Muir, K.W. unpublished data.
47. Niecke, E.; Engelmann, M.; Zorn, H.: Krebs, B.; Henkel, G. Angew. Chem. Intl. Edn. 1980, 19, 710.

RECEIVED June 12, 1981.

Recent Developments in the Chemistry of Dicoordinated Phosphorus Radicals and Cations

S. G. BAXTER, A. H. COWLEY, R. A. KEMP, S. K. MEHROTRA, and J. C. WILBURN[1]

Department of Chemistry, The University of Texas at Austin, Austin, TX 78712

Phosphide anions, phosphinyl radicals, and phosphenium cations are two-coordinate phosphorus species which are related by gain or loss of one electron:

$$R_2P^- \underset{+1e^-}{\overset{-1e^-}{\rightleftharpoons}} R_2P^{\cdot} \underset{+1e^-}{\overset{-1e^-}{\rightleftharpoons}} R_2P^+$$

Phosphide anions are well documented and are used widely as reagents. However, much less is known concerning phosphinyl radicals and phosphenium cations.

Phosphinyl Radicals as Ligands

Interesting work by Lappert, Goldwhite, and co-workers (1,2) has established that phosphinyl radicals can be stabilized by means of bulky substituents such as $(Me_3Si)_2CH$. We have found that these radicals constitute a new class of ligand intermediate in character between phosphenium cations and phosphide anions.

Upon treatment of $[\{(Me_3Si)_2CH\}_2P]^{\cdot}$ (I) (1,2) with $Fe_2(CO)_9$ in n-hexane at ambient temperature, the initial blood-red color of the solution darkened rapidly. Evacuation of the volatiles produced a purple-black solid radical of composition $[\{(Me_3Si)_2CH\}_2PFe(CO)_4]^{\cdot}$ (II). The magnetic moment of II was found to be 0.88 μ_B at 300K and the 70 eV mass spectrum exhibited a parent peak at m/e 517.

Despite their paramagnetism, it is possible to record limited NMR data for radicals I and II: I: $^{31}P\ \{^1H\}$ (s, +404 ppm); II: ^1H Me_3Si (s, δ 0.28), $^{13}C\ \{^1H\}$ Me_3Si (br s, δ 3.0), $^{31}P\ \{^1H\}$ (s, +300 ppm). The fact that ^1H and ^{13}C NMR spectra are detectable for II but not for I suggests that the unpaired electron resides in the $Fe(CO)_4$ moiety of II. Two additional

[1] Current address: Allied Chemical, Buffalo Research Laboratory, Buffalo, NY 14210

0097–6156/81/0171–0391$05.00/0

lines of evidence support this view. First, in contrast to that of I (for which a_P = 9.63 and a_H = 0.64 mT, and g = 2.009) (1,2), the ESR of II comprises a narrow singlet (g = 2.004) with no perceptible hyperfine coupling to ^{31}P or ^{1}H. Second, the Mössbauer spectrum of II exhibits unresolved hyperfine coupling; moreover, in contrast to (phosphenium)$Fe(CO)_4$ complexes (3) the isomer shift (-0.03 mm/sec) and quadrupole splitting (0.52 mm/sec) for the two intense lines of II are significantly removed from the linear (Collins-Pettit) (4) plot of isomer shifts vs. quadrupole splittings typical of diamagnetic $LFe(CO)_4$ complexes. Assuming the phosphinyl ligand occupies an axial site of a locally trigonal bipyramidal geometry at Fe, the P-Fe σ-bond in II will be formed by interaction of the phosphorus lone pair with the a_1 MO of the C_{3v} $Fe(CO)_4$ fragment (5,6), and delocalization of the unpaired electron can occur via interaction of the singly-occupied P(3p) orbital with the Fe($3d_{xz}$) component of the 1e MO. If the latter interaction is sufficiently strong, the singly-occupied MO would be a π*(CO) orbital. The fact that the CO stretching frequencies of II (2040, 1930, 1920, and 1880 cm^{-1}) are ~100 cm^{-1} lower than those of typical phosphenium ion complexes (7) is consistent with the idea of one-electron transfer into the $Fe(CO)_4$ moiety.

The reaction of I with $Co_2(CO)_8$ in toluene resulted in the production of a purple solid radical of composition $[\{(Me_3Si)_2CH\}_2PCo_2(CO)_7]^{\cdot}$ (III). The 70 eV mass spectrum of III does not exhibit a parent peak; the highest m/e peaks appear at 495 and 464 and are attributable to $[\{(Me_3Si)_2CH\}_2PCo_2(CO)]^+$ and $[\{(Me_3Si)_2CH\}_2PCo(CO)_2]^+$, respectively. It is not possible to record ^{1}H and $^{13}C\{^{1}H\}$ NMR spectra for III; however, the $^{31}P\{^{1}H\}$ NMR spectrum consists of a broad singlet at +420 ppm. The magnetic moment of III is 1.82 μ_B at 300K. The pattern of CO stretching frequencies (2070, 2035, 2010, 1990, and 1975 cm^{-1}) is very similar to that of phosphine and phosphite complexes of composition $R_3PCo_2(CO)_7$ (8), thus suggesting a comparable structure for III. The ESR spectrum of III in toluene consists of a 15-line pattern, indicating delocalization of the unpaired electron from the phosphinyl radical onto the $Co_2(CO)_7$ moiety (^{59}Co, I = 7/2, natural abundance 100%).

Pentamethylcyclopentadienyl-Substituted Phosphenium Ions

Since the heavier group 4A elements can engage in pentahapto bonding with cyclopentadienyl and pentamethylcyclopentadienyl groups, we were prompted to investigate the bonding in isoelectronic compounds featuring P^+ and As^+.

Treatment of $(Me_5C_5)PCl_2$ (9) with Me_3SiNMe_2 affords $(Me_5C_5)(Me_2N)PCl$ (IV) which was characterized by elemental analysis, mass spectroscopy (parent peak, m/e 245), and NMR: ^{1}H Me_2N (d, δ 2.58, J_{PNCH} = 11.3 Hz), Me_5C_5 (br s, δ 1.9); $^{13}C\{^{1}H\}$ Me_2N (d, δ 41.51, J_{PNC} = 17.5 Hz), C_5Me_5 (br, δ 138.6); $^{31}P\{^{1}H\}$ (s, 144.8 ppm). The Me_5C_5 ring of IV is bonded in the monohapto manner, since on cooling to -40°C the 200 MHz ^1H NMR spectrum of the pentamethylcyclopentadienyl methyl protons exhibits three resonances: Me_a (d, 3H, δ 1.37, J_{PCCH_a} = 6.65 Hz), Me_b (d, 6H, δ 1.77, J_{PCCCH_b} = 6.63 Hz), Me_c (s, 6H, δ 1.83).

Treatment of IV with a stoichiometric quantity of Al_2Cl_6 in CH_2Cl_2 solution at -78°C, followed by warming to room temperature affords a red-brown solution. The presence of the $AlCl_4^-$ anion as the sole aluminum-containing species was evidenced by the presence of a sharp singlet (w_h ~ 9 Hz, δ = 103 ppm) (10) in the ^{27}Al NMR, thus indicating the formation of the phosphenium ion, $[(Me_5C_5)(Me_2N)P]^+$ (V). ^1H Me_2N (d, 6H, δ 3.12, J_{PNCH} = 7.8 Hz), C_5Me_5 (d, 15H, δ 2.14, J_{PCCH} = 2.6 Hz); ^{13}C {^1H} Me_2N (d, δ 43.04, J_{PNC} = 12.9 Hz), C_5Me_5 (s, δ 10.76), C_5Me_5 (d, δ 130.4, J_{PC} = 11.8 Hz). Several pieces of spectroscopic evidence lead to the conclusion the Me_5C_5 ring of V is not bonded in the monohapto manner. First, the ^1H and ^{13}C spectra of V indicate that the ring and Me carbons of the Me_5C_5 moiety are equivalent; moreover, the equivalence of the methyl groups persists to -100°C and -80°C in ^1H and ^{13}C NMR experiments, respectively. Second, the ^{31}P chemical shift of V (111.0 ppm) is 33.8 ppm upfield (i.e. shielded) compared to that of the phosphorus(III) chloride precursor, IV. In all cases reported to date, phosphenium ion formation via halide ion abstraction from precursor phosphorus(III) halides has been accompanied by a downfield ^{31}P NMR chemical shift in excess of 100 ppm. (11-14) The upfield shift is attributed to multihapto bonding between P^+ and the Me_5C_5 ligand. Support for this suggestion is provided by the fact that ~100 ppm upfield ^{11}B NMR chemical shifts have been observed (15) when pentahapto boron cations, $[\eta^5\text{-}Me_5C_5)BX]^+$ are formed via X^- abstraction from the corresponding monohapto boron dihalides, $(\eta^1\text{-}Me_5C_5)BX_2$.

The foregoing NMR observations on V are consistent with a static pentahapto structure, or with tri- and dihapto structures with low barriers to migration. MNDO calculations (16,17) on $[(Me_5C_5)(Me_2N)P]^+$ reveal the following: (i) the global minimum is the dihapto structure, (ii) the penta- and trihapto structures do not correspond to minima, and (iii) the barrier to circumannular migration of the Me_2NP moiety in the dihapto structure is very low (<2 kcal/mole).

Acknowledgment

The authors are grateful to the National Science Foundation (Grant CHE79-10155) and the Robert A. Welch Foundation for generous financial support, and to Professor R. L. Collins for the Mössbauer spectrum.

Literature Cited

1. Gynane, M. J. S.; Hudson, A.; Lappert, M. F.; Power, P. P.; Goldwhite, H. J. Chem. Soc. Chem. Commun. 1976, 623.

2. Gynane, M. J. S.; Hudson, A.; Lappert, M. F.; Power, P. P.; Goldwhite, H. J. Chem. Soc. Dalton Trans. 1980, 2428.
3. Baxter, S. G.; Collins, R. L.; Cowley, A. H.; Sena, S. F. J. Am. Chem. Soc. 1981, 103, 714.
4. Collins, R. L.; Pettit, R. J. Chem. Phys. 1963, 39, 3433.
5. Elian, M.; Hoffmann, R. Inorg. Chem. 1975, 14, 1058.
6. Hoffmann, R.; Chen, M. M. L.; Elian, M.; Rossi, A. R.; Mingos, D. M. P. Inorg. Chem. 1974, 13, 2666.
7. Montemayer, R. G.; Sauer, D. T.; Fleming, S.; Bennett, D. W.; Thomas, M. G.; Parry, R. W. J. Am. Chem. Soc. 1978, 100, 2231.
8. Szabó, P.; Fekete, L.; Bor, G.; Nagy-Magos, Z.; Markó, L. J. Organomet. Chem. 1968, 12, 245.
9. Jutzi, P.; Saleske, H.; Nadler, D. J. Organomet. Chem. 1976, 118, C8.
10. Akitt, J. W. Annual Reports on NMR Spectroscopy 1972, 5A, 465.
11. Maryanoff, B. E.; Hutchins, R. O. J. Org. Chem. 1972, 37, 3475.
12. Fleming, S.; Lupton, M. K.; Jekot, K. Inorg. Chem. 1972, 11, 2534.
13. Thomas, M. G.; Schultz, C. W.; Parry, R. W. Inorg. Chem. 1977, 16, 994.
14. Cowley, A. H.; Cushner, M. C.; Lattman, M.; McKee, M. L.; Szobota, J. S.; Wilburn, J. C. Pure & Appl. Chem. 1980, 52, 789.
15. Jutzi, P.; Seufert, A. Angew. Chem. Int. Ed. Engl. 1977, 16, 330.
16. Dewar, M. J. S.; Thiel, W. J. Am. Chem. Soc. 1977, 99, 4899.
17. Dewar, M. J. S.; McKee, M.; Rzepa, H. J. Am. Chem. Soc. 1978, 100, 3607.

RECEIVED July 7, 1981.

^{31}P NMR Investigations on Dicoordinated Phosphorus Compounds in P(II) = C–P(III) Systems

R. APPEL, V. BARTH, H. KUNZE, B. LAUBACH, V. PAULEN, and F. KNOLL

Inorganic Department of Chemistry, University of Bonn, Gerhard-Domagk-Strasse 1, D-5300 Bonn 1, FRG

A number of compounds having the P(II)=C-P(III) backbone structure (phosphino methylene phosphanes) have recently (1,2) been prepared by our group via condensation and migration of trimethysilyl groups at low temperatures.

$$R\ RP(SiMe_3)_2 + \begin{matrix} R'N{=}CCl_2 \\ \\ O{=}CCl_2 \end{matrix} \longrightarrow \begin{matrix} RP{=}C(NR'SiMe_3){-}P(R)SiMe_3 \\ + 2\ ClSiMe_3 \\ RP{=}C(OSiMe_3){-}P(R)SiMe_3 \end{matrix}$$

(SiMe$_3$ of the lower product ← Z)

	1	2	3	4	5
R =	CH_3	$C(CH_3)_3$	C_6H_5	C_6H_5	C_6H_5
R' =	--	------	---	C_6H_4X	YCO

4		a	b	c	d	e	f	g	h	i	j	k
X=	H	o-Cl	m-Cl	p-Cl	o,m Cl_2	o-F	p-F	p-Br	o-Me	p-Me	o-OMe	o-CF_3

5	a	b
Y=	OC_2H_5	C_6H_5

2		a	b	c	d	e	f	g
Z =	$SiMe_3$	H	Cl	CH_3	nC_4H_9	CH_3CO	C_6H_5CO	ClCO

The phosphino methylene phosphanes so formed occur as yellow oils with little or no tendency towards crystallization. In order to obtain suitable materials for x-ray structure studies, we converted two compounds to convenient derivatives.

0097–6156/81/0171–0395$05.00/0

Crystal Structure

The (benzoyl-trimethyl-silylamino(phenyl-trimethyl-silyl-phosphano) methylene) phenylphosphane was heated in acetonitrile to 80-90°C. A cyclic product was formed, which readily crystallized and turned out to be of a Z-structure according to the CIP rules:

```
                 N(COC6H5)SiMe3                       NSiMe3
                /                                    / \
      P=C                          ------->     P=C    C(C6H5)OSiMe3
     /    \                                    /   \  /
C6H5       P(C6H5)SiMe3                   C6H5      PC6H5

      A    B          5                        A    B          6
```

Since the ring was formed by silyl migration, we believe the configuration with respect to the P=C double bond did not change. This assumption is supported by the ^{31}P nmr data described in the next section.

Transformation of another compound was done by treatment with more ioscyanide dichloride, giving 1,3,4,6-tetraphospha-1,5-hexadiene.

```
C6H5        N(C6H5)SiMe3                          C6H5       N(C6H5)SiMe3
     \     /                 + C6H5NCCl2               \     /
2     P=C                  ----------------->           P=C
           \                 - NCC6H5                        \
            P(C6H5)SiMe3     - 2 ClSiMe3                      PC6H5
                                                   A B        |               7
4     A     B                                                 PC6H5
                                                             /
                                                          P=C
                                                         /    \
                                                     C6H5      N(C6H5)SiMe3
```

The x-ray structure of the meso-form of the product (3) indicates an E-E configuration. Again we presume the same orientation in the starting material on the basis of ^{31}P nmr shift comparison.

NMR Investigations

Two particular structures and the starting compounds indicate a stronger ^{31}P nmr shift for the A phosphorus to lower field in case of an E configuration. The J(AB) coupling is not diagnostic of an E or Z structure.

		Z 6	(Z) 5	EE 7	(E) 3
^{31}P δ (A)	ppm	+112.7	+155.3	+258.0	+213.8
^{31}P δ (B)	ppm	+ 72.3	- 38.6	- 12.3	- 39.0
J(AB)	Hz	22.1	160.0	24.6	78.9
^{1}H δSiCH	ppm	+ 0.3	0.0	- 0.4	------
$^{5}J(PH)$	Hz	0.9	0.3	0.0	------

^{31}P (85% H_3PO_4), ^{1}H (TMS), frequency scale

	E/Z	$^{31}P\delta A$	$^{31}P\delta B$	J(AB)	$^{1}H\delta$ SiMe	$^{5}J(PH)$	structure predicted
1	10%	+192.0	-90.0	23.5			E
1	90%	+153.0	-65.0	29.5			Z
2	2%	+276.5	-30.0	30.0			E
2	98%	+231.0	-11.0	9.9	+0.4	2.0	Z
2a		+232.0	- 6.5	47.0	+0.3	0.3	Z
2c		+222.0	+ 3.0	5.9			Z
2d		+223.3	+15.8	0.0			Z
2e		+241.5	+47.5	6.5			Z
2f		+235.5	+39.6	13.0			Z
2b		+252.1	+106.0	374/422 (t°)			?
2g		+255.0	+61.0	11.5			?
3		+164.0	-37.0	73.0			?
4		+213.8	-39.0	78.9			E
4c		+221.9	-39.0	77.7			E
4f		+200.9	-39.9	82.5			E
4g		+220.4	-34.4	77.8	-0.2	0.0	E
4i		+203.6	-40.0	80.8			E
4b	70%	+228.8	-38.8	70.0	-0.13	0.0	E
4b	30%	+142.4	-41.0	36.8	+0.19	2.4	Z
4a		+139.8	-40.9	39.1	+0.21	2.2	Z
4d		+143.0	-40.6	36.6	+0.12	2.5	Z
4e		+157.4	-40.2	71.9	+0.06	2.0	Z
4h		+135.6	-40.7	41.5			Z
4j		+141.1	-41.4	62.0			Z
4k		+137.0	-39.3	35.9			Z
5a	50%	+243.7	-30.3	81.0			E
5a	50%	+180.3	-41.2	204.0			Z
5b		+155.3	-38.6	160.0			Z

δ ppm, J Hz, ^{31}P (85% H_3PO_4), ^{1}H (TMS)

Low Temperature Spectra

On lowering the temperature we observed in the ^{31}P nmr spectra of 4a an additional splitting of the signals, giving two pairs of doublets showing temperature dependent intensities due to inversion and or rotation hindrance.

Furthermore we detected at -80°C two new pairs of doublets in the P-$SiMe_3$ region of the spectrum, giving for the first time evidence of a phosphaguanidine, the silatautomer of 4a R'N=C(P(R)$SiMe_3$)$_2$. The reversible low temperature pattern is of the predicted type for a mixture of the two d/l diastereomeric pairs of compounds.

Results and Discussion

With respect to the two structures known so far, the more positive ^{31}P nmr shift pointed towards an E configuration. This seemed to be supported by a series of compounds prepared with differently substituted isocyanide dichlorides, showing the more space requiring groups to move opposite to the substituent on the methylene phosphorus. Moreover, because of easy cis-migration, only the Z compounds seemed to be capable of forming diastereomeric modifications of the silatantomer.

To be on the safe side, we added a third structure determination after being successful in growing crystals of 4a. This was completed shortly before we left for the conference and it turned out to be a E structure instead of the predicted Z configuration. The determination was redone using a different crystal and in addition a ^{31}P nmr of the solid compound was kindly run by Dr. Forster (Bruker Company). The structure was confirmed, the ^{31}P nmr spectra in solution and in the solid phase showed nearly the same shifts of close inspection of the three crystal structures showed the nitrogen to be planar, but only two structures show the nitrogen and its ligands to be coplanar with the methylene phosphorus, thus optimizing orbital interactions.

	Structure	Coplanarity	^{31}P δ(A) ppm
4a	E	yes	+132.8
6	Z	yes	+115.7
7	EE	no	+258.0

It is therefore not the proposed E or Z configuration which governs the strong shift differences seen in the ^{31}P nmr spectra, but rather the possible interaction of the Π-system with other suitable orbitals.

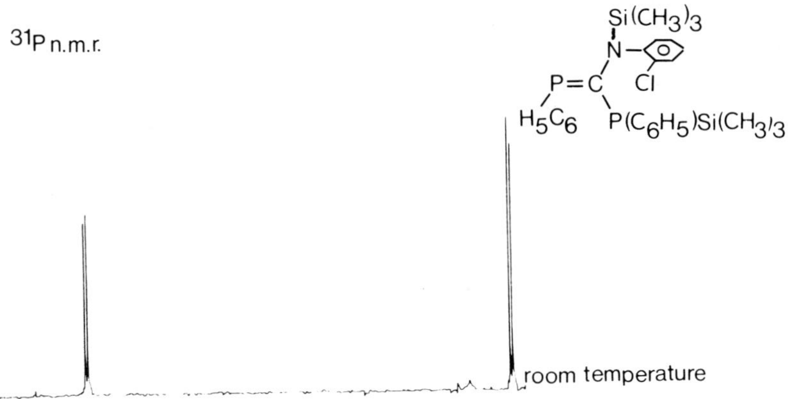

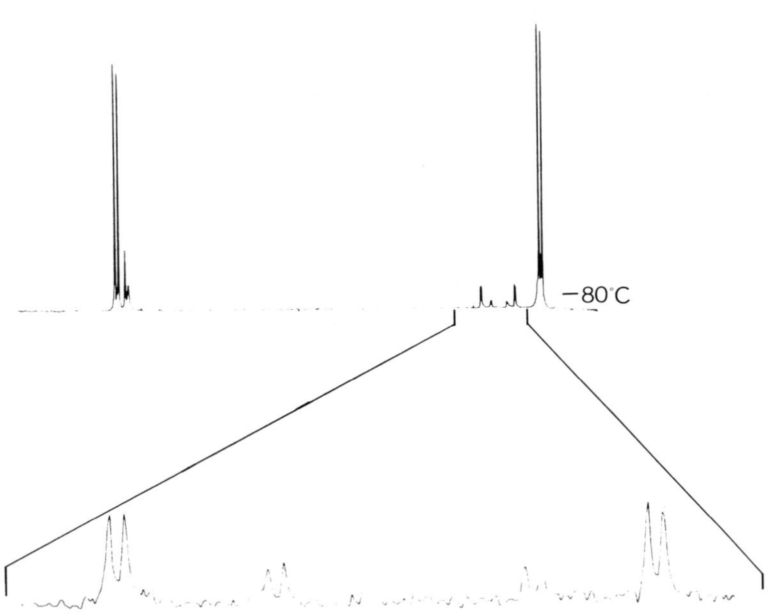

Figure 1. ^{1}H NMR spectra of 4a.

Literature Cited

1. Appel, R.; Barth, V. Angew. Chem. 1979, 91, 497, Angew. Chem. Int. Ed. Engl. 1979, 18, 469.
2. Appel, R.; Laubach, B. Tetrahedron Lett. 1980, 2497.
3. Appel, R.; Barth, V.; Knoll, F.; Ruppert, I. Angew. Chem. 1979, 91, 936, Angew. Chem. Int. Ed. Engl. 1979, 18, 873.
4. Becker, G.; Mundt, O. Z. Anorg. Allg. Chem. 1980, 462, 130.
5. Becker, G.; Mundt, O. Z. Anorg. Allg. Chem. 1978, 443, 53.

RECEIVED July 7, 1981.

Synthesis and Properties of Phosphaalkenes

T. A. VAN DER KNAAP, T. C. KLEBACH, F. VISSER, R. LOURENS, and F. BICKELHAUPT

Vakgroep Organische Chemie, Vrije Universiteit, De Boelelaan 1083, 1081 HV Amsterdam, The Netherlands

Three years ago, it was shown that phosphaalkenes (1), which are not substituted at the P=C double bond by hetero atoms and owe their thermal stability mainly to steric hindrance around this bond can be obtained by base-induced elimination of HCl from 2 (1). The predominant role of steric hindrance was obvious from the in-

$$R^1\underset{\displaystyle Cl}{P}\text{-}CHR^2R^2 \xrightarrow[-HCl]{base} R^1P{=}CR^2R^3 \qquad \begin{array}{l} \mathbf{1a}:\ R^1 = Mes,\ R^2 = R^3 = C_6H_5 \\ \mathbf{1b}:\ R^1 = R^2 = R^3 = Mes \end{array}$$

2   1

fluence of the groups R on stability: 1 with R^1 = mesityl (= Mes) or 2,6-dimethylphenyl resulted in stable, isolable compounds, while those with R^1 = 2-methylphenyl or phenyl were too unstable for isolation. For this reason it was expected that trimesitylphosphaethene (1b) would be a particularly stable phosphaalkene. However, in this case the amount of the steric hindrance is apparently so large as to prevent the formation of 1b. Even 2b (R^1=R^2=R^3=mesityl) could not be obtained in the usual way by treatment of 3 with HCl; instead, the interesting phosphonium salt 4 (^{31}P-NMR: δ=31.9 ppm, $^1J_{PH}$ = 574 Hz) was formed, providing, as a spin-off, insight into the mechanism of transformations of the type $R_2PX + HY \rightarrow R_2PY$. The primary product 4 in this case is presumably for steric reasons prevented from further reaction via phosphoranes such as 5 to form 2a.

$$\underset{\mathbf{3}}{Mes\text{-}\underset{\displaystyle NEt_2}{P}\text{-}CHMes_2} \xrightarrow{HCl} \underset{\mathbf{4}}{\left[Mes(H)(NEt_2)\overset{\oplus}{P}\text{-}CHMes_2\right] Cl^{\ominus}} \rightleftarrows \underset{\mathbf{5}}{Mes\text{-}P(Cl)(H)(NEt_2)\text{-}CHMes_2} \not\rightarrow \xrightarrow{-HNEt_2} \mathbf{2b}$$

0097-6156/81/0171-0401$05.00/0

However, 2b could be synthesized by reacting 6 with dimesitylmethyl potassium; elimination of HCl could not be achieved under a variety of conditions, probably due to the inaccessibility of the proton in 2b for base.

$$\underset{6}{MesPCl_2} + KCHMes_2 \longrightarrow 2b \xrightarrow[\times]{-HCl} \underset{1b}{MesP{=}CMes_2}$$

Similarly, attempted elimination of HCl from 7 to furnish 8 was unsuccessful.

$$\underset{7}{Cl_2P\text{-}CHMes_2} \xrightarrow[\times]{-HCl} \underset{8}{ClP{=}CMes_2} \xrightarrow{?} 1b$$

With respect to the structure of phosphaalkenes, considerable progress has been achieved by the X-ray crystal structure determination of 1a (2) and of its $Cr(CO)_5$ complex 9 (3). Both the structural data and the close similarity of spectral data confirm that 1a as such and as a ligand in 9 has essentially the same structure of a planar, non-delocalized phosphaethene (Table 1).

Table 1. Selected spectral and structural data of 1a and 9.

$2,4,6\text{-}(CH_3)_3C_6H_2\text{-}P(X){=}C(C_6H_5)_2$

1a: X = lone pair

9 : X = $Cr(CO)_5$

Compound	NMR: P=C (δ in ppm) ^{31}P	^{13}C	Bond distances (in pm) P=C	P-C	P-Cr	bond angle (in °) C=P-C
1a	233.06	193.37 J_{PC}=43.5Hz	170	184	-	108.7
9(4)	237.3	190.94 J_{PC}=32.3Hz	168	182	236	109.8

The P=C bond lengths of 170 and 168 pm, respectively, are clearly shorter than those of delocalized systems which typically range from 172 to 176 pm. In view of the well-known tendency of tricoordinate phosphines towards small bond angles, it is not surprising to find the C=P-C bond angle in 1a (108.70°) to be more acute than expected for pure sp^2-hybridization, while in the first row analog of 1a, N-mesitylbenzophenone-imine, the C=N-C angle is 120.8° (4). The structural data of 1a imply a relatively high s-character in the lone pair at phosphorus, intermediate between those in tertiary phosphines and λ^3-phosphorins. In line with this interpretation the IR-spectrum (ν_{CO}=1955 cm^{-1}) and the structural data of 9 (Cr-C_{trans}:187 pm, C-O_{trans} 113 pm) show 1a to be a ligand of intermediate basicity and π-acceptor strength.

Finally, a program was initiated to investigate the chemical reactivity of 1a. Earlier experiments had shown (1) that the polarity of the P=C bond in polar reactions is reversed as compared to that of the N=C bond in imines; this is further exemplified by the following reactions:

1a $\xrightarrow{HCl}$ 2a $\left(\xrightarrow{\text{excess HCl}} 6 + CH_2(C_6H_5)_2\right)$ (1)

$\xrightarrow{H_2O}$ (Mes-P(OH)-CH$(C_6H_5)_2$) ——→ Mes-P(=O)(H)-CH$(C_6H_5)_2$

$\xrightarrow{H_2O_2}$ Mes-P(=O)(OH)-CH$(C_6H_5)_2$

$\xrightarrow[CH_3ONa]{CH_3OH}$ Mes-P(OCH3)-CH$(C_6H_5)_2$ (1)

$\xrightarrow[\text{2.) } O_2]{\text{1.) EtOH}}$ Mes-P(=O)(OEt)-CH$(C_6H_5)_2$

$\xrightarrow[\text{2.) } D_2O]{\text{1.) n-BuLi}}$ Mes-P(n-Bu)-CD$(C_6H_5)_2$

Remarkable are the reactions with O_2, S_8, and Br_2, in which formally the P=C bond of 1a is cleaved:

1a $\xrightarrow{O_2}$ polymer + O=C$(C_6H_5)_2$

$\xrightarrow{S_8}$ polymer + S=C$(C_6H_5)_2$

$\xrightarrow{Br_2}$ MesPBr$_2$ + Br$_2$C$(C_6H_5)_2$

Reactions with dienes and carbonyl compounds did either not occur below 150°C (e.g. 2,3-dimethylbutadiene, cyclopentadiene, tetrachloro-α-pyron; acetaldehyde), or were accompanied by decomposition (2,3-dicarbomethoxybutadiene, hexachlorocyclopentadiene, 1,3-diphenylisobenzofuran; acrolein). However, 1,3-dipoles reacted readily to give well-defined addition products:

1a + Mes-C≡N-O ($\oplus$ $\ominus$) $\xrightarrow[25^oC,\ rapid]{C_6H_{12}}$ Mes-C(=N-O)-C(C_6H_5)(C_6H_5)-P(Mes) **10**

1a + $C_6H_5N_3$ $\xrightarrow{80^o}$
- $\xrightarrow[Th]{C_6H_6}$ Mes-P(=N~C_6H_5)=C(C_6H_5)(C_6H_5) **11**
- $\xrightarrow[24h]{CS_2}$ C_6H_5-N(N=N)-C(C_6H_5)(C_6H_5)-P(Mes) **12** + 8% **11**

12 $\xrightarrow{H_2O, O_2}$ Mes-P(=O)(NHC_6H_5)-CH$(C_6H_5)_2$ **13**

1a + $(C_6H_5)_2CN_2$ $\xrightarrow[80^oC,\ 24h]{C_6H_6}$ N=N-C(C_6H_5)(C_6H_5)-C(C_6H_5)(C_6H_5)-P(Mes) **14**

The structure assignments of 10 ($\delta^{31}P$: 9.3 ppm; $\delta^{13}C$: 98.8 ppm ($^1J_{PC}$ = 30 Hz), P-C-O; 160.6 ppm ($^1J_{PC}$ = 45 Hz), P-C=N) and 14 ($\delta^{31}P$: 47.36 ppm; $\delta^{13}C$: 47.75 ppm ($^1J_{PC}$ = 14.7 Hz); 64.06 ppm ($^2J_{PC}$ = 20.5 Hz) follow from the spectral data. In the reaction between 1a and phenyl azide, the strong solvent dependence of both the rate and the product formation is remarkable. In C_6H_6 (as in $CDCl_3$), the reaction is fast, affording the imino-ylid 11 as the sole product ($\delta^{31}P$: 18.8 ppm; $\delta^{13}C$: 68.32 ppm ($^1J_{PC}$ = 166.5 Hz); 68.26 ppm ($^1J_{PC}$ = 166.5 Hz); ca. 1:1 mixture of cis/trans isomers). In CS_2, the reaction is slow; besides 8% 11, the major product is the cycloadduct 12 ($\delta^{31}P$: 140.8 ppm; $\delta^{13}C$: 61.1 ppm ($^1J_{PC}$ = 49 Hz)); the structure of 12 is further confirmed by its conversion to 13. Once formed, 11 is not converted to 12; thus, both compounds are obviously formed by different mechanisms.

Literature Cited

1. Klebach, Th.C.; Lourens, R.; Bickelhaupt, F.; J. Am. Chem. Soc. 1978, 100, 4886.
2. Stam, C.H.; personal communication.
3. Klebach, Th.C.; Lourens, R.; Bickelhaupt, F.; Stam, C.H.; Van Herk, A.; J. Organometal. Chem. 1981, 210, 211.
4. Bokkers, G.; Kroon, J.; Spek, A.L.; Acta Cryst. B 1979, 35, 2351.

RECEIVED July 7, 1981.

Routes to Dicoordinated Phosphorus Compounds

K. ISSLEIB, H. OEHME, H. SCHMIDT, and G.-R. VOLLMER

Department of Chemistry, Martin-Luther-University, Weinbergweg 16, 4022 Halle Saale, FRG

Derivatives of trivalent phosphorus of coordination number two are available from stable or also unisolated intermediates by elimination of small stable molecules XY according to the general

$$X\text{-}C\text{-}P\text{-}Y \rightarrow \;>C=P- \; + \; XY$$

scheme where XY may be hydrogen halide, water, alcohol, alkoxysilane, chlorosilane, ether and elemental hydrogen. Examples of the application of this concept are the various possibilities of formation of 1,3-benzazaphospholes, 1,3-benzthiaphospholes, carbosilylated phosphaalkenes, dialkylaminoalkylidene-phosphines and other compounds of that type.

1,3-Benzazaphospholes

These compounds (1) are synthesized from primary o-aminophenylphosphine and various cyclization reagents, from secondary o-aminophenylphosphines under elimination of an ether and by oxidation or thermal treatment of 1,3-benzazaphospholenes (Scheme 1). In these cyclizations the NH-benzazaphospholes are formed exclusively. This underlines the C=P double bond system in these special cases to be favored over the C=N bond. A PH isomer was never observed.

Scheme 1.

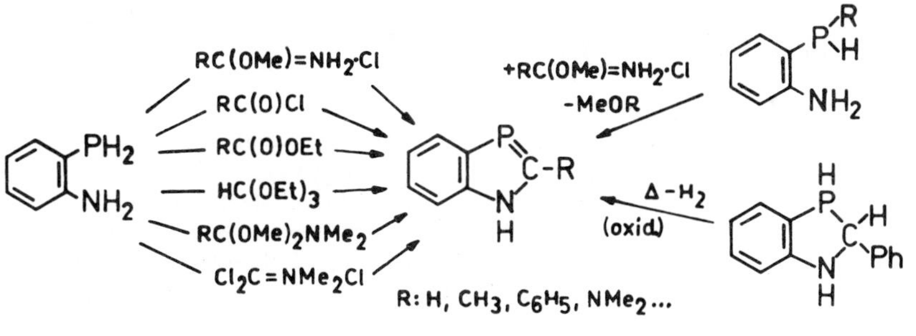

0097–6156/81/0171–0405$05.00/0

1,3-Benzthiaphosphates

o-Mercaptophenylphosphine is indifferent towards imidoester-hydrochlorides or ortho esters. By use of N,N-dimethyl carboxylic acid amide acetals, however, 1,3-benzthiaphospholes (2) could be prepared. The 2-phenyl derivative is made more conveniently by heating a mixture of the phosphine and benzaldehyde to 120°C. This reaction is a further example of an aromatization through elimination of molecular hydrogen (Scheme 2).

Scheme 2.

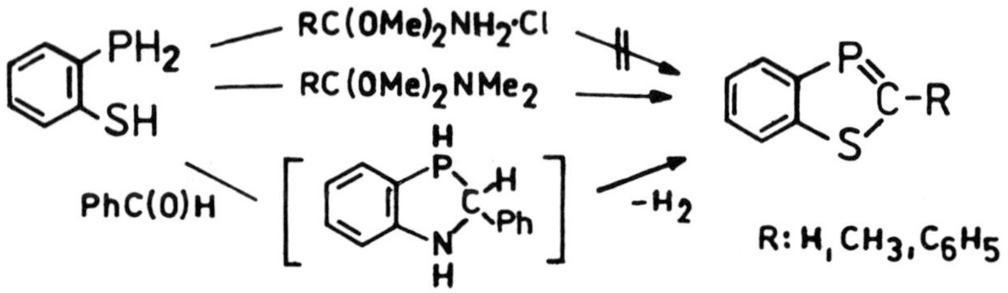

$RC(OMe)_2NH_2{\cdot}Cl$; $RC(OMe)_2NMe_2$; PhC(O)H; $-H_2$; R: H, CH_3, C_6H_5

1,3-benzazaphospholes and 1,3-benzthiaphospholes as well as the 2-t-butyl-1,3-benzoxaphosphole (3) are phosphorus aromatics of extraordinary stability. Their structures have been proved based on their nmr, uv and mass spectra (e.g., ^{31}P shifts at +70 to +80 ppm, ^{13}C shifts (C-2) at +159 to 214 ppm).

The reactivity of the compounds is rather limited. For example, benzazaphospholes are not attacked by dilute aqueous acids and bases, oxygen, sulfur or alkyl halides. The coordinating abilities, metalation and interaction of the so formed ambident anion with acyl halides is demonstrated in Scheme 3 (4).

Scheme 3.

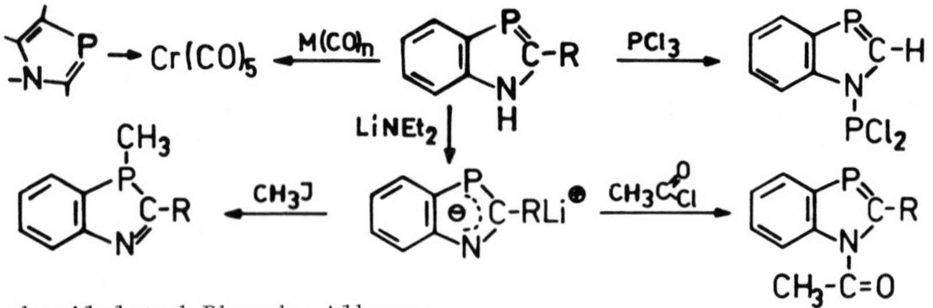

$Cr(CO)_5$; $M(CO)_n$; PCl_3; $LiNEt_2$; CH_3J; CH_3COCl; PCl_2; CH_3-C=O

Carbosilylated Phospha-Alkenes

Continued research on N-silyl-phospha-amidines and N,N-bis-silylated phosphaguanidines (5) led us to carbosilylated phosphaalkenes. The various routes to trimethylsilylmethyl-chlorophosphines and their conversion into carbosilylated phosphaalkenes is given in Scheme 4 (6). Interaction of these products with hydrogen chloride leads to Si_2CH-PRCl and Si-CH_2-PRCl, respectively. By HCl abstraction from these products, the parent phosphaalkene can be regenerated but mono-carbosilylated phosphaalkenes can also be produced. When R is CH_2-Si as a consequence of elimination of either HCl or trimethylchlorosilane, different phosphorus derivatives are obtained. A noteworthy characteristic of carbosilylated phospha-alkenes are their P-nmr signals which appear at extremely low fields (+376 to +438 ppm).

Scheme 4.

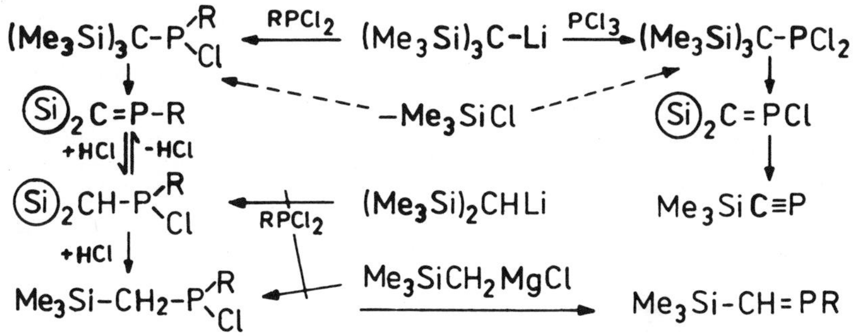

<u>Dialkylamino Alkylidene Phosphines</u>

In the synthesis of specially substituted methylene diphosphines, made from secondary phosphines and carbonyl derivatives (7), a carbenium ion adjacent to trivalent phosphorus as the transition state has been discussed. The transfer of this reaction principle to primary phosphines and suitable carbonyl compound revealed a further pathway to derivatives of dicoordinated phosphorus (8). Aromatic phosphines react with carboxylic acid amide acetals under elimination of alcohol giving dialkylamino-alkylidene phosphines (Scheme 5). A modification of the synthesis

Scheme 5.

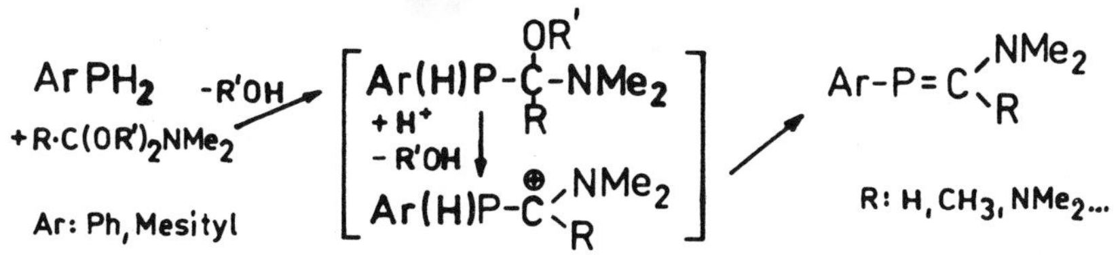

of D passing through the identical unisolated intermediates, is the interaction of alkali metal phosphides (MPHR) with dialkylamino alkoxy carbenium fluoroborates (Scheme 6).

The application of further suitable carbenium salts offers a wide field of synthetic routes to derivatives of dicoordinated phosphorus.

Scheme 6.

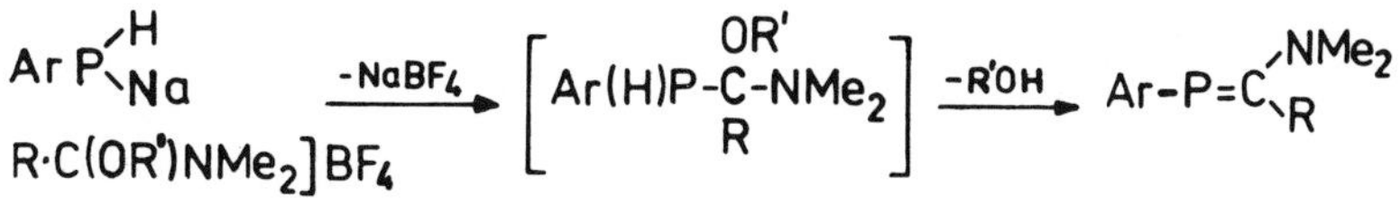

The *o*-phenylendiphosphine which is easily available by reduction of the appropriate benzene phosphonates give similar results on reaction with amide acetales or other cyclization reagents.

Finally *bis*-monometalated o-phenylenediphosphine is the basic starting material for further condensations which are in progress.

Literature Cited

1. Issleib, K.; Vollmer, R.-G.; Oehme, H.; Meyer, H. *Tetrahedron Letters* 1978, 441.
2. Issleib, K.; Vollmer, R.-G. *Tetrahedron Letters* 1980, 3483.
3. Heinicke, J.; Tzschach, A. *Z. Chem.* 1980, *20*, 342.
4. Issleib, K.; Vollmer, R.-G. *Z. Anorg. Allg. Chem.* in press.
5. Issleib, K.; Schmidt, H.; Meyer, H. *J. Organomet. Chem.* 1978, *160*, 47 and 1980, *192*, 33.
6. Issleib, K.; Schmidt, H.; Wirkner, Chr., *Z. Anorg. Allg. Chem.* in press.
7. Oehme, H.; Leissring, E. *Z. Chem.* 1979, *19*, 416.
8. Oehme, H.; Leissring, E.; Meyer, H. *Tetrahedron Letters* 1980, 1141.

RECEIVED June 30, 1981.

Reactions of 2,4,6-Tri(*t*-butyl)phenyllithium with Phosphorus Halides

MASAAKI YOSHIFUJI, ICHIRO SHIMA, and NAOKI INAMOTO

Department of Chemistry, Faculty of Science, The University of Tokyo, Hongo, Tokyo 113, Japan

Sterically hindered compounds have been of interest in view of their protecting nature of reactive sites or molecules (1, 2, 3). During the course of our studies of phosphorus-containing reactive intermediates (4 - 8) we have found the following interesting reactions of 2,4,6-tri-t-butylphenyllithium (I) with some phosphorus halides.

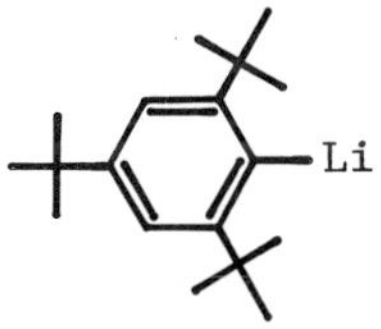

(I)

The reaction of I with $P(O)Cl_3$ gave a very crowded molecule, bis(2,4,6-tri-t-butylphenyl)phosphinic chloride (II, mp 210 - 210.5° C) in 49 % yield, together with 2,4,6-tri-t-butylchlorobenzene (III, 29 % yield) (9), while the reaction of I with $P(S)Cl_3$ gave mainly III in 59 % yield. Compound II has two very bulky 2,4,6-tri-t-butylphenyl groups attached geminally on one element. The chlorine atom of II resists toward the further nucleophilic attack and II is quite insensitive to moisture. The X-ray analysis of II

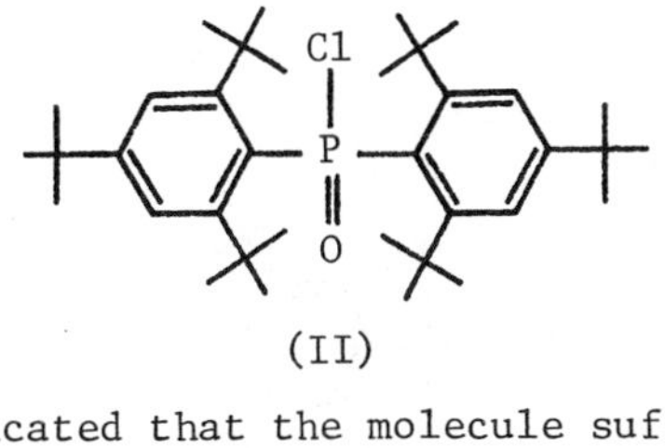

(II) 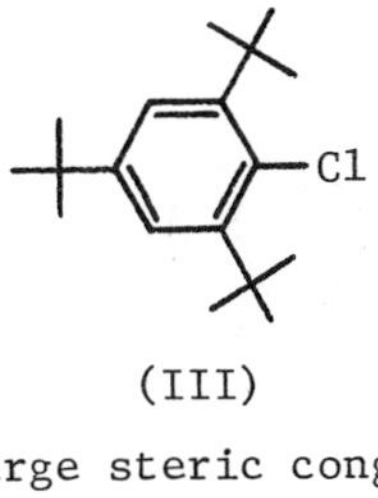 (III)

indicated that the molecule suffers a large steric congestion by the four ortho t-butyl groups to force it to take boat-shaped benzene rings (10). Such distortions have previously found only in the case of bridged benzene rings (11). The *ab initio* calculation

0097-6156/81/0171-0409$05.00/0

of the boat-benzene skeletons of II indicated that the molecule II contains 61.4 kcal/mol higher energy than that with flat benzene rings. More interestingly the *ab initio* calculation of boat benzene suggested that boat-shaped benzene is more stable than envelope or chair-formed benzene, probably due to the overlap of 1,4-electrons (12).

The reaction of I with PCl_3 gave 2,4,6-tri-t-butylphenylphosphonous dichloride (IV) in high yield, whereas the Friedel-Crafts type of reaction of 1,3,5-tri-t-butylbenzene and PCl_3 in the presence of aluminum chloride failed to give IV, but the reaction involving the migration of one of the t-butyl groups to the phosphorus atom afforded t-butyl-3,5-di-t-butylphenylphosphinic chloride (V, mp 132.5 - 133.5°C, 86 %) after hydrolysis (13).

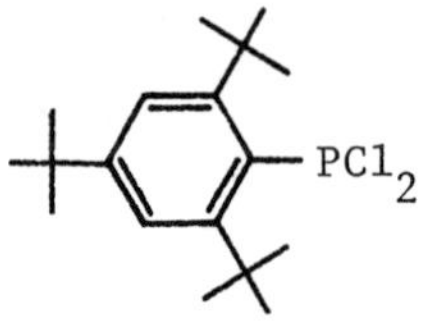

PCl_2 (IV)

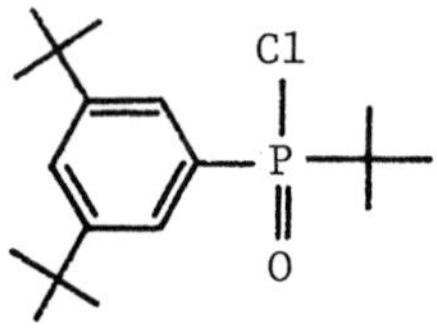

Cl P O (V)

Compound IV in refluxing toluene in the presence of pyridine or $P(S)Cl_3$ gave a cyclic product, 5,7-di-t-butyl-1-chloro-3,3-dimethyl-1-phosphaindan (VI, 64 %) or its 1-sulfide (VII, 68 %) respectively.

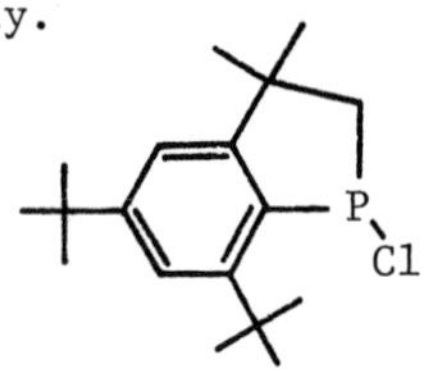

P Cl (VI)

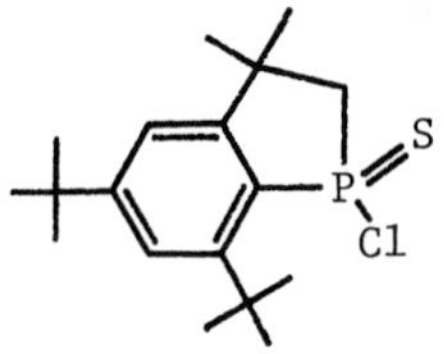

P S Cl (VII)

Compound IV in THF was allowed to react with magnesium to give surprisingly bis(2,4,6-tri-t-butylphenyl)diphosphene (VIII, mp 175 - 176 °C (decompn.)) in 54 % yield as stable orange-red crystals, which might be a dimerization product of 2,4,6-tri-t-butyl-

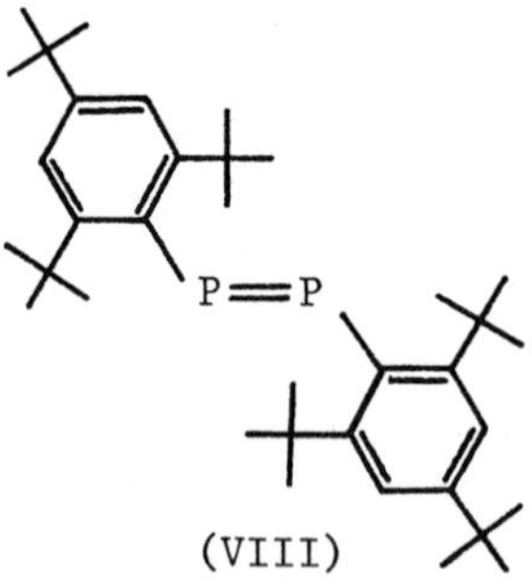

P=P (VIII)

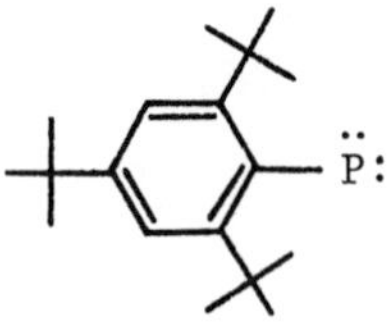

P: (IX)

phenylphosphinidene (IX) as a reactive intermediate. Compound VIII is the first isolated molecule with a localized P=P bond, in which steric hindrance makes an important contribution to the stabilization (14). Diphenyldiphosphene or phosphobenzene was an erroneous structural formula reported in 1877 (15) and about 70 years later it turned out to represent cyclopolyphosphines $(PhP)_n$ ($n = 4,5,6$) by molecular weight determination (16) and X-ray analysis (17, 18). Since then it has been thought that "phosphobenzene" never exists as a stable molecule (19). However, Compound VIII is fairly stable to air and moisture and can be characterized by ^{1}H, ^{13}C, and ^{31}P NMR, mass , UV, and IR spectral analysis and also by molecular weight determination. ^{1}H NMR (CCl_4) δ 7.30 (s, 4H, arom.), 1.45 (s, 36H, *o*-t-Bu), 1.35 (s, 18H, *p*-t-Bu); ^{31}P NMR (C_6D_6) δ -59.0 ppm (from external 85% H_3PO_4); mass spectrum *m/e* calcd: 552.4012; found; 552.4012; UV (CH_2Cl_2) λmax (ε) 284 (15660), 340 (7690), 460 (1360) nm; molecular weight (vapor phase osmometric): 552.1 (in C_6H_6). Anal. ($C_{36}H_{58}P_2$) C, H. The X-ray crystallographic analysis of VIII indicated that the structure is similar to that of anti-azobenzene and that the bond distance between the two phosphorus atoms in VIII (2.034(2) Å) is shorter than the average value observed in cyclopolyphosphines (14).

Compound VIII gave 2,4,6-tri-t-butylphenylphosphonic chloride (X) by the reaction with chlorine in carbon tetrachloride followed by hydrolysis in almost quantitative yield.

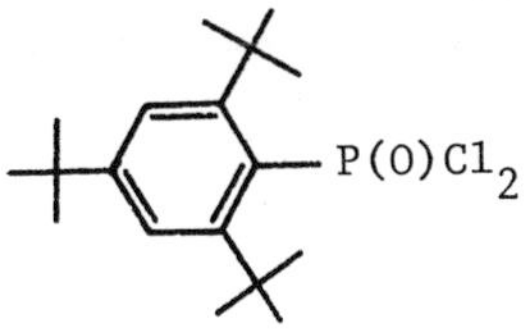

$P(O)Cl_2$

(X)

Such sterically hindered and unusual phosphorus compounds described here might serve as good model compounds for the theoretical studies as well as studies in organic or inorganic chemistry.

Acknowledgments

This work was supported in part by the Scientific Research Grant of the Ministry of Education, Science, and Culture of Japan (Nos. 384029, 464163, and 543008).

Literature Cited

1. Klebach, Th. C.; Lourens, R.; Bickelhaupt, F. J. Am. Chem. Soc. 1978, 100, 4886.
2. Baudler, M.; Saykowski, F. Z. Naturforsch. 1978, 33b, 1208.
3. Dimroth, K.; Mach, W. Angew. Chem. Internat. Ed. Engl. 1968, 7, 460.

4. Yoshifuji, M.; Nakayama, S.; Okazaki, R.; Inamoto, N. J. Chem. Soc. Perkin I 1973, 2065.
5. Nakayama, S.; Yoshifuji, M.; Okazaki, R.; Inamoto, N. J. Chem. Soc. Perkin I 1973, 2069.
6. Nakayama, S.; Yoshifuji, M. ; Okazaki, R.; Inamoto, N. Bull. Chem. Soc. Jpn. 1975, 48, 546.
7. Nakayama, S.; Yoshifuji, M.; Okazaki, R.; Inamoto, N. Bull. Chem. Soc. Jpn. 1975, 48, 3733.
8. Nakayama, S.; Yoshifuji, M.; Okazaki, R.; Inamoto, N. Bull. Chem. Soc. Jpn. 1976, 49, 1173.
9. Yoshifuji, M.; Shima, I.; Inamoto, N. Tetrahedron Lett. 1979, 3963.
10. Yoshifuji, M.; Shima, I.; Inamoto, N.; Hirotsu, K.; Higuchi, T. Angew. Chem. Internat. Ed. Engl. 1980, 19, 399.
11. Coulter, C. L.; Trueblood, K. N. Acta Crystallog. 1963, 16, 667.
12. Yoshifuji, M.; Shima, I.; Inamoto, N.; Aoyama, T. Tetrahedron Lett. in press.
13. Yoshifuji, M.; Okazaki, R.; Inamoto, N. J. Chem. Soc. Perkin I 1972, 559.
14. Yoshifuji, M.; Shima, I.; Inamoto, N.; Hirotsu, K.; Higuchi, T. to be submitted.
15. Köhler, H.; Michaelis, A. Ber. dtsch. chem. Ges. 1877, 10, 807.
16. Kuchen, W.; Grünewald, W. Chem. Ber. 1965, 98, 480.
17. Daly, J. J.; Maier, L. Nature 1964, 203, 1167.
18. Daly, J. J.; Maier, L. Nature 1965, 208, 383.
19. Emsley, J.; Hall, D. "The Chemistry of Phosphorus": Harper & Row, Publishers: London, 1976, p. 461.

RECEIVED July 7, 1981.

Synthesis of New Dicoordinated Phosphorus Compounds with a P(III)=N Bond

C. MALAVAUD, L. LOPEZ, T. N'GANDO M'PONDO, M. T. BOISDON, Y. CHARBONNEL, and J. BARRANS

ERA CNRS N° 926, Laboratoire des Hétérocycles du Phosphore et de l'Azote, Université Paul Sabatier, 118 Route de Narbonne, 31 062 Toulouse Cédex, France

In recent years, more and more varied dicoordinated phosphorus compounds have been isolated ; this has meant the appearance of a new and very interesting branch of phosphorus chemistry.

Although the number of dicoordinated phosphorus compounds with a $-P_{III}=C$ bond is fairly large, compounds with a $-P_{III}=N$ bond are still few in number since only one heterocycle, namely 1,2,4,3-triazaphosphole (1), and two types of acyclic compounds, namely the diazaphosphapropenes (2) and the tetrazaphosphapentenes (3), are known :

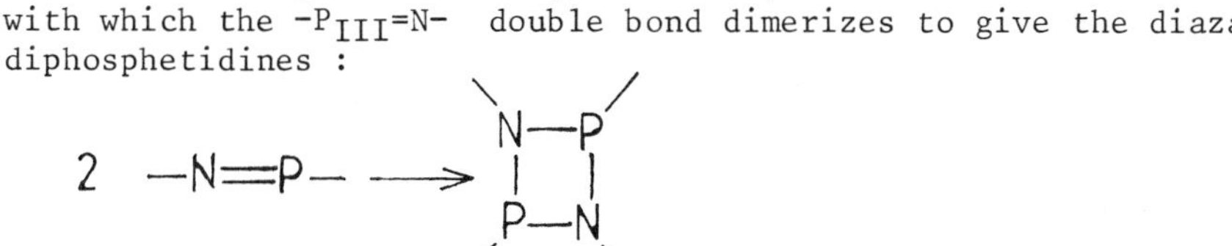

The rarity of these compounds, in comparison with the variety of usable nitrogen-containing reactants, seems due to the ease with which the $-P_{III}=N-$ double bond dimerizes to give the diazadiphosphetidines :

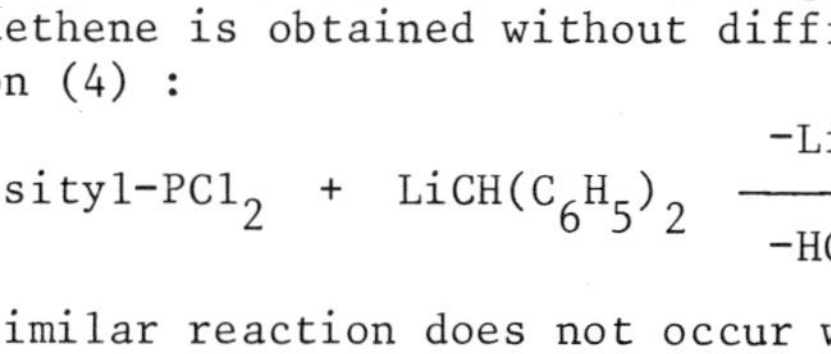

It is for this reason that although 2,2-diphenyl-1- mesitylphosphaethene is obtained without difficulty from the following reaction (4) :

$$\text{Mesityl-PCl}_2 + \text{LiCH}(C_6H_5)_2 \xrightarrow[-\text{HCl}]{-\text{LiCl}} \text{Mesityl-P=C}(C_6H_5)_2$$

a similar reaction does not occur when the diphenylmethane is replaced by heavily substituted amines. With t-Butylamine, for example, an analogous reaction, at room temperature gives diazadiphosphetidine.

0097-6156/81/0171-0413$05.00/0

Many other aliphatic or aromatic amines produce the same reaction. However if N-alkyl (or aryl)2-aniline is made to react with tris-dimethylaminophosphine, tricoordinated phosphorus compounds are obtained. Their analyses and mass spectra show that they are tetramers (5) : when the energy of the ionizing electron beam is nearly 70 eV the mass spectra of the oligomers show an intense $[M]^+$ peak which corresponds to the monomer ion (n = 1) ; but when the spectra are obtained by field desorption m.s. the molecular ion peak corresponds to a tetramer (n = 4) ; one sees also smaller peaks for ions n = 3, 2 or 1.

These oligomers, when heated in various solvents are in equilibrium with the monomer : a dicoordinated phosphorus compound:

[Scheme: oligomer (benzodiazaphosphole, N-R)$_n$ ⇌ (Δ) n monomer (dicoordinated N=P benzodiazaphosphole, N-R)]

The percentage of dicoordinated phosphorus compound present depends equally on the substituent R, on the solvent and on the temperature. It has been possible to prepare these oligomers from monosubstituted aliphatic or aromatic diamines, from amidrazones, from semicarbazides, from substituted thiosemicarbazides, from 2-aminophenol, from 2-aminothiophenol and even from the monoimines of orthophenylediamine (Table I). It seems that these oligomers are generally the thermodynamic products of the reaction, except perhaps for the amidrazones, and the 1-phenyl semicarbazide in which case they would be kinetic products (the thermodynamic products being the triazaphospholes).

For instance, when the following amidrazone iPr-C(NH_2)=N-NHMe reacts with $P(NMe_2)_3$ in refluxing benzene, one obtains only the corresponding 1,2,4,3-triazaphosphole, but if the reaction is carried out in refluxing toluene with $P(NEt_2)_3$ one obtains 20% of an oligomer ($\delta^{31}P$ = 79). This oligomer depolymerised with BF_3 gives the known complexed triazaphosphole.

If these oligomers are made to react with Lewis acids (BF_3 , $AlCl_3$, $ZnCl_2$ etc...) the dicoordinated phosphorus compound is obtained quantitatively :

[Scheme: $[R^1\{N,X\}P]_4$ —L.A.→ $4R^1\{N(\to L.A.),X\}P$ —base→ $R^1\{N,X\}P$ —Δt→ oligomer]

These compounds react with the amines to give the corresponding dicoordinated phosphorus compounds, a number of which may remain in the free state for a few hours before once again oligomerizing. Particular notice should be given to the synthesis of the diazaphospholes

TABLE I

$\left(C_6H_4\langle{}^{X}_{N}\rangle P-\right)_4$	OLIGOMER $\delta^{31}P$	OLIGOMER F °C	P_{II} COMPLEXE-BF_3 $\delta^{31}P$	P_{II} COMPLEXE-BF_3 F °C	FREE P_{II} $\delta^{31}P$
X					
NH	80	75	222		238
N(sBu)	81	218	226,7	104	232
NPr	86	194	226	147	
N(iPr)	84	255	225,6	132	232
N Benzyl	86	204	224,8	122	
NC_6H_5			205,9		
S	92	135	160	191	
N=CHØ	104	95	236		
N=CHØCl	102		268		
O			262 ($AlCl_3$)		
$\left(CH_2CH_2\langle{}^{N(R)}_{N(R)}\rangle P-\right)_4$					
H					
Pr	99,5	135	285	78	
iPr	99	148	284		
X=C(NHØ)(HN-NH$_2$) O	85	155	232		
X=C(NHØ)(HN-NH$_2$) S	80	190	248		
O=C(NH$_2$)(HN-NHØ)	94	175	229		242
iPr-C(=NH)(N(Me)-NH$_2$)	79		246		258

$\delta^{31}P$ (réf. : 85% H_3PO_4 aq.sol.) and melting points of some oligomers and their complexes with BF_3 (6,7)

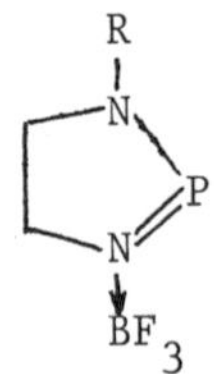

which are not stabilized either by electron delocalization, or by steric hindrance : the basicity of the unsubstituted nitrogen atom is simply suppressed by complexation.

This then is a method of synthetizing new various dicoordinated phosphorus compounds, such as 5-oxo -1,2,4,3-triazaphosphole derivatives, by depolymerizing complexation ; its further study in now under way.

Of course these compounds react in the same way as the dicoordinated phosphorus compounds with P_{III}=N bond previously obtained in the laboratory ; in particular they give addition reactions with alcohols and amines, and spirophosphoranes with diols or amines while the oligomer is degraded by these reactants (6).

Another interesting reaction is the action of imines of amidrazones with hexamethyl phosphorus triamide : we have obtained the three isomers of the monosubstituted 1,2,4,3-triazaphosphole (8) :

$$P(NMe_2)_3 + R\text{-}C(NH_2)\text{=N-N=}CHR^1 \longrightarrow \text{(R, H, N, N, N, P ring)}$$

the study of their chemical properties are in progress.

It is known that dicoordinated phosphorus compounds can be obtained by the action of Lewis bases, such as triethylamine or DBU, on chlorophosphines :

$$-\underset{Cl}{P}-\underset{H}{C}- \xrightarrow[-\ HCl]{Base} -P=C-$$

This dehydrohalogenation can also be obtained by the action of a Lewis (7) : for example, BF_3 reacts with 2-chloro-4,5-benzo-1,3,2-diazaphospholane according to the reaction scheme :

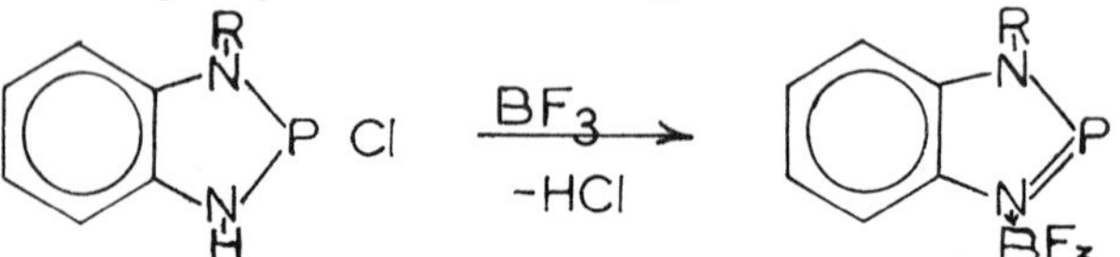

The benzodiazaphosphole is obtained, which was previously prepared by depolymerization of the oligomer. Analogous results are obtained in the case of 2-chloro-4,5 -benzo 1,3,2-oxazaphospholane but a stronger Lewis acid such as $AlCl_3$ must be used.

Litterature cited :

(1) Y. Charbonnel; J. Barrans ; Tetrahedron, 1976, 32, 2039
(2) E. Niecke ; W. Flick ; Angew Chem., 1973, 85, 586.
(3) O. Scherer ; G. Schnabl ; Chem. Ber., 1974, 107, 2123.
(4) Th.C. Klebach ; R. Lourens ; F.Bickelhaupt ; J. Am. Chem.Soc., 1978, 100, (15) 4886.
(5) C. Malavaud ; M.T. Boisdon ; Y. Charbonnel ; J. Barrans ; Tetrahedron Letters, 1979, 5, 447.
(6) C. Malavaud ; Thèse d'Etat, 1980, Toulouse;
(7) Th. N'Gando M'Pondo ; Thèse d'Etat, 1981, Toulouse.
(8) A. Schmidpeter ; H. Tautz ; Z. Naturforsch. 1980, 35b, 1222-8.

RECEIVED June 30, 1981.

COMPOUNDS WITH PENTACOORDINATED AND HEXACOORDINATED PHOSPHORUS

Cyano Anions of Dicoordinated, Tricoordinated, Tetracoordinated, Pentacoordinated, and Hexacoordinated Phosphorus

ALFRED SCHMIDPETER and FRANZ ZWASCHKA

Institut für Anorganische Chemie der Universität München, Meiserstrasse 1, D-8000 München 2, FRG

WILLIAM S. SHELDRICK

GBF, Mascheroder Weg 1, D-3300 Braunschweig-Stöckheim, FRG

Cyanoligands may change the properties of the center which they are attached to quite definitely. In particular they reduce its basicity and participate in an eventual anionic charge. This is illustrated by the pseudohalide ions $E(CN)_n^-$, n = (7 - group number of E), in accord with the "cyano displacement principle". They are found in the first short period for n = 1 to 3 and also in the second short period for n = 1:

$$\begin{array}{cccc} C(CN)_3^- & N(CN)_2^- & OCN^- & F^- \\ & \mathbf{P(CN)_2^-} & SCN^- & Cl^- \end{array}$$

The dicyanophosphide ion which continues this series has now been synthesized by reacting phosphorus tricyanide with dialkylphosphites (R = CH_3, C_2H_5):

$$P(CN)_3 + HPO(OR)_2 + Et_3N \rightarrow Et_3NH^+P(CN)_2^- + NCPO(OR)_2$$

$$P(CN)_3 + Na,KPO(OR)_2 \rightarrow Na,K^+P(CN)_2^- + NCPO(OR)_2$$

While however neither its triethylammonium salt nor its alkali salts are stable in solution, the salts of crown ether-alkali complexes are stable as such and in solution.

In the reaction which we have found, phosphorus tricyanide behaves as an interpseudohalogen, $(NC)_2P$-CN, with "positive cyanogen" which is taken over nucleophilically by the dialkylphosphite anion. With reverse polarity however, phosphorus tricyanide reacts with the dialkylphosphite in the absence of a base:

$$P(CN)_3 + HPO(OR)_2 \rightarrow HCN + (NC)_2P\text{-}PO(OR)_2$$

The dicyanophosphide ion is bent, as is the isoelectronic molecule $S(CN)_2$. A transfer of charge into the cyano groups is expected from charge density calculation and is demonstrated by N-coordination to the cation in the crystal (Figure 1). It is also in accord with the observed wider CPC-angle (95°) and shorter PC-distance (170 pm) and the lower CN and higher PC stretching fre-

0097–6156/81/0171–0419$05.00/0

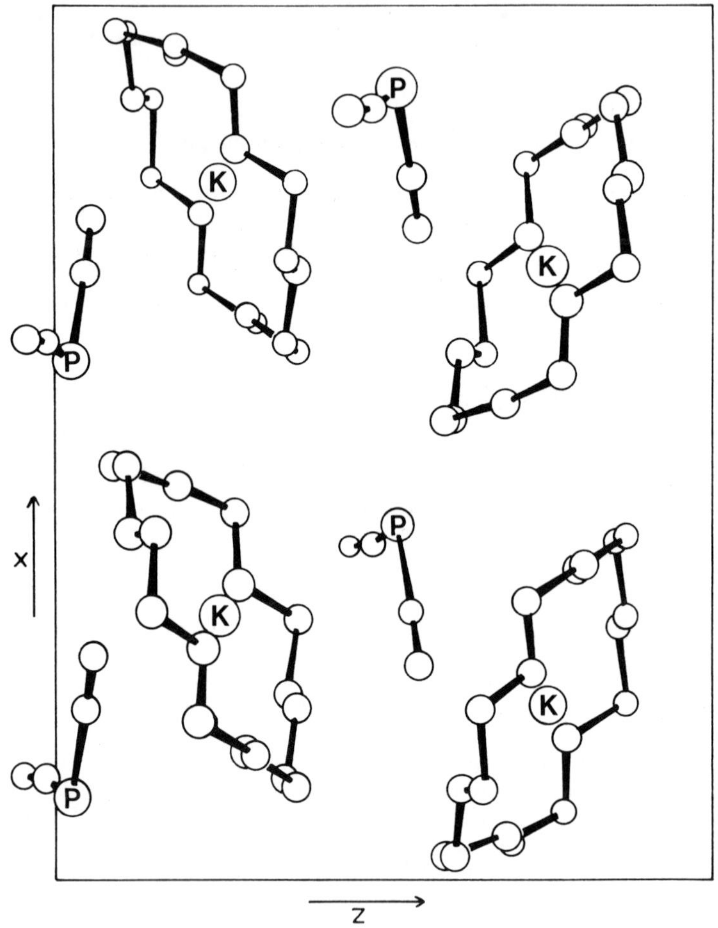

Figure 1. [18]Crown-6-KP(CN)₂.

quencies as compared with those of $P(CN)_3$ and $CH_3P(CN)_2$. The dicyanophosphide ion could thus react ambivalently, but the nucleophilicity of phosphorus outweighs by far that of nitrogen.

$P(CN)_2^-$ may be used to introduce this group by nucleophilic substitution. This reaction has been used e.g. to prepare the dicyanophosphino derivatives of heterocyclic cations, which in turn may be reduced by phosphite again. The resulting compounds are benzimidazolones and quinolones in which the PCN group as a novel pseudochalcogen takes the place of oxygen.

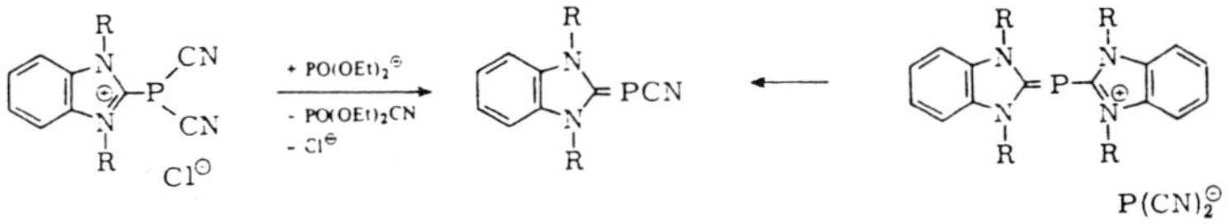

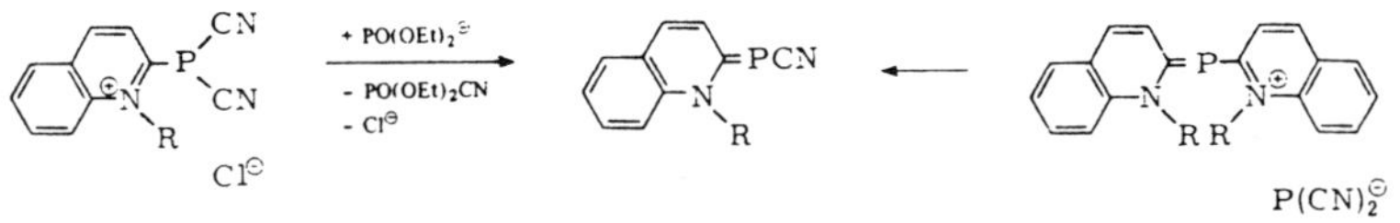

They occupy a connecting position between the historic phosphamethinecyanine cations synthesized by Dimroth as the first -P=C- compounds, and the dicyanophosphide anion (and they show indeed intermediate spectroscopic and structural data, Figure 2). In a metathetic reaction these ions afford salts which have the same composition as the above compounds. They are stable as crystals but surprisingly rearrange in solution even at room temperature e.g. on attempted recrystallization. The exchange could proceed via a diphosphetane betaine.

A typical reaction of halide ions is their combination with a halogen molecule X_2 to give the linear trihalide ions X_3^-. In accord with its postulated pseudohalide character, the dicyanophosphide ion adds bromine and iodine at room temperature to give anions of X_3^- type with $P(CN)_2$ as central member. The crown ether-sodium salts of these hypervalent anions, dicyanodihalophosphates (III), can be isolated in crystalline form.

$$P(CN)_2^- + X_2 \rightarrow P(CN)_2X_2^- \quad (\rightarrow P(CN)_2X + X^-)$$

Surprisingly the reaction does not lead just to the dicyanohalophosphanes (which is the case for X = Cl). The ions with X = Br or I do not dissociate in solution as is apparent from the characteristic upfield shift of the ^{31}P-NMR signal (-165, -172).

Addition of cyanogen bromide or iodide instead of halogen gives the tricyanohalophosphates(III), X = Br, I.

$$P(CN)_2^- + XCN \rightarrow P(CN)_3X^- \leftarrow P(CN)_3 + X^-$$

In accord with their stability mobile equilibria lying far to the side of the products are established from phosphorus tricyanide and crown ether sodium chloride, bromide or iodide, from which the phosphates(III) can be isolated as crystals.

In spite of the interest in hypervalent phosphorus(III) compounds as substitution intermediates, little is known about their geometry. It is proposed to be ψ-trigonal bipyramidal with the lone pair equatorial and the entering nucleophile and the leaving group opposite to each other in apical positions. No structure determinations of a hypervalent, four-coordinate phosphorus(III) species has so far been done, however, to provide a model for this transition-state geometry. We can now present the structures of the anions $P(CN)_2Br_2^-$ (Figure 3), $P(CN)_3Cl^-$ (Figure 4), $P(CN)_3Br^-$ (Figure 5), and $P(CN)_3I^-$. While the former two show the expected ψ-trigonal bipyramidal coordination with cyano ligands in the equatorial positions and an almost linear Br-P-Br and Cl-P-CN arrangement, respectively, the latter two are dimeric, showing an unexpected ψ-octahedral coordination of P(III).

The tetracyanophosphate(III) would be the concluding member in the series $P(CN)_nX_{4-n}^-$. It forms indeed from phosphorus tricyanide and a phosphonium cyanide, but decomposes above -10^oC:

$$3\ P(CN)_3 + 3\ CN^- \rightarrow 3\ P(CN)_4^- \rightarrow P(CN)_2^- + P_2C_{10}N_{10}^{2-}$$

The two salts formed in this decomposition can be isolated and separated. A structure determination shows the dianion (Figure 6) to consist of a three-coordinate dicyano- and a six-coordinate pentacyano-phosphorus connected by a C_3N_3 unit. Whether carbon or nitrogen is the bridging atom therein is still open to discussion.

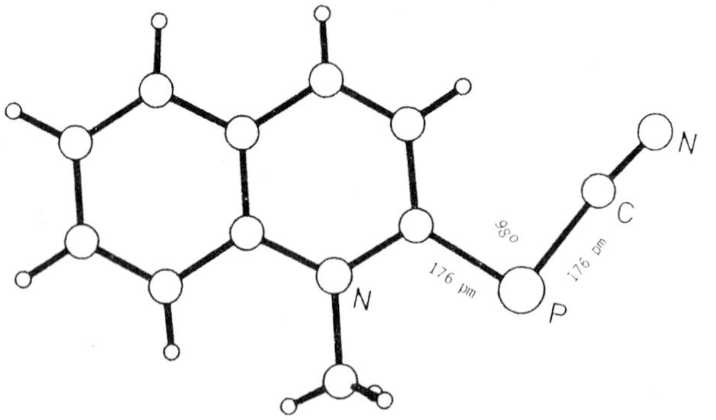

*Figure 2. Molecular structure of cyanophosphinidene-*N*-methylquinolone.*

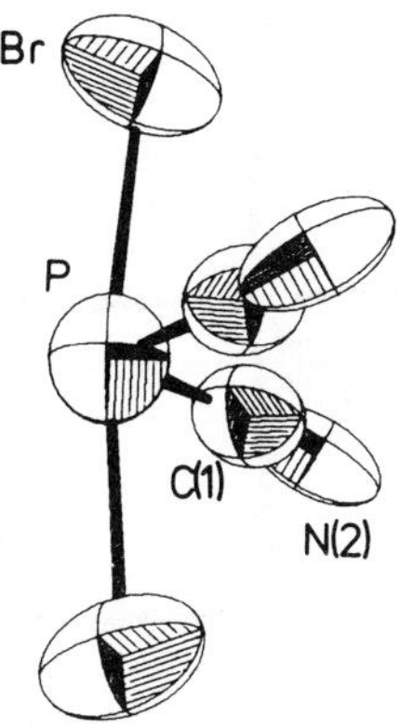

Figure 3. Molecular structure of the anion $P(CN)_2Br_2^-$.

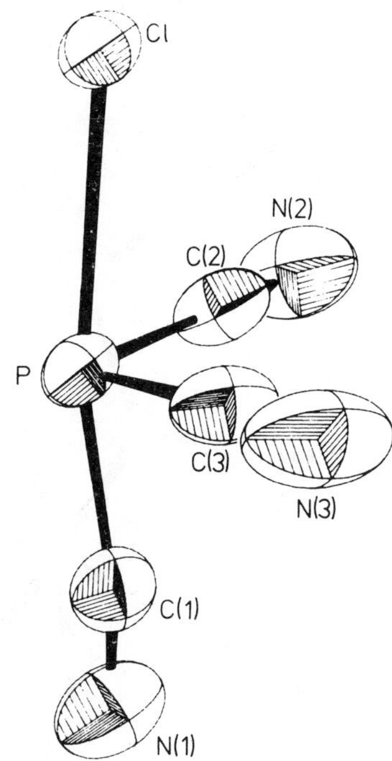

Figure 4. Molecular structure of the anion $P(CN)_3Cl^-$.

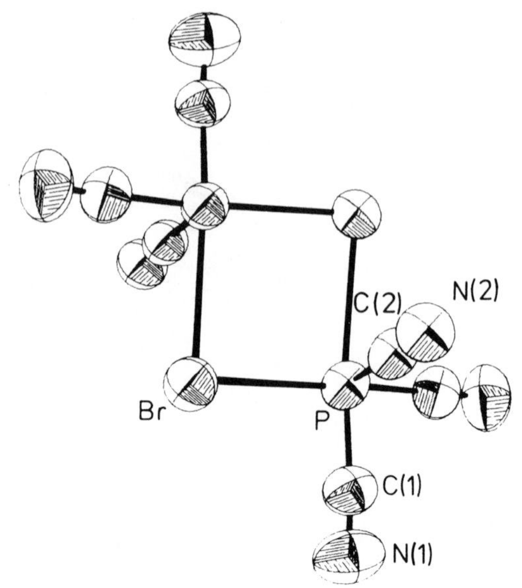

Figure 5. Molecular structure of the anion $[P(CN)_3Br^-]_2$.

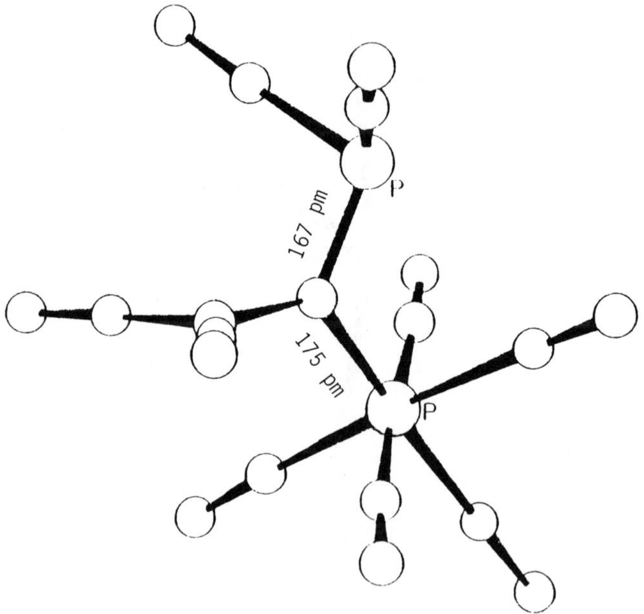

Figure 6. Molecular structure of the anion $P_2C_{10}N_{10}^{2-}$.

RECEIVED June 30, 1981.

87

Preparation, Reactions, and Structures of Some *N,N′*-Dimethylurea-Bridged Phosphorus Compounds

NORBERT WEFERLING and REINHARD SCHMUTZLER

Lehrstuhl B für Anorganische Chemie der Technischen Universität, Pockelsstrasse 4, D-3300 Braunschweig, FRG

As against a wealth of compounds involving directly bonded λ^3P phosphorus atoms there is a relative scarcity of compounds with phosphorus atoms of different oxidation number and/or coordination number directly linked together[1,2].

This is especially true of the $\lambda^5P\lambda^3P$ linkage of which Roesky and his co-workers have reported a few examples only as recently as 1979/1980 [3,4]. Likewise, a first example of a compound with λ^5P-λ^5P directly bonded has been reported by Richman and Holmes only in 1980 [5]. We have undertaken attempts at the synthesis of bifunctional phosphorus compounds involving both P-P bonded systems as well as with phosphorus functions separated by other groups, especially the N,N'-dimethylurea group, MeNC(:O)NMe [4-6]. The bulk of the synthesis work was based on oxidative addition reactions with either hexafluoroacetone (HFA; 1) or tetrachloro-o-benzoquinone (TOB; 2) at the two novel P(III) compounds, Cl(Me)PN(Me)-C(:O)N(Me)P(Me)Cl, 3, and the diphosphine 4 which were obtained according to eq. (1) and (2):

$$Me(Me_3Si)NC(:O)N(SiMe_3)Me \ + \ 2\ MePCl_2 \longrightarrow$$
$$\longrightarrow Cl(Me)PN(Me)C(:O)N(Me)P(Me)Cl \ + \ 2\ Me_3SiCl \qquad (1)$$
$$3\ ;\ \delta_P\ +119$$

$$3 + \begin{cases} \xrightarrow{Bu^n_3P} [Bu^n_3PCl]^+Cl^- \\ \xrightarrow[Me_3SnH]{NEt_3} \left.\begin{matrix}[HNEt_3]^+Cl^- \\ Me_3SnCl\end{matrix}\right\} \end{cases} + \ \text{cyclo-}[MeN\text{-}C(=O)\text{-}NMe\text{-}P(Me)\text{-}P(Me)] \qquad (2)$$
$$4\ ;\ \delta_P\ +20.1$$

Attempted oxidative addition of 1 at 3 which was expected to produce a N,N'-dimethylurea-bridged diphosphorane, led to an unusual reaction and product 5 with simultaneous elimination of one

0097–6156/81/0171–0425$05.00/0

phosphorus as $MePCl_2$ (eq. 3)

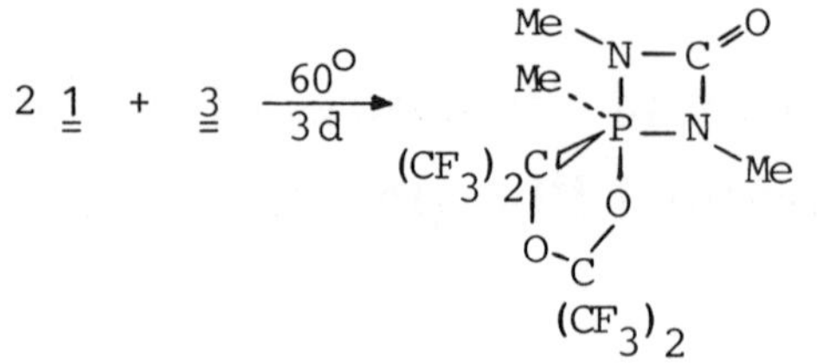

$$2\ \underline{\underline{1}} + \underline{\underline{3}} \xrightarrow[3\,d]{60^{\circ}} \underline{\underline{5}} + MePCl_2 \qquad (3)$$

5 ; δ_P -32.9

The reaction may proceed by oxidative addition of 2 equivalents of 1 at a λ^3P atom of 3, producing a monophosphorane intermediate in which attack at the λ^5P atom by the nitrogen atom γ to P, followed by nucleophilic attack of Cl^- at λ^3P may occur (eq. 4)

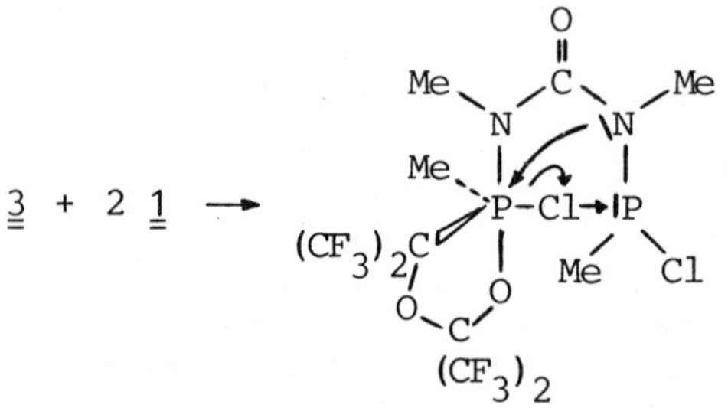

$$\underline{\underline{3}} + 2\ \underline{\underline{1}} \longrightarrow [\text{intermediate}] \longrightarrow MePCl_2 + \underline{\underline{5}} \qquad (4)$$

Alternatively, oxidative addition of 1 may occur at a λ^3P atom of 1,3-dimethyl-2-methyl-1,3,2λ^3-phosphadiazetidin-4-one, 6 resulting from the thermolysis of 3 in accord with eq. (5) for which we have obtained ^{31}P n.m.r. evidence,

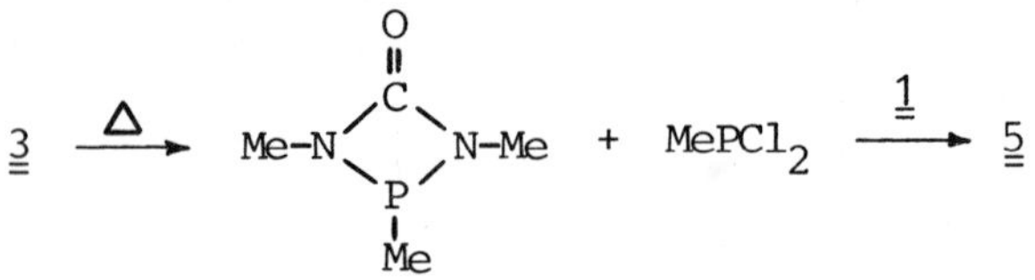

$$\underline{\underline{3}} \xrightarrow{\Delta} \text{Me-N(C=O)N-Me(P-Me)} + MePCl_2 \xrightarrow{\underline{\underline{1}}} \underline{\underline{5}} \qquad (5)$$

This latter possibility, however, is thought to be less likely, in view of observations on the reaction of 3 with two equivalents of TOB, 2 which occurs smoothly at temperatures as low as 0° C, producing the novel spirophosphorane 6, closely analogous to 5.

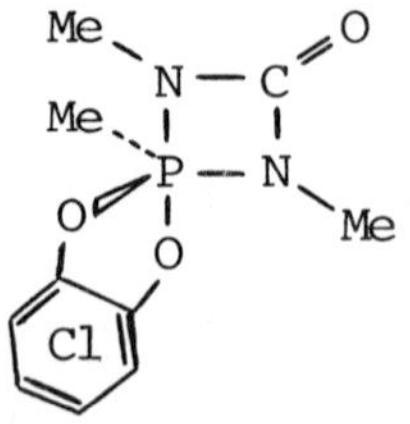

6 ; δ_P -24.4

The formation of 5 is noteworthy also, in as much as in the reactions of P(III) compounds with HFA, carbon-carbon condensation of the latter with formation of the 1,3,2λ^5 dioxaphospholane system is normally observed [8]. Formation of the 1,4,3λ^5 dioxaphospholane ring system as in 5 has rarely been reported; furthermore, products of this type are known to undergo decomposition into the starting compounds [9,10] or, in the presence of hydrogen on the carbon atom α to λ^5-P, to rearrange directly into 1,2λ^5-oxaphosphetanes [11,12]. We have observed by ^{19}F and ^{31}P n.m.r. that 5 is unchanged over the entire temperature region investigated (room temperature to 160°C). The structure of 5 has been established by a single crystal X-ray diffraction study (W.S. Sheldrick); the geometry at phosphorus is relatively close to trigonal-bipyramidal.

Reaction of 4 with four equivalents of 1 was found to produce a complicated mixture of products; no evidence for oxidative addition of 1 at both λ^3P atoms of 4 with retention of the P-P bond was found. When TOB was employed, however, a clean reaction in accord with eqn. (6) was observed.

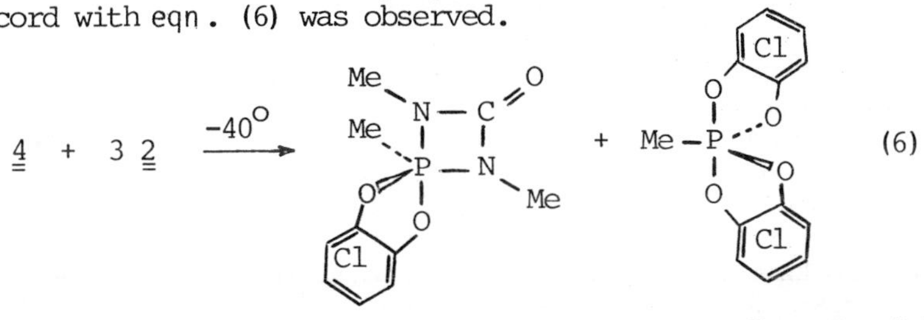

$$\underline{4} + 3\,\underline{2} \xrightarrow{-40^{\circ}} \underline{6} + \underline{7} \qquad (6)$$

6 ; 7 ; δ_P -8.5

A stable intermediate λ^5Pλ^3 bonded compound, 8 is formed when 2 and 4 are allowed to react in an exact 1:1 molar ratio in ether between -40 to 0°C. Elimination of methylphosphinidene, MeP (undergoing oxidative addition with 2) apparently is preferred to reaction of 8 with further 2 to give a λ^5Pλ^5P bonded diphosphorane. The λ^3P atom in 8 does, however, exhibit reactions typical of P(III), with retention of the P-P bond, such as shown below

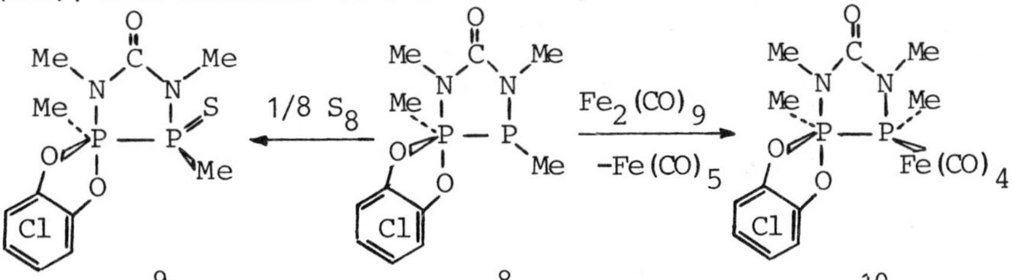

$$\underline{9} \xleftarrow{1/8\ S_8} \underline{8} \xrightarrow[-Fe(CO)_5]{Fe_2(CO)_9} \underline{10}$$

9	8	10
$\delta(\lambda^5P)$ -16.5	$\delta(\lambda^3/\lambda^5P)$ +3.5	$\delta(\lambda^5P)$ -8.4
$\delta(\lambda^4P)$ +54.9		$\delta(\lambda^4P)$ 108.4
$^1J(PP)$ 232		$^1J(PP)$ 46

Single crystal X-ray diffraction studies of both 9 and 10 (J.W. Gilje; D. Schomburg; W.S. Sheldrick) have confirmed that the P-P bond in these compounds has remained intact. Both compounds represent a novel structural feature, $\lambda^5P\lambda^4P$, in phosphorus chemistry. Compound 10, in addition, is noteworthy as it combines both a nonmetallic and a metallic five-coordinate center in the same molecule.

Literature cited

1. L. Maier, in "Organic Phosphorus Compounds" (eds. G.M. Kosolapoff and L. Maier); Wiley-Interscience, New York, London, Sydney, Toronto 1972; Vol. 1, pp. 1, 289. I.F. Lutsenko and M.V. Proskurnina, Uspekh. Khim., 47, 1648 (1978).
2. A.H. Cowley (Editor), Compounds containing Phosphorus-Phosphorus Bonds; Dowden, Hutchinson + Ross, Inc., Stroudsburg, Pa., 1973.
3. H.W. Roesky, K. Ambrosius, and W.S. Sheldrick, Chem. Ber., 1979, 112, 1365.
4. H.W. Roesky, K. Ambrosius, M. Banek, and W.S. Sheldrick, Chem. Ber., 1980, 113, 1847.
5. J.E. Richman and R.R. Holmes, J. Am. Chem. Soc., 1980, 102, 3955.
6. R.E. Dunmur and R. Schmutzler, J. Chem. Soc. (A), 1971, 1289; R.K. Harris, J.R. Woplin, R.E. Dunmur, M. Murray, and R. Schmutzler, Ber. Bunsenges. Phys. Chem., 1972, 76, 44; O. Schlak, R. Schmutzler, R.K. Harris, E.M. McVicker, and M.I.M. Wazeer, Phosphorus and Sulfur, in press.
7. ^{31}P N.m.r. shifts (ext. 85% H_3PO_4 ref.) to high field are given with a negative sign.
8. E.g. F. Ramirez, C.P. Smith, J.F. Pilot, and A.S. Gulati, J. Org. Chem., 1968, 33, 3387.
9. V.N. Volkovitskii, I.L. Knunyants, and E.G. Bykhovskaya, Zh. Vses. Khim. Obshch., 1973, 18, 112 (C.A. 78, 148035w (1973)).
10. R.K. Oram and S. Trippett, J. Chem. Soc. Perkin I, 1973, 1300.
11. F. Ramirez, C.P. Smith, and J.F. Pilot, J. Am. Chem. Soc., 1968, 90, 6726.
12. J.A. Gibson, G.V. Röschenthaler, K. Sauerbrey, and R. Schmutzler, Chem. Ber., 1977, 110, 3214.

RECEIVED June 30, 1981.

A Stable Monocyclic Triarylalkoxyhydridophosphorane

A 10-P-5 Species with an Apical P–H Bond

MICHAEL R. ROSS[1] and J. C. MARTIN

Roger Adams Laboratory, University of Illinois, Urbana, IL 61801

Spectroscopic data presented in an earlier report (1) were interpreted in terms of an apical P-H bond disposition and a trigonal bipyramidal (TBP) structure for a monocyclic triarylhydridophosphorane (**1b**). We have an X-ray crystallographic solution of the structure of apical hydridophosphorane **1a** as well as the structure of its diaryldialkoxy spirobicyclic analogue, **2**. Table I lists important bond lengths and angles for the two species, and, for comparison, the phosphatrane of Verkade (2), **3**, and equatorial hydridophosphorane **9**.

Phosphorane **1a** has a $^1J_{PH}$ value (269 Hz in $CDCl_3$) and an infrared P-H stretching frequency (2100 cm^{-1}, $CHCl_3$) that are significantly smaller than the corresponding values for **2** ($^1J_{PH}$ = 733 Hz, ν_{P-H} = 2430 cm^{-1}) (1) and other equatorial hydridophosphoranes (3). These observations are reconciled with an apical P-H disposition for **1a** in solution on the basis of a correlation between $^1J_{PH}$ and ligand electronegativity (Equation 1). This was developed using data for a large number of hydridophosphoranes

$$^1J_{PH} = 306\ [\sigma_I(\text{equatorial}) + 0.505\sigma_I(\text{apical})] + 595 \qquad (1)$$

which could, with some assurance, be assigned TBP structures with equatorial P-H bonds. Apical hydridophosphoranes **1a**, **1b**, and **3** (3, 4) all have $^1J_{P-H}$ values much lower than Equation 1 predicts.

The apical P-H bond length of 1.35(3) Å for **1a** is, within experimental error, the same as that reported for the equatorial P-H bond of **2**, 1.32(3) Å, and for the apical P-H bond of **3**, 1.349(71) Å (2). All are shorter than the sum of phosphorus and hydrogen covalent radii, 1.40 Å (4), and considerably shorter than P-H bond lengths in a variety of hydridophosphines and in the PH_4^+ cation, 1.41-1.45 Å (5). The two apical P-O bonds for **2** (average value = 1.745 Å are in the usual range of bond lengths for such species (6). The P-O bond for **1b** (1.825(3) Å is longer.

[1] Current Address: Monomer Process Research Department, Rohm and Haas Company, Research Laboratories, Spring House, PA 19477

0097–6156/81/0171–0429$05.00/0

Table I. Bond Lengths (Å) and Angles (Deg) in Hydridophosphoranes

Bond Length, Angle	Compound			
	1a[a]	2[a]	3[b]	9[c]
a	1.825(3)	1.743(2)	1.986(5)	1.721(2)
b	1.35(3)	1.748(3)	1.35(7)	1.701(3)
c	1.825(3)	1.32(1)	1.577(3)	1.36(3)
∠ab[d]	176(1)	178.47(3)	172.2(48)	177.4(1)
∠cd	112.7(1)	114.5(11)	120.0(1)	115(1)
∠ac	90.7(1)	86.1(11)	87.6(1)	91(1)
∠ad	--	88.09(14)	87.1(2)	90.8(1)
∠de	123.6(2)	127.6(2)	119.3(3)	123.7(1)
∠ae	85.5(1)	88.48(14)	87.6(1)	92.6(1)
∠ce	123.7(2)	117.9(11)	120.0(1)	121(1)

[a]Reported herein. [b]Reference (3). [c]Reference (10). [d]The O-P-H angle for **1a** is bent toward the t-Bu aryl group; the N-P-H angle for **3** is bent toward one equatorial oxygen; **2** and **9** have O-P-P angles bent away from the equatorial hydrogen.

The phosphorus atom lies 0.036 Å below the equatorial plane of **1b**, toward the apical hydrogen. This represents a small displacement along the pathway from the ideal TBP toward **4**, with P-O bond lengthening and P-H bond shortening.

Addition of trifluoromethanesulfonic (triflic) acid to a $CDCl_3$ solution of **1a** (**1b**) gives alkoxyhydridophosphonium salt **5a** (**5b**). Hellwinkel's (7) bicyclic phosphorane **8** behaves in an analogous manner. An increase of about 300 Hz in the value of $^1J_{PH}$ upon formation of **5a,b** from **1a,b** is consistent with a smaller degree of phosphorus s-orbital character in the apical P-H bond of the phosphorane (p/2 hybridization) than in the sp^3 phosphonium P-H bond (8). The conversion of **2** to **7** causes only the small change (17 Hz) in $^1J_{P-H}$ expected for a smaller change in s-orbital character (ca. sp^2 to sp^3).

In our preliminary communication (1) we reported a pKa of 11.7±0.1 for **1b**. Utilizing the same ^{31}P NMR technique we find pKa = 10.3±0.2 for **2**. While the structure of the conjugate base of **2**, phosphoranide **10** can be unambiguously assigned (see chemical shift models **11**, **12**, and **13**) (11) the same structural assessment is not as satisfying for the conjugate base of **1a** (or **1b**). Phosphine derivatives **14**-**20** (12) have ^{31}P NMR chemical shifts similar to those seen for the conjugate bases of **1a** (-9.92 ppm) and **1b** (-11.1 ppm). Weak P-O bonding would not, however, be expected (13) to lead to a significant upfield ^{31}P shift if the bond is a sufficiently weak one. Estimates of pKa values for the equilibrium between open-chain tautomer of **1** and its conjugate open-chain base **21b** can be used with the measured pKa value for **1** and the

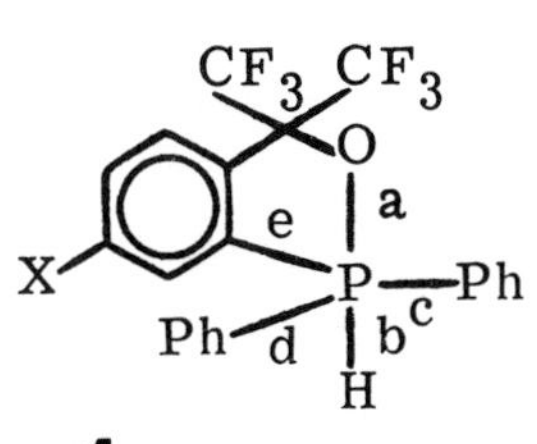

1a, X = t-Bu
1b, X = H

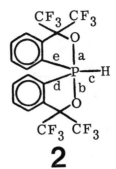

2

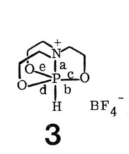

BF_4^-

3

4

$CF_3SO_3^-$

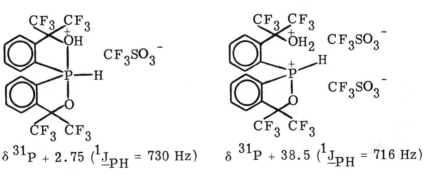

5a, X = t-Bu, δ ^{31}P + 10.2 ($^1J_{PH}$ = 569 Hz)
5b, X = H, δ ^{31}P + 11.7 ($^1J_{PH}$ = 560 Hz)

$CF_3SO_3^-$

δ ^{31}P + 2.75 ($^1J_{PH}$ = 730 Hz)

6

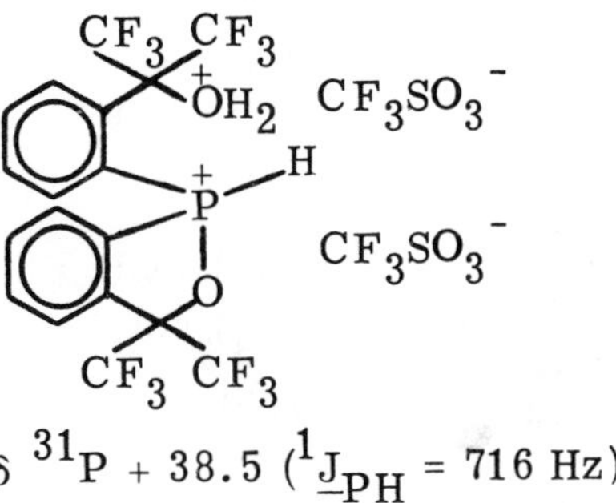

δ ^{31}P + 38.5 ($^1J_{PH}$ = 716 Hz)

7

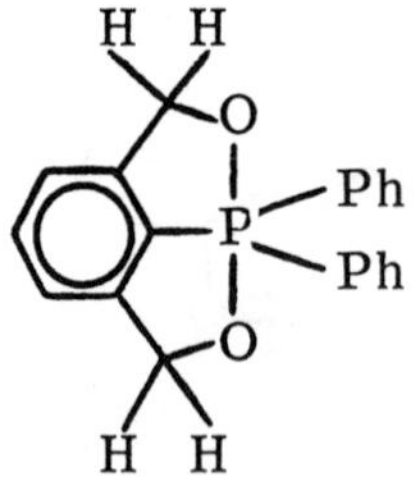

8, $\delta\ ^{31}P$ -25.9

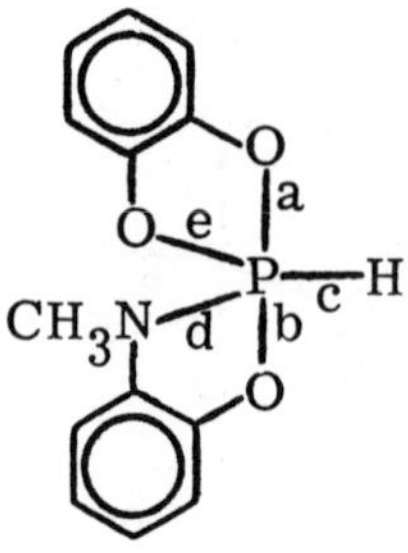

9

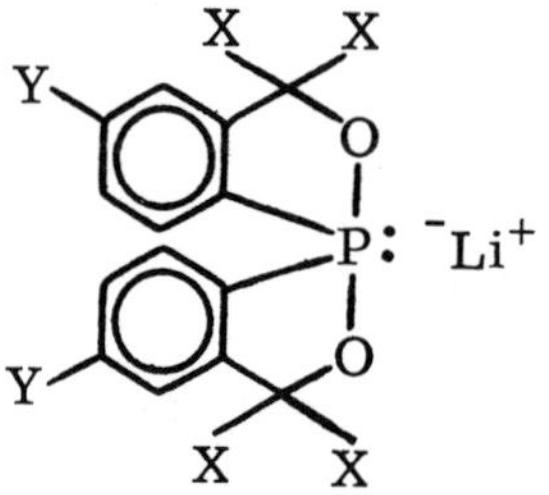

10, X = CF_3, Y = H, $\delta\ ^{31}P$ -18.5
11, X = CF_3, Y = CH_3, $\delta\ ^{31}P$ -17.9
12, X = CH_3, Y = H, $\delta\ ^{31}P$ -35.0

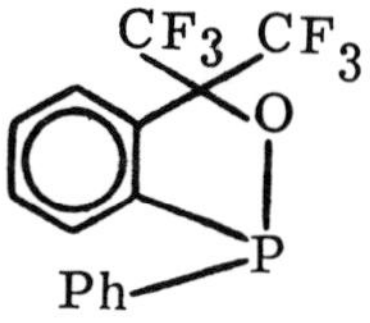

13, $\delta\ ^{31}P$ + 113

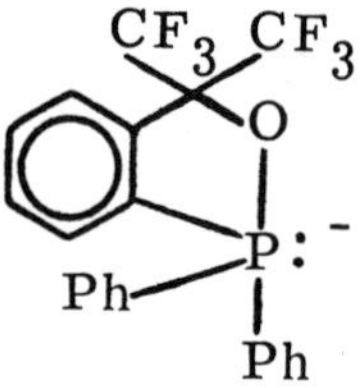

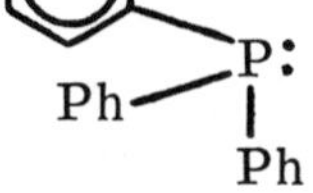

	X	$\delta\ ^{31}P$
14	-C$(CH_3)_2OCH_2$Ph,	+21.9
15	-C$(CH_3)_2$OH,	-11.1
16	-CH_2OH,	-16.0
17	-CO_2H,	-5.0
18	-OCH_3,	-13.5
19	-HC(CH_3)OH,	-17.0
20	-CF_3,	-10.9

a or **b**

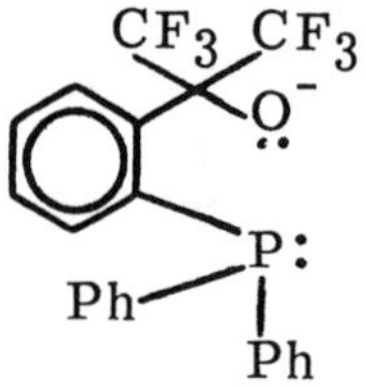

21
($\delta\ ^{31}P$ - 9.92 ppm)

experimental lower limit for the free energy difference between **1** and its unobserved open-chain isomer to determine that the lower limit to the energy difference between the true conjugate base of **1** (either **21a** or **21b**) and tautomer **21b** is zero. There is therefore no direct evidence requiring that we postulate energetically significant P-O bonding in **21**. On the other hand, since we have only a lower limit for the energy difference between **1** and its open-chain isomer, we cannot rule out some P-O bonding in **21**. Further work will be required to remove this ambiguity.

This research was supported in part by a grant (CA13963) from the National Cancer Institute.

Literature Cited

1. Ross, M. R.; Martin, J. C. J. Am. Chem. Soc. 1981, 103, 1234.
2. Milbrath, D. S.; Verkade, J. G. J. Am. Chem. Soc. 1977, 99, 6607. Clardy, J. C.; Milbrath, D. S.; Springer, J. P.; Verkade, J. G. J. Am. Chem. Soc. 1976, 98, 623.
3. Brazier, J. F.; Houalla, D.; Loenig, M.; Wolf, R. Top. Phosphorus Chem. 1976, 8, 99.
4. Pauling, L. "The Nature of the Chemical Bond"; Cornell University Press: Ithaca, N.Y., 1973; p 224-228.
5. Corbridge, D. E. C. Top. Phosphorus Chem. 1966, 3, 91.
6. Holmes, R. R.; Deiters, J. A. J. Am. Chem. Soc. 1977, 99, 3318.
7. Hellwinkel, D.; Krapp, W. Chem. Ber. 1978, 111, 13.
8. Hoffmann, R.; Howell, J. M.; Muetterties, E. L. J. Am. Chem. Soc. 1972, 94, 3047.
9. Musher, J. I. Angew. Chem., Int. Ed. Engl. 1969, 8, 54.
10. Clark, T. E.; Day, R. O.; Holmes, R. R. Inorg. Chem. 1979, 18, 1653.
11. Granoth, I.; Martin, J. C. J. Am. Chem. Soc. 1979, 101, 4623.
12. Granoth, I.; Martin, J. C. J. Am. Chem. Soc. 1981, 103, in press (compound **14**); Landvatter, E.; Rauchfuss, T. B.; Wrobluski, D. A., private communication (compounds **16**, **19**); Granoth, I.; Alkabets, R.; Shirin, E.; private communication (compound **15**); Wrobluski, D. A.; Rauchfuss, T. B. Inorg. Synth., submitted (compound **17**); McEwen, W. E.; Shiau, W.-I.; Yeh, Y.-I.; Schulz, D. N.; Pagilagan, R. A.; Levy, J. B.; Symmes, C. Jr.; Nelson, G. O.; Granoth, I. J. Am. Chem. Soc. 1975, 97, 1787 (compound **18**); Miller, G. R.; Yankowsky, A. W.; Grim, O. S. J. Chem. Phys. 1969, 51, 3185 (compound **20**).
13. Granoth, I.; Martin, J. C. J. Am. Chem. Soc., in press.

RECEIVED July 1, 1981.

Monocyclic Phosphoranide and Phosphoranoxide Anions

P(V) Oxyphosphorane Carbanion — P(IV) Ylide Alkoxide Tautomerism

ITSHAK GRANOTH, RIVKA ALKABETS, EZRA SHIRIN, YAIR MARGALIT and PETER BELL

Israel Institute for Biological Research, Ness-Ziona 70450, Israel

Pentacoordinate hydroxyphosphoranes are likely intermediates or transition states in substitution reactions at tetracoordinate phosphorus (1). Recently, stable hydroxyphosphoranes (2, 3) and their conjugate bases - metal phosphoranoxides (4, 5) have been isolated. Spectroscopic evidence (4 - 7) for equilibria between P(IV) compounds and hydroxyphosphoranes have been reported. Observation (8) and isolation (9) of P(IV) TBP phosphoranide species have also been announced. All these phosphoranes are stabilized by several features dominated by their spirobicyclic nature.

Monocyclic Phosphoranide Anion. The intramolecular oxidative addition of hydroxyalkyl phosphites, which gives P-H phosphoranes, is well known (10). Some P-H phosphoranes are so stable that the open-chain P(III) tautomers cannot be detected spectroscopically or even by attempted H_2O_2 oxidation (8). Thus, it is surprising to find no evidence for an equilibrium between phosphine alcohol 1 and its closed-ring tautomer phosphorane 2. Phosphine 1 is quaternized by alkyl halides giving phosphonium halides such as 3. These in turn are converted to alkoxyphosphoranes, such as 4, by NaH (Scheme I).

Phosphine 1 shows the expected ^{31}P NMR δP-11.1 (THF), typical (11) of ortho substituted triphenylphosphine. Addition of $LiAlH_4$ to this solution gives molecular hydrogen and two signals in the ^{31}P NMR spectrum at -11.1 and -30.3. This observation suggests that deprotonation of 1 leads to a relatively slow equilibrium of phosphine alkoxide 5 (major component) and phosphoranide 6. This mixture and alkyl halides give phosphoranes such as 4.

The instability of phosphorane 2 is surprising because 2 accommodates four features which are known to stabilize hypervalent compounds (12). The central phosphorus atom in the TBP 2 bears two highly apicophilic and three poorly apicophilic ligands (2). The phosphorus atom is contained in a five-membered ring linking an apical oxygen to an equatorial aromatic ring carbon (12). The gem-dimethyl conformational effect favors a closed ring structure. The increased stability of phosphoranide 6, com-

0097-6156/81/0171-0435$05.00/0

Scheme I

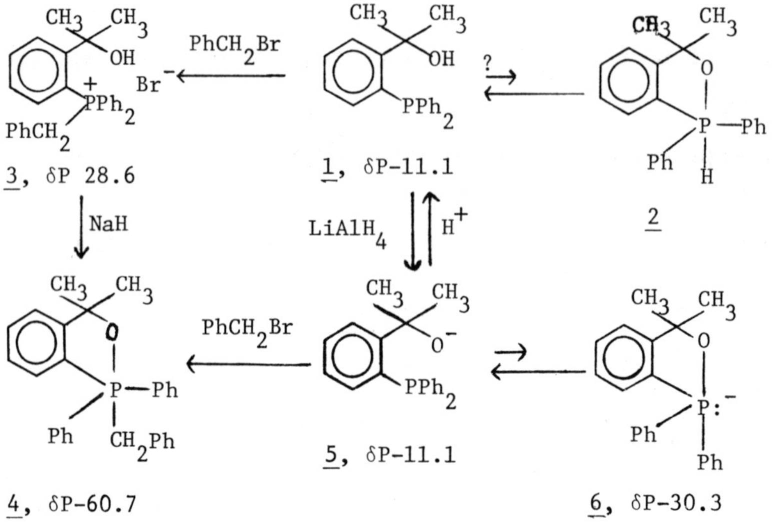
PhCH2Br
?
3, δP 28.6
1, δP-11.1
2
NaH
LiAlH4
H+
PhCH2Br
4, δP-60.7
5, δP-11.1
6, δP-30.3

pared with its conjugate acid 2, is analogous to the enhanced stability of reported phosphoranoxide anions (3,4,5). This may result from the increased electronegativity difference between apical and equatorial ligands in these TBP bases, compared with their respective conjugate acids.

Phosphorane-Ylide Tautomerism. The lability of α protons in alkylphosphoranes is not known. Deuterium exchange of the benzyl protons is not observed, even in the presence of NaOD in D_2O-CD_3SOCD_3. Deprotonation of 4 in THF by CH_3Li at room temperature is fast. This red solution is shown by variable temperature ^{31}P NMR to contain an equilibrium mixture of phosphorane 7 and ylide 8 (Scheme II). This mixture and CH_3I give phsophorane 9.

Scheme II

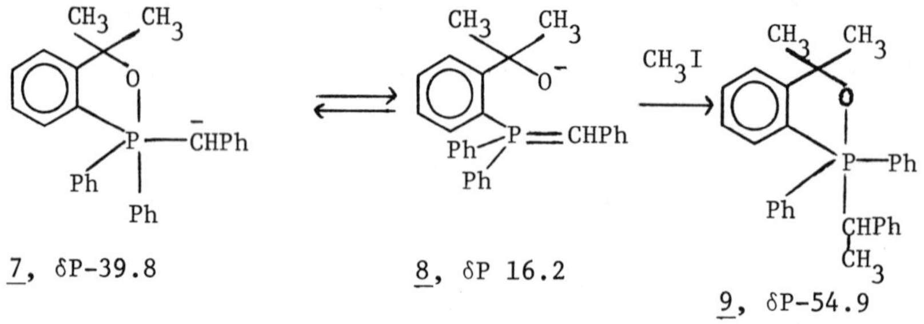
CH3I
7, δP-39.8
8, δP 16.2
9, δP-54.9

Monocyclic Phosphoranoxide Anion. The ^{31}P NMR chemical shift value and line width of 10 is solvent and pH sensitive, suggesting the equilibria shown in Scheme III. The structure of 11 is confirmed by preparation and characterization of the analogous stable BF_4^- salt. Deprotonation of 10 by NaH in THF is much faster than that of its para isomer. This deprotonation is followed by broadening, gradual upfield shift and resharpening of the ^{31}P NMR signal from 38.5 to -31.0 ppm. The latter, somewhat broad line is consistent with structure 14 in equilibrium with a small concentration of 13.

Scheme III

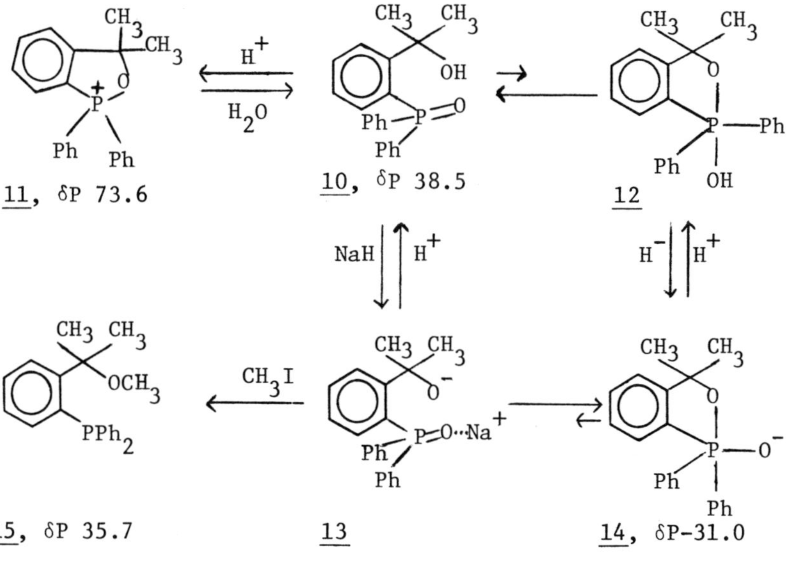

11, δP 73.6 ⇌ (H^+ / H_2O) 10, δP 38.5 ⇌ 12

10: NaH ↓ ↑ H^+ 13; 12: H^- ↓ ↑ H^+ 14

15, δP 35.7 ← (CH_3I) 13 ⇌ 14, δP-31.0

The equilibrating 14 and 13 react surprisingly fast with CH_3I, giving ether 15 as the sole product. The para isomer of 13 does not react with CH_3I in THF at room temperature during 24 hours. The extraordinary rapid formation of 15, presumably from 13, may result from the increased reactivity of 13 enabled by intramolecular solvation of the metal by the phosphine oxide oxygen. Alternative mechanisms, such as a methoxyphosphorane intermediate, cannot be ruled out at this stage.

Literature Cited

1. Westheimer, F.H. Acc. Chem. Res. 1968, 1, 70.
2. Segall, Y.; Granoth I. J. Am. Chem. Soc. 1978, 100, 5130.
3. Granoth, I.; Martin, J.C. J. Am. Chem. Soc. 1979, 101, 4618.

4. Granoth, I.; Martin, J.C. J. Am. Chem. Soc. 1978, 100,5229.
5. Munoz, A.; Garrigues, B; Koenig, M. J.C.S. Chem. Comm. 1978, 219.
6. Gallagher, M.; Munoz, A.; Gence, G.; Koenig, M. J.C.S. Chem. Comm. 1976, 321.
7. Ramirez, F.; Nowakowsky, M; Marecek, J.F. J. Am. Chem.Soc. 1977, 99, 4515.
8. Granoth, I.; Martin, J.C. J. Am. Chem. Soc. 1979, 101, 4623.
9. Garrigues, B.; Koenig, M.; Munoz, A. Tetrahedron Lett. 1979, 4205.
10. Burgada, R. Bull. Soc. Chim. France 1975, 407.
11. Grim, S.O.; Yankowsky, A.W. Phosphorus and Sulfur 1977, 3, 191.
12. Martin, J.C.; Perozzi, E.F. J. Am. Chem. Soc. 1974, 96, 3155.

RECEIVED July 7, 1981.

90

Selectivity in Reactions of Tricyclic Phosphatranes

D. VAN AKEN, I. I. MERKELBACH, J. H. H. HAMERLINCK, P. SCHIPPER, and H. M. BUCK

Eindhoven University of Technology, Department of Organic Chemistry, The Netherlands

Five-coordinated phosphorus compounds are known to adopt the trigonal bipyramidal (TBP) configuration, in which the apical positions are preferred by electron-withdrawing groups, whereas electron-donating groups are situated in equatorial positions (1). In addition, the apical bonds are longer and weaker than the equatorial bonds originating apical entry and departure of groups (phosphorylation) (2). Little attention has been paid to reactions in which external nucleophiles discriminate between pseudo-equatorial and pseudo-apical carbon atoms in a TBP configuration. The observation of such selective reactions is hampered by the occurrence of pseudo-rotation which brings about ligand exchange in the TBP (3). In order to obtain a definite answer with respect to equatorial vs. apical reactivity for nucleophiles we have synthesized the rigid tricyclic phosphatranes 1-4 in which a transannular N → P bond brings phosphorus in a TBP configuration. Previously the characterization of the related compound 5 was published by Verkade et al. (4a).

$[RXP(OCH_2CH_2)_3N]^+BF_4^-$ $\qquad$ $[HP(OCH_2CH_2)_3N]^+BF_4^-$

1 R = Me; X = O $\qquad$ 5

2 R = Et; X = O

3 R = Me; X = S

4 R = Et; X = S

The molecular constraint which precludes ligand exchange unambiguously ensures that the tricyclic compounds undergo nucleophilic attack exclusively at the pseudo-equatorial carbon atom, irrespective of the nature of the carbon atom situated in the apical ligand of the TBP. The compounds 1-4 were obtained by alkylation of the chalcogen atom of the corresponding bicyclic (thio)phosphate, 6 and 7, with trialkyloxonium tetrafluoroborate at -78 °C (5).

$X = P(OCH_2CH_2)_3N$

6 X = O

7 X = S

0097-6156/81/0171-0439$05.00/0

All experimental evidence indicates that the structure of the O(S) alkylated compounds is a tricyclic cage (4b). In solution NMR confirms the presence of a N → P bond (4b). To verify the geometry, the structure of 4 was determined crystallographically. The configuration of phosphorus is indeed TBP with O-P-O angles near 120°, O-P-N angles of 85-87°, and O-P-S angles of 94-95°. The S-P-N angle is 178°. The N-P bond length is 2.05 Å which is only slightly longer than the corresponding distance of 1.99 Å found in 5 (4a). With aqueous OH^-, compounds 1-4 react exclusively at the pseudo-equatorial carbon atom leading to the monocyclic products 8-11.

$RXP(:O)(OCH_2CH_2)_2NCH_2CH_2OH$

8 R = Me; X = O
9 R = Et; X = O
10 R = Me; X = S
11 R = Et; X = S

In all products the ^{31}P couplings show that the bond of the exocyclic group linked to phosphorus is preserved in 8, $J(POCH_3)$ = 11 Hz and in 10, $J(PSCH_3)$ = 16 Hz. In our opinion, the preference for pseudo-equatorial nucleophilic attack is an intrinsic property of the electronic configuration of the TBP. The dp π bond (P = O), which is formed in the reaction is already developed to some extent for the equatorial oxygen atoms in the TBP owing to back-donation (1). As a result, some electron density is transferred from the pseudo-equatorial carbon atoms to phosphorus, rendering them more acceptable for nucleophilic attack. Since five-coordination is very wide-spread for group IV and V elements as well as for transition metals, our model for pseudo-equatorial attack of nucleophiles might have a generalized feature.
In addition, we will present the mechanistic aspects concerning the electron-entry on four-coordinated phosphorus compounds. There are indications promoting the concept that a TBP is involved with the odd electron in an equatorial position. This presumption is based on the temperature-dependent ESR spectra of the radicals derived from 5. X-irradiation of a single crystal of 5 at liquid nitrogen temperature (77 K) yields the spectrum of radical 13 exclusively with $a_{P\parallel}$ = 1120 G, $a_{P\perp}$ = 930 G, while ^{14}N splitting is not resolved. From these values one calculates a_P^{iso} = 993 G, indicating the phosphorus 3s spin density is 0.27, while the anisotropic contribution places 0.61 of the spin density in its 3p orbital, giving a total spin density of 0.88 on phosphorus. Rotation on the crystallographic c-axis shows two orientations with an angle between their $a_{P\parallel}$ components of 70°. Raising the temperature, these signals disappear irreversibly at 193 K, and those of 12 become apparent (6). Again on rotation two orientations are present which are perpendicular. The anisotropic phosphorus coupling constants $a_{P\parallel}$ = 888 G, $a_{P\perp}$ = 753 G, indicate a spin density of 0.21 in the phosphorus 3s orbital and 0.43 in its 3p orbital. An additional hyperfine coupling is observed due to ^{14}N, a_N = 22 G, which value is nearly isotropic, indicating a spin

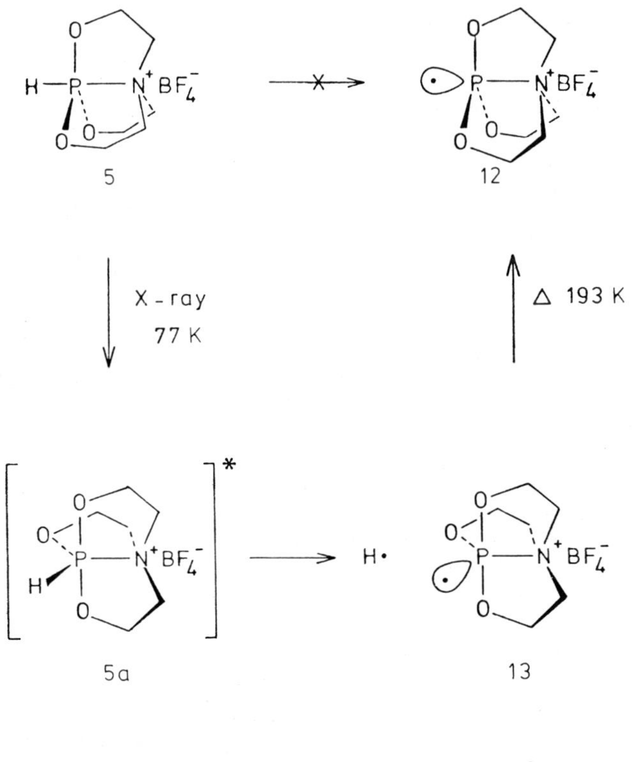

H—P—N$^+$ BF_4^-

5

12

X - ray
77 K

Δ 193 K

5a

H•

13

density of 0.05 in the 2s orbital of nitrogen. In the plane perpendicular to the crystal mounting axis (c-axis) the directions of $a_{P\parallel}$ of 12 and 13 differ 35° or the complementary angle of 55° which cannot be assigned uniquely, due to the presence of two orientations for both radicals in the unit cell (4a). The irreversible arrangement of 13 to 12 at higher temperatures exhibits the latter species as thermodynamically more stable. This may be conceivable on the basis of release of ring strain by the five-membered ring in structure 13 which spans two equatorial positions. However, the driving force for the intermediate formation of structure 13 is less obvious. Its geometry can only be formed prior to P-H bond scission 5 → 5a since at the radical stage only the reverse reaction was observed. Apparently, during X-irradiation 5 absorbs energy in a photochemical way, to produce 5a in the excited state. Subsequently, the latter undergoes P-H bond scission with retention of configuration to produce 13. These processes may be understood on the basis of balancing electronic energy versus strain energy. Apparently, structure 5a with the hydrogen in an equatorial position is electronically favoured over 5 with hydrogen in an apical position. In contrast, the strain energy in 5a is enhanced with respect to that of 5. The latter factor may be less important for 5a in the excited state, since in the excited state the bond lengths are increased. Therefore, in this particular situation the electronic factor has become dominant. In this way the irreversible retro-arrangement of 13 to 12 is also consistent. These experiments strongly suggest that the initial state for the capture of an electron by a four-coordinated phosphorus may lead to a TBP configuration via an equatorial entry of the electron.

The authors thank Dr. J.C. Schoone, University of Utrecht, The Netherlands, for collecting the X-ray data.
This work was supported by the Netherlands Foundation for Chemical Research (SON) with financial aid from the Netherlands Organization for the Advancement of Pure Research (ZWO).

Literature Cited

1. Muetterties, E.L.; Schunn, R.A. Quart. Rev. Chem. Soc. 1966, 20, 245. For a review, see: Luckenbach, R. "Dynamic Stereochemistry of Penta-coordinated Phosphorus and Related Elements"; Thieme, G., Stuttgart, 1973.
2. Marquarding, D.; Ramirez, F.; Ugi, I.; Gillespie, P. Angew. Chem. 1973, 85, 99.
3. Ugi, I.; Ramirez, F. Chem. Brit. 1972, 8, 198; Musher, J.I. J. Chem. Educ. 1974, 51, 94.
4. a. Clardy, J.C.; Milbrath, D.S.; Springer, J.P.; Verkade, J.G. J. Am. Chem. Soc. 1976, 98, 623.
   b. Milbrath, D.S.; Verkade, J.G. ibid. 1977, 99, 6607.
5. Murray, M.; Schmutzler, R.; Gründemann, E.; Teichmann, H. J. Chem. Soc. (B) 1971, 1714.
6. Hamerlinck, J.H.H.; Schipper, P.; Buck, H.M. J. Am. Chem. Soc. 1980, 102, 5679.

RECEIVED June 30, 1981.

The Perfluoropinacolyl Group: A Stabilizing Substituent for Unusual Phosphites and Phosphoranes

GERD-VOLKER RÖSCHENTHALER, and RAINER BOHLEN
Fachbereich 3 der Universität, 28 Bremen 33, FRG

WERNER STORZER
Lehrstuhl B für Anorganische Chemie der Technischen Universität, 33 Braunschweig, FRG

The perfluoropinacolyl moiety $OC(CF_3)_2C(CF_3)_2O$ stabilizes higher valence states of main group elements (1) because of its high group electronegativity and bidentate character. The investigations of Ramirez, Trippett, and Knunyants show that pentacovalent phosphorus compounds containing this grouping are readily obtained by oxidative cyclisation reacting phosphorus(III) compounds with hexafluoroacetone. We were successful in synthesizing fluorophosphoranes in a similar manner (2, 3):

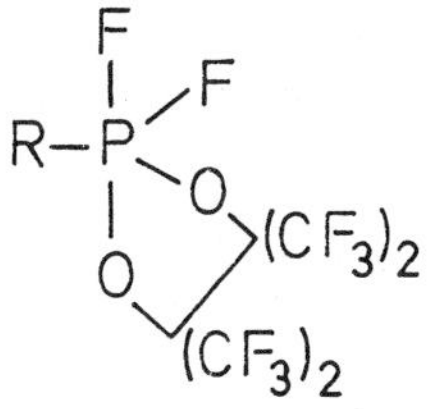

R = $N(SiMe_3)_2$, OMe, N=PF_2Ph

Two fluorines attached to phosphorus instead of the perfluoropinacolyl moiety would cause rapid decomposition. These findings prompted us to try the synthesis of phosphites, cyclic and bicyclic phosphoranes by a step by step procedure using dilithium perfluoropinacolate 1 and phosphorus halogenides as starting materials:

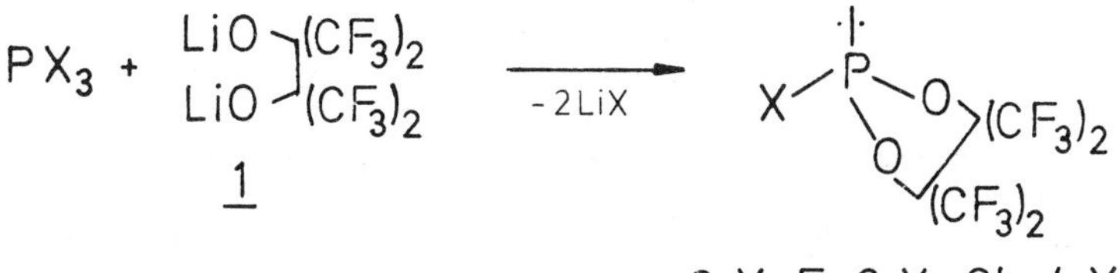

$$PX_3 + \underset{\mathbf{1}}{LiOC(CF_3)_2C(CF_3)_2OLi} \xrightarrow{-2LiX} X\text{-}P\langle OC(CF_3)_2C(CF_3)_2O \rangle$$

2: X = F; 3: X = Cl; 4: X = Br

0097–6156/81/0171–0443$05.00/0

The chlorophosphite 3 is a precursor for several substitution reactions yielding surprisingly stable phosphorus(III) compounds (Scheme 1):

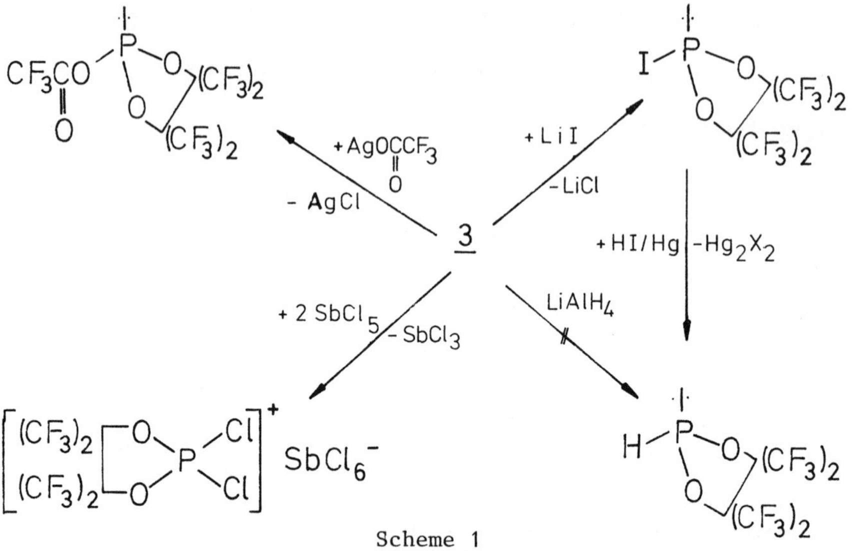

Scheme 1

The phosphites 3 and 4 are oxidized by chlorine and bromine to furnish stable covalent phosphoranes. No Arbuzov reaction was observed. The trichlorophosphorane 5 can be reacted with 1 again to form a spirophosphorane 7 which is hydrolyzed without ring cleavage to give an extremely stable hydroxyphosphorane or a phosphoraneoxide anion:

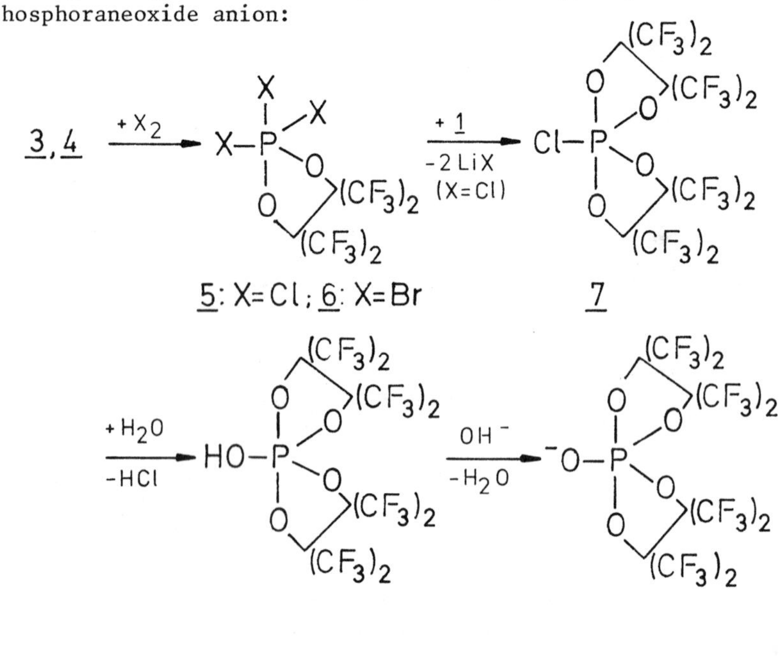

5: X=Cl; 6: X=Br

7

Aminophosphites are transferred to aminospirophosphoranes via a similar reaction route:

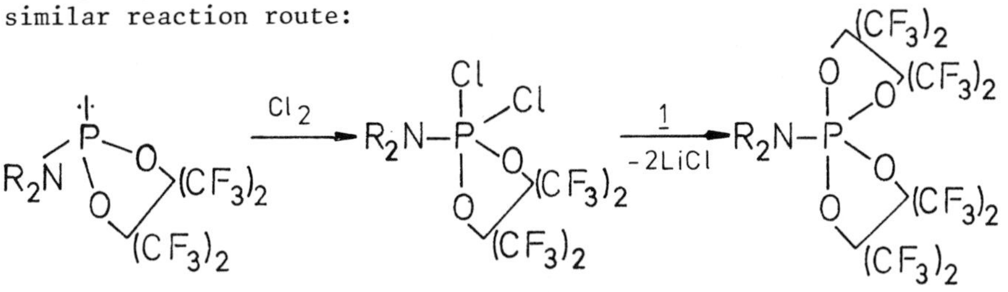

A nitrogen-bridged diphosphorane is obtained by reacting the phosphite precursor with chlorine. The diphosphorane decomposes slowly to form 5 and a diazadiphosphetidine.

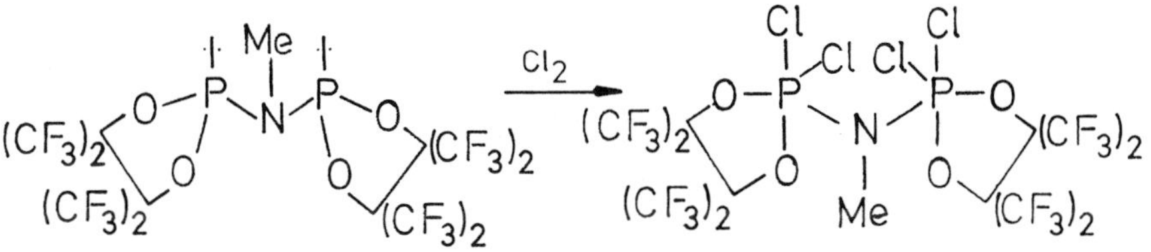

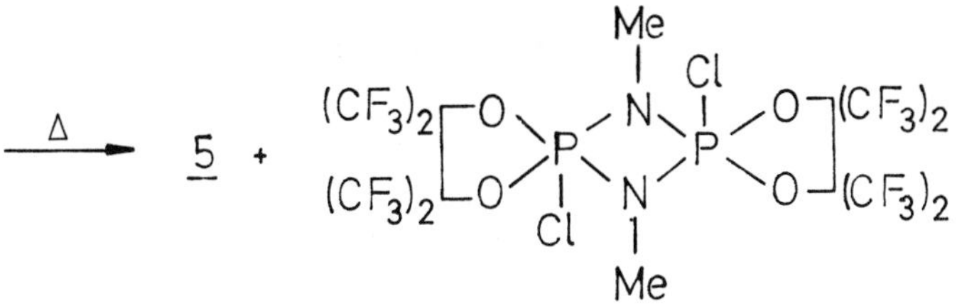

A very interesting chemistry develops from the aminophosphite 8 having a H_2N group bonded to phosphorus, the first stable species of that type, which is synthesized by reacting 3 with lithium amide. Compound 8 easily adds hydrogen fluoride to form a aminofluorohydridophosphorane. Chlorine yields a dichlorophosphorane 9 which decomposes upon heating to give a cyclic phosphazene. Substitution of the two chlorines in compound 9 by the perfluoropinacolyl grouping yields a aminospiro system. Two remarkable compounds are found during the reaction of 8 with ammonium perfluoropinacolate and hexafluoroacetone. In the first case a hydridospirophosphorane, in the latter a phosphorane containing a four and five membered ring can be isolated(Scheme 2).

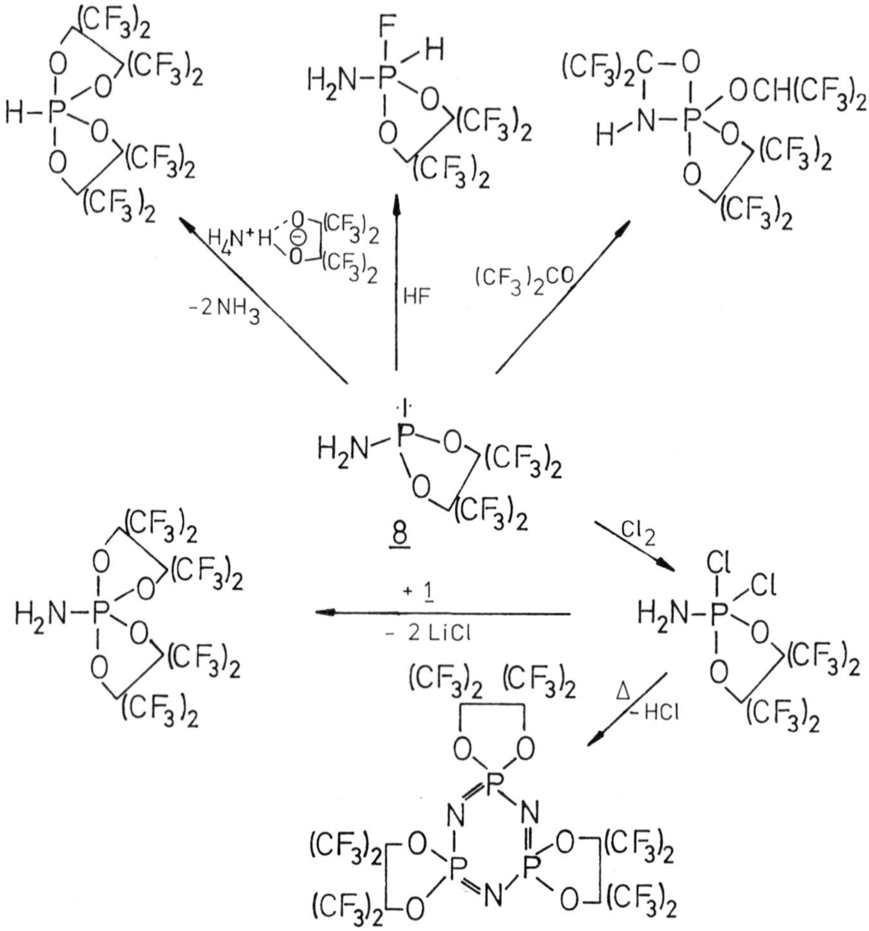

Scheme 2

Deutsche Forschungsgemeinschaft is thanked for financial support, Professor Schmutzler for access to research facilities, and Daikin Company, Osaka, Japan for the generous gift of hexafluoroacetone.

1. Allen, M.; Janzen, A. F.; Willis, C. J. Can. J. Chem. 1968, 46, 3671.
2. Gibson, J. A.; Röschenthaler, G.-V.; Schmutzler, R. J. Chem. Soc. Dalton Trans. 1975, 918.
3. Röschenthaler, G.-V.; Gibson, J. A.; Schmutzler, R. Chem. Ber. 1977, 110, 611.

RECEIVED June 30, 1981.

92

Tartaric Acid in Phosphorus Chemistry: Phosphor Emetics and Oligomers

A. MUNOZ, L. LAMANDÉ, M. KOENIG, and R. WOLF

ERA N° 926 Associée au CNRS, Laboratoire des Hétérocycles du Phosphore et de l'Azote, Université Paul Sabatier, 118 Route de Narbonne, 31062 Toulouse, France

The complex of tartaric acid and antimony (emetic) was described three centuries ago. Nevertheless, the structure of this compound has been elucidated these last fifteen years by X-ray diffraction (1). In fact, emetic presents a binuclear cyclic structure. Many authors mentioned similar complex with transition metals (vanadium (2), chromium (3)) or metalloids (arsenic (4), bismuth (5)). Emetic with phosphorus was not mentioned. Nevertheless, tartaric acid or alkyl tartrates has been utilized in phosphorus chemistry : tartaric acid reacts with trialkyl phosphites giving heterocyclic phosphites (6). Starting from alkyl tartrates, we prepared spirophosphoranes with a P-H bond and sixcoordinated compounds (7). With unprotected tartaric acid, many possibilities appear : condensation as a diol, as a di(α-hydroxyacid), or even as a β-hydroxyacid.

In fact, natural tartaric acid (R,R) reacts easily at room temperature with phosphorus trichloride. The main products of the reaction are spirophosphoranes with a P-H bond formed from the α-hydroxyacid linkage (scheme (A)).

```
                HO-CO     CO-OH                 ┌ -O-CO     CO-O    H    ┐
n PCl3 + n        |         |     - 3n HCl      │   |         |    \ |    │  (A)
                HO-CH ─── CH-OH  ──────────►    │   |         |     >P<   │
                                                └ -O-CH ─── CH-O /        ┘ n
```

We isolated two kinds of compounds :

- After six hours (or less) of condensation, we obtained a white powder, soluble in THF, whose nmr spectrum and elemental analysis are consistent with the oligomer 1a.

- When condensation time is higher than six hours, we isolated a white powder insoluble in THF, whose spectrum, elemental analysis and molecular weight are consistent with the cyclic dimer 2a.

0097–6156/81/0171–0447$05.00/0

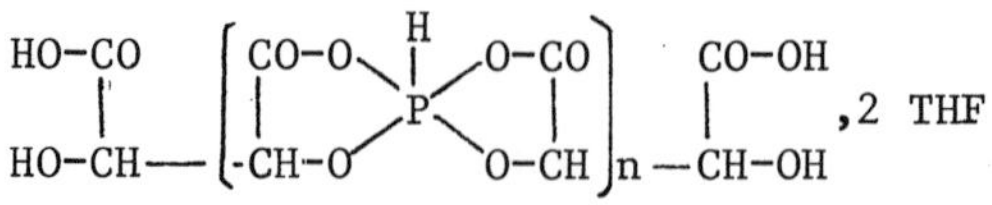

1a : 2 < n < 6

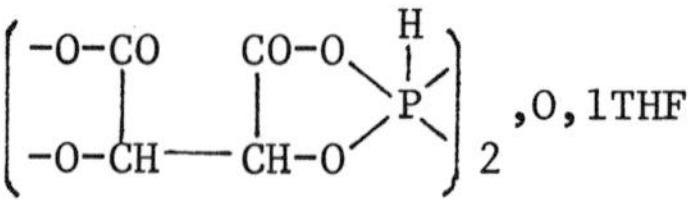

2a

Starting with these compounds, we synthetised the corresponding oligomer or dimer hydroxyphosphoranes and adducts with an hexacoordinated phosphorus atom, according to reactions (B) and (C).

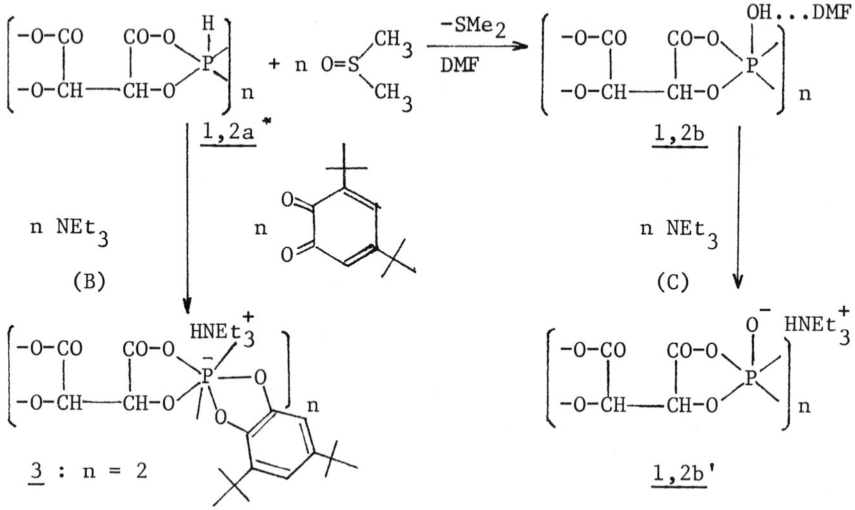

$-SMe_2$ DMF

1,2a*

1,2b

n NEt_3 (B)

n NEt_3 (C)

3 : n = 2

1,2b'

* for oligomers terminal groups omitted.

On the other hand, condensed spirophosphoranes 1a and 2a are deprotonated, under mild conditions, by triethylamine, giving an equilibrium between phosphoranide and phosphite anions, quite similar to that observed previously with monomolecular phosphoranes (scheme (D)) (8).

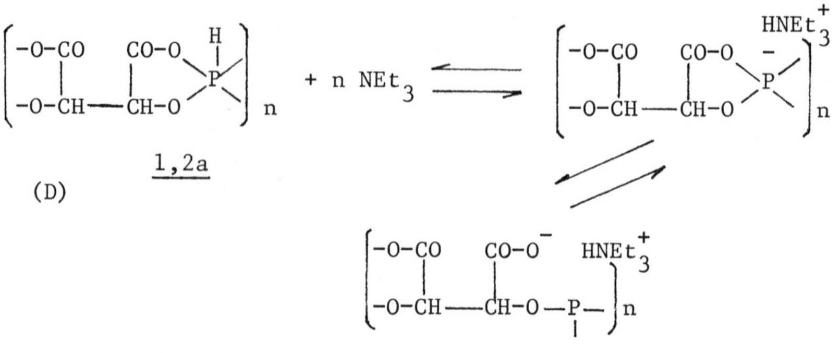

+ n NEt_3

1,2a

(D)

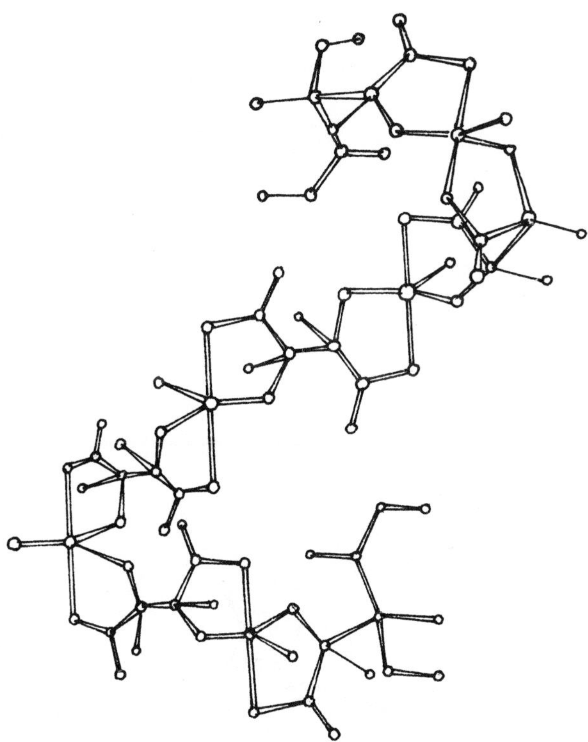

Figure 1. Oligomers 1a and 1b.

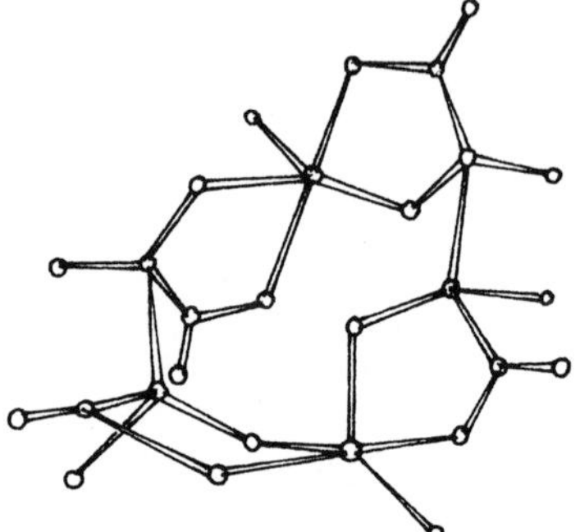

Figure 2. Dimers 2a and 2b.

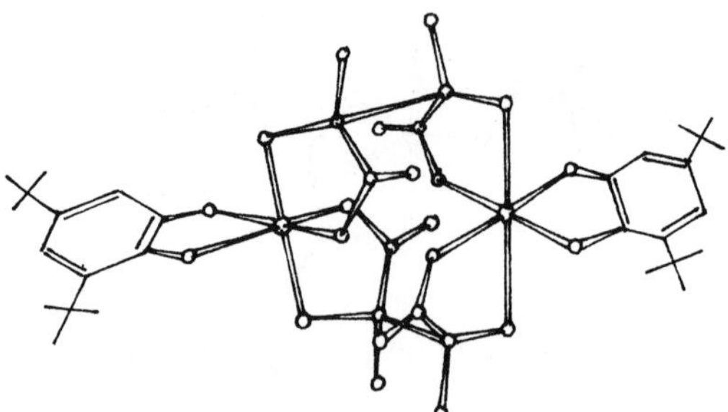

Figure 3. Hexacoordinated adduct 3.

Analysis of nmr spectra allowed us to determine the absolute configuration of the phosphorus atom of pentaco-ordinated compounds 1a, 2a and 1b, 2b. Protons P-O-CH-CH-O-P and exocyclic substituents P-H and P-O X (X = H...DMF, $HNEt_3^+$) are in a cis position. The more likely structure of oligomers 1a and 1b is a sequence of TBP drawn up as helix (fig. 1), while dimers 2a and 2b should present an emetic structure with a pentaco-ordinated phosphorus atom (fig. 2). All these compounds manifest strong optical activity.

These results complete the chemistry of the tartaric acid through elements of periodic classification. If adducts between phosphorus and tartaric acid have been lastly discovered, they offer, in return, a larger co-ordination variety than emetics which exhibit exclusively tetraco-ordination. It is interesting to notice that real phosphorus emetic, spirophosphoranide anion, is rather unstable in contrast with the arsenic or antimony homologues, while pentaco-ordinated or sixco-ordinated compounds, which are stable, are unknown for these metalloids.

Concerning phosphorus chemistry, the synthesis of condensed compounds, optically active, with pentaco-ordinated phosphorus adopting helicoidal or macrocyclic structures, represents, in our opinion, significant progress. Thus, hydroxyphosphoranes 1b, 2b are tautomers of hydroxyphosphoric esters, similar to phosphoric esters of natural polyhydroxylated compounds.

LITERATURE CITED

1. KIOSSE, G.A. ; GOLOVASTIKOV, N.I.; BELOV, N.V. Soviet. Phys. Doklady. 1964, 9, 198.
2. ORTEGA, R.B.; CAMPANA, C.F.; TAPSCOTT, R.E. Acta Cryst. 1980, B 36, 1786, and references there in.
3. KAIZAKI, S.; HIDAKA, J.; SHIMURAI, Y. Bull. Soc. Chim. Jap. 1967, 40, 2207
4. SCHLESSINGER, G.; Inorg. Synth. 1970, 12, 267, and references there in.
5. ROSENHEIM, A.; VOGELSANG, W. Z. Anorg. Chem. 1906, 48, 205
6. SAMITOV, YU.YU.; MUSINA, A.A.; GURARII, L.I.; MUKMENEV, E.T.; ARBUSOV, B.A. Bull. Acad. Sci. USSR 1975, 24, 1407, and references there in.
7. KOENIG, M.; MUNOZ, A.; GARRIGUES, B.; WOLF, R. Phosphorus and Sulfur 1979, 6, 435
8. GARRIGUES, B.; KOENIG, M.; MUNOZ, A. Tetrahedron Letters 1979, 4205

RECEIVED July 7, 1981.

Nucleophilic Substitution at Pentacoordinated Phosphorus

Addition-Elimination Mechanism

A. SKOWROŃSKA, J. STANEK-GWARA, and M. NOWAKOWSKI

Polish Academy of Sciences, Centre of Molecular and Macromolecular Studies, 90-362 Łódź, Boczna 5, Poland

There is a growing realization that hexacoordinate phosphorus species may play an important role in organophosphorus chemistry (1-8). The intriguing problem of the extent to which the hexacoordinate compounds are involved as intermediates in the substitution at pentacoordinate phosphorus atom still remains unresolved.

Recently we undertook the study of nucleophilic displacement in chlorophosphoranes containing one or two catechol bidentate ligands with trimethylsilyl azide or tetraalkylammonium azide (9). We found that this process proceeds via individually distinct hexacoordinate species.

The present investigation was undertaken in an effort to generalize the above findings. Low temperature ^{31}P NMR spectroscopy has been used for the detection and identification of hexacoordinate compounds and for the elucidation of subsequent reactions. Reaction between hexafluoroisopropoxyspirophosphorane 1 and 1,2 mole of phenol in the presence of triethylamine in methylene chloride solution at -80°C led to the hexacoordinate adduct 2 accompanied by some phenoxyspirophosphorane 4 and traces of compounds 5. After warming of the reaction mixture to -70°C the amounts of 2 and 4 increased to 76% and 20% respectively. At -40°C the hexacoordinate adduct 3 appeared and the amount of 4 decreased in favour of 5. At room temperature the spectrum indicated hexacoordinate adducts 3 and 6 as main products. The full picture of the above reaction is presented in Scheme I. The chemical shifts related to 85% phosphoric acid of compounds 2, 3, 5, 6 are in the highfield range of -104 ppm to -109 ppm which must be associated with hexacoordinate phosphorus derivatives. The structure of hexacoordinate anions 2 and 3 is in addition supported by the doublet due to the splitting of phosphorus by the methine proton of $R_f = CH(CF_3)_2$ group. There is no indication for formation of a hexacoordinate structure containing two OR_f groups when an excess of phenol is present. Among two pairs of hexacoordinate isomeric structures 2, 3 and 5, 6, the 2 and 5 seems to be kinetically and 3, 6 thermodynamically controlled.

0097-6156/81/0171-0453$05.00/0

Scheme I

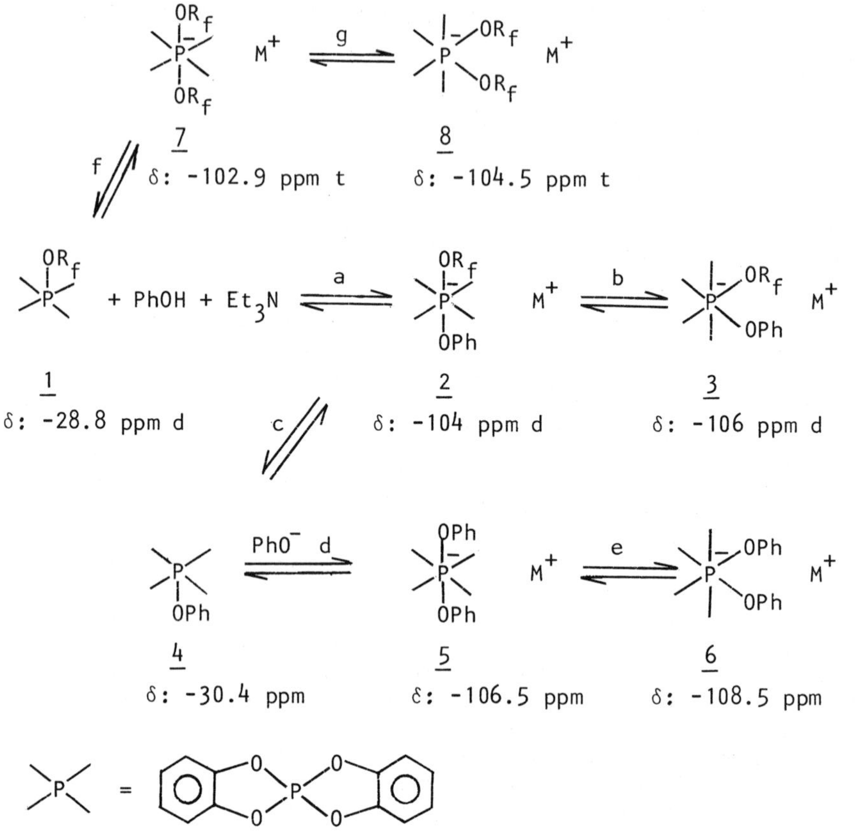

7 (δ: -102.9 ppm t) ⇌(g) 8 (δ: -104.5 ppm t)

1 (δ: -28.8 ppm d) + PhOH + Et_3N ⇌(a) 2 (δ: -104 ppm d) ⇌(b) 3 (δ: -106 ppm d)

4 (δ: -30.4 ppm) ⇌(PhO^- d) 5 (δ: -106.5 ppm) ⇌(e) 6 (δ: -108.5 ppm)

$R_f = (CF_3)_2CH-$

$M = Et_3NH$

On the basis of the work of Ramirez (10), Trippett (11) and Dillon (12) the more stable hexacoordinate isomers are of cis-structure and consequently we assigned the structures of 2, 5 as trans and those of 3, 6 as cis. The elimination step (c) leading to the phosphorane 4 also involves the less stable isomer 2. The rate of elimination seems in this case to be higher than that of isomerisation of 2 (reaction b) into the more stable hexacoordinate anion 3 (b). The latter does not undergo elimination to 4 under experimental conditions employed. These observations indicate that nucleophilic displacement at pentacoordinate phosphorus in 1 implies a trans-relationship of nucleophile to leaving group and, therefore, should proceed with inversion of configuration at the phosphorus atom.

Associative nucleophilic displacement at P(5) has been observed in other reactions of the phosphorane and nucleophiles described in Scheme II.

Scheme II

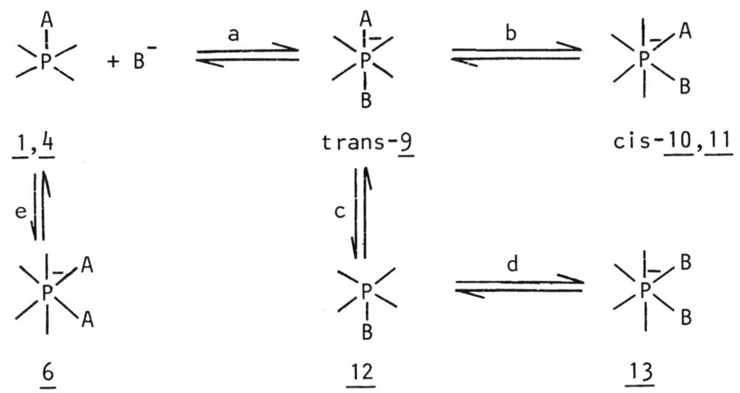

1,9,10, A = R_fO; 4,6,11, A = PhO; B = N_3

^{31}P NMR chemical shifts, δ ppm: 6 -108.1; 9 -102.2 d; 10 -107.9 d; 11 -108.5; 12 -27; 13 -113.

Treatment of spirophosphorane 1 with 1.1 equiv. of tetraalkylammonium azide in CH_2Cl_2 solution at -95°C gave the P(6) trans-adduct 9 as a major product (47%), accompanied by cis-adducts 10 (14%), 13 (10%) and unchanged spirophosphorane 1. At -20°C the amount of 10 increased to 82% with parallel decrease of adduct 9 (4%) and 1 disappeared.

The ^{31}P NMR spectrum of a solution of spirophosphorane 4 in CH_2Cl_2, treated at -95° with tetraalkylammonium azide showed the presence of P(6) cis-adducts 11 (55%), 13 (12%), traces of spirophosphorane 12 (4%) and unchanged spirophosphorane 4. At -30°C the spectrum showed signals due to 11 (47%), 13 (25%) and 6 (23%). Formation of hexacoordinate anion 13 provide indirect proof for the elimination process c giving the phosphorane 12.

Further studies on phosphorane 1,4,15 brought out additional significant information. The reactions of 4 with phenol and 15 with diethylphosphoric acid in the presence of triethylamine in CH_2Cl_2 solution give the octaedral P(6) adducts of trans configurations 5, 16. Adducts 5, 16, the only products at -100°C, gradually rearrange into cis-isomers 6, 17 being the only species at -50°C. When potassium hexafluoroisopropanolate was allowed to react with the phosphorane 1 in the presence of 18-crown ether at -80°C, the first formed trans-adduct 7 slowly isomerises at room temperature into adduct 8. The ^{31}P NMR chemical shifts of 5, 6 and 7, 8 are very close to those observed in the experiment described in Scheme I.

Scheme III

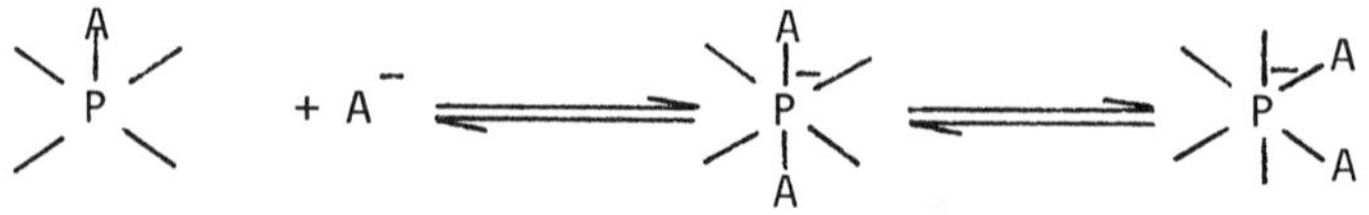

A | P + A^- ⇌ A | P | A ⇌ P(A)(A)

1,4,15 trans-5,7,16 cis-6,8,17

4,5,6, A = PhO; 1,7,8, A = R_fO; 15,16,17, A = $OP(O)(OEt)_2$

^{31}P NMR chemical shifts,δppm: 5 -106.5; 6 -108.5; 7 -102.6 t, J_{P-H} 16.1 ± 0,48 Hz; 8 -105.5 t, J_{P-H} 17.09 ± 0,48 Hz; 16 -12.8 d, -106.2 t, J_{P-O-P} 29 ± 2.44 Hz; 17 -13.75 d, -114.7 t, J_{P-O-P} 22 ± 2.44 Hz.

Further work is required to generalise the above findings.

Literature Cited:

1. Trippett Stuart, Ed.; Organophosphorus Chemistry (Specialist Periodical Report); The Chemical Society, London, vol. 4-10, 1973-79, ch. 2.
2. Ramirez F., Tasaka K., Desai N.B., Smith C.P., J.Am.Chem.Soc., 1968, 90, 751.
3. Ramirez F., Ugi I., Marquarding D., Angew.Chem. Int. Ed. Engl., 1973, 13, 91.
4. Kluger R., Covitz F., Dennis E., Williams L.D., Westheimer F.H. J.Am.Chem.Soc., 1964, 91, 6066.
5. Archie W.C. Jr., Westheimer F.H., J.Am.Chem.Soc., 1973, 95, 5955.
6. Ramirez F., Marecek J.F., Tetrahedron Lett., 1977, 967.
7. Aksnes G., Eide A.I., Phosphorus 1974, 4, 209.
8. Aksnes G., Khall F.Y., Majewski P.J., Phosphorus and Sulfur 1977, 3, 157.
9. Skowrońska A., Pakulski M., Michalski J., J.Am.Chem.Soc., 1979, 101, 1979.
10. Sarma R., Ramirez F., McKeever B., Marecek J.F., Prasad V.A.V., Phosphorus and Sulfur, 1979, 5, 323.
11. Font Friede J.J.H.M., Trippett S., J.C.S.Chem.Comm., 1980, 157.
12. Dillon K.B., Platt A.W.G., Waddington T.C., J.C.S. Chem.Comm., 1979, 889.

RECEIVED July 2, 1981.

NEW PHOSPHORUS LIGANDS AND COMPLEXES (INCLUDING CATALYTIC PROPERTIES)

Metal Chelates of Aminoalkylphosphonic Acids

Stabilities, Properties, and Reactions

ARTHUR E. MARTELL

Department of Chemistry, Texas A&M University, College Station, TX 77843

Stabilities

Alkylphosphonate groups are monodentate donors of intermediate hardness that may replace carboxylate groups in chelating agents such as nitrilotriacetic acid (NTA), ethylenediaminetetraacetic acid (EDTA), and diethylenetriaminepentaacetic acid (DTPA). The corresponding ligands nitrilotrimethylenephosphonic acid (NTP), **1**, ethylenediaminetetramethylenephosphonic acid (EDTP), **2**, and diethylenetriaminepentamethylenephosphonic acid (DTPP), **3**, form metal chelates that are generally more stable than those of the analogous aminopolycarboxylates. The stability constant data in Table I (1) for representative alkaline earth and transition metal chelates of these ligands show mixed effects. The alkaline earth NTP chelates are more stable than those of NTA, while the reverse is the case for EDTP and EDTA. For both NTP and EDTP, the transition metal chelates are considerably more stable than

Table I
Comparison of Stabilities of Metal Chelates of Aminopolyacetate and Aminopolymethylenephosphonate Ligands*

	Log K_f^+					
Metal Ion	NTP	NTA	EDTP	EDTA	HEDTP	HEDTA
Mg^{2+}	7.2	5.47	8.43	8.83		
Ca^{2+}	7.5	6.39	9.36	10.61	5.62	8.2
Co^{2+}	14.4	10.38	17.11	16.26		
Ni^{2+}	11.1	11.50	16.38	18.52		
Cu^{2+}	17.4	12.94	23.21	18.70	16.29	17.5
Zn^{2+}	16.4	10.66	18.76	16.44		

* $t = 25.0°C$, $\mu = 0.10$ (KNO_3); + $\text{Log } K_f = [ML^{-(n-2)}]/[M^{2+}][L^{-n}]$ where H_nL represents the ligand.

0097–6156/81/0171–0459$05.00/0

those of the aminopolycarboxylate analogs, except for Ni(II), which shows a curious and thus far unexplained reversal. The metal chelates of hydroxyethylethylenediaminetriacetic acid (HEDTA) seem to be more stable than those of hydroxyethylethylenediaminetrimethylenephosphonic acid (HEDTP), 4, but the lack of data for the latter precludes a more complete comparison for these ligands.

For metal ions of higher charge, the phosphonates are clearly superior to the corresponding aminoacetates (2). An example of this effect may be found in the comparison of the stabilities of the Fe(III) chelates of N,N'-bis(o-hydroxybenzyl)-N,N'-ethylenediaminedimethylphosphonic acid (HBEDPO), 5, and the corresponding diacetic acid ligand N,N'-bis(o-hydroxybenzyl)-N,N'-ethylenediaminediacetic acid (HBED), 6. Although the latter was specifically designed to be an effective chelating ligand for iron(III), the iron(III) chelates of the former ligand are about 4-5 orders of magnitude more stable. Although only qualitative evidence is generally available for the chelates of aminopolyalkylphosphonic acids of tripositive and tetrapositive metal ions, it appears that in such complexes phosphonate donor groups are increasingly more effective, relative to carboxylate donor groups, as the charge of the metal ions increase. The difference in stabilities of the metal chelates of ligands 1-6 and the corresponding ligands containing aminoacetate functions are both enthalpic and entropic in origin (2,3). The higher charge of the phosphonate group leads to increased mutual coulombic repulsion when more than one phosphonate donor is coordinated to a metal ion, an effect that is increasingly compensated for as the charge of the central metal ion increases. A related effect involves hydration energy of the free ligands that is lost on formation of the metal chelates, which is compensated for by the energy of coordinate bond formation, which increases with the charge of the metal ion. The entropy effects resulting from the release of solvated water molecules from both the anion and cation favor metal chelate formation. This effect increases rapidly with the charge of the central metal ion.

Insolubility of Protonated Chelates

The usefulness of aminophosphonic acids as sequestering agents for highly charged metal ions is somewhat impaired by the strong tendencies of the metal chelates of these ligands to form highly protonted species that tend to precipitate in neutral or weakly acid solution. Thus metal chelates of the ligand EDTP in which the four phosphonate groups are coordinated to a metal ion, have four negative oxygen donors that are not coordinated to the metal. These chelates readily combine with hydrogen ions, forming protonated chelates of the type $MH_nL^{(8-n-m)^-}$, where m is the charge of the metal ion and n may vary from 1-4. The reduction of ionic charge and intermolecular hydrogen bonding resulting from extensive protonation tends to induce precipitation of the metal chelates. Thus the iron(III) chelate of EDTP is insoluble in the weakly acid to neutral pH range, while at higher pH the chelate species become more soluble as the result of proton neutralization and increased negative charge of the complex.

Thermal Degradation in Solution

Another problem encountered in the industrial applications of aminopolymethylenephosphonic acid chelating agents is their hydrolytic instability at high temperature in aqueous solution. It has been found (4), for example, that NTP decomposes about 100 times more rapidly than NTA at temperatures above 200°C in aqueous solution. NMR evidence indicates that NTP undergoes hydrolysis to aminomethylphosphonic acid, **11**, and hydroxymethylphosphonic acid, **10**, as indicated by the following reaction scheme. This type of hydrolytic instability is much different from the decomposition route of NTA, which has been found to occur through a decarboxylation process. The high basicity of the NTP anion would be expected to increase the tendency toward hydrolysis by increasing the population of protonated species such as **7** at high pH.

7 → 8 + 9

9 + HOH → 10 (HOCH$_2$—PO$_3^{2-}$)

8 + H^+ → ^-O_3P—CH_2—N^+H_2—CH_2—PO_3^{2-} → 11 ($^-O_2P(O)$—CH_2—NH_2) + 9

Polyphosphates as Ligands

It is interesting to note that condensed phosphate ligands such as tripolyphosphate (TPP), ATP, and ADP contain negative donor oxygens that are harder bases than those of the phosphonates, and have characteristically different metal ion affinities. The stabilities of the alkaline earth chelates of these ligands are accordingly higher relative to those of the transition metal ions than the relative stabilities of the phosphonates described in Table I. For example the stability constants of the Mg(II) and Ca(II) chelates of TPP are $10^{5.67}$ and $10^{5.20}$, respectively, while those of Co(II) and Ni(II) are only $10^{6.94}$ and $10^{6.75}$, respectively. Similarly, the stability constants of the Mg(II) and Ca(II) chelates of ATP are $10^{4.06}$ and $10^{3.77}$, while those of Co(II) and Ni(II) are only $10^{4.63}$ and $10^{5.02}$, respectively. The relative stabilities of the Mg(II) and Ca(II) chelates of these ligands is $Mg^{2+} > Ca^{2+}$, which is a reversal of the order observed for the corresponding chelates of the aminopolycarboxylate and aminopolyphosphonate ligands.

Pyridoxal-Catalyzed Dephosphonylation

Because of their widespread occurrence in biological systems the phosphonate analogs of natural amino acids such as aminomethylphosphonic acid (AP), 2-aminoethylphosphonic acid (AEP), and 2-amino-3-phosphonopropionic acid (APP), their reactions are of interest as models for biological processes. These aminophosphonic acids form metal chelates having structures analogous to those of the corresponding aminocarboxylic acids. They also form Schiff bases (7) which in the form of their dipolar ions, or in the form of their metal chelates, are more stable (7) than those formed by the analogous aminocarboxylic acids (glycine, β-alanine, and aspartic acids, respectively). The metal chelates of the Schiff base of APP have been found to undergo an interesting and novel dephosphonylation reaction (8), which required a prior transamination step. The products of the reaction are alanine and inorganic phosphate, while the pyridoxal is regenerated, thus providing the basis for a catalytic process.

Literature Cited

1. Martell, A. E.; Smith, R. M. "Critical Stability Constants", Vol.1, Plenum Press, New York, 1974.
2. Martell, A. E. Pure & Appld. Chem., 1978, 50, 813.
3. Pitt, C. G.; Martell, A. E."Inorganic Chemistry in Biology and Medicine"; American Chemical Society, Washington, D.C., 1980 pp.279-312.
4. Martell, A. E.; Motekaitis, R. J.; Fried, A. R.; Wilson, J. S.; MacMillan, D. T. Can. J. Chem., 1975, 53, 3471.
5. Smith, R. M.; Martell, A. E. "Critical Stability Constants", Vol.4, Plenum Press, New York, 1976, p.63.
6. Smith, R. M.; Martell, A. E. "Critical Stability Constants", Vol.2, Plenum Press, New York, 1975.
7. Langohr, M.; Martell, A. E. J. Inorg. Nucl. Chem., 1978, 40, 149.
8. Martell, A. E.; Langohr, M. J.C.S. Chem. Commun., 1977, 342.

RECEIVED June 30, 1981.

Transition VIB Metal π-Complexes of λ^3- and λ^5- Phosphorins and Some of Their Reactions

K. DIMROTH

Fachbereich Chemie, Philipps-Universität, Hans-Meerwein-Strasse, D-3550 Marburg, FRG

In spite of the aromaticity of λ^3-phosphorins 1 (1) the chemistry of 1 is quite different from the pyridines (2). The parent λ^3-phosphorin 1a (3) is too sensitive for extensive chemical studies. Therefore Märkl and we have investigated the chemical reactions of 2,4,6-trisubstituted λ^3-phosphorins such as 1b (4) and 1c. Two reasons are mainly responsible for the different chemical properties of pyridines and λ^3-phosphorins: i) The two coordinate nitrogen is more electronegative than the two coordinate phosphorus; ii) In all λ^3-phosphorins phosphorus is the reactive moiety which partakes therefore in chemical reactions. Reason i) is the cause that phosphorins are electron rich aromatic heterocycles. This is proven by physical methods, e.g. by PE-spectra (5,6) and by reactions with $M(CO)_6$ (M=Cr, Mo, W) which give rise to $\eta^6\pi^6$ $M(CO)_3$ complexes 2 (7), whereas in phenyl substituted pyridines the phenyl substituents are complexed (8). Point ii) can be illustrated by the reaction with nucleophiles. Whereas pyridines add the nucleophile to C-2 of the ring, λ^3-phosphorins add nucleophiles (Me^-, Ph^-, OMe^-) to the phosphorus atom, producing λ^4-phosphorin anions 3 which, by addition of electrophiles (Me^+, Et^+) to the phosphorus atom, afford λ^5-phosphorins 4 (9). ^{13}C NMR spectroscopy clearly shows that λ^5-phosphorins 4, containing also a planar heterocyclic ring (10), are phosphorus ylids with a delocalized negative charge in the pentadienyl part of the ring (11,12).

The pyridine nitrogen atom easily adds hard (H^+, CH_3^+) and soft ($Cr(CO)_5$, $HgAc_2$) acids. In contrast λ^3-phosphorins add soft acids only, giving P-complexes (13,14). Complexes of the type λ^3-phosphorin $P{\rightarrow}Hg(Ac)_2$ are important intermediates for the synthesis of 1,1-dialkoxy- and many other λ^5-phosphorins (e.g. 4b) in the presence of nucleophiles (ROH, ArOH, R_2NH) in a redox process yielding Hg (15). Another important reaction of the λ^3-phosphorins is the addition of two radicals ($\cdot NPh_2$, $\cdot OR$, $\cdot Cl$) to the phosphorus atom (16,17). The 1,1-substituents, especially in the 1,1-dichloro-λ^5-phosphorins 4c, can be easily exchanged for many nucleophiles (Me^-, OR^-, SR^-, F^-, etc.). This leads to a large number of λ^5-phosphorins 4 with varying 1,1-substituents at phosphorus. The remainder of the phosphorin ring is unaltered (17,2).

0097-6156/81/0171-0463$05.00/0

1 $\xrightarrow[-3CO]{+Cr(CO)_6}$ **2** $\xrightarrow{+R^{1\ominus}}$ **6** $\xrightarrow{+R^{1'\oplus}}$ **7**

1 $\xrightarrow{+R^{1\ominus}}$ **3** $\xrightarrow{+R^{1'\oplus}}$ **4** $\xrightarrow[-3CO]{+Cr(CO)_6}$ **5**

1 a) $R^2=R^4=R^6=H$

b) $R^2=R^4=R^6=Ph$

c) $R^2=R^4=R^6=$*t*-But

4 a) R^1, $R^{1'}$ =Alkyl, aryl

b) R^1, $R^{1'}$ =OCH_3

c) R^1, $R^{1'}$ =Cl

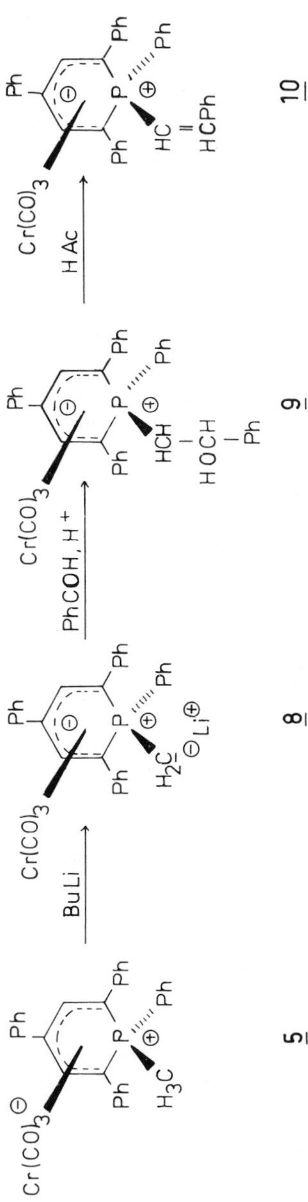
BuLi
PhCOH, H+
H Ac
5
8
9
10

Boat-like $\eta^5\pi^6$ $Cr(CO)_3$ complexes 5 (18) are formed (19,20) by treating 2,4,6-substituted λ^5-phosphorins with $Cr(CO)_6$. ^{13}C NMR spectroscopy confirms a phosphorus ylid structure, and the ^{13}C-^{31}P coupling constants of the 1,1-substituents allow a determination of their exo/endo configuration in agreement with X-ray analysis (21), the larger substituent always being in *exo* position. λ^5-Phosphorin $Cr(CO)_3$ complexes 5 and 7 also can be synthesized from λ^3-phosphorin $Cr(CO)_3$ complexes 2. Nucleophiles such as Me^-, Et^-, Ph^-, OCH_3^-, add to the *exo* position at the phosphorus atom giving anions 6 which add electrophiles (Me^+, Et^+) to the *endo* position producing the same or stereoisomeric λ^5-phosphorin $Cr(CO)_3$ complexes 5 or 7 (22,23). Mild oxidation of the λ^5-phosphorin $Cr(CO)_3$ complexes allows regeneration of the λ^5-phosphorins 4.

The ylid character of λ^5-phosphorins and their $Cr(CO)_3$ complexes again is evident when one or both groups on phosphorus are CHR_2 as one can abstract a proton giving a carbanion. Reaction with electrophiles (e.g. D^+, CH_3^+, and RCHO) causes side chain addition. No Wittig olefination is found with aldehydes. Instead a 1(2'-hydroxy) product 9 is formed which can be dehydrated to the λ^5-phosphorin derivative 10.

Literature cited

1. Ashe III, A. J. *J. Acc. Chem. Res.* 1978, 11, 153.
2. Dimroth, K. *Top. Curr. Chem.* 1973, 38, 1.
3. Ashe III, A. J. *J. Am. Chem. Soc.* 1971, 93, 3293.
4. Märkl, G. *Angew. Chem. Engl.* 1966, 5, 846.
5. Oehling, H.; Schäfer, W.; Schweig, A. *Angew. Chem. Engl.* 1971, 10, 656.
6. Batich, C.; Heilbronner, E.; Hornung, V.; Ashe III, J. A.; Clark, D. T.; Cobley, U. T.; Kilcast, D.; Scalan, I. *J. Am. Chem. Soc.* 1973, 95, 928.
7. Deberitz, J.; Nöth, H. *Chem. Ber.* 1970, 103, 2541.
8. Deberitz, J.; Nöth, H. *J. Organomet. Chem.* 1973, 61, 271.
9. Märkl, G.; Lieb, F.; Merz, A. *Angew. Chem. Engl.* 1967, 6, 87.
10. Thewalt, U. *Angew. Chem. Engl.* 1969, 8, 769.
11. Bundgaard, T.; Jakobsen, H. J.; Dimroth, K.; Pohl, H. H. *Tetrahedron Lett.* 1974, 3179.
12. Dimroth, K.; Berger, S.; Kaletsch, H. *Phosphorus and Sulfur* 1981 in press.
13. Deberitz, J.; Nöth, H. *Chem. Ber.* 1973, 106, 2222.
14. Kanter, H.; Dimroth, K. *Tetrahedron Lett.* 1975, 541.
15. Dimroth, K.; Städe, W. *Angew. Chem. Engl.* 1968, 7, 881.
16. Dimroth, K.; Hettche, A.; Kanter, H.; Städe, W. *Tetrahedron Lett.* 1972, 835.
17. Kanter, H.; Mach, W.; Dimroth, K. *Chem. Ber.* 1977, 110, 395.
18. Debaerdemaeker, T. *Acta Crystallogr., Sect. B* 1979, 35, 1686.
19. Lückoff, M.; Dimroth, K. *Angew. Chem. Engl.* 1976, 15, 503.

20. Dimroth, K.; Lückoff, M.; Kaletsch, H. Phosphorus and Sulfur 1981 in press.
21. Dimroth, K.; Berger, S.; Kaletsch, H. Phosphorus and Sulfur 1981 in press.
22. Dimroth, K.; Kaletsch, H. Polish J. Chem. 1981, 6 in press.
23. Dimroth, K.; Kaletsch, H. Angew. Chem. 1981 in press.

RECEIVED July 7, 1981.

96

Synthesis of Transition Metal Phosphoranides

Conversion of Bicyclic Phosphoranes into Phosphoranides and Phosphane Adducts with Transition Metal Derivatives

J. G. RIESS, F. JEANNEAUX, P. VIERLING, and J. WACHTER[1]

Laboratoire de Chimie Minérale Moléculaire (ERA 473), Université de Nice, 06034 Nice, France

A. GRAND

Laboratoire de Chimie, Département de Recherches Fondamentales, Centre d'Etudes Nucléaires de Grenoble 85X, 38041 Grenoble, France

Phosphoranide anions **1** - i.e. species based on pentacoordinated phosphorus having a lone pair as one of its five substituents - were proposed by Wittig and Maercker as early as 1967, to act as intermediates or transition states in nucleophilic substitutions at tricoordinated phosphorus (1). Shortly afterwards in 1969, Hellwinkel brought indirect evidence for the formation of phosphoranide **3** in equilibrium with the carbanion **4**, since the action of a base on phosphorane **2** gave, after acidic treatment, a mixture of **5** and **6** in proportions that depend on the experimental conditions (2).

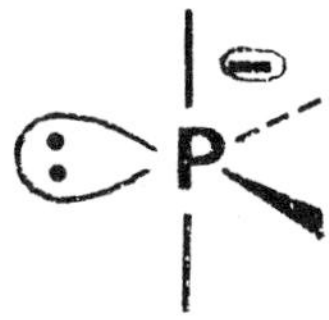

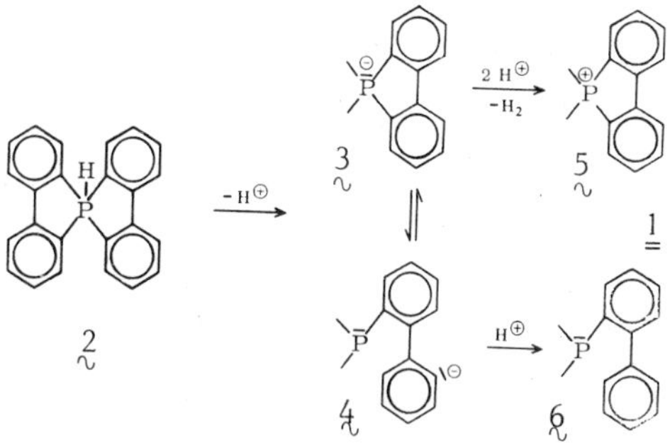

[1] Current address: Universität Regensburg, Fachbereich Chemie, 8400 Regensburg, Universitätsstrasse 31, FRG

0097-6156/81/0171-0469$05.00/0

Similarly, Savignac et al obtained the alkylated compound 8 when treating phosphorane 7 with NaH, $NaNH_2$ or $BuLi/R_2NH$, then with MeI, which strongly suggests a phosphoranide intermediate (3).

1) $NaNH_2$ 2) IMe 7 8 2

On the other hand, evidence *against* an intermediate hypervalent tetracoordinated phosphorus anion in nucleophilic substitutions at phosphorus in tertiary phosphines was put forward by Kyba: the fact that the substitution reaction 3 occurs with complete inversion of configuration at phosphorus was interpreted to mean that it proceeds without even one pseudorotation of 9, which makes the passage through such an intermediate unlikely (4).

RLi 9 3

It was only in 1978 that the first *direct* observation of a phosphoranide lithium salt was made by Granoth and Martin (5). By allowing the phosphonium trifluoromethane sulfonates 10 to react with $LiAlH_4$ they obtained new compounds which were characterized by NMR ($\delta\,^{31}P$ = −35 ppm) and assigned structure 11. Further evidence was gained by protonation, which afforded phosphoranes 12; the latter reaction could be reversed under the action of $LiAlH_4$:

$LiAlH_4$ H^+ $LiAlH_4$ 10 11 12 4

More recently, Munoz et al assigned a signal at +78.4 ppm to the presence of phosphoranide anion 14 in a dynamic equilibrium mixture 5, obtained by treating 13 with triethylamine or pyridine (6) :

Base 13 $(Base\text{-}H)^+$ $(Base\text{-}H)^+$ 14 $(Base)^+$ 5

Transition metal compounds having a phosphoranide *ligand* remained unknown. The approach we chose to accede to this new class of compounds was to provoke the intramolecular addition of an ionic fifth substituent on the phosphorus atom of a phosphane previously coordinated to the transition metal. The anionic site was created by abstraction of a proton from a secondary amine.

The *aminophosphoranide* molybdenum adduct 16 was obtained in 66% yield as yellow crystals by allowing LiMe to react with the cationic adduct 15 in a thf/ether solution (3:1) at -20°C. Infrared monitoring of the reaction showed that the ν(CO) vibrations of 15 at 1850 and 1978 cm^{-1} had completely disappeared after 30 mn, while two new absorptions had developed at 1855 and 1945 cm^{-1}. The evolution of methane was ascertained by IR. It appears indefinitely stable at room temperature and melts (with decomp) at 145°C.

15 16 17

Compound 16 exhibits a single resonance in the $^{31}P\{^1H\}$NMR spectrum at 43.9 ppm, which is an unusual location when compared to the 185-200 ppm range usually found for Mo-P(III) adducts (7). The ^1H spectrum shows a single sharp signal for the C_5H_5 protons at 5.32 ppm in $CDCl_3$.

The most prominent features of the molecule as established by X-ray diffraction are : the 5-connected character of the metal-bound phosphorus atom, i.e. the presence of the original phosphoranide ligand; the shortest Mo(II)-P bond reported so far (2.38Å), 0.07-0.14Å shorter than those found in complexes having the Cp-Mo(II)PR_3 pattern; the presence of the hitherto unknown Mo<P,N cycle (Mo-N: 2.23Å) which is all the more remarkable in view of the low basicity expected from a P-bond nitrogen atom; one of the longest P-N bonds known (1.91Å). Also very unusual is the location of the oxygen atoms in the equatorial sites and of the phenyl group in an apical site of the almost perfect bipyramidal arrangement of the substituents on phosphorus.

Similar behavior was observed with the tungsten analog of 16, giving in 90% yield an orange-yellow crystalline compound ($\delta^{31}P$ = 26.4 ppm; J_{P-W} = 232 Hz) whose spectral and analytical data are consistent with the formulation.

Upon heating at 60°C for 4 h, phosphoranide 16 converts into the isomeric stable compound 17 which also contains a bipyramidal phosphoranide ligand. But in 17 the two additional electrons needed by the metal to complete its 18-electron valence shell are provided by an oxygen instead of the nitrogen atom, thus giving the new Mo<O,P cycle.

Phosphoranide 17 shows a singlet at 23.8 ppm in the ^{31}P NMR. The prominent features of its crystal structure (9) are : as in 16 a very short P-Mo bond (2.37Å); a very long P-O(Mo) bond (1.893Å) compared to the other P-O bond (1.653Å); and the fact that the oxygen atoms recover their usual more favorable apical positions, which may be the driving force of this conversion.

When compound 18, the iron analog of 15, was treated under similar conditions with LiMe, it did not yield a phosphoranide, but exhibited another original behavior: it underwent the abstraction of a proton, but also the cleavage of the Fe-N bond, with formation of a P-N bond (which converts the bidentate cyclic ligand into a monodentate, through P, bicyclic one), and the migration of the phenyl group from phosphorus to iron, to yield compound 19!

18 19

The structure of 19 was established by X-ray crystallography. It shows a short Fe-C(phenyl) σ-bond (2.04Å) when compared to that found in $(\eta^5\text{-Cp})Fe(CO)(PPh_3)(\eta^1\text{-}C_6H_5)$ (2.14Å) (10) and the shortest P-Fe bond observed so far (2.105Å) which may arise from a high π-accepting capability and/or a small cone-angle of the constrained bicyclic ligand (11).

That these transformations can be reversed under the action of an acid is even more surprising, and seems to have no precedent in the literature. This new behavior probably entails the assistance of the N-H group. It is the action of an acid or of a base on that group that triggers the redistribution of bonds about phosphorus and iron, probably in a synchronous pair of 1,2 shifts at the P-Fe bond, that is perhaps uniquely attributable to the transannular relationship of N and P in this flexible ligand.

Literature Cited

1 Wittig, G.; Maercker, A. Organometal. Chem. 1967, 8, 91.
2 Hellwinkel, D. Chem. Ber. 1969, 102, 528.
3 Savignac, P.; Richard, B.; Leroux, Y., Burgada, R. J. Organometallic Chem. 1975, 93, 331.
4 Kyba, E.P. J. Am. Chem. Soc. 1979, 98, 4805.
5 Granoth, I.; Martin, J.C. J. Am. Chem. Soc. 1978, 100, 7434; 1979, 101, 4623.
6 Garrigues, B.; Koenig, M.; Munoz, A. Tetrahedron Lett. 1979 p.4205.
7 Wachter, J.; Jeanneaux, F.; Riess, J.G. Inorg. Chem. 1980, 19, 2169.
8 Reisner, M.G.; Bernal, I.; Brunner, H.; Doppelberger, J. J. Chem. Soc. Dalton Trans. 1979, 1664.
9 Wachter, J.; Mentzen, B.F.; Riess, J.G. Angew. Chem. 1981, 93, 299.
10 Semion, V.A.; Struchkov, Yu.T. Zh. Struckt. Khim. 1969 p. 88.
11 Grec, D.; Hubert-Pfalzgraf, L.G.; Riess, J.G.; Grand, A. J. Am. Chem. Soc. 1980, 102, 7133.

RECEIVED July 1, 1981.

Secondary Phosphino Macrocyclic Ligands

EVAN P. KYBA and HEINZ H. HEUMÜLLER

Department of Chemistry, The University of Texas at Austin, Austin, TX 78712

The use of secondary phosphines as ligands for transition metals has led to some interesting complexes (2,3). Our interest in unusual macrocyclic ligand systems led us to the synthesis of the first macrocycles (VI, VII, Scheme 1) which contain secondary-phosphino ligating sites.

In order to prepare these macrocycles it was necessary to synthesize the previously unknown o-bis(phosphino)benzene (IV). Lithium aluminum hydride reduction of the o-bis(phosphonate) III gave IV in 50% yield (^{31}P NMR, δ -123.8 ppm, J_{PH} = 207 Hz). The phosphonate III could be obtained in modest yields by the photo-activated nucleophilic aromatic substitution by sodium diethyl-phosphite on o-chloroiodobenzene in liquid ammonia solution (4).

Recently we have developed a more general approach to molecules exemplified by III. Thus the Diels-Alder cycloaddition of alkyne II and α-pyrone, followed by aromatization by loss of carbon dioxide, led to the isolation of III (72%) (5). Alkyne II was obtained in high yields, in two steps from dichloroacetylene and triethylphosphite via Arbuzov-type reactions (5). Since the intermediate chloroalkyne phosphonate I was isolable (90%), phosphorus nucleophiles other than triethylphosphite could be used to give unsymmetrical alkyne diphosphoryl species. We have demonstrated this approach by the reaction of I with Ph_2POEt and $PhP(OEt)_2$ (5).

Treatment of II in THF with two equivalents of n-butyllithium generated V, which gave VI and VII upon high dilution reactions (6) in THF with bis(3-methanesulfonyloxypropyl)sulfide and bis(3-chloropropyl)methylamine, respectively in yields of about 80%.

Both macrocycles were highly air sensitive, colorless, distillable viscous oils, and both were isolated as a mixture of two

0097-6156/81/0171-0473$05.00/0

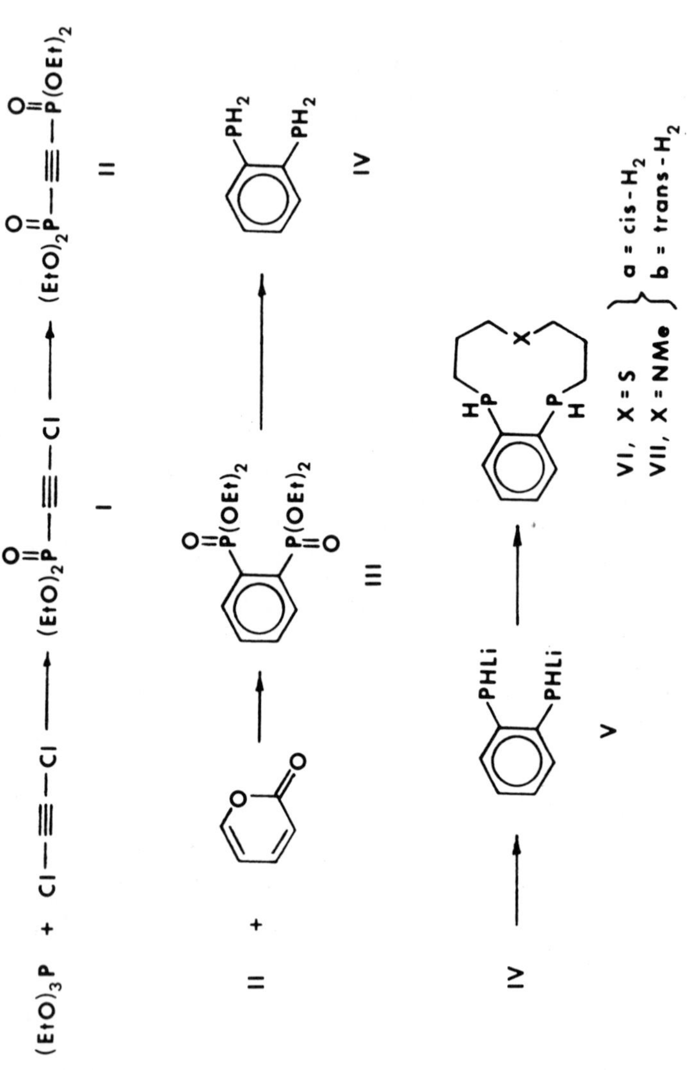

Scheme 1.

$(EtO)_3P$ + Cl—≡—Cl ⟶ $(EtO)_2P(O)$—≡—Cl (I) ⟶ $(EtO)_2P(O)$—≡—$P(O)(OEt)_2$ (II)

II + (2-pyrone) ⟶ o-$C_6H_4[P(O)(OEt)_2]_2$ (III) ⟶ o-$C_6H_4(PH_2)_2$ (IV)

IV ⟶ o-$C_6H_4(PHLi)_2$ (V) ⟶ VI, VII

VI, X = S; VII, X = NMe; a = cis-H_2, b = trans-H_2

isomers. Thus VI in C_6D_6 exhibited two singlets in the proton-decoupled ^{31}P NMR spectrum at δ -58.1 and -64.0 ppm in an area ratio of 62:38, and similarly, VII in C_6D_6 gave singlets at -55.1 and -61.5 ppm (area ratio 69:31). The major isomers of VI and VII show no fluxionality at temperatures as low as -90°C in toluene solution as observed by ^{31}P NMR spectroscopy. In contrast, the minor isomers exhibited coalescence temperatures at ca. -60°C, and with further cooling, two peaks arose (low temperature limiting ^{31}P NMR absorptions at δ -70.2 and -60.0 ppm) for VI and three peaks at δ -74.1, -54.6, and -44.5 ppm for VII.

Treatment of a benzene solution of VI with an excess of nickel chloride in methanol resulted in the precipitation of a yellow powder. This was recrystallized from methanol to give air-stable yellow-orange crystals (60%) with the stoichiometry (VI)·$NiCl_2$·3MeOH (by combustion analysis). When this recrystallized complex was treated with excess aqueous sodium cyanide in the presence of C_6D_6 at room temperature for 10 min VI was regenerated, but now the ^{31}P NMR absorptions at δ -58.1 and -64.0 ppm were in an area ratio of 95:5. Similar treatment of VII with nickel chloride led to a dark oily precipitate which required considerable manipulation to partially purify it. Treatment with sodium cyanide as above led to the regeneration of VII also enriched in the major isomer (85:15).

Examination of Dreiding and Corey-Pauling-Koltung molecular models reveals that the trans-ligand (b) cannot chelate a metal ion without severe steric interactions while the cis-(a) can do so easily. The fact that the nickel (II) complexation-cyanide decomplexation sequence leads to enrichment of the major isomers of VI and VII leads us to postulate that the major isomers are the cis-(a) species. We have equilibrated VIa and VIb thermally to a ratio of 58:42; thus the macrocyclization gives a non-equilibrium mixture favoring the cis isomer, analogous to PPh and AsMe macrocyclizations that we have described recently (1,7,8).

Acknowledgement.

We are most grateful to the Air Force Office of Scientific Research (AFOSR-79-0090) and the Robert A. Welch Foundation (F-573) for generous support of this work.

Literature Cited.

1. For the previous paper in the series, Phosphino-macrocycles, see Kyba, E.P.; Davis, R.E.; Hudson, C.W.; John, A.M.; Brown, S.B.; McPhaul, M.J.; Liu, L.-L.; Glover, A.C. J. Am. Chem. Soc. 1981, 103, 0000.
2. Stelter, O. in "Topics in Phosphorus Chemistry", Volume 9, Griffith, E.J. and Grayson, M., Editors, Wiley Interscience, New York, 1972, p. 433.

3. Booth, G. in "Organic Phosphorus Compounds", Volume 1, Kosolapoff, G.M. and Maier, L., Editors, Wiley Interscience, New York, 1972, p. 433.
4. Bard, R.R.; Bunnett, J.F.; Traber, R.P. J. Org. Chem. 1979, 44, 4918.
5. Kyba, E.P.; Rines, S.P.; Owens, P.O.; Chou, S.-S.P.; Tetrahedron Lett. 1981, 22, 1875.
6. Kyba, E.P.; Chou, S.-S.P. J. Org. Chem. 1981, 46, 860.
7. Kyba, E.P.; John, A.M.; Brown, S.B.; Hudson, C.W.; McPhaul, M.J.; Harding, A.; Larsen, K.; Niedzwiecki, S.; Davis, R.E. J. Am. Chem. Soc. 1980, 102, 139.
8. Kyba, E.P.; Chou, S.-S.P. J. Am. Chem. Soc. 1980, 102, 7012.

RECEIVED June 30, 1981.

98

Dicoordinated and Tricoordinated Acyclic Phosphazenes as Complex Ligands

O. J. SCHERER, H. JUNGMANN, and R. KONRAD

University of Kaiserslautern, Paul-Ehrlich-Strasse, 6750 Kaiserslautern, FRG

Starting with the phospha(III)azene RR'N-P=NR, R=$(CH_3)_3C$, R'=$(CH_3)_3Si$, and $Pt(COD)_2$ the platinum(0) complex PtL_3 (A) with three phospha(III)azene ligands can be synthesized in high yield.

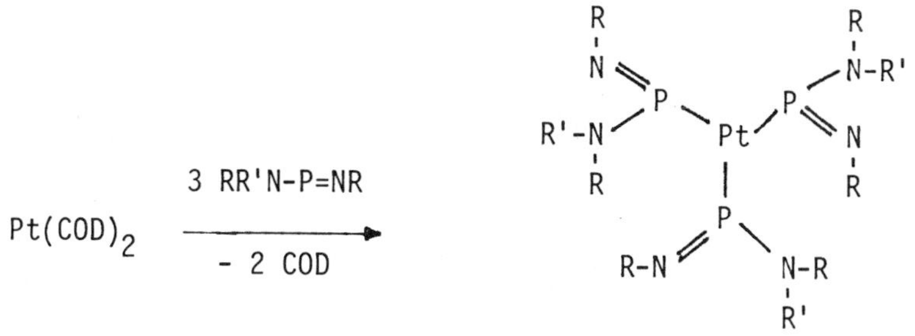

$$Pt(COD)_2 \xrightarrow[-\ 2\ COD]{3\ RR'N-P=NR} A$$

R=$(CH_3)_3C$, R'=$(CH_3)_3Si$

A

DNMR studies show that at ambient temperature the $(CH_3)_3Si$-groups of A are dynamic. This intramolecular 1.3-exchange can be stopped at about - 20° C.

PtL_3 (A) is a very useful starting material for the synthesis of a variety of new platinum(0) complexes containing phospha(III)-azenes as ligands (see scheme).

0097–6156/81/0171–0477$05.00/0

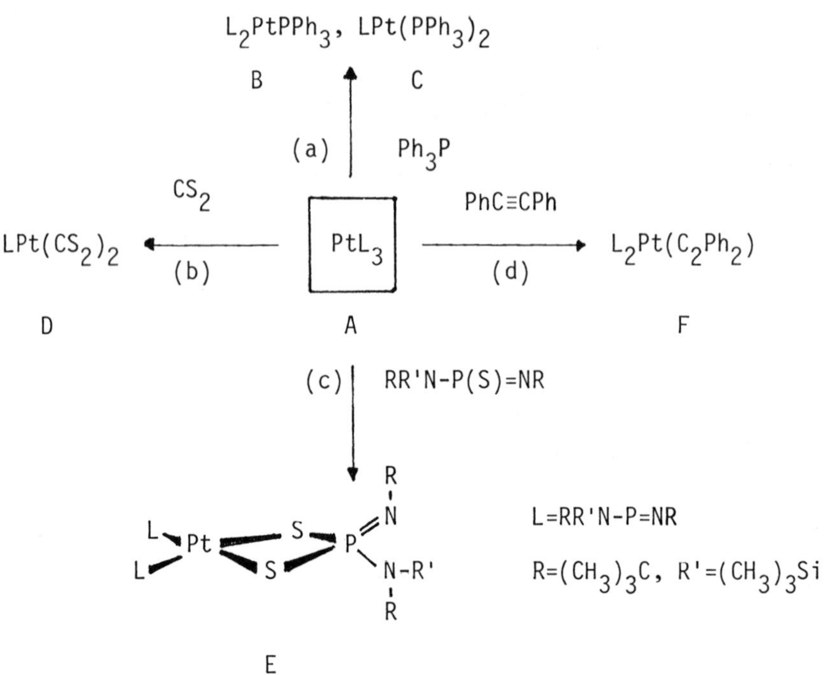

According to reaction (a) mixed substituted platinum(O)complexes of the hitherto unknown type PtL_2L' and $PtLL'_2$ (where L and L' are phosphorus ligands) can be isolated and characterized by nmr spectroscopy. Instead of Ph_3P other R_3P derivatives and Ph_3As can be used. Complex D is very labile. IR data give evidence for a σ-coordination of the CS_2 ligands. With the double ylide RR'N-P(S)=NR (reaction (c)) the four-membered chelate complex E with pentavalent phosphorus of coordination number four is formed. Compound F is an example of the well characterized alkyne complexes (here with the phospha(III)azene ligand L).

All platinum phospha(III)azene complexes (A → F) show dynamic behaviour for L. As described for A the 1.3 - $(CH_3)_3Si$-group exchange within the ligand L can be stopped at low temperature (ca. - 20 to - 30^0 C) for B → F. It is interesting to note that for E even at - 80^0 C the $(CH_3)_3Si$-group exchange of the R'-group is fast on the nmr time scale. All compounds with two phospha-(III) azene ligands L at about - 60^{0} C show hindered rotation around the Pt-L bond with formation of "cis" and "trans" isomers.

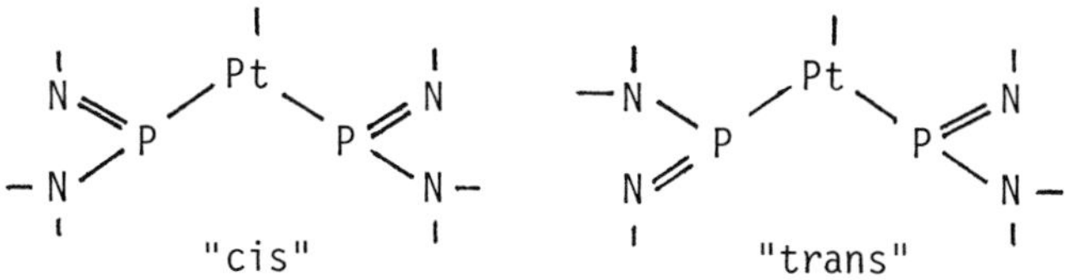

"cis" "trans"

With B as starting material all reactions of the scheme can be done in the same way with formation of $(Ph_3P)Pt(CS_2)_2$, $L(Ph_3P)Pt(C_2Ph_2)$ and $L(Ph_3P)PtS_2P(=NR)(NRR')$ (L=RR'N-P=NR, $R=(CH_3)_3C$, $R'=(CH_3)_3Si$).

Compound C with RR'N-P(S)=NR forms G,

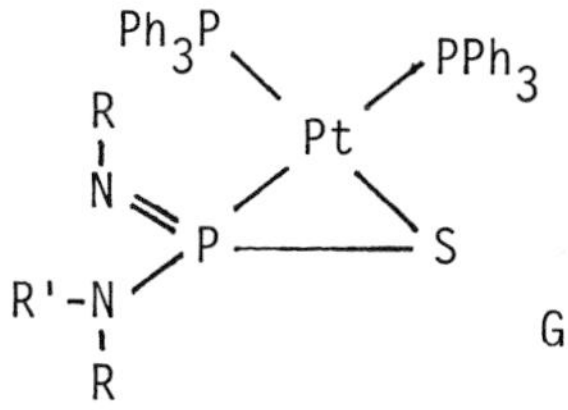

G

a complex with a derivative of monomeric metathiophosphoric acid as η^2-ligand [1].

Contrary to the free ligand RR'N-P(S)=NR, the $(CH_3)_3Si$-group of G can be hydrolyzed with formation of $(Ph_3P)_2PtSP(=NR)(NHR)$ ($R=(CH_3)_3C$); a platinum complex with the unknown RHNP(S)=NR as a side-on coordinated ligand.

[1] O. J. Scherer and H. Jungmann, Angew. Chem. 91, 1020 (1979), Intern. Edit. 18, 953 (1979).

RECEIVED June 30, 1981.

Metal Complexes of Amino(cyclophosphazenes)

V. CHANDRASEKHAR and S. S. KRISHNAMURTHY

Department of Inorganic and Physical Chemistry, Indian Institute of Science, Bangalore-560 012, India

M. WOODS

Department of Chemistry, Birkbeck College, University of London, Malet Street, London WC1E 7HX, England

Metal complexes of cyclophosphazenes have aroused considerable interest from both structural and biomedical points of view(1). The coordination of cyclophosphazenes to metal ions occurs in different ways: (a) through the skeletal nitrogen atoms, e.g., $N_4P_4(NHMe)_8.PtCl_2$(1); (b) through a skeletal and an exocyclic nitrogen atom, e.g., $W(CO)_4.N_4P_4(NMe_2)_8$(1); (c) through one skeletal nitrogen atom while another is protonated, e.g., $N_4P_4Me_8H.CuCl_3$(1); (d) through the phosphorus atom in the ring, e.g., $[N_3P_3Ph_4(CH_3)H]_2.PdCl_2$(2). Instances are also known where a protonated phosphazene species functions merely as a counter ion without any direct interaction with the metal(1). The factors that are responsible for the divergent behaviour of the cyclophosphazene ligands towards the metal ions are poorly understood and clearly more work is needed in this area. A study of the metal complexes of the (amino)cyclophosphazenes $N_3P_3R_6$[R = NHMe(I), NMe_2(II)], $N_4P_4R_8$[R = NHMe(III), R =NMe_2(IV)] and the spirocyclic phosphazene, $N_3P_3(HNCH_2CH_2NH)(NMe_2)_4$(V), is undertaken with a view to rationalising the ligating ability of the cyclophosphazenes in terms of electronic and steric factors associated with them.

Experimental Procedure

A solution of the metal chloride(hydrated or anhydrous) and the cyclophosphazene in the required stoichiometry(metal:ligand 1:1 or 1:2) in methanol or methyl ethyl ketone was heated under reflux for

0097-6156/81/0171-0481$05.00/0

4-8 hrs in the presence of dimethoxy propane. The solution was filtered and the solvent was removed from the filtrate *in vacuo* to obtain a solid. The solid was washed several times with petrol-benzene mixture (1:1) to remove excess ligand(if any). The complexes were crystallised from acetone or methanol. In the case of mercuric chloride complexes, addition of the warm solutions of the metal salt and the cyclophosphazene in methanol resulted in the precipitation of the complexes. The following complexes were isolated: I.2$HgCl_2$(VI) m.p. 230°(d), III.2$HgCl_2$(VII) m.p. 225°, III.$NiCl_2$(VIII) m.p. 192°-195°, V.2$HgCl_2$(IX)m.p.198°, $(V)_2$.$NiCl_2$(X) m.p. 186°, $(V)_2$.$CoCl_2$(XI) m.p.176°-180°. The dimethylamino derivatives(II and IV) did not form complexes(UV spectral evidence).

Discussion

The nickel chloride complex(X) of the spirocyclic phosphazene(V) is diamagnetic suggesting a square planar coordination around nickel. The P=N stretching vibration for the complex appears as two **split** bands(1215 and 1185 cm^{-1}) compared to a single band at 1200 cm^{-1} observed for the ligand. The phosphorus chemical shifts move upfield compared to those of the ligand; the spiro phosphorus atom is the one most affected [complex: δ_{PR_2} 21.9; $\delta_{P(spiro)}$ 25.5; $^2J(P\text{-}N\text{-}P)$ 31.6 Hz; ligand: δ_{PR_2} 26.7; $\delta_{P(spiro)}$ 35.5; $^2J(P\text{-}N\text{-}P)$ 40.0 Hz]. The proton resonances move downfield upon complexation, the most affected being the N-H and the N-CH_2 protons (complex: $\delta_{N\text{-}CH_3}$ 2.7; $\delta_{N\text{-}CH_2}$ 3.52: $\delta_{N\text{-}H}$ 4.7; ligand: $\delta_{N\text{-}CH_3}$ 2.6; $\delta_{N\text{-}CH_2}$ 3.34; $\delta_{N\text{-}H}$ 2.2). The complex exhibits a low molar conductance in MeCN. These results are consistent with the structure shown in Figure 1. The cobalt chloride complex(XI) of the spirocyclic phosphazene(V) has a magnetic moment of 4.6 B.M. and a high molar conductance in MeCN(280 mhos). Its electronic spectrum in acetone or MeCN is characteristic of a tetrahedral Co(II) complex(3) λ_{max}(CH_3CN): 692(895), 635(sh), 590(358). It is likely that one exocyclic and one endocyclic nitrogen atom from each ligand molecule are involved in coordination.

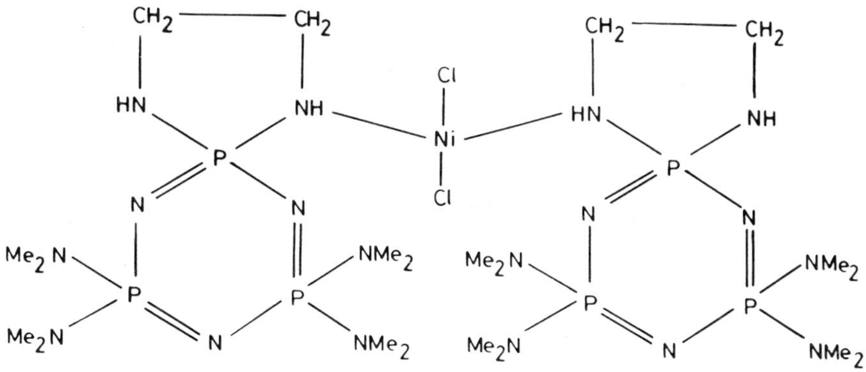

Figure 1. Possible structure of the nickel chloride complex of the spirocyclic phosphazene, $N_3P_3(NH\ CH_2CH_2NH)(NMe_2)_4$.

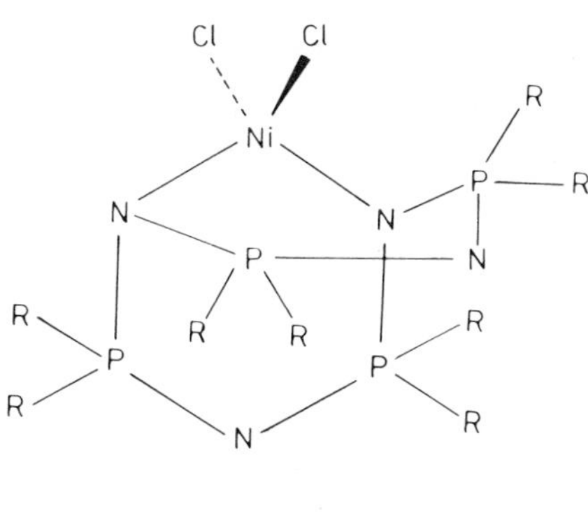

R = NHMe

Figure 2. Possible structure of the nickel chloride complex of $N_4P_4(NHMe)_8$.

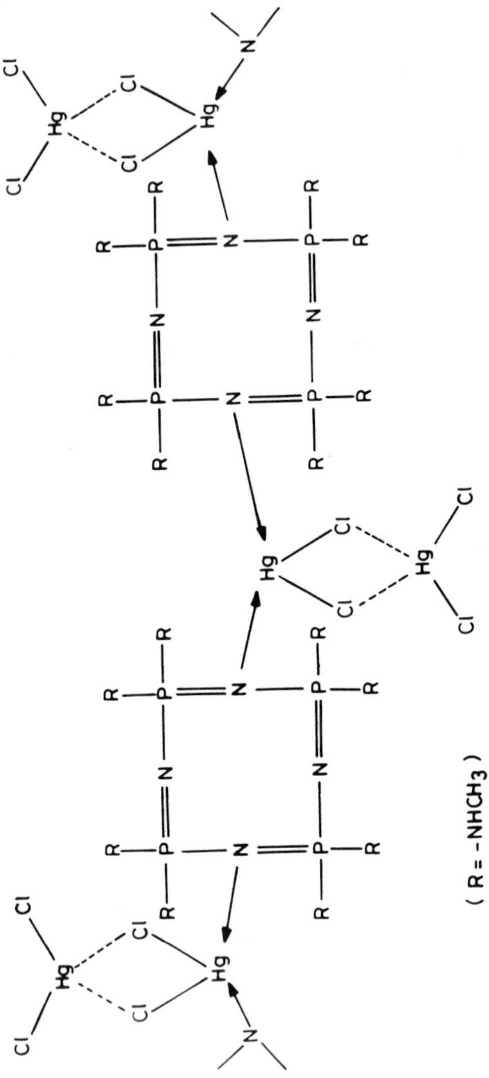

Figure 3. *Possible structure of the mercuric chloride complex of $N_4P_4(NHMe)_8$.*

The nickel chloride complex(VIII) of $N_4P_4(NHMe)_8$ is paramagnetic (3.2 B.M.). The phosphorus NMR spectrum shows a single line which shifts upfield by 4.0 δ compared to the ligand. A tetrahedral structure is proposed for this complex in which the metal is coordinated to two antipodal nitrogen atoms of the ligand (Figure 2).

The extreme insolubility and high melting points of the mercuric chloride complexes suggest that they are polymeric as illustrated for $N_4P_4(NHMe)_8 \cdot 2HgCl_2$ (VII)(Figure 3). A number of polymeric mercury complexes are known(4,5).

X-ray crystallographic studies reveal that both ring protonation and complexation affect the phosphazene in a similar way but the effect of protonation is more pronounced(1). For the hydrochloride adducts of amino(cyclophosphazenes), the proton NMR shifts move downfield whereas phosphorus chemical shifts are markedly upfield to those observed for the free bases. The ring P=N stretching frequency also undergoes an upward shift for the hydrochloride adducts(1). Similar trends are observed for the metal complexes prepared in the present study. The mode of coordination of cyclophosphazenes to transition metal ions mainly depends upon the ring size and the nature of the substituent. The rigid six-membered cyclotriphosphazenes have low propensity for forming metal complexes and clearly steric effects predominate.

Acknowledgement

The authors thank the University Grants Commission and the Council of Scientific and Industrial Research, New Delhi and the Overseas Development Administration, U.K. for support.

Literature Cited

1. Krishnamurthy,S.S.; Sau,A.C.; Woods,M. Adv. Inorg. Radiochem., 1978, 21, 41.
2. Schmidpeter,A.; Blanck,K.; Hess,H.; Riffel,H. Angew. Chem. Int. Edn. Engl., 1980, 19, 650.
3. Cotton,F.A.; Goodgame,D.M.L.; Goodgame,M. J. Am. Chem. Soc., 1961, 83, 4690.
4. Wirth,T.H.; Davidson,N. J.Am.Chem.Soc., 1964, 86, 4325.
5. Cheung,K.K.; Sim,G.A. J.Chem.Soc.,1965, p.5988.

RECEIVED June 30, 1981.

100

Structural and Magnetic Investigation on Transition Metal Complexes with Tripodal Polytertiary Phosphines

L. SACCONI

Istituto di Chimica Generale ed Inorganica, Università di Firenze, Istituto CNR, 39, Via J. Nardi 50132 Firenze, Italy

Tripodal tritertiary phosphines as 1,1,1-tris(diphenyl--phosphinomethyl)ethane, triphos, and tris(2-diphenylphosphino--ethyl)amine, np_3, are capable: a) to act as tridentate or tetradentate ligands; b) to promote several reactions with formation of several types of metal complexes which are presented.

By reaction of cobalt(II) aquacation with triphos and HSR (R = H, Me) different dinuclear metal complexes are formed with formulae $[(triphos)Co(\mu\text{-}SR)_2Co(triphos)]^{2+}$ and $[(triphos)Co(\mu\text{-}S)_2Co(triphos)]^{o,+,2+}$. These dinuclear complexes contain from 32 to 34 valence electrons. Those with 34 electrons present antiferromagnetic behavior. These properties are explained in terms of a molecular orbital treatment of the type suggested by Hoffmann and Burdett for unpuckered dinuclear metal complexes.

The ligands triphos or/ and np_3 in presence of compounds of iron, cobalt, nickel, rhodium, iridium and palladium, by reaction with THF solutions of white phosphorus, P_4, or yellow arsenic, As_4, form mononuclear or dinuclear sandwich complexes containing the cyclo-triphosphorus or cyclo-triarsenic units which behave as 3π-electrons rings.

The metal complexes formed have the general formulae $[(triphos)M(\eta^3\text{-}P_3)]$ (M = Co, Rh, Ir) and $[(np_3)Co(\eta^3\text{-}P_3)]$; $[(triphos)M'(\eta^3\text{-}P_3)M''(triphos)]^{+,2+,3+}$ (M = Co, Ni, Rh, Ir, Pd). The complexes can be also of the ethero-metal type.

An iron complex can be formed by using the ethyl derivative of triphos, p_3Et_6: $[(triphos)Co(\eta^3\text{-}P_3)Fe(p_3Et_6)]^{2+}$. Many of the complexes here presented are paramagnetic. The number of valence electrons range from 30 to 34. This unprecedented magnetic behavior can be accounted for by a molecular orbital treatment of the type suggested by Hoffmann.

0097–6156/81/0171–0487$05.00/0

In the case of "simple sandwich" complexes the X-ray diffraction studies have shown that the M-P(triphos), M-P(cyclo) and P-(cyclo)-P(cyclo) distances reveal effects of both the increase in the principal quantum number and the "lantanide contraction". For "triple decker sandwich" complexes the M-M distances increase whereas the P(cyclo)-P(cyclo) distances decrease with increasing number of valence electrons.

With the ligand np_3 one can obtain the complexes of general formula $[(np_3)M(\eta\text{-}P_4)]$ (M = Ni, Pd) which contain an intact tetrahedral molecule P_4 σ-bonded to the metal.

Also the complexes formed with the ligand triphos, cobalt(II) aquacations and CS_2 are of some interest. Among these compounds there is the CS_2-bridged dinuclear diamagnetic complex $[(triphos)Co(\eta\text{-}CS_2)Co(triphos)]^{2+}$. By using the ligand p_3Et_6 the unexpected complex complex $[(p_3Et_6)Fe\ S_2C(PEt_2\text{-}CH_2)_2C(CH_3)_2]^{2+}$ is obtained. This is the first fivecoordinated complex of iron(II) in a singlet state.

The tertiary phosphine PEt_3, in the presence of nickel(II) aquacations and H_2S, form diamagnetic tri- and enneanuclear dications having formulae $[Ni_3(\mu^3\text{-}S)_2(PEt_3)_6]^{2+}$ and $[Ni_9(\mu^3\text{-}S)_6(\mu^4\text{-}S)_3]^{2+}$. The first complex is formed by a trigonal bipyramidal kernel of two apical sulfur atoms and three equatorial nickel ions. The structure of the second compound consists of nine nickel ions located at the apices of two confacial octahedra. The six phosphines are bonded to the six external nickel ions. Six sulfur atoms lie practically in the planes of the two external triangular faces of the two octahedra, triply bridging all the nickel ions. The three internal sulfur atoms are quadruply bridging the nickel ions.

The contributions of all the coworkers are greatly acknowledged.

RECEIVED July 7, 1981.

101

The Use of Alkylaminobis(difluorophosphines) as Ligands to Stabilize Novel Binuclear Complexes

J. H. KIM, K. S. RAGHUVEER, T. W. LEE, L. NORSKOV-LAURITZEN, V. KUMAR, M. G. NEWTON, and R. B. KING

Department of Chemistry, University of Georgia, Athens, GA 30602

Alkylaminobis(difluorophosphines), $RN(PF_2)_2$ (R = CH_3, C_6H_5, etc.) can function as bidentate ligands in binuclear metal complexes by bridging a bonded pair of metal atoms to form a five-membered chelate ring (I). These bridges appear to hold together metal-metal bonds so that they survive chemical transformations which cleave unbridged metal-metal bonds. This property gives the transition metal coordination chemistry of the $RN(PF_2)_2$ ligands a richness which appears to rival that of the most versatile unsaturated hydrocarbon ligands in transition metal organometallic chemistry such as cyclopentadienyl and cyclooctatetraene.

In 1980 we published a survey (1) of our major results in this area as of late 1979. These results include extensive work on binuclear $CH_3N(PF_2)_2$ complexes of cobalt (2, 3, 4, 5) and nickel (6). This paper summarizes our more recent results in this area with particular emphasis on binuclear complexes of chromium, molybdenum, and tungsten as well as some new results on iron carbonyl derivatives.

Chromium, Molybdenum, and Tungsten Derivatives

Several compounds of the type $[RN(PF_2)_2]_nM_2(CO)_{11-2n}$ (M = Cr, Mo, and W; R = CH_3 and C_6H_5; n = 3, 4, and 5) have been prepared which, at least formally, are derivatives of the unknown binuclear metal carbonyls $M_2(CO)_{11}$. Thus pyrolysis of the molybdenum complexes $RN(PF_2)_2Mo(CO)_4$ (R = CH_3 and C_6H_5) at 100-120°C results in extensive rearrangement to give the yellow binuclear complexes $[RN(PF_2)_2]_3Mo_2(CO)_5$ (R = CH_3 and C_6H_5) shown by X-ray diffraction in the case R = C_6H_5 to have structure II (M = Mo). The chromium and tungsten analogues $[RN(PF_2)_2]_3M_2(CO)_5$ (M = Cr and W) can be obtained by photolysis of the corresponding metal hexacarbonyls with $CH_3N(PF_2)_2$ in a 1 to 1.5:1 ligand/metal mole ratio. Pyrolysis or photolysis of mixture of the $RN(PF_2)_2$ ligands and $M(CO)_6$ (M = Mo and W) in a 2 to 2.5:1 ligand/metal mole ratio gives mixtures of the yellow binuclear complexes $[CH_3N(PF_2)_2]_4M_2(CO)_3$ and $[CH_3N(PF_2)_2]_5M_2CO$ in the case of $CH_3N(PF_2)_2$ and the yellow binuclear complexes

0097-6156/81/0171-0489$05.00/0

$[C_6H_5N(PF_2)_2]_4Mo_2(CO)_3$ and $[C_6H_5N(PF_2)_2]_5Mo_2(CO)_2$ in the case of $C_6H_5N(PF_2)_2$. The complex $[CH_3N(PF_2)_2]_4Mo_2(CO)_3$ has been shown by X-ray diffraction to have structure III (M = Mo, R = CH_3). The compounds II and III are rare examples of structurally characterized binuclear molybdenum carbonyl complexes containing bridging carbonyl groups. In II where the two metal atoms are equivalent the carbonyl bridge is symmetrical whereas in III where the two metal atoms are non-equivalent the carbonyl bridge is unsymmetrical.

The addition reactions of the $RN(PF_2)_2$ ligands to the metal-metal triple bonds in the cyclopentadienylmolybdenum carbonyls $[R'_5C_5Mo(CO)_2]_2$ (IV: R' = H and CH_3) have also been studied. Thus reactions of IV with $RN(PF_2)_2$ under relatively mild conditions results in the simple adducts, $RN(PF_2)_2[Mo(CO)_2C_5R'_5]_2$ of apparent structure V. In addition reaction of $[(CH_3)_5C_5Mo(CO)_2]_2$ with $CH_3N(PF_2)_2$ under more vigorous conditions also results in CO loss to give $[(CH_3)_5C_5Mo(CO)(PF_2)_2NCH_3]_2$ formulated tentatively as VI because its infrared spectrum indicates the presence of bridging carbonyls but the absence of terminal carbonyls.

Iron Carbonyl Derivatives

Reactions of iron carbonyls with $CH_3N(PF_2)_2$ give products of the type $[RN(PF_2)_2]_nFe_2(CO)_{9-2n}$ which may be regarded as derivatives of $Fe_2(CO)_9$. The following compounds of this type have been reported in previous papers:

(1) $CH_3N(PF_2)_2Fe_2(CO)_7$ (n = 1): This orange complex is formed from the reaction of $Fe_2(CO)_9$ with $CH_3N(PF_2)_2$ in a 1:1 mole ratio; a more major product of this reaction is $CH_3N(PF_2)_2[Fe(CO)_4]_2$, a compound with two isolated $Fe(CO)_4$ groups without an iron-iron bond (6).

(2) $[CH_3N(PF_2)_2]_2Fe_2(CO)_5$ (n = 2): This orange complex is formed by the photolysis of $Fe(CO)_5$ with $CH_3N(PF_2)_2$ in a 1:1 mole ratio in diethyl ether or pentane (6, 7). An alternate method for preparing this compound involves the reaction of $Fe_3(CO)_{12}$ with $CH_3N(PF_2)_2$; an intermediate in this reaction is yellow $[CH_3N(PF_2)_2Fe(CO)_3]_2$ containing two square pyramidal iron atoms and no iron-iron bond (6, 7).

(3) $[CH_3N(PF_2)_2]_4Fe_2CO$ (n = 4): This is formed by the ultraviolet irradiation of $Fe_3(CO)_{12}$ with excess $CH_3N(PF_2)_2$ in diethyl ether (8). The infrared spectrum of this yellow complex indicated a terminal carbonyl group rather than the expected carbonyl group. Since this observation was initially somewhat puzzling, the structure of this complex was determined by X-ray diffraction. This revealed the unexpected structure VII in which one of the four $CH_3N(PF_2)_2$ ligands has undergone phosphorus-nitrogen bond cleavage to form separate PF_2 and CH_3NPF_2 units which function as bridging and terminal ligands, respectively.

The final member of this series, orange $[CH_3N(PF_2)_2]_3Fe_2(CO)_3$ (n = 3), has now been isolated in low yield from the photochemical reaction of $Fe_3(CO)_{12}$ with $CH_3N(PF_2)_2$. The infrared spectrum of

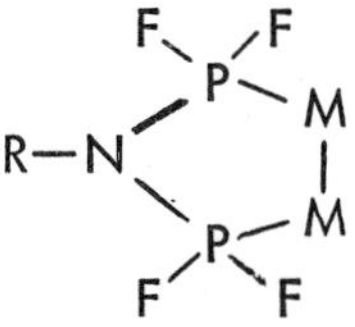

I

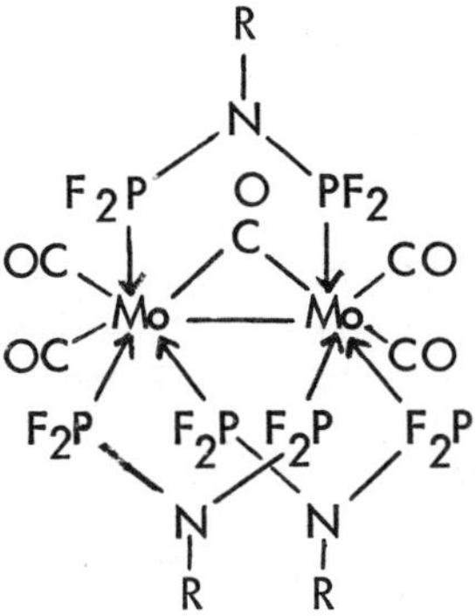

II

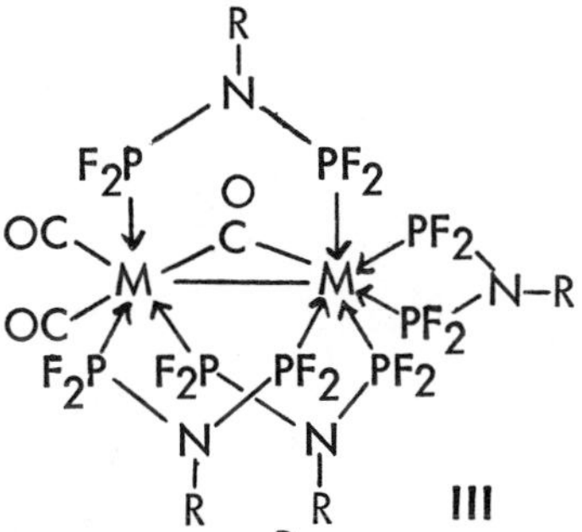

III

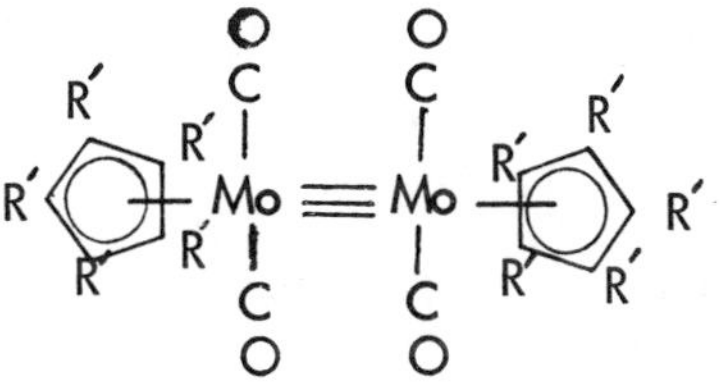

IV

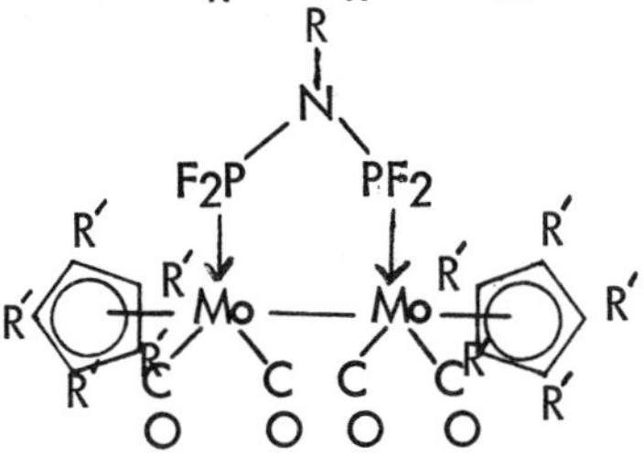

V

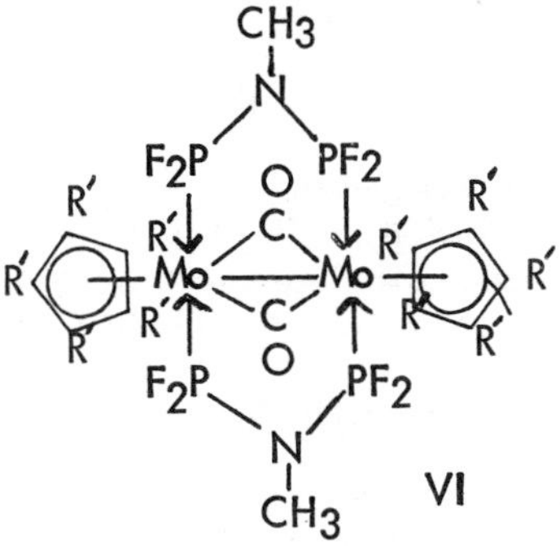

VI

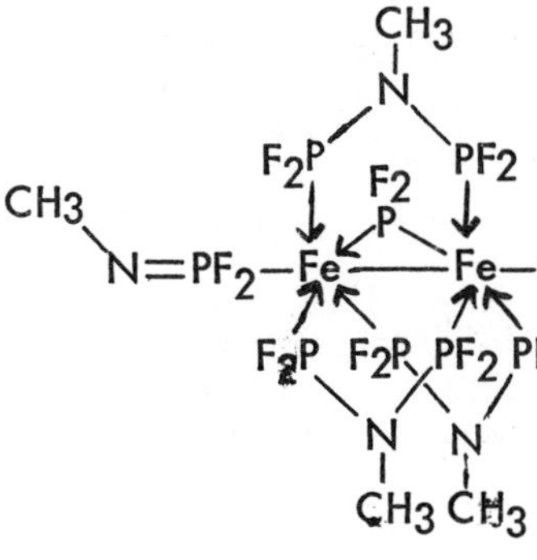

VII

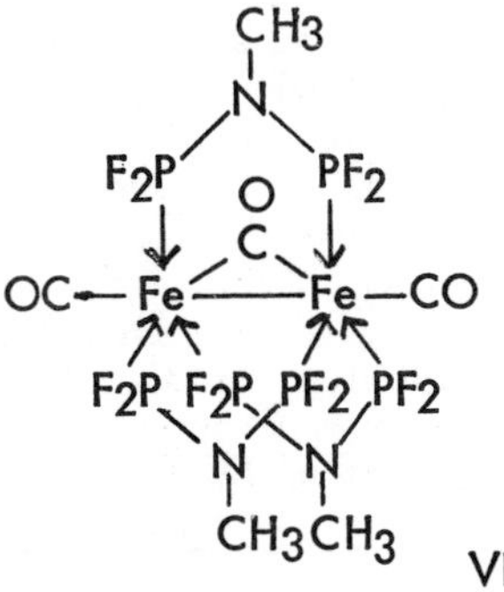

VIII

this complex exhibits a terminal $\nu(CO)$ frequency at 1977 cm.$^{-1}$ and a bridging $\nu(CO)$ frequency at 1776 cm.$^{-1}$. This suggests structure VIII analogous to that reported (2,3) for $[CH_3N(PF_2)_2]_3Co_2(CO)_2$ but with an extra bridging carbonyl group.

Reinvestigation of the reaction between $Fe_2(CO)_9$ and $CH_3N(PF_2)_2$ using a larger $CH_3N(PF_2)_2/Fe_2(CO)_9$ ratio has also been productive. The initial product from the reaction of $Fe_2(CO)_9$ with excess $CH_3N(PF_2)_2$ at room temperature appears to be the yellow volatile liquid $CH_3N(PF_2)_2Fe(CO)_4$ of structure IX in which only one of the two PF_2 groups of the $CH_3N(PF_2)_2$ ligand is bonded to the iron atom. The infrared $\nu(CO)$ spectrum of $CH_3N(PF_2)_2Fe(CO)_4$ suggests a mixture of the axially and equatorially substituted isomers with the equatorially substituted isomer (IX) predominating. Hydrolysis of the PF_2 group in $CH_3N(PF_2)_2Fe(CO)_4$ (IX) occurs easily to give $CH_3NHPF_2Fe(CO)_4$. The photolysis of $Fe_2(CO)_9$ with excess $CH_3N(PF_2)_2$ in diethyl ether gives $[CH_3N(PF_2)_2]_2Fe(CO)_2$ indicated by its n.m.r. spectrum to have structure X containing one monodentate and one bidentate $CH_3N(PF_2)_2$.

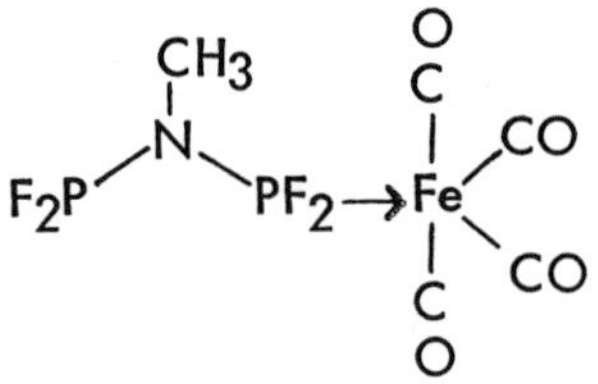

IX

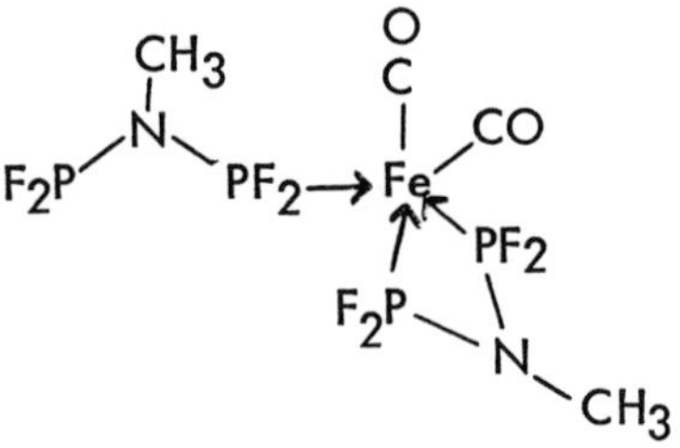

X

Acknowledgment

Portions of this research were supported by the Air Force Office of Scientific Research (Grant AFOSR-75-2869) and the National Science Foundation (Grant CHE-77-5991).

Literature Cited

1. King, R. B. Acc. Chem. Res. 1980, 13, 243.
2. Newton, M. G.; King, R. B.; Chang, M.; Pantaleo, N.S.; Gimeno, J. Chem. Commun. 1977, 531.
3. King, R. B.; Gimeno, J.; Lotz, T. J. Inorg. Chem. 1978, 17, 2401.
4. Newton, M. G.; Pantaleo, N. S.; King, R. B.; Lotz, T. J. Chem. Commun. 1978, 514.
5. Chaloyard, A.; El Murr, N.; King, R. B. J. Organometal. Chem. 1980, 188, C13.
6. King, R. B.; Gimeno, J. Inorg. Chem. 1978, 17, 2390.
7. Newton, M. G.; King, R. B.; Chang, M.; Gimeno, J. J. Am. Chem. Soc. 1977, 99, 2802.
8. Newton, M. G.; King, R. B.; Chang, M.; Gimeno, J. J. Am. Chem. Soc. 1978, 100, 326.

RECEIVED July 7, 1981.

New Aspects of the Coordination Chemistry of Carbonyl Phosphines

EDITH F. LANDVATTER and THOMAS B. RAUCHFUSS

School of Chemical Sciences, University of Illinois, Urbana, IL 61801

The oxidative addition reaction is one of the major juncture points uniting inorganic chemistry, organic chemistry, and catalysis. Continued interest in this reaction has been tempered by the recognition that the simpler hydrocarbons are not often viable substrates. Within the context of monometallic platinum metal chemistry, several approaches have been employed in an attempt to compensate for this deficiency. Thus the DuPont group has demonstrated (1) the activation of a number of hydrocarbon substrates using platinum metal complexes of the basic trialkylphosphines. Using an alternative approach we have exploited the chelate effect as a means of enhancing the propensity of a substrate towards oxidative addition (2), eqn. 1.

$$C_6H_4(PPh_2)(XH) + L_nMPPh_3 \rightarrow C_6H_4(PPh_2)(X)M(H)L_n + PPh_3 \quad (1)$$

This approach is a relatively unexplored one but one which is capable of yielding novel compounds such as acylhydrido complexes from the scission of aldehydic C-H bonds. In an effort to delineate its scope, we have examined some of the quantitative aspects of the chelate assisted oxidative addition in comparison with the more conventional versions of this reaction.

Conversion of Vaska's complex, trans-$IrCl(CO)(PPh_3)_2$, 1, to its iridium(III) derivatives via oxidative addition is characterized by a bleaching of its 440 nm MLCT band. Thus, by measuring the diminution of the intensity of this absorbance (ε = 765), one can determine the degree to which a given substrate effects oxidative addition (3).

Employing this spectrophotometric method we surveyed the reactivity of 1 towards simple and chelating organic substrates; the chelating agents employed in this study were the monosubstituted triphenylphosphines (4):

0097-6156/81/0171-0493$05.00/0

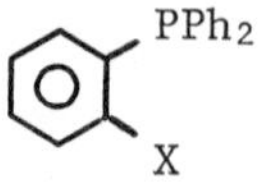

$X = CO_2H$, PCO_2H
$X = OH$, POH
$X = CHO$, PCHO

All data were collected on toluene solutions (28°C) which were 5×10^{-4} M in iridium; the results of a representative experiment are shown in Figure 1. It proved necessary to use the simple substrates in considerable (20X) excess in order to detect the oxidative addition reaction; using the measured equilibrium constant, we then calculated the [Ir(III)]/[Ir(I)] (R_{HX}) which would exist if the initial concentrations of substrate and 1 were both 5×10^{-4} M. Those calculated values are shown in Table I, together with our results for the chelating substrates.

Table I. [Ir(III)]/[Ir(I)] Values for the Reaction of trans-$IrCl(CO)(PPh_3)_2$ (5×10^{-4}M) with (XH, Y-substituted benzene) (5×10^{-4}M)

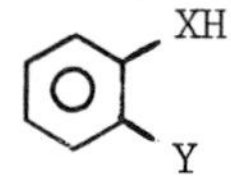

XH	Y=H	$Y=PPh_2$
CO_2H	5.60×10^{-4}	1.34
OH	8.82×10^{-5}	1.76
CHO	$< 4 \times 10^{-5}$	1.03

As expected, the trend in the R values for PhCHO, PhOH, and $PhCO_2H$ followed that for their aqueous acidities. Benzoic acid and phenol proved to be feebly acidic towards 1 while absolutely no interaction between benzaldehyde and 1 was observed even with 10^4-fold excess of this substrate. Chelation has a large impact on the reactivity of these functional groups towards 1. Thus, the R_{PCO_2H} was 10^3 greater than that for the parent benzoic acid. Nonetheless, the effectiveness of PCO_2H as a precursor to a stable iridium(III) hydride is limited since on a preparative scale the product of the reaction of equimolar quantities of 1 and PCO_2H consists of a mixture which contains considerable quantities of the iridium(I) complex ($\nu_{CO} = 1960$ cm^{-1}) wherein the carboxylic acid group is pendant. The chelating phenol, POH, is comparable to PCO_2H in its R value; the enhancement factor, R_{POH}/R_{PhOH}, is significantly greater than that for the benzoic acids.

The most unusual substrate in this study was the phosphine aldehyde, PCHO. This substrate adds extremely effectively to 1, yielding an R_{PCHO} of 1×10^2. The significance of this result lies in the recognition that its enhancement factor, i.e., R_{PCHO}/R_{PhCHO}, is *ca.* 10^6 while similar enhancement factors for the

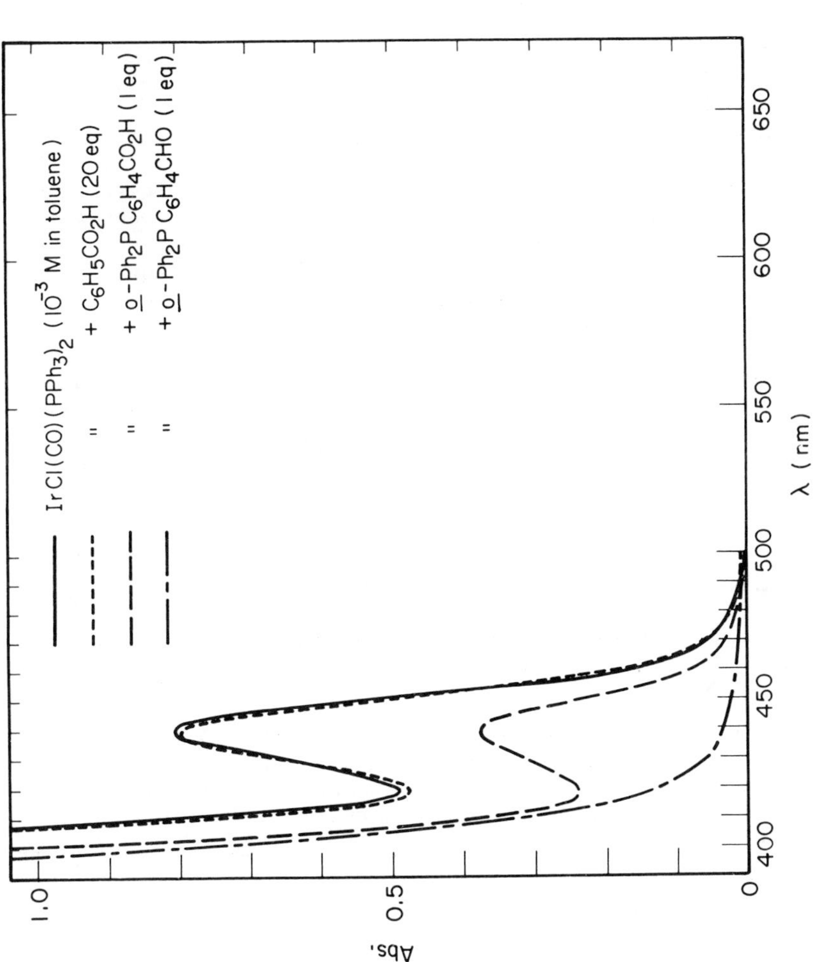

Figure 1. Bleaching of 445-nm band of Vaska's complex by selected oxidative addition substrates at 28°C.

other substrates are ca. 10^3. A conceivable explanation for this observation is based on a kinetic effect, i.e., PhCHO adds very slowly to 1 and our R_{PhCHO} does not represent an equilibrium value. We reject this explanation since it is not clear why benzaldehyde would be slow to effect oxidative addition to 1 while other nonpolar substrates, e.g., H_2, add readily.

It is generally recognized that the ability of many substrates to oxidatively add to low-valent group VIII metal complexes is related to their electrophilicity as judged by their aqueous acidity or their ability to participate in Menschutkin-type reactions. Another class of effective substrates is characteristically nonpolar and notoriously unreactive towards conventional nucleophiles; H_2 and GeH_4 are typical examples here. The present study indicates that such nonpolar substrates, as represented by the formyl group, are more amenable to chelate enhancement than are the polar type of substrates. This phenomenon has been observed previously and is the basis for some of the classic examples of the cyclometallation reaction, such as those involving $IrCl(PPh_3)_3$ and $Ru(Me_2PCH_2CH_2PMe_2)_2$ (5).

An additional result of our study focuses on the importance of the donor substituent on the chelating substrate. By comparison of the affinities of various iridium(I) complexes for phosphino and arsino substituted aldehydes as outlined in Scheme 1, we were able to isolate the analytically pure acyl hydrides for three of the four permutations.

Scheme 1

trans-$IrCl(CO)(PPh_3)_2$ + PCHO → $IrCl(PCO)H(CO)(PPh_3)$ + PPh_3
trans-$IrCl(CO)(AsPh_3)_2$ + PCHO → $IrCl(PCO)H(CO)(AsPh_3)$ + $AsPh_3$
trans-$IrCl(CO)(AsPh_3)_2$ + AsCHO → $IrCl(AsCO)H(CO)(AsPh_3)$ + $AsPh_3$
trans-$IrCl(CO)(PPh_3)_2$ + AsCHO → no reaction

Spectrophotometrically, we could detect no reaction between 1 and AsCHO. This result demonstrates that the identity of the neutral donor component of the bifunctional substrate is crucial for the success of our methodology. This observation substantiates the mechanism outlined in Scheme 2 for the chelate assisted oxidative addition reaction.

Scheme 2

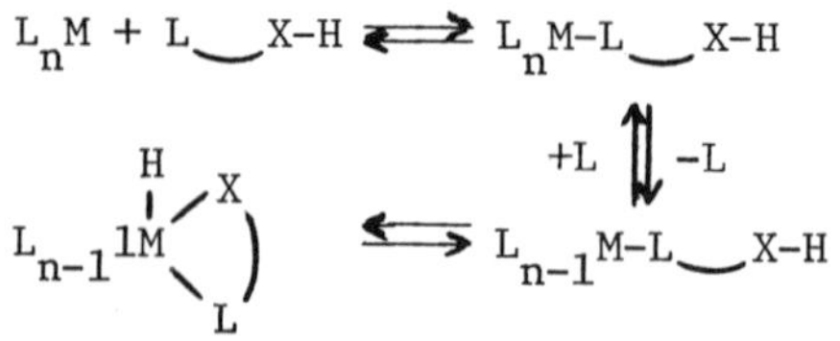

$$L_nM + L\frown X\text{-}H \rightleftharpoons L_nM\text{-}L\frown X\text{-}H$$

$$\text{+L} \upharpoonleft\downharpoonright \text{-L}$$

$$L_{n-1}\overset{H}{\overset{|}{1M}}(X)(L) \rightleftharpoons L_{n-1}M\text{-}L\frown X\text{-}H$$

Acknowledgment

Acknowledgment is made to the donors of the Petroleum Research Fund, administered by the American Chemical Society, for support of this research.

Literature Cited

1. Tulip, T. H.; Thorne, D. L. *J. Am. Chem. Soc.* 1981, *103*, 2448-2450 and references therein.
2. Rauchfuss, T. B. *J. Am. Chem. Soc.* 1979, *101*, 1045-1047.
3. Deeming, A. J.; Shaw, B. L. *J. Chem. Soc. (A)* 1969, 1802-1804.
4. Rauchfuss, T. B. *Platinum Met. Rev.* 1980, *24*, 95-99.
5. Bruce, M. I. *Angew. Chem. Int. Ed. Eng.* 1977, *16*, 73-86.

RECEIVED September 15, 1979.

New Chiral Aminophosphine Ligands: Application to Catalytic Asymmetric C–C Bond Formation

GÉRARD BUONO, CHRISTIAN TRIANTAPHYLIDES, and GILBERT PEIFFER

Laboratoire de l'Ecole de Chimie de Marseille et des Organophosphorés, Université d'Aix-Marseille III, Rue H. Poincaré, 13397 Marseille, France

ANDRÉ MORTREUX and FRANCIS PETIT

Laboratoire de Chimie Organique Appliquée, ERA CNRS 458 Ecole Nationale Supérieure de Chimie de Lille, BP 40, 59650 Villeneuve d'Ascq, France

Chiral phosphines have been widely used to prepare low-valent transition metal complexe catalysts for enantioselective organic transformations. In many studies, catalytic processes for asymmetric C-H induction give almost quantitative optical yields (1). Several approaches have been envisaged in the preparation of chiral phosphines : (i) The phosphorus atom is the chiral center (2). (ii) The diphenylphosphino groups are connected by a chiral link (3). (iii) Asymmetric centers are located both on phosphorus atom and organic chain (4). Often the chiral phosphine is the chelating diphosphine to decrease conformational mobility (5). Asymmetric C-C bond forming reactions are of great significance for the synthesis of optically active compounds (6), however the use of chiral transition metal catalysts for such reactions has been much less successful (1). We describe in this work the synthesis of chiral aminophosphines obtained from natural amino acids. This synthetic approach allows the synthesis of mono bi and multidentate chiral ligands in which two achiral phosphorus centers are connected by a chiral link. These ligands can be used in the asymmetric codimerization of cyclohexa-1,3-diene with ethylene catalysed by nickel complexes (7).

Synthesis of chiral aminophosphines from natural aminoacids.

Chiral aminoacids were generally used for configurational correlation and more recently for reagent preparation in order to induce asymmetric synthesis (8). Several chiral N and N,N' substituted diamines, diamino alcohols and acylphosphide are readily obtained from aminoacids by a method outlined in the scheme I. Among the different known methods of peptide bond formation (9) only the azide one is appropriate to the coupling step when the amino protecting group is an acetyl or a formyl. Chiral starting materials were the N-formyl and the N-acetyl methyl esters of optically pure amino acids :
(R)-phenylglycine R_1 = Ph, (S)-phenylalanine R_1=$PhCH_2$ (S)-Methionine R_1= CH_2CH_2SMe, (S)-valine R_1=$(CH_3)_2CH$, etc. The activated azide compounds(2) were prepared as generally described via hydrazides

0097–6156/81/0171–0499$05.00/0

Scheme I

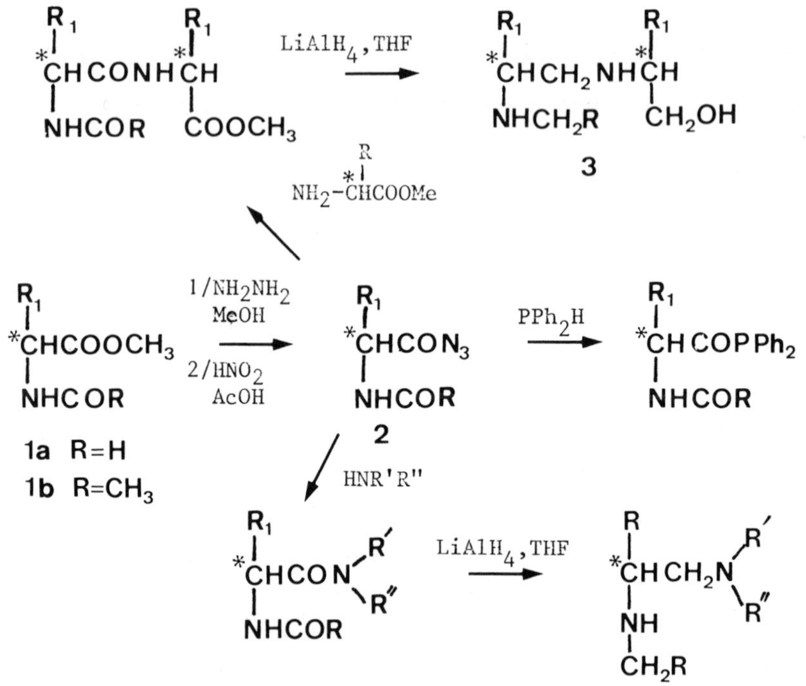

and suitable condensation of amine compounds gave amides without racemization (60 % yield from compounds 1a, 1b). The diamide compounds were converted into the diamino ones by reduction with $LiAlH_4$ in THF (90% yield). From these compounds we can synthesize different bidentate and tridentate chiral heterophosphines as shown in scheme II by action of the chlorodiphenylphosphine and triethylamine.

Scheme II

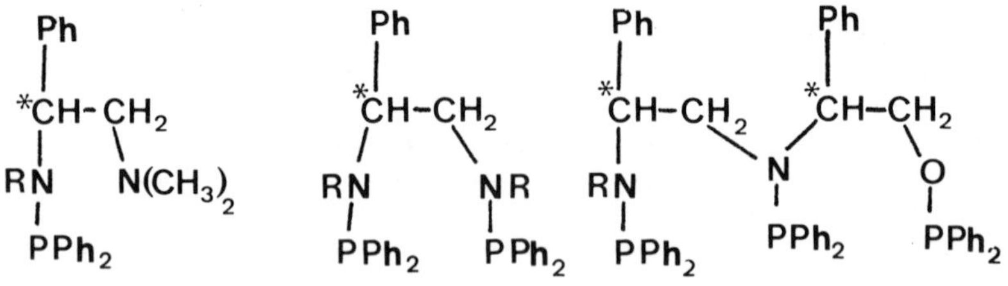

Moreover, the diamino alcohol 3 resulting from (R) phenylglycine gives with hexamethylphosphoroustriamide only one chiral trico-ordinated compound 4.

(Scheme III)

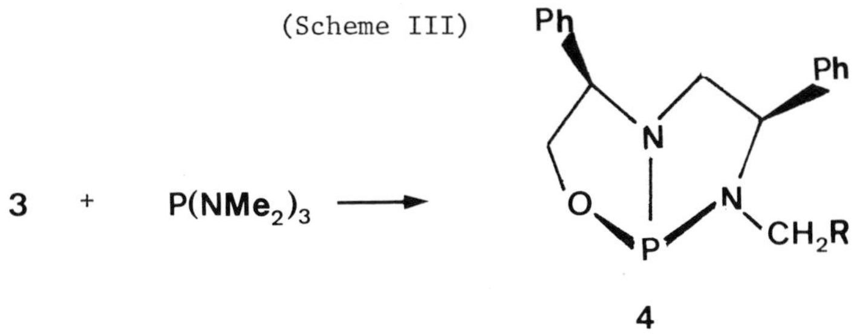

3 + $P(NMe_2)_3$ ⟶ 4

Asymmetric C-C bond formation catalysed by metallic complexes with chiral amino-phosphines as ligands.

Catalytic reactions in which an asymmetric carbon is created have been less intensely studied than asymmetric hydrogenation(1). Before enantioselectivity analysis, we endeavored to increase the regioselectivity of the expected products. For instance by varying the nature of the substituents (R may be alkyl or aryl) at the nitrogen atom of the ligands $Ph_2PNR_1R_2$ a variation in regioselectivity to normal and branched aldehyde was observed during hydroformylation of 1-hexene with $(PPh_2NR_1R_2)_2$ catalysts (10). A good correlation has been shown between the acceptor properties of the aminophosphine ligands and the selectivity for linear aldehyde formation. Recently it has been reported that chiral aminophosphines presented a high enantioselectivity in catalytic hydrogentation (11). We have also observed interesting results in regio and enantioselectivity in the asymmetric codimerization of cyclohexa-1,3 diene with ethylene catalysed by chiral aminophosphine nickel complexes (7), (Scheme IV).

Optimization of the catalytic system (-70°C, aminophosphine/Ni=10) gave (+)-(S)-3-vinylcyclohexene with 73,5% optical purity in 87% chemical yield when (-)-R-N-methyl-N-(1-phenyl-1-ethylamino)diphenylphosphine was used.

Scheme IV

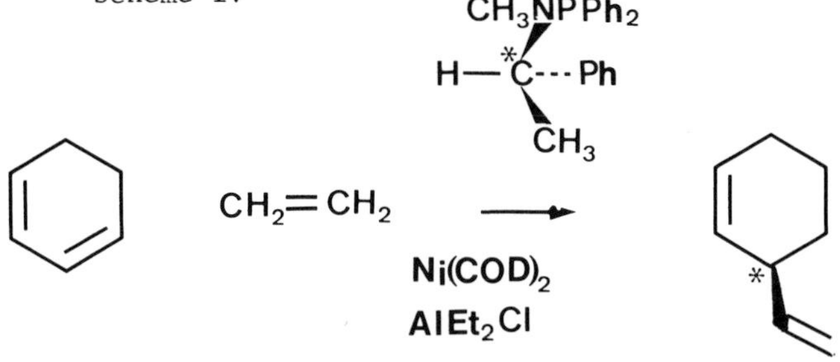

$CH_2{=}CH_2$ ⟶ (CH_3NPPh_2–*C(H)(Ph)(CH_3); $Ni(COD)_2$; $AlEt_2Cl$)

Literature cited

1. Kagan, H.B. ; Fiaud, J.C. "Topics in Stereochemistry" ; Vol.10, Eliel, E.L. ; Allinger, N.L. ; New York, N.Y.,1978, p 175-261.
2. Horner, L. ; Stegel, H. Phosphorus. 1972, 1, 199, 209 ; Knowles, W.S. ; Sabacky, M.J. ; Vineyard, B.D. Chem.Technol. 1972, 591 ; Knowles, W.S. ; Sabacky, M.J. ; Vineyard, B.D. ; Weinkauff, D.J. J.Am.Chem.Soc. 1975, 94, 2569.
3. Aguiar, A.M. ; Morrow, C.J. ; Morrison, J.D. ; Burnett, R.E. ; Masler, W.F. ; Bhacca, N.S. ; J.Org.Chem. 1976, 41, 1545.
4. Valentine, D. ; Blount, J.F.; Toth, K. J.Org.Chem. 1980, 45, 3691 ; Valentine, D. et al. J.Org.Chem. 1980, 45, 3698, 3703.
5. Kagan, H.B. ; Langlois, N. ; Dang, T. J.Organomet.Chem. 1975, 90, 353 ; 1975, 91, 105 and papers cited therein ; Fryzuk, M.D. Bosnich, B. J.Am.Chem.Soc. 1977, 99, 6262 ; 1978, 100, 5491 ; Achiwa, K. J.Am.Chem.Soc. 1976, 98, 8265.
6. Meyers, A.I. Acc.Chem.Res. 1978, 11, 375 ; Apsimon, J.W. ; Seguin, R.P. Tetrahedron. 1979, 35, 2797.
7. Buono, G. ; Peiffer, G. ; Mortreux, A. ; Petit, F. J.C.S., Chem.Commun. 1980, 937.
8. Enders, D. ; Eichenauer, H. Tetrahedron Lett. 1977, 191 ; Enders, D et al. J.Am.Chem.Soc. 1979, 101, 5654 ; Meyers, A.I.; Poindexter, G.S. ; Brich, Z. J.Org.Chem. 1978, 43, 892.
9. Gross, E. ; Meienhofer, J. "The peptides", Vol. 1, Academic Press, New York, N.Y., 1979, p 197 - 239.
10. Grimblot, J. ; Bonnelle, J.B. ; Vacher, C. ; Mortreux, A. ; Petit, F. ; Peiffer, G. J. Mol. Catal. 1980, 9, 357.
11. Fiorini, M. ; Giongo, G.M. ; Marcoti, F. ; Marconi, W. J. Mol Catal. 1976, 1, 451 ; Pracejus, G. ; Pracejus, H. Tetrahedron Lett. 1977, 3497 ; Hanaki, K. ; Kashiwabara, K. ; Fujita, J. Tetrahedron Lett. 1978, 489.

RECEIVED July 7, 1981.

^{31}P NMR Studies of Catalytic Intermediates in Triphenylphosphine Rhodium Complex Hydroformylation Systems

ALEXIS A. OSWALD, JOSEPH S. MEROLA, and EDMUND J. MOZELESKI — The Corporate Research Science Laboratories, P.O. Box 45, Linden, NJ 07036

RODNEY V. KASTRUP — The Analytical and Information Division of Exxon Research and Engineering Company, P.O. Box 121, Linden, NJ 07036

JOHN C. REISCH — The Intermediates Technology Division of Exxon Chemical Company, P.O. Box 241, Baton Rouge, LA 70821

Phosphorus-31 NMR was found to be a powerful tool in studies correlating the structure of triphenylphosphine rhodium complexes with their behavior as catalysts (1,2). In a previous study (2), the rate of dissociation of the tris(triphenylphosphine) rhodium carbonyl hydride (Formula I, in Figure 1, i.e., tris-TPP complex) low temperature, low pressure hydroformylation catalyst, discovered by Wilkinson (3), was determined in the presence of excess TPP in aromatic solvents at varying temperatures. The dissociation product, i.e., bis-(triphenylphosphine) rhodium carbonyl hydride (II), was proposed (2) as a key catalytic intermediate in the selective terminal hydroformylation of n-1-olefins as outlined by the catalyst cycle of Figure 1 using commercial TPP complex plus excess TPP catalyst systems (4,5,6). In the present work, further catalytic intermediates and their equilibria were studied (Figure 1). These studies were carried out primarily at low temperatures using a JEOL FX900 spectrometer, under varying partial pressures of reactant gases, particularly H_2 and CO mixtures, in NMR pressure tubes. Chemical shifts were calculated with reference to a 1M solution of aqueous H_3PO_4.

For most of the present studies, the tris-TPP complex was generated from dicarbonyl acetylacetonato rhodium (III, Acac complex) and TPP by low pressure hydrogenation via the scheme shown in Figure 1. The first step of this sequence is known (3). The complete conversion to the tris-TPP complex under ambient conditions was developed as a facile and quantitative, preparative method during the course of the present work.

The Acac complex was usually added to provide a 1% mixture in an appropriate solution of TPP in a previously nitrogenated, stirred, 9 to 1 volume mixture of toluene and perdeuterobenzene. One of the CO ligands was immediately displaced by TPP as indicated by the instant gas evolution. After removing the CO in vacuum, the resulting solutions of the intermediate (IV) were hydrogenated to obtain the tris-TPP complex. The 2,4-pentanedione by-product did not compete in the presence of excess TPP and H_2 for coordinating the rhodium.

0097-6156/81/0171-0503$05.00/0

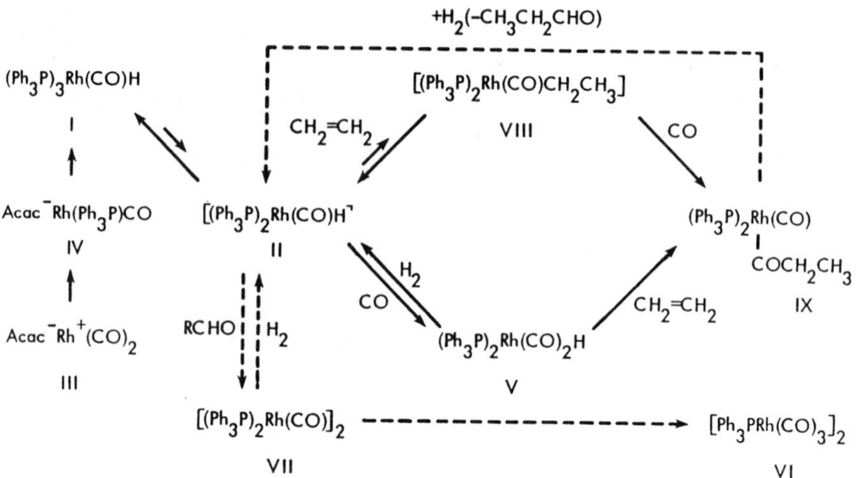

Figure 1. Catalytic intermediates and their relationship to the proposed catalytic cycle of selective terminal hydroformylation.

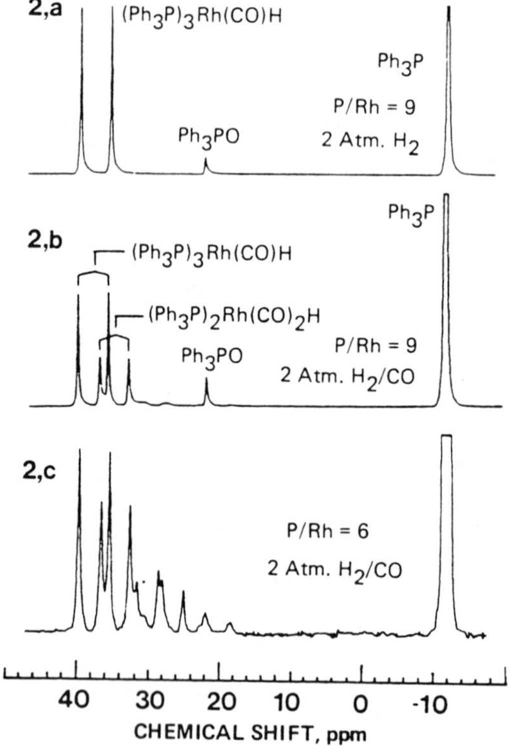

Figure 2. The effect of H_2 and H_2/CO on 1% toluene solutions of Acac $Rh(CO)_2$ plus TPP as shown by ^{31}P NMR at −90° C.

In the second figure (2a), a ^{31}P NMR spectrum at -90° of such a tris-TPP complex is shown in the presence of 6 moles free TPP per mole complex under about 2 Atm hydrogen pressure after equilibration. In a series of experiments similar mixtures containing different ratios of the starting Acac complex and TPP, were placed under about 2 Atm pressure of a 1 to 1 mixture of H_2 and CO, and equilibrated by mechanical shaking. The NMR spectra were then determined at -90°. At P/Rh ratios of 9 and 6, the tris-TPP complex was the dominant species (Figure 2, b and c). At a P/Rh ratio of 3.3, these species became minor components.

In another series of experiments, a 1% solution of the tris-TPP complex, (I), plus excess TPP to provide a P/Rh ratio of 9, was placed under H_2 and then ^{13}CO pressure. The results were similar. They are illustrated by Figure 3. When 2 Atm of H_2 was followed by 2 Atm CO, the mixture contained predominantly the presumed dicarbonyl hydride (Figure 1, V and Figure 3,a). The use of a high ratio of H_2 to CO (2 Atm to 0.3 Atm), resulted mainly in the tris-TPP complex (Figure 3,b). On the other hand, 0.3 Atm CO alone resulted mainly in an unidentified complex, presumably a tricarbonyl dimer (Figure 1, VI and Figure 3,c). Apparently, the partial pressures of not only the CO but of the H_2 as well have a major effect on the equilibria between the different complex species.

When a 1% n-valeraldehyde solution of the Acac complex (III), plus varying amounts of TPP, were pressured to 2 Atm with a 1 to 1 mixture of H_2/CO, the initial complex equilibria were about the same as those previously found in the aromatic solvent. As is shown by Figure 4, when the P/Rh ratio was 15, the dominant Rh compound at 35°C was the tris-TPP complex (I). At a P/Rh ratio of 6, the presumed dicarbonyl hydride (V) was the major complex. When the P/Rh ratio was only 3.3, major amounts of the rhodium were in the form of fast exchanging more highly carbonylated species.

However, it was found that in aldehyde solution the tris-TPP complex is largely converted to a presumed dimer (VII) when stored for 16 hours (Figure 5,a). It appears that the same dimer is also formed quantitively from the Acac complex under N_2 at room temperature (Figure 5,b). These reactions could be reversed to provide the tris-TPP upon pressuring the solution by 2 Atm H_2 (Figure 5,c).

In Figure 1, the observed reactions are indicated in the context of a proposed catalytic scheme for terminal hydroformylation. The reaction pathway involving the alkyl phosphine intermediate (VIII) is the most likely. Overall, the results of the NMR studies provide consistent explanations for the process parameters of selective hydroformylation, particularly of the low pressure continuous product flashoff process (5,6). It was shown that, in contrast to prior indications (3), the tris-phosphine complex (I) is a remarkably stable and favored species in the presence of excess phosphine and H_2. This complex (I) is postulated to have a key role in the reversible generation and

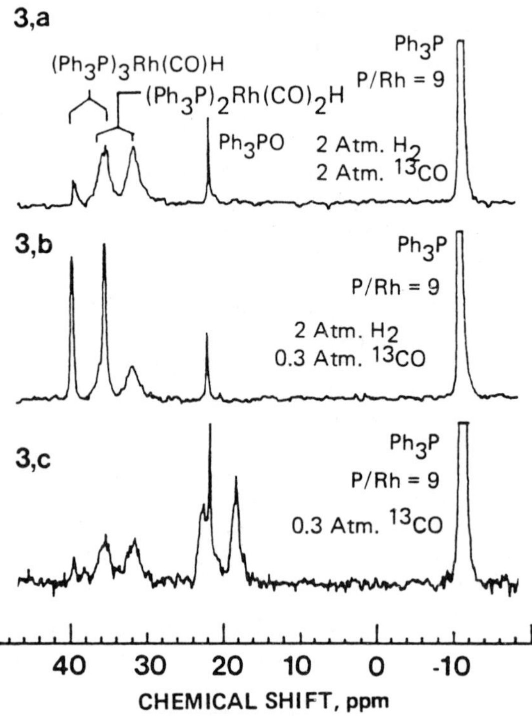

Figure 3. The effect of $H_2/^{13}CO$ and ^{13}CO on 1% toluene solutions of $(Ph_3P)_3Rh(CO)H$, plus TPP as shown by ^{31}P NMR at −60° C.

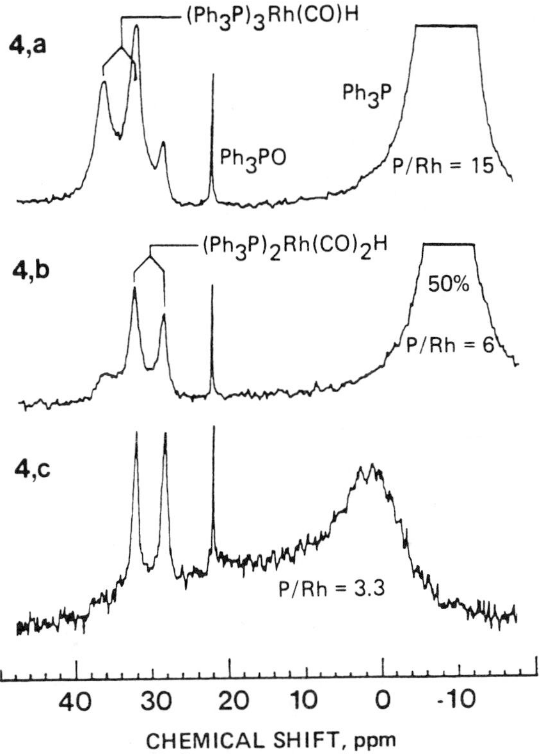

Figure 4. The effect of H_2/CO on 1% valeraldehyde solutions of $AcacRh(CO)_2$ plus TPP as shown by ^{31}P NMR at −35° C.

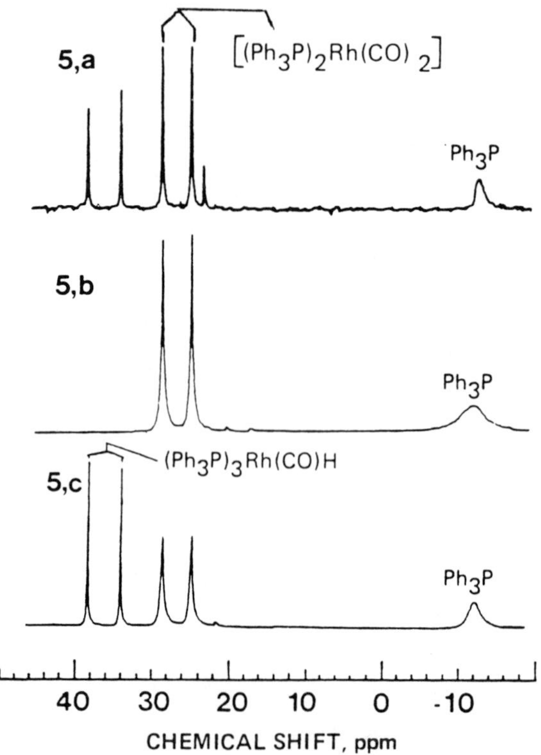

Figure 5. Complexes derived in 1% aldehyde solutions (a) from $(Ph_3)_3Rh(CO)H$, (b) from $AcacRh(CO)_2$ plus 3TPP, and (c) from $AcacRh(CO)_2$ 3TPP under 2 atm. H_2 as shown in 1% i-*butyraldehyde solution by ^{31}P NMR at −60° C.*

stabilization of the reactive, key catalytic trans-bis-phosphine complex intermediate, II.

Literature Cited:

1. Tolman, C. A., Meakin, P. Z., Lindner, D. L., Jesson, J. P., J. Amer. Chem. Soc., 1974, 96, 2762.
2. Advances in Chemistry, Editors Alyea, E. C., and Meek, D. W., American Chemical Society, 1981, 196, 78, paper of "^{31}P NMR Studies of Equilibria and Ligand Exchange in Triphenylphosphine Rhodium Complex and Related Chelated Bis-Phosphine Rhodium Complex Hydroformylation Catalyst Systems," by Kastrup, R. V., Merola, J. S., and Oswald, A. A., in press.
3. Bonati, F., and Wilkinson, G., J. Chem. Soc. A., 1964, 3159; Yagupsky, G., Brown, C. K., and Wilkinson, G., J. Chem. Soc. A, 1968, 2660 and Chem. Commun., 1969, 1244; Evans, D., D., Osborn, J. A., Wilkinson, G., J. Chem. Soc. A, 1968, 3133; Yagupsky, M., Brown, C. K., Yagupsky, G., and Wilkinson, G., J. Chem. Soc. A, 1970, 941; Yagupsky, G., Brown, C. K., Wilkinson, G., J. Chem. Soc. A, 1979, 1392; Brown, C. K., Wilkinson, G., J. Chem. Soc. A, 1970, 2753.
4. Pruett, R. L., Smith, J. A., J. Org. Chem., 1969, 34, 327.
5. Advances in Organometallic Chemistry, Editors: Stone, F. G. A., West, R., Academic Press, Inc., 1979, 17, 1. Chapter on Hydroformylation by Pruett, R. L.
6. New Synthesis with Carbon Monoxide, Editor Falbe, J., Springer-Verlag, Berlin-Heidelberg-New York, 1980, 1, Chapter on Hydroformylation, Oxo Synthesis Roelen Reaction by Cornils, B.

RECEIVED July 7, 1981.

REACTION MECHANISMS INVOLVING ORGANIC AND INORGANIC PHOSPHORUS COMPOUNDS

New Data on the Mechanism of the Perkow–Arbuzov Reaction

LÁSZLÓ TŐKE, IMRE PETNEHÁZY, and GYŐNGYI SZAKÁL — Department of Organic Chemical Technology, Technical University Budapest, 1521 Budapest, Műegyetem, Hungary

HARRY R. HUDSON and LUBA POWROZNYK — Department of Chemistry, The Polytechnic of North London, Holloway Road, London, N7 8DB, England

CHRISTOPHER J. COOKSEY — Department of Chemistry, University College London, 20 Gordon Street, London WC1H OAJ, England

In spite of many previous studies on the mechanisms by which trialkyl phosphites interact with α-halogenocarbonyl compounds, the reactive intermediates which lead to ketophosphonate (Arbuzov reaction) and to vinyl phosphate (Perkow reaction) have in no cases been clearly identified. It is generally believed (1), however, that the Arbuzov product 4 results from initial attack by phosphorus at the α-carbon atom, whereas the Perkow product 7 is formed by initial attack at the carbonyl carbon atom, followed by migration of phosphorus from carbon to oxygen (Scheme 1).

Scheme 1

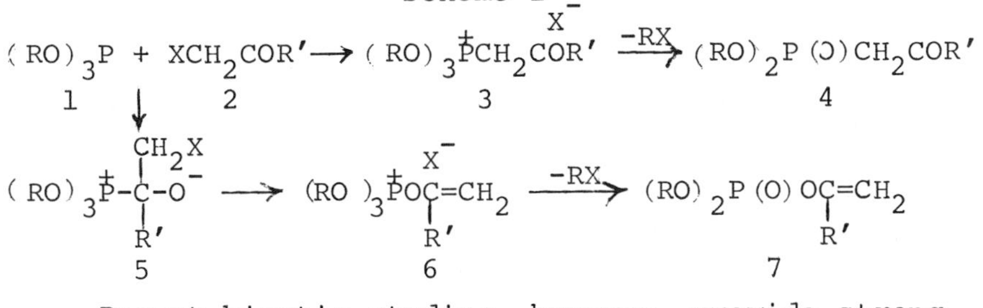

$$\underset{1}{(RO)_3P} + \underset{2}{XCH_2COR'} \longrightarrow \underset{3}{(RO)_3\overset{+}{P}CH_2COR'\ X^-} \xrightarrow{-RX} \underset{4}{(RO)_2P(O)CH_2COR'}$$

$$\downarrow$$

$$\underset{5}{(RO)_3\overset{+}{P}-\underset{R'}{\overset{CH_2X}{C}}-O^-} \longrightarrow \underset{6}{(RO)_3\overset{+}{P}O\underset{R'}{C}=CH_2\ X^-} \xrightarrow{-RX} \underset{7}{(RO)_2P(O)O\underset{R'}{C}=CH_2}$$

Recent kinetic studies, however, provide strong evidence for the involvement of a common first intermediate (2).

We have now carried out crossover experiments in which the reaction of trimethyl phosphite with a mixture of differently substituted chloro- and bromoacetophenones (e.g. $C_6H_5COCH_2Cl$ and p-$MeC_6H_4COCH_2Br$) affords a mixture of products whose composition shows that halogen exchange has occurred. In other words, the ratio of Perkow to Arbuzov products is that which would be expected if $C_6H_5COCH_2Br$ and p-$MeC_6H_4COCH_2Cl$ were also present in the reaction mixture (5 mol% in benzene, 8 mol% in chlorobenzene, 23 mol% in acetonitrile).

0097–6156/81/0171–0513$05.00/0

As direct halogen exchange between the ketones was excluded by separate experiments the results are best explained by assuming that an ionic exchange reaction occurs, involving halide ion from a quasiphosphonium intermediate (Scheme 2).

Scheme 2

$$C_6H_5COCH_2Br + Cl^- \rightleftharpoons C_6H_5COCH_2Cl + Br^-$$

Confirmation of this effect was obtained by carrying out experiments in the presence of added nucleophiles (in the form of triethylbenzylammonium chloride, or tetrabutylammonium bromide) which modified both the product ratio and the overall reaction rate.

To obtain more information on the nature of the quasiphosphonium intermediates involved in these systems we have studied the reactions of sterically hindered neopentyl esters by means of ^{31}P nmr spectroscopy. Trineopentyl phosphite and α-bromoacetophenone gave rise to a peak at +41 ppm due to the ketophosphonium intermediate 3 (R = Me_3CCH_2; R′= Ph; X = Br) within half an hour of mixing the reactants in acetone-d_6 at 27 ^oC (^{31}P nmr shifts are relative to 85% H_3PO_4 downfield positive). Peaks due to the ketophosphonate 4 +19 ppm and the vinyl phosphate 7 (-7 ppm) were also observed (compound 4 and 7 have satisfactory elemental analysis and spectroscopic data). The concentration of the intermediate reached a maximum after about two hours when it was precipitated from acetone solution by the addition of anhydrous ether to give white crystals of trineopentyloxy(phenacyl)phosphonium bromide, identified by elemental analysis and nmr spectroscopy (^{31}P δ+41, in $CDCl_3$). When redissolved in acetone-d_6, deuterochloroform, acetic acid, or acetic acid-acetone mixtures, the intermediate decomposed to yield ketophosphonate 4 but none of the vinyl phosphate 6 (Perkow product). Nor was the course of reaction affected by the addition of chloride ion or of α-chloroacetophenone in acetonitrile.

A proposal (3) that vinyl phosphate might be formed from the ketophosphonium intermediate, after rearrangement *via* a four-membered cyclic phosphorane 8 is therefore excluded by our results under these reaction conditions.

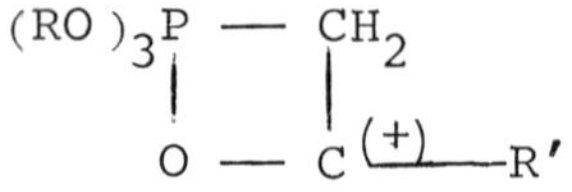

$$\begin{array}{ccc} (RO)_3P & — & CH_2 \\ | & & | \\ O & — & C^{(+)}—R' \end{array}$$

8

We also found that the vinyl phosphate was the major reaction product in the initial stages of the reaction of trineopentyl phosphite with α-bromoacetophenone in acetone-d_6, but that the product ratio changed in favour of the ketophosphonate as the reaction proceeded. The results are consistent with a reaction scheme in which the ketophosphonium and vinyloxyphosphonium intermediates are formed by parallel pathways, either directly from the reactants, of from a common first intermediate X and in the light of our previous kinetic studies (2) the latter is indicated (Scheme 3).

Scheme 3

$$1 + 2 \longrightarrow X \begin{cases} \nearrow 3 \longrightarrow 4 \\ \searrow 6 \longrightarrow 7 \end{cases}$$

It is clear also that the vinyloxyphosphonium intermediate 6 is much less stable than the ketophosphonium intermediate 3, although this difference may only be apparent when the alkoxy group (R) is sterically hindered as in the case of neopentyl.

Intermediates derived from dineopentyl phenylphosphonite and from neopentyl diphenylphosphinite were found to be considerably more stable. Reaction with α-bromoacetophenone gave the corresponding Arbuzov intermediates 9, 10 as crystalline solids, the diphenylphosphinite derivative 10 being sufficiently

$$Ph\overset{+}{P}(OR)_2CH_2COPh\ Br^- \quad (9) \qquad Ph_2\overset{+}{P}(OR)CH_2COPh\ Br^- \quad (10)$$

$$Ph_2P(OR)OC(:CH_2)Ph\ Cl^- \quad (11) \qquad (R{=}Me_3CCH_2)$$

stable for X-ray crystal structure determination to be carried out. In deuterochloroform at 100°C (sealed tube), the intermediates were again shown to decompose to yield the corresponding Arbuzov products exclusively, the ^{31}P nmr signals for 9 (δ +67.1) and for 10 (δ +68.3) being replaced by single peaks at +32.1 and +28.2 ppm, respectively. The presence of phosphorus-bonded CH_2 groups in the products was confirmed by ^1H nmr spectroscopy (J for P-CH_2 15-20 Hz); signals due to vinyl protons were not detectable. From neopentyl diphenylphosphinite we were also able to obtain the first example of a crystalline Perkow intermediate 11 (δ +55.9) by reaction with α-chloroacetophenone in chloroform. Although rather unstable, it could be stored at 0°C and was shown to decompose to the corresponding vinyl ester (δ +29.7) when dissolved in deuteterochloform ($t_{1/2}$ ca. 40 minutes, at room temperature) The presence of vinyl protons in the intermediate and in its decomposition product was indicated by complex multiplets in the ^{1}H nmr spectrum centred at 5.45 and 5.1 ppm, respectively.

By carrying out the reaction of trineopentyl phosphite with α-bromoacetophenone in the presence of acetic acid (10% in acetone-d_6) we have also obtained evidence for the transient existence of the vinyloxyphosphonium species. Under these conditions, we observed the rapid appearance of three transient intermediates (^{31}P α +92, +40, and -7.3 ppm), although vinyl phosphate was the almost exclusive final product. The peak at -7.3 ppm disappeared most rapidly, with the simultaneous development of a corresponding peak at -7.2 ppm due to vinyl phosphate, and is tentatively assigned to the vinyloxyphosphonium bromide. It is known that tetraalkoxyphosphonium ions have ^{31}P chemical shifts which are not far removed from those of the corresponding phosphate esters (4).

The identities of the peaks at +40 and +92 ppm are uncertain. One likely species to be present in acid conditions is the protonated form of the betaine 5. Such intermediates are assumed to be formed in acid conditions and to give rise to the formation of α-hydroxyphosphonates, $(RO)_2P(O)CH_2X(R')OH$ (1). It is also likely that the betaine 5 is the first common intermediate (Scheme 3) in the formation of Arbuzov and Perkow products, these being formed by a 1,2-migration of the phosphite moiety to the α-carbon atom or to oxygen, respectively. We detected no signal at +13 ppm which would be expected if the bromo(trineopentyloxy) phosphonium intermediate had been formed, although such intermediates are known to be very short-lived. Nor was the corresponding phosphorobromidate detected at -8.5ppm (5).

Literature cited

1. Chopard,P.A.; Clark,V.M.; Hudson,R.F. and Kirby,A. Tetrahedron 1965, 21, 1961.
2. Tőke,L.; Petneházy,I. and Szakál,Gy. J.Chem.Soc.Res. (S) 1978, 155.
3. Marquading,D.; Ramirez,F.; Ugi,I. and Gillespie,P. Angew.Chem.Int.Edn. 1973, 12, 91.
4. Finley,J.H.; Denney,D.Z. and Denney,D.B. J.Amer. Chem.Soc. 1969, 91, 5826.
5. Michalski,J.; Pakulski,M and Skowronska,A. J.C.S. Perkin I. 1980, 833.

RECEIVED July 7, 1981.

Structure and Reactivity of Quasiphosphonium Intermediates

H. R. HUDSON, A. T. C. KOW, and K. HENRICK

Department of Chemistry, The Polytechnic of North London, Holloway Road, London N7 8DB, England

New quasiphosphonium halides derived from neopentyl diphenylphosphinite and dineopentyl phenylphosphonite are reported and it is shown that the attachment of phenyl groups to phosphorus provides enhanced stability. The products, $Ph_2P(OR)MeX$ (X = Cl, Br, or I) and $PhP(OR)_2MeX$ (X = Br or I) (R = Me_3CCH_2), are resistant to moist air and can be handled in the open laboratory for the purpose of X-ray diffraction studies (1). The rate of initial reaction between the ester and an alkyl halide increases with the number of phenyl substituents on phosphorus; from the diphenylphosphinite an adduct with chloromethane can be prepared.

^{31}P chemical shifts (downfield from 85% H_3PO_4) of +72 ppm for $Ph_2P(OR)MeX$ (X = Cl, Br, or I) and of +74 ppm for $PhP(OR)_2MeX$ (X = Br or I) confirm the phosphonium structure in solution. Although phosphorus (V) intermediates have been shown to occur in reactions of certain phosphorus (III) esters with halogens (2), their role in Michaelis-Arbuzov reactions involving alkyl halides is less certain. In no cases have phosphorus (V) intermediates been detected by ^{31}P nmr and the stereochemical evidence for their involvement (3) appears to require further investigation (4). In solvents such as chloroform, quasiphosphonium intermediates exist essentially as ion-pairs (5).

To gain information on the kinetics and mechanism of the product-forming stage of the Michaelis-Arbuzov reaction, we have followed the decomposition of intermediates by ^1H nmr spectroscopy. In $CDCl_3$ the reactions follow excellent first-order kinetics (Table I) and are in accord with rate-determining collapse to products within the undissociated ion-pair (Figure 1). Synchronous

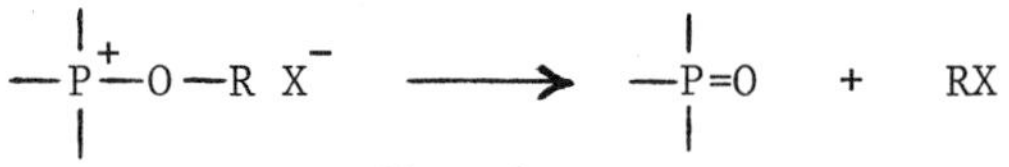

$$-\overset{|}{\underset{|}{P}}{}^{+}-O-R\ X^{-} \longrightarrow -\overset{|}{\underset{|}{P}}=O \quad + \quad RX$$

Figure 1.

attack by halide ion and alkyl-oxygen fission are indicated by absence of rearrangement in the neopentyl group and although the process may be considered to be of the S_N2-type it is not

0097–6156/81/0171–0517$05.00/0

Table I. First-order decompositions in $CDCl_3$

Compound	t/°C	$10^7 k_1/s^{-1}$		
		Cl	Br	I
$Ph_2P(OR)MeX$	33	1.5	1.54	2.75
	60	92	74	105
$PhP(OR)_2MeX$	33	-	114	148
	60	-	4670	5260
$(RO)_3PMeX$	33	-	670	1100

bimolecular, as kinetically separate species are not involved. A similar process has been reported for the decomposition of neopentyloxytriphenylphosphonium chloride in carbon tetrachloride(6).

First order kinetics are also consistent with a sequence which involves rate-determining dissociation of the ion-pair (Figure 2, $k_{-1} << k_2$). This can be excluded, at least for neopentyl derivatives, since their stability depends on the relatively slow

$$\text{—}\overset{|}{\underset{|}{P^+}}\text{—OR } X^- \underset{k_{-1}}{\overset{k_1}{\rightleftharpoons}} \text{—}\overset{|}{\underset{|}{P^+}}\text{—OR} + X^-$$

$$X^- \curvearrowright R\text{—}O\text{—}\overset{|}{\underset{|}{P^+}}\text{—} \xrightarrow{k_2} XR + O\text{=}\overset{|}{\underset{|}{P}}\text{—}$$

Figure 2.

rate at which the neopentyl group is attacked by nucleophiles. Entropies of activation are also not far from zero, being slightly positive in most cases but slightly negative for the decomposition of $Ph_2P(OR)MeI$ (Table II).

Table II. Activation parameters*

Compound	$\Delta H^{\ddagger}$/kcal mol^{-1}			$\Delta S^{\ddagger}$/cal K^{-1} mol^{-1}		
	Cl	Br	I	Cl	Br	I
$Ph_2P(OR)MeX$	30.3	28.5	26.7	9.06	3.44	-1.44
$PhP(OR)_2MeX$	-	27.2	26.2	-	7.71	4.84

*Decomposition in $CDCl_3$

A second alternative which accords with first order kinetics consists in the formation of a low steady-state concentration of dissociated ions, followed by rate-determining attack of halide on the quasiphosphonium ion ($k_2 << k_{-1}$). Although the extent of dissociation in $CDCl_3$ is expected to be small and cannot be determined directly, it may be estimated by the Fuoss equation which permits calculation of K_e from the mean ionic radius of the ions and the dielectric constant of the medium (7). For the present purpose we have used the minimum interionic distance as determined by X-ray crystallography for the bromides and we have based our calculation on the published dielectric constant for $CHCl_3$ which, it is assumed, will not differ significantly from that for $CDCl_3$. The K_e values so obtained have been used to calculate k_2 for the dealkylation step and to derive the appropriate activation parameters. Although $\Delta H^{\ddagger}$ values do not differ significantly from those derived from k_{obs} the entropies of activation ($\Delta S^{\ddagger}$) are large and positive and are inconsistent with a conventional S_N2 reaction (Table III).

Table III. Activation parameters assuming S_N2 interaction of dissociated ions ($CDCl_3$)

Compound	t/°C	$10^{12}K_e$	$\Delta H^{\ddagger}$	$\Delta S^{\ddagger}$
$PhP(OR)_2MeBr$	33	7.97	30.1	68.0
	60	5.41		
$Ph_2P(OR)MeBr$	33	22.7	38.4	52.5
	60	15.6		

Rates of decomposition do not vary much with the halide ion. The reactivity order falls between that which is usally observed in the gas phase or in aprotic media (Cl>Br>I) and that which is observed in hydroxylic solvents (Cl<Br<I). Weak solvation by chloroform is proposed. Variations from first-order decomposition may occur in more ionizing media, for example CD_3CN or $PhNO_2$, in which dissociation leads to the incursion of a second-order component.

The increase in stability of quasiphosphonium intermediates $Ph_nP(OR)_{n-3}MeX$, with increasing n, is attributed mainly to inductive effects of the substituent groups. Phosphorus-carbon bond lengths for Ph - P and Me - P in the intermediates do not support the view that mesomeric release from phenyl to phosphorus is involved (1). On this basis, the stability should increase with decreasing n, if p_π-d_π interaction between oxygen and phosphorus is a significant factor (8), whereas exactly the opposite is observed.

The possibility that R-O fission may occur in advance of attack by halide ion has been demonstrated in the reactions of dineopentyl phenyl phosphite or neopentyl diphenyl phosphite with alkyl halides. In these cases the Michaelis-Arbuzov intermediates cannot be isolated but the neopentyl halides which are formed are contaminated with varying amounts of rearrangement product (t-pentyl halide) (Table IV). These results may explain the formation of mixtures of isomeric halides (9,10,11) in certain applications of Rydon's method for alkyl halide preparation (12). S_N1 fission of the R-O bond appears to be encouraged by the electron-attracting effect of phenoxy-substituents (Figure 3).

Table IV. Formation of rearranged alkyl halides (%) in Michaelis-Arbuzov reactions of neopentyl esters

Reactants		t-halide	Reactants		t-halide
$(RO)_2(PhO)P$	MeI	0	$(RO)(PhO)_2P$	MeI	0
"	MeBr	0.1	"	MeBr	4.4
"	$PhCH_2Cl$	0	"	$PhCH_2Cl$	19.3
			"	*MeBr	50

*In presence of PhOH (1 mol equiv).

Figure 3.

The phenol which is formed as a by-product in these systems also appears to have a significant effect on the mechanism since it increases the proportion of rearranged products obtained. For example, it was found that the interaction of neopentyl diphenyl phosphite with methyl bromide yielded pentyl bromides containing 4.4% of the tertiary isomer, whereas the percentage rearrangement rose to 50 in the presence of one molecular equivalent of phenol.

Literature cited

1. Henrick, K.; Hudson, H. R.; Kow, A. Chem. Comm., 1980, 226.
2. Skowronska, A.; Mikolajczak, J.; Michalski, J. Chem. Comm., 1975, 986.
3. Bodkin, C. L.; Simpson, P. JCS Perkin II, 1972, 2049.
4. Adamcik, R. D.; Chang, L. L.; Denney, D. B. Chem. Comm., 1974, 986.
5. Hudson, H. R.; Rees, R. G.; Weekes, J. E. Chem. Comm., 1971, 1279; JCS Perkin I, 1974, 982.
6. Jones, L. A.; Sumner, C. E., Jnr.; Franzus, B.; Huang, T. T. S.; Snyder, E. I. J. Org. Chem., 1978, 43, 2821.
7. Robinson, R. A.; Stokes, R. H. "Electrolytic Solution"; Butterworth (London), 1959, p 400.
8. Nesterov, L. V.; Kessel, A. Ya.; Samitov, Yu. Yu.; Musina, A. A. Zhur. obshchei Khim., 1969, 39, 1179.
9. Karabatsos, G. J.; Orzech, C. E.; Meyerson, S. J. Am. Chem. Soc., 1964, 86, 1994.
10. Gerrard, W.; Hudson, H. R.; Parrett, F. W. Nature, 1966, 211, 740.
11. Hudson, H. R. J. Chem. Soc.(B), 1968, 664.
12. Landauer, S. R.; Rydon, H. N. J. Chem. Soc., 1953, 2224.

RECEIVED June 30, 1981.

Reactions of Triorganosilyl Halides with Esters of Tricoordinated and Tetracoordinated Phosphorus

J. CHOJNOWSKI, M. CYPRYK, J. MICHALSKI, and L. WOZNIAK

Polish Academy of Sciences, Centre of Molecular and Macromolecular Studies, 90-362 Łódź, Boczna 5 Poland

To get a broader concept of the nucleophilic reactivity of phosphorus acid derivatives towards organosilanes we studied mechanism of reactions of triorganosilyl halides with esters of tricoordinate and tetracoordinate phosphorus having the general formulae:

$$\begin{matrix} X \\ RO \\ RO \end{matrix} \Big\rangle P \qquad\qquad \begin{matrix} X \\ RO \\ RO \end{matrix} \Big\rangle P{=}Y$$

X=OAlk, OAr, $OSiR_3$, O^-, SR, halogen and some other groups

Y=O, S, Se, R=alkyl or aryl

The reaction between a triorganosilyl halide and trialkyl phosphite takes a course which departs from the Arbuzov scheme since it leads to silyl esters of alkylphosphonic acids rather than to silylphosphonates (1, 2). The mechanism of this process was not understood although evidence for the possible transient formation of triorganosilyl phosphites was presented (2). Recently we noticed that the reaction of alkyl phosphite with trimethylsilyl iodide involved various intermediates which were stable enough at lower temperatures to be identified and their formation and disappearance can be followed by ^{31}P NMR spectroscopy. Therefore the investigation of this process gives a deeper insight into the mechanism of the reaction of silyl halides with oxyesters of tricoordinated phosphorus.

^{31}P NMR studies over a large temperature range revealed that two separate stages can be distinguished in this process. In the first stage, which can be conveniently followed in the temperature range -20°-0°C using methylene chloride as a solvent, only products of tricoordinated phosphorus are formed. The second stage including transformation of these tricoordinate P intermediates to tetracoordinate phosphorus products (trimethylsilyl esters of alkylphosphonic acid) in principle does not occur at a significant rate in this temperature range.

0097-6156/81/0171-0521$05.00/0

Initially the process leads to almost exclusive formation of dimethyl iodophosphite which reacts in a consecutive step with the other product trimethylmethoxysilane to form dimethyl(trimethylsilyl) phosphite and methyl iodide. The results are explained by four-centered mechanism involving the nucleophilic attack on phosphorus according to the following scheme:

$$(MeO)_3+Me_3SiI \rightleftharpoons \left[\begin{array}{l} \quad MeO \\ \quad\;\; | \\ MeO-P\cdots I \\ \quad\;\; \vdots \qquad \vdots \\ \quad MeO\cdots SiMe_3 \end{array}\right]^{\ddagger} \rightleftharpoons (MeO)_2PI+Me_3SiOMe$$

(1)

$$(MeO)_2PI+Me_3SiOMe \longrightarrow \left[\begin{array}{l} \quad\;\; OMe \\ \quad\;\;\; | \\ MeOP\cdots OSiMe_3 \\ \quad\;\; \vdots \qquad \vdots \\ \quad\;\; I\cdots Me \end{array}\right]^{\ddagger} \longrightarrow (MeO)_2P(OSiMe_3)+MeI$$

The possibility of the nucleophilic attack of phosphorus on silicon resulting in the formation of a phosphonium or phosphorane intermediate was excluded by a careful analysis of the rate of formation of both phosphite products as well as the analysis of the analogous reaction of $(MeO)_2P(OSiMe_3)$ with Me_3SiI which was studied separately.

The first reaction of scheme 1 is reversible, which was proved by studying the reaction of diphenyl iodophosphite with trimethylmethoxysilane.

In the case of mixed alkyl aryl esters the iodosilane reacts almost exclusively with the alkyl group. Similarly in the case of mixed alkyl silyl esters, the alkoxyl group is replaced by iodine.

$$(2) \qquad \begin{array}{c} X \\ \diagdown \\ AlkO-P \\ \diagup \end{array} + Me_3SiI \rightleftharpoons \begin{array}{c} X \\ \diagdown \\ PI \\ \diagup \end{array} + Me_3SiOAlk$$

$$X=ArO,\ Me_3SiO$$

On the other hand, the reaction of the silyl iodide with alkyl or aryl chlorophosphites leads first to the full displacement of chlorine by iodine.

$$(3) \qquad \begin{array}{c} Cl \\ RO \end{array}\!\!\gtrdot P + Me_3SiI \longrightarrow \begin{array}{c} I \\ RO \end{array}\!\!\gtrdot P + Me_3SiCl$$

The reaction 3 may serve as a convenient method of synthesis of alkyl or aryl iodophosphites.

In a sharp contrast to phosphites, alkyl esters of tetracoordinate phosphorus react with triorganosilyl halides (at least with bromide and iodide) according to a mechanism which is similar to the Arbuzov scheme. The interaction of a phosphoryl group with silicon leads to the ionization of the silicon-halogen bond and the formation of the phosphonium salt intermediate. The salt undergoes decomposition involving dealkylation of the phosphonium ion by the counter-ion (3).

$$(4)\quad >P(=O)OR' + R_3SiX \underset{}{\overset{fast}{\rightleftharpoons}} >P^{+}(OR')(OSiR_3)\ X^{-} \xrightarrow{slow} >P(=O)(OSiR_3) + R'X$$

X=Cl, Br, I R'=alkyl R=alkyl or aryl

The reaction provides an easy way for removal of an alkyl group from esters of phosphorus and it has found wide application in the synthesis of bioactive acids of phosphorus (4). It may be also used for the selective displacement of the alkyl group by trimethylsilyl group in the presence of some reactive groups bonded to phosphorus like halogen, amine, SR etc. (5).

The fast reversible formation of the phosphonium salt also deserves attention since it is the first step in important reactions in organosilicon chemistry. As a consequence of the particularly high susceptibility of the silicon in the phosphonium ion to nucleophilic attack, phosphoryl compounds catalyze nucleophilic attack in substitution at silicon (6).

In order to obtain more information about the fast step of the reaction 4, the silyl ester group exchange (reaction 5) was studied.

$$(5)\quad (PhO)_2P(=O)OSiMe_3 + Me\alpha NpPhSiCl \underset{k_{-1}}{\overset{k_1}{\rightleftharpoons}} (PhO)_2P(=O)OSiMe\alpha NpPh + Me_3SiCl \qquad K$$

The kinetics and equilibrium of the reaction 5 in carbon tetrachloride was investigated by ^{31}P NMR spectroscopy. The process follows the kinetic law for a second order reversible reaction (first order in both the ester and the silyl halide). Values of rate and equilibrium constants 25°C and of activation parameters are: $k_1=0.064\ dm^3mol^{-1}s^{-1}$, $k_{-1}=0.048\ dm^3mol^{-1}s^{-1}$, $K=1.33$, $\Delta H^{\ddagger}=21\ kJ\ mol^{-1}$, $\Delta S^{\ddagger}=-39$ e.u. These results are in accord with a phosphonium intermediate mechanism.

Reaction 5 is accompanied by intermolecular (6) and intramolecular (7) silyl group migration.

(6) $>P(=O)\overset{(1)}{-}OSiMe_3 + >P(=O)\overset{(2)}{-}OSiMe\alpha NpPh \rightleftharpoons >P(=O)\overset{(1)}{-}OSiMe\alpha NpPh + >P(=O)\overset{(2)}{-}OSiMe_3$

(7) $>P(=O)-OSiR_3 \rightleftharpoons >P(=O)-OSiR_3$

Reaction 6 was detected using esters having a different phosphorus moiety. This reaction may also be responsible for the disappearance of optical activity in the system during reaction 5 when the optically active isomer of MeαNpPhSiCl is used. The racemization follows an irregular optical rotation-time curve and proceeds more slowly than reaction 5, giving evidence of the stereoselective course of reaction 5. The intramolecular migration 7 must lead to the racemization of esters with the center of chirality on the phosphorus atom and it is the reason for the failure to isolate the optically active isomer having $P(O)OSiMe_3$ structure. It does not seem, however, to change configuration at the silicon atom.

Literature Cited

1. Malatesta, L.; Gazz. Chim. Ital. (1950), 80, 527.
2. Bugerenko, E. F.; Tchernyshev, E. A.; Popov, E. M.; Izv. Acad. Nauk SSSR, ser. khim.(1966), 1391.
3. Borecka, B.; Chojnowski, J.; Cypryk, M.; Michalski, J.; Zielińska, J.; J. Organometal. Chem. (1979), 171, 17.
4. McKenna, C. E.; Higa, M. T.; Cheung, N. H.; McKenna, M. C.; Tetrahedron Lett. (1977), 155.
5. Chojnowski, J.; Cypryk, M.; Michalski, J.; Synthesis (1978), 777.
6. Chojnowski, J.; Cypryk, M.; Michalski, J.; J. Organometal. Chem. (1978), 161 C 31-35.

RECEIVED July 7, 1981.

Halogenolysis of the Phosphorus–Sulfur Bond in Thioloesters of Organic Phosphorus Thioacids

B. KRAWIECKA, J. MICHALSKI, and E. TADEUSIAK

Polish Academy of Sciences, Centre of Molecular and Macromolecular Studies, Boczna 5, 90-362 Łódź, Poland

The reaction of organic phosphorus thioloesters with halogenating agents has been known for twenty years. This reaction has been shown to proceed via different pathways depending on the reaction medium (1-3). The reaction in non-aqueous solvents has been applied successfully to the synthesis of optically active phosphino, phosphono and phosphorochloridates III (4-7). Such reactions are of importance in phosphorus stereochemistry since they provide excellent access to this class of compounds. Their course is strongly dependent on reaction conditions, and inversion, retention and racemization have been observed. Rational interpretation of the experimental facts was not available.

We have recently demonstrated (8) that, in the case of a model with a sterically crowded phosphorus atom, intermediate products containing two phosphorus atoms bonded through an oxygen-bridge are formed. This allowed for the first time a rational mechanistic explanation to be given of the stereochemical changes involved using the following scheme.

$RR'P(O)SR''$ (I) $\xrightarrow{Cl_2}$ $RR'P(O)\overset{+}{S}(Cl)R''\ \ Cl^-$ (II)

I + II → $RR'(R''S)\overset{+}{P}\text{-}O\text{-}P(O)R'R\ \ Cl^-$ (IV) → $RR'P(O)Cl$ (III) + I

IV ⇌ $R'RCl\overset{+}{P}\text{-}O\text{-}P(O)R'R\ \ {}^-SR''$ (V)

V $\xrightarrow{-R''SCl,\ +Cl_2}$ $R'RCl\overset{+}{P}\text{-}O\text{-}P(O)R'R\ \ Cl^-$ (VI) → $RR'P(O)Cl$ (III racemic)

$R = Bu^t$, $R' = Ph$, $R'' = Me$

0097–6156/81/0171–0525$05.00/0

The scheme provides the possibility of an unambiguous interpretation of the reaction stereochemistry. The reaction between I + II occurs with inversion of configuration at the phosphoryl center and with retention at the phosphonium center. The decomposition of the salt IV into phosphinochloridate III and starting phosphinothiolate I involves a second inversion at the phosphoryl center which consequentially results in retention of configuration. Racemization arises instead, from the decomposition of the salt VI, which leads to one molecule with unchanged configuration and another with inverted. Evidence supporting this reaction mechanism was based on ^{31}P NMR spectroscopy and on the independent synthesis of the transient salts IV, V and VI.

The aim of the present investigation was to generalize the above findings for other models, including those without steric hindrance at the phosphorus center. Halogenolyses were performed with elemental chlorine, sulfuryl chloride, or elemental bromine.

Participation of transient species containing two phosphorus atoms was demonstrated for the thiolates I: R = Bu^t, R' = OMe, R" = Me; R = Et, R' = OEt, R" = Me; R = R' = Me_3CCH_2O , R" = Me; R = R' = OEt, R" = CH_2CH_2Cl.

When the starting thiolate bears an alkoxy group e.g. R = Bu^t, R' = OMe, R" = Me dealkylation is also observed. Use of sulfuryl chloride as the chlorinating agent in toluene solution results in demethylation of the methoxy group in the initially formed product VII, while the use of elemental chlorine gives puzzling results. Although demethylation occurs it does so only after ligand exchange of the S-methyl group with Cl in product VIII.

R = Bu^t, R' = R" = Me

The first step in all above mentioned reactions was assumed to be formation of the halogenosulfonium salt of the type II. We can now support this statement experimentally by the following observations: The stability of the halogenosulfonium salt depends strongly on the nature of Y. For example, when X = Y = Br this salt is stable in many cases, even at ambient temperature, whereas when X = Y = Cl it is not observed above -50°C. This difference is due to lower nucleophilicity of Br^- toward phosphorus. Relatively stable salts (IX) are formed when Y^- is of low nucleophilicity such as $SbCl_6^-$, $SbBr_6^-$, Br_3^-.

$$\mathrm{R(R')P(=O)-S^{+}(X)-R''}\quad Y^-$$

IX: R = Bu^t, R' = Ph, R" = Me

(a) X = Y = Cl
(b) X = Cl, Y = $SbCl_6$
(c) X = Cl, Y = $HgCl_3$
(d) X = Y = Br
(e) X = Br, Y = $SbBr_6$
(f) X = Br, Y = Br_3

The ^{31}P chemical shifts (downfield from 85% H_3PO_4) of sulfonium salts IX with various Y^- are very close: (a) δ = +86.1 ppm (temp. -55°C), (b) δ = +90.1 ppm, (c) δ = +86.8 ppm, (d) δ = +86.7 ppm, (e) δ = +89.2 ppm, (f) δ = +92.2 ppm. This suggests that all these salts have "true" sulfonium structure with little interaction within the ion pair involved. It is of interest to note that IX (X = Y = Br) reacts with cyclohexene to give 1,2-dibromocyclohexane and the starting thiolate. Sulfonium salts IX react readily with external nucleophiles of high P-nucleophilicity e.g. water and alcohols. The reaction between I (R = Bu^t, R' = Ph, R" = Me) and elemental bromine in methanol solution leads to the ester (X) with predominant inversion of configuration (3).

$$\mathrm{Bu^tPhP(O)SMe} \xrightarrow[\mathrm{MeOH}]{\mathrm{Br_2}} \mathrm{Br^-}\;\; \mathrm{MeOH} \rightarrow \mathrm{Bu^tPhP(O)-S^{+}(Br)-Me} \longrightarrow \underset{\mathrm{X}}{\mathrm{Bu^tPhP(O)OMe}}$$

Partial racemization of the ester obtained can be explained by the intermediate formation of a salt of type V which undergoes methanolysis with inversion of configuration at the phosphoryl center. This must lead to the ester X with net retention of configuration for reasons discussed in detail above. We have excluded the alternative pathway involving preliminary formation of t-butyl phenyl phosphinobromidate XI followed by reaction of XI with methanol.

$$\mathrm{Bu^tPhP(O)SMe} \xrightarrow{\mathrm{Br_2}} \underset{\mathrm{XI}}{\mathrm{Bu^tPhP(O)Br}} \longrightarrow \underset{\mathrm{X}}{\mathrm{Bu^tPhP(O)OMe}}$$

The rates of both reactions are considerably lower than that of I with bromine in methanol discussed above.

An important aspect of this paper is the demonstrated facile formation of the P-O-P system by nucleophilic attack of phosphoryl oxygen on the four-coordinated phosphorus center. This result confirms the reaction schemes proposed earlier for a number of reactions, important from the synthetic point of view of organophosphorus compounds (9-13) and polymers (14).

Acknowledgement. This work was supported by the Polish Academy of Sciences.

Literature Cited

1. Saville, B. Chem.Ind. (London) 1956, 660.
2. Stirling, C.J.M. J.Chem.Soc. 1957, 3597.
3. Cooper, D.B.; Hall, C.R.; Harrison, J.H.; Inch, T.D. J.Chem. Soc., Perkin Trans. 1 1977, 1969.
4. Michalski, J.; Ratajczak, A. Rocz.Chem. 1963, 37, 1185.
5. Michalski, J.; Mikołajczyk, M.; Omelańczuk, J. Tetrahedron Lett. 1968, 3565.
6. Krawiecka, B.; Skrzypczyński, Z.; Michalski, J. Phosphorus 1973, 3, 177.
7. Hall, C.R.; Inch, T.D. J.Chem.Soc., Perkin Trans. 1 1979, 1104.
8. Krawiecka, B.; Michalski, J.; Tadeusiak, E. J.Am.Chem.Soc. 1980, 102, 6584.
9. Kosolapoff, G.M.; Watson, R.M. J.Am.Chem.Soc. 1951, 73, 4101.
10. Kosolapoff, G.M. Science 1948, 108, 485.
11. Toy, A.D.F. J.Am.Chem.Soc. 1949, 71, 2268.
12. Simpson, P.; Zwierzak, A. J.Chem.Soc., Perkin Trans. 1 1975, 201.
13. Aaberg, T.; Gramstad, T.; Husebye, S. Tetrahedron Lett. 1979, 2263.
14. Vogt, W. Macromol.Chem. 1973, 163, 89.

RECEIVED July 2, 1981.

Isotope Effects in Amination Reactions of Chlorocyclophosphazenes

J. M. E. GOLDSCHMIDT, R. HALEVI, and E. LICHT

Department of Chemistry, Bar-Ilan University, Ramat-Gan 52100 Israel

The general substitution reactions of hexahalogenocyclotriphosphazenes, $N_3P_3X_6$ (X=F,Cl,Br; Y=Substituent)

$$N_3P_3X_6 + n\ HY \longrightarrow N_3P_3X_{6-n}\ Y_n (I) + nHX$$

can yield up to 12 products. Three of these arise from n = 1, 5 and 6 and the other 9 from n = 2, 3 and 4, for each of which gem-, cis- and trans- isomers can exist. As an example of these reactions, amination of the hexachloro-compound, $N_3P_3Cl_6$ has been particularly extensively investigated, mostly to isolate and characterize the products formed (1). In reactions which can lead to isomers, mixtures of them are generally formed, though typically the quantity of one of them greatly exceeds that of the rest. In this behavior, called stereoselectivity, the operation of characteristic substitution patterns was revealed. Attempts to rationalize these isomeric preferences mechanistically have been widespread and many factors have been considered. Because the amination reactions have been demonstrated to be kinetically controlled (2), our experiments aimed at elucidating the stereoselectivity mechanistically have focused on detailed kinetic studies.

The general mechanism that has been proposed (3,4) for the first stage of substitution of $N_3P_3Cl_6$ by amine leading to (I) with X = Cl, Y = NR_2 (R = alkyl or H) and n = 1 is

$$N_3P_3Cl_6 + HNR_2 \rightleftharpoons N_3P_3Cl_6 \cdot HNR_2 \xrightarrow[\text{(base)}]{-H^+} N_3P_3Cl_6NR_2^- \xrightarrow[\text{(r.d)}]{-Cl^-} N_3P_3Cl_5NR_2$$

This proposal rests on a combination of 3 chief experimental findings. (1) The order of the reactions (5-9) is never less than one with respect to the concentration of the phosphazene and the amine requiring that both of these be involved in the rate-determining, or a prior, step. (2) The existence of a second-order term in the concentration of amine in non-polar solvents arising from base catalysis by amine (7). (3) The effect of the leaving group, the rate of reaction increasing in the order F < Cl < Br (5). Corroboratorive evidence for the proposed mechanism comes from the values of the activation parameters measured in THF in which most of our

0097-6156/81/0171-0529$05.00/0

TABLE I
Kinetic Data in T.H.F. for Reactions

$$N_3P_3Cl_6 + 2HNR_1R_2 \longrightarrow N_3P_3Cl_5NR_1R_2 + H_2NR_1R_2{}^+Cl^-$$

R_1	R_2	k_2 at 30°C $dm^3mol^{-1}sec^{-1}$	ΔH* $kJ\ mol^{-1}$	ΔS* $JK^{-1}mol^{-1}$	Refs.
Me	H	23.2±0.6	2.9±1.7	-205±6	9
Me	Me	11.5±0.2	7.1±2.1	-197±8	3
Piperidine		1.33±0.02	12.1±2.1	-188±6	4
Bu^t	H	0.0163±0.0003	28.8±1.7	-184±6	4

work was performed because of the simple overall second-order invariably found in it because the solvent acts as a base (3). Some typical values of ΔH* and ΔS* appear in Table I. The values of ΔH* are always rather low whilst values of ΔS* are so very low that it is always rate-controlling. The bulk of the ΔH* is ascribed to the first pre-equilibrium whilst ΔS* is largely associated with the solvation of the Cl^- ion being formed in the rate-determining step. This crude division of the activation parameters is supported by the data on the steric effects of the alkyl groups of the nucleophile (Table I). The reduction in the rate of reaction with increasing size of the alkyl groups results only from increases in ΔH*, the values of ΔS* being virtually constant, this latter fact reflecting the common final step in all the reactions (4). There being disagreement on the deprotonation step, it also having been considered concerted with Cl^- ion departure (3), we carried out comparative studies using piperidine and N-deuteropiperidine, (Table II). The combination of a second-order term in piperidine concentration (in toluene) with the absence of a measurable isotope effect in both solvents studied

TABLE II
Rate Constants for Amines and N-Deuterated Analogues in Reactions (10)

$$N_3P_3Cl_6 + 2H(D)NR_1R_2 \longrightarrow N_3P_3Cl_5NR_1R_2 + H_2(D_2)NR_1R_2{}^+Cl^-$$

R_1	R_2	Solvent	T°C	$k_2{}^a$ HNR_1R_2	$k_3{}^b$	$k_2{}^a$ DNR_1R_2	$k_3{}^b$
Piperidine		T.H.F.	30	7.41 ±0.19	-	7.40 ±0.18	-
Piperidine		Toluene	0	2.13±1.3 $\times10^{-3}$	1.71 ±0.5	2.16±0.6 $\times10^{-3}$	1.7 ±0.3
Bu^t	H	T.H.F.	30	0.0163 ±0.0003	-	0.0163 ±0.0002	-

[a] $k_2 = dm^3\ mol^{-1}\ sec^{-1}$ [b] $k_3 = dm^6\ mol^{-2}\ sec^{-1}$

is interpreted as supporting the separation of the fast pre-equilibrium deprotonation from the Cl^- ion departure step.

Geminal and non-geminal (cis and trans) isomers can be formed in the second stage of the substitution, leading to (I) with X = Cl, Y = NR_2 and n = 2. At this stage of substitution geminal isomers are only obtained with primary amines (except azidirine) and this is accommodated by the following proposed conjugate-base

$$P_2Cl_4N_3P{\scriptstyle\begin{matrix}NHR\\ Cl\end{matrix}} \xrightleftharpoons[\text{(base)}]{-H^+} P_2Cl_4N_3P{\scriptstyle\begin{matrix}NR^-\\ Cl\end{matrix}} \xrightarrow[\text{(r.d)}]{-Cl^-} P_2Cl_4N_3P{=}NR \xrightarrow[\text{(fast)}]{H_2NR} P_2Cl_4N_3P(NHR)_2$$

mechanism. Although kinetically unconfirmed, this mechanism is supported by the enhanced gem-isomer formation observed on addition of base (11).

With respect to the 2 non-geminal isomers which are the exclusive products formed with secondary amines, and which are produced in varying quantities together with the gem-isomer with primary amines, the general mechanism of their formation resembles that of the first stage of substitution, but the reaction is slower and proceeds at the 'unsubstituted' phosphorus atom because of mesomeric charge transfer from the substituent amino-group (3). With all amines the trans-isomers predominate in the non-geminal reactions (12). Three effects have been advanced to explain trans-preference (1) The 'cis-effect' (13), (2) the steric effect (14) and (3) the substituent solvating effect (SSE) (9,15). To test these proposals we performed a kinetic study in THF of cis- and trans-isomer formation at different temperatures for the reactions shown in Table III and evaluated the values of the activation parameters (Table III).

TABLE III

Activation Parameters for Reactions (15)

$$N_3P_3Cl_5NMe_2 + 2HNMe_2 \longrightarrow \frac{\text{Cis}}{\text{Trans}}\}N_3P_3Cl_4(NMe_2)_2 + H_2NMe_2^+Cl^-$$

	Cis-Isomer	Trans-Isomer	Units
ΔH*	8.8 ± 0.5	28.5 ± 1.3	kJ mol^{-1}
ΔS*	-239 ± 8	-159 ± 8	JK^{-1} mol^{-1}

These values can only be reconciled with SSE in which the following transition state structure which is patently sterically impossible for the reaction that leads to the cis-isomer is postulated. This is formed by reprotonation on the amino-substituent after prior deprotonation of the primary product formed. The formation of an H^+Cl^- ion pair, rather than a 'bare' Cl^- as in the cis-reaction, in the rate-determining step accounts for the higher values of ΔS* in the trans-reaction compared with the cis-reaction. Supporting evidence for this proposal comes from the results of experiments (Table IV) in which the ratios of non-geminal isomeric products using piperidine and N-deuteropiperidine were determined. The average ratio $(trans/(cis + trans))_D/$

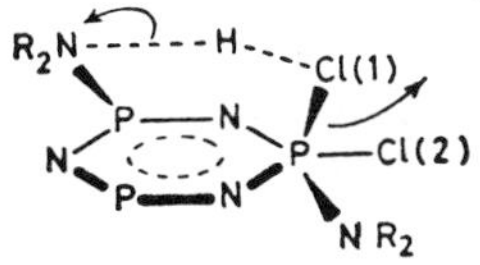

$(trans/(cis + trans))_H$ is equal to 0.95. The reduced amount of trans-isomer found with the deuterium compound adds support for the SSE mechanism involving a H-bonded transition state structure. However a full study of the kinetics of the cis- and trans- reactions at various temperatures using normal and N-deuteropiperidine needs to be carried out to validate this conclusion.

TABLE IV
N-Deuterium Effect on Ratio of Trans to Total Non-Geminal Isomers in Reaction

$$N_3P_3Cl_5NC_5H_{10} + 2H(D)NC_5H_{10} \longrightarrow N_3P_3Cl_4(NC_5H_{10})_2 + H_2(D_2)NC_5H_{10}^{+}Cl^{-}$$

	1 = HNC_5H_{10}	2 = DNC_5H_{10}	
Expt.	$\left(\frac{Trans}{Cis+Trans}\right)_H$ x100	$\left(\frac{Trans}{Cis+Trans}\right)_D$ x100	$^2/_1$
1	82.6±1.6	79.7±1.1	0.964
2	79.0±2.7	73.6±1.5	0.932
3	75.7±1.1	72.7±2.0	0.959
4	72.3±0.4	70.7±3.7	0.978
5	75.6±2.1	71.5±2.2	0.946

Literature Cited

1. Krishnamurthy, S.S.; Sau, A.C.; Wood, M. Adv. Inorg. Chem. Radiochem. 1978, 24, 41.
2. Friedman, N.; Goldschmidt, J.M.E.; Sadeh, U.; Segev, M. J. Chem. Soc. Dalton Trans. 1981, 103.
3. Goldschmidt, J.M.E.; Licht, E. J. Chem. Soc. A 1971, 2429.
4. Goldschmidt, J.M.E.; Licht, E. J. Chem. Soc. Dalton Trans. 1979, 1012.
5. Moeller, T.; Kokalis, S.G.J. Inorg. Nucl. Chem. 1963, 25, 1397.
6. Krishnamurthy, S.S.; Sundaram, P.M. Inorg. Nucl. Letters 1979, 15, 367.
7. Capon, B.; Hills, K.; Shaw, R.A. J. Chem. Soc. 1965, 4059.
8. Eliahu, S. M.Sc. Dissertation, Bar-Ilan University, 1969.
9. Goldschmidt, J.M.E.; Licht, E. J. Chem. Soc. Dalton Trans. 1972, 728.
10. Licht, E. Ph.D. Dissertation, Bar-Ilan University, 1971.
11. Gabay, Z.; Goldschmidt, J.M.E. J. Chem. Soc. Dalton Trans. 1981 (in press).
12. Biran, Z.; Goldschmidt, J.M.E. J. Chem. Soc. Dalton Trans. 1979, 1017.
13. Keat, R.; Shaw, R.A. J. Chem. Soc. A 1966, 908.
14. Schmutz, J.L.; Allcock, H.R. Inorg. Chem. 1975, 14, 2433.
15. Goldschmidt, J.M.E.; Goldstein, R. J. Chem. Soc. Dalton Trans. 1981 (in press).

RECEIVED July 7, 1981.

Zwitterionic σ-Complexes: Their Role as Intermediates in Phosphorylation of Aromatics by Phosphorus Compounds

Y. GOLOLOBOV and P. ONYS'KO

The Institute of Organic Chemistry, Academy of Sciences of the Ukrainian SSR, Kiev, USSR

Two types of reactions were known for P^{III} compounds with nitroaromatics - oxidation of P^{III} and substitution of a nitro-group by a phosphoryl moiety (1).

We have found a new direction for the interaction between the above-mentioned reagents in DMSO, namely the attack of the unsubstituted carbon atom in trinitrobenzene (TNB) by P^{III}. This leads to new σ-complexes, which are stabilized without a dissociated cation in contrast to the "ordinary" Meisenheimer σ-complexes.

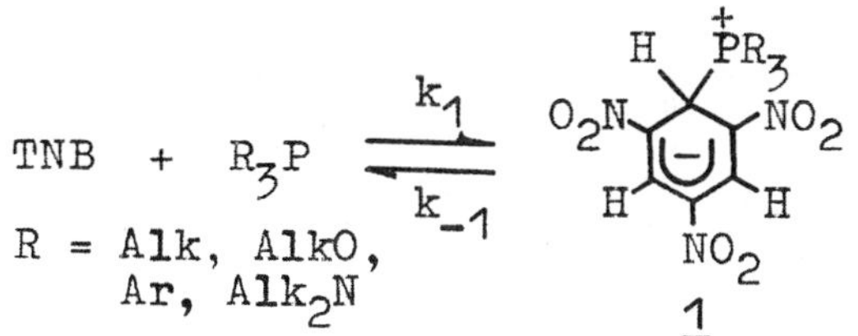
TNB + R3P
k1
k-1
R = Alk, AlkO,
Ar, Alk2N
H
PR3
O2N
NO2
H
H
NO2

1

The formation of 1 seems sufficiently common to such nucleophilic reagents as P^{III}.

We have found some evidences in support of this type of interaction from spectroscopic studies of the complexes like 1.

The alternative ionic scheme of the reaction

$$TNB + R_3P \nrightarrow (O_2N)_3C_6H_2^- + \overset{+}{P}HR_3$$

is rejected by coupling constant $^2J_{HCP}$ 12-21 Hz which is a typical value for the formation of P-σ-complexes of type 1.

The substituents on the phosphorus affect the reaction equilibrium constant $K = k_1/k_{-1}$ (Table I)(k_1 in $M^{-1}s^{-1}$, k_{-1} in s^{-1}, K in M^{-1})(2).

0097-6156/81/0171-0533$05.00/0

Table I

Compound of P^{III}	$k_1 \cdot 10^3$	$k_{-1} \cdot 10^4$	K
$(EtO)_3P$	1.81	0.26	71
$(EtO)_2PPh$	3.0	2.8	11
$EtOPPh_2$	4.8	11.8	4.5

The σ-complexes' stability decreases when attaching phenyl groups to the phosphorus atom. It should be noted that PPh_3 does not form a σ-complex with TNB, but it does with the more electrophilic $C_6H_3(SO_2CF_3)_3$-symm.

σ-Complexes of TNB with $(AlkO)_2P(O)H$

Owing to effective solvation in DMSO, dialkyl phosphites form with TNB comparatively stable complexes 2 in this solvent.

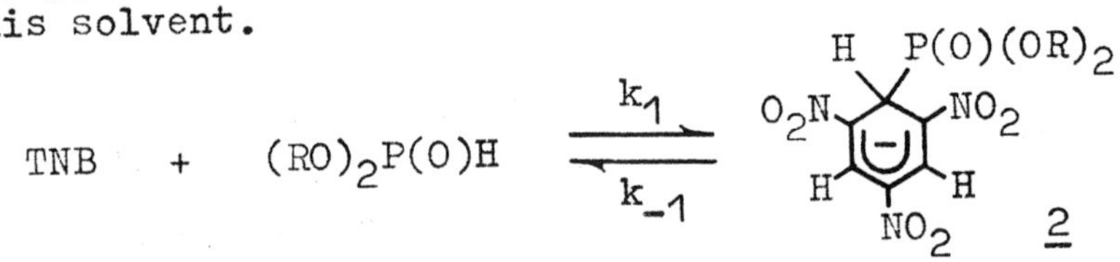

$$\text{TNB} + (RO)_2P(O)H \underset{k_{-1}}{\overset{k_1}{\rightleftharpoons}} \mathbf{2}$$

(2: σ-complex bearing H and $P(O)(OR)_2$ at C-1; O_2N, NO_2, H, H, NO_2 ring substituents)

The complexation rate of $(EtO)_2P(O)H$ in DMSO ($k_1 = 5 \cdot 10^{-5}$, $k_{-1} = 3 \cdot 10^{-5}$) is considerably lower than that of $(EtO)_3P$ (Table I). However in the presence of Et_3N σ-complexing is a practically irreversible reaction for $(EtO)_2P(O)H$.

σ-Complexes of P^{III} with derivatives of TNB

We have found that $(AlkO)_3P$ and $(AlkO)_2P(O)H$ in the presence of Et_3N form stable σ-complexes on the C-3 atom in 3 and 4. This is peculiar to the nature of P-nucleophilic reagents while charged nucleophiles as a rule form more thermodynamically stable C-1 complexes (3).

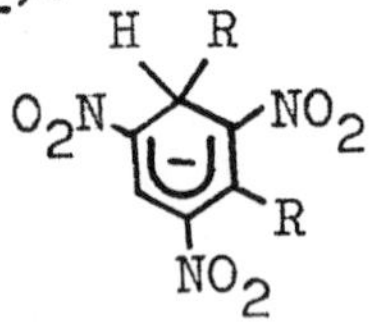

(Structure: σ-complex with H, R at C-1; O_2N, NO_2, R′, NO_2 ring substituents)

3 R = $\overset{+}{P}(OAlk)_3$, R' = MeO, $P(O)(OMe)_2$

4 R = $P(O)(OAlk)_2$, R' = Me, MeO, $P(O)(OMe)_2$

It should be noted that the derivatives of TNB both with donor and acceptor groups in the aromatic ring have lower reaction rates as compared with TNB, which is apparently due to steric hindrance of the substituents.

Nucleophilic phosphorylation of halogenated TNB's

1,3,5-Trinitrochloro(fluoro)benzene has been phosphorylated by P^{III} compounds either by nucleophilic attack at Cl or through σ-complex formation:

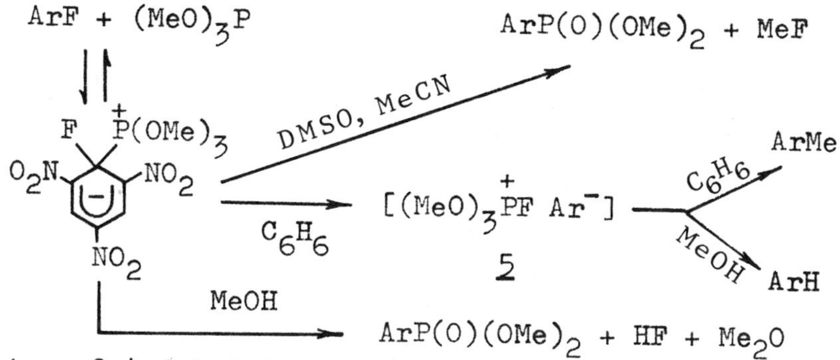

Ar = 2,4,6-trinitrophenyl

It follows from this scheme that the use of a protonic solvent (MeOH) as a test for nucleophilic attack at Cl is not absolutely reliable, since ionic pairs of type 5 may be formed by an alternative path.

σ-Complexes 1 and 2 were obtained in solution (DMSO, MeCN) and purified by column chromatography. They can be kept at 20° for a few hours, after which they convert slowly to corresponding phosphonates.

$$\underline{1} \text{ or } \underline{2} \ (R = AlkO) \xrightarrow{DMSO} ArP(O)(OAlk)_2$$

Nucleophilic phosphorylation of 5-nitropyrimidines

2-R-5-Nitropyrimidines 6 do not form stable σ-complexes with $(RO)_3P$ or $(RO)_2P(O)H$ without bases. Stable complexes 7 are formed only with strongly nucleophilic $(RO)_2P(O)H$ in the presence of Et_3N.

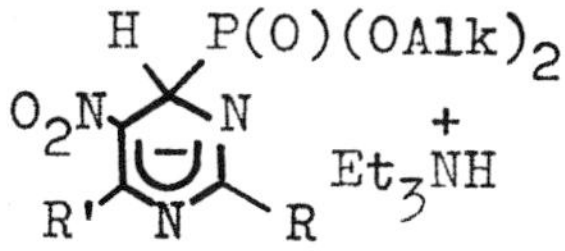

7 R = H, MeO, Ph; R' = H

9 R = H, MeO; R' = MeO

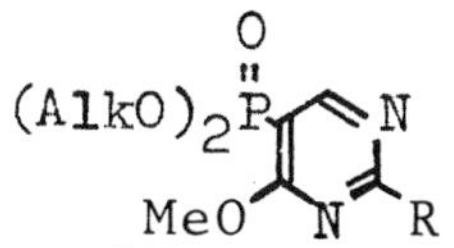

10

R = H, MeO

In all cases formation of complexes was detected only on the nonsubstituted C-4 (C-6) atom of the heterocycle (contrary to many other nucleophiles).

The reaction of 2-R-4-methoxy-5-nitropyrimidines 8 with $(RO)_2P(O)H$ and Et_3N takes place very unusually, since it leads to the substitution of the 5-NO_2 group which is usually inert. σ-Complexes 9 are formed rapidly in the reaction and then rearrange to phosphonates 10.

Despite the fact that trialkyl phosphites do not form stable complexes with 5-nitropyrimidines, they easily react with 6-R-4-chloro-5-nitropyrimidines (R= MeO, NH_2, Cl) to give the corresponding pyrimidinyl-4-phosphonates or (for R = Cl) pyrimidinyl-4,6-diphosphonates.

In 2,4-dichloro-5-nitropyrimidines both chlorines can be substituted but Cl in position 4 is the first to be substituted, which agrees with a preferential formation of σ-complexes to the C-4 atom of 5-nitropyrimidines.

It is quite possible that P-nucleophiles are well suited to accomplish nucleophilic aromatic substitution either through σ-complexes formation or through halogen attack.

Literature Cited

1. Cadogan, J.I.J. Quart.Rev. 1968, 22, 222.
2. Gololobov, Yu.G.; Onys'ko, P.P.; Prokopenko, V.P. Dokl. Akad. Nauk SSSR 1977, 237, 105.
3. Strauss, M.J. Chem.Revs 1970, 70, 667.

RECEIVED July 7, 1981.

Use of X-Ray Structural Results on Phosphorus Compounds in Modeling Reaction Mechanisms

ROBERT R. HOLMES and JUDITH C. GALLUCCI
Department of Chemistry, University of Massachusetts, Amherst, MA 01003

JOAN A. DEITERS
Department of Chemistry, Vassar College, Poughkeepsie, NY 12601

During the past four years we have determined the crystal structure of over thirty cyclic containing phosphorus compounds and have established that their structures in the solid state and in solution in general are similar (1,2). Recent x-ray structures in the bis(biphenylylene) series (3,4) show steric effects as the size of R increases.

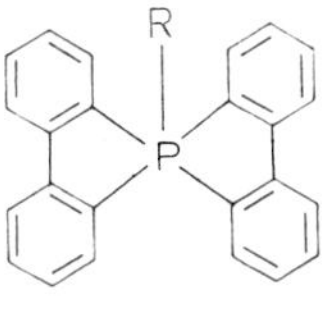

R

8-$(CH_3)_2$N-1-Np

1-Np

CH_3

Ph

Np = naphthyl

The structure of 8-Me_2N-1-Np derivative (3) is at a point 2/3 the way from a trigonal bipyramid (R equatorial) toward the rectangular pyramid with one of the rings in the unique apical-basal orientation. Dimeric phosphoranes exhibiting unusual bridgehead structures are the difluoro and dichloro triazaphospholes (5). Their structures are half-way between the idealized trigonal bipyramid and square pyramid and exist in the cis-facial arrangement (relative to the trigonal bipyramid).

X = F, Cl

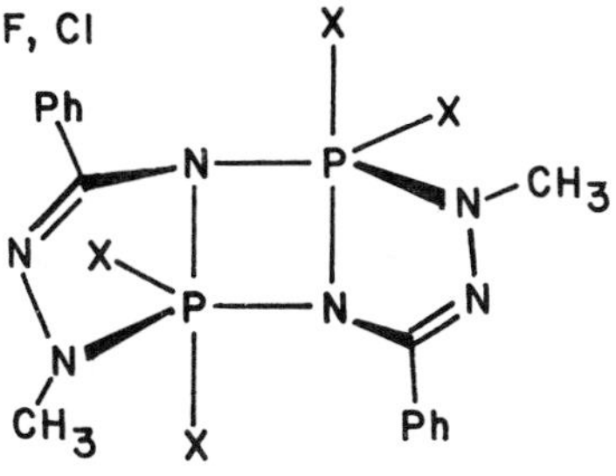

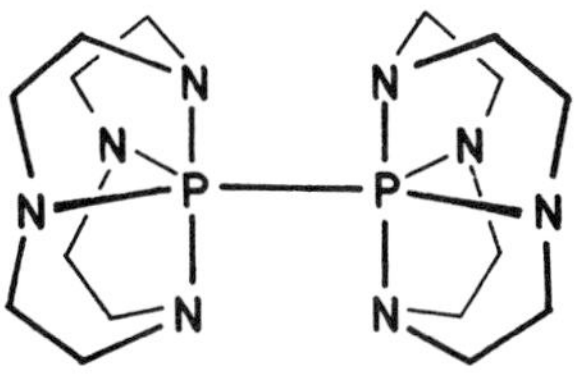

0097–6156/81/0171–0537$05.00/0

The first x-ray structure of a P(V)-P(V) derivative, the dimeric cyclen phosphorane $[(NCH_2CH_2)_4P]_2$ (6) above, shows a normal P-P single bond length (2.264(2) Å).

Using the results of x-ray studies as a structural base, we have parameterized a molecular mechanics program (7) to include terms specific for pentacoordinate phosphorus and have shown that computer simulated structures compare favorably with those obtained by x-ray diffraction.

In general, phosphorane structures form a series which may be placed incrementally along a C_{2v} coordinate (Berry coordinate) connecting the idealized trigonal bipyramid (TP) with the square (or rectangular) pyramid (RP) (8,9). In most cases these structural distortions can be interpreted in terms of inherent molecular properties, ie., substituent electronegativity (or apicophilicity), ring strain effects, steric and pi bonding considerations (9). It is determined, for example, that like atoms in a five-membered ring and ring unsaturation are required to form a RP., cf 1 and 2.

1 (TP→RP 97%)

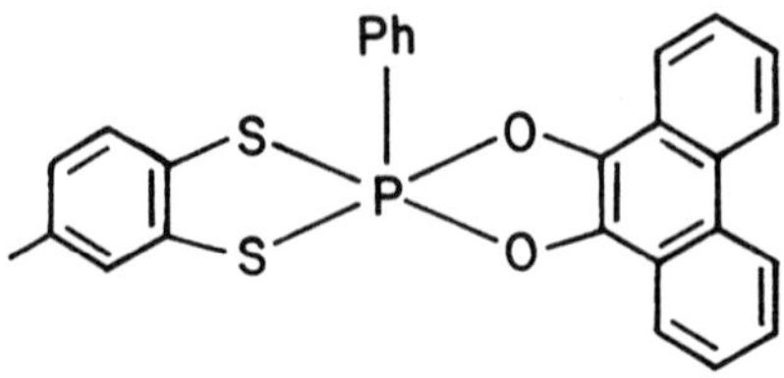

2 (TP→RP 40%)

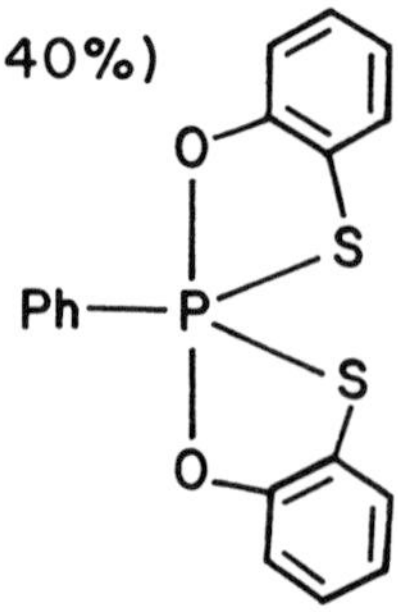

When short intermolecular contacts are apparent in the crystallographic data, their presence is due to either intermolecular hydrogen bonding or steric interactions caused by bulky substituents (9). For some of these situations, we simulated both the isolated molecule and the molecular structure perturbed by neighboring molecules in the unit cell to establish that this is the case (10,11).

With this background data, modeling of a reaction coordinate for a phosphorus compound proceeding by an associative mechanism through a pentacoordinated transition state becomes attractive. We have simulated the steric course of the alkaline hydrolysis of chiral five- (12) and six- (13) membered cyclic phosphonium salts, whose reaction kinetics and product stereochemistries had been studied previously by Marsi and coworkers (14,15). For this purpose, we determined the absolute configuration of the phospholanium iodide 3 (12), and the x-ray structures of the related phosphorinanium salts, 4 and 5 (13).

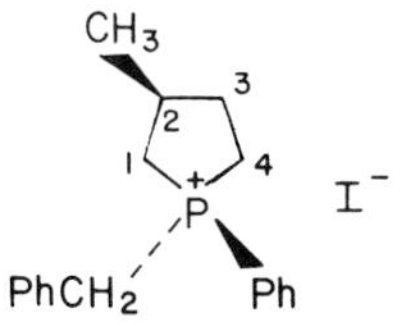

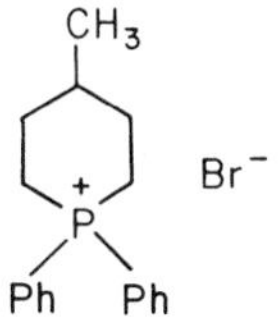

Molecular mechanics modeling (13) of hydroxide attack on the chiral trans benzyl derivative 6 indicated that steric control during transition state formation is a principal factor in the reduced amount of inverted product formed (15) compared to that for the cis isomer 7.

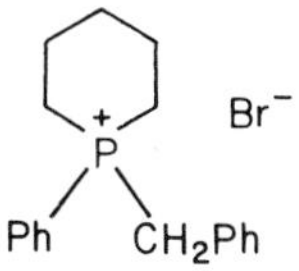

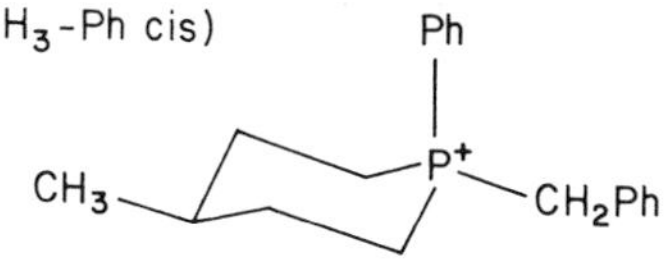

For the chiral phenylphospholanium salts, modeling (13) shows, in agreement with the initial suggestion of Marsi (14), that steric control in the ground state is important in accounting for the decreased amount of inversion at phosphorus for 8 compared to that for 9. In 8, in-line displacement of the methoxy group results in greater steric interference between the approaching hydroxide ion and the ring methyl group.

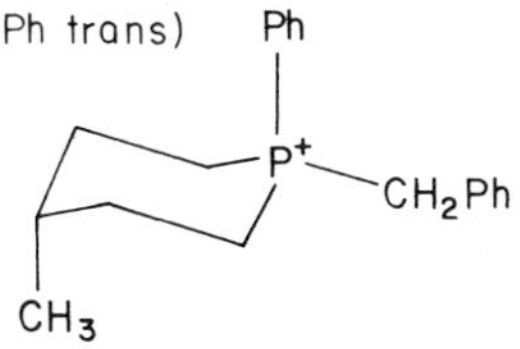

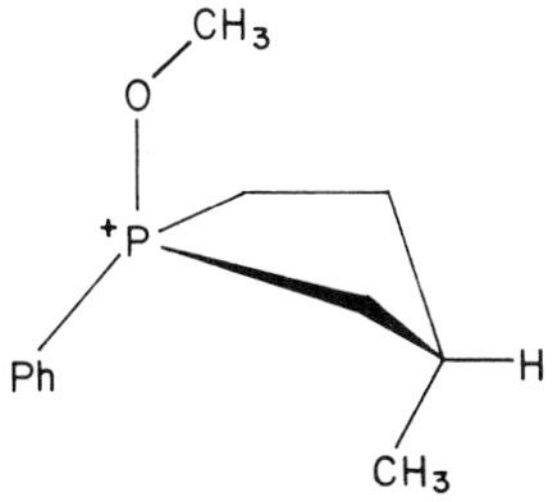

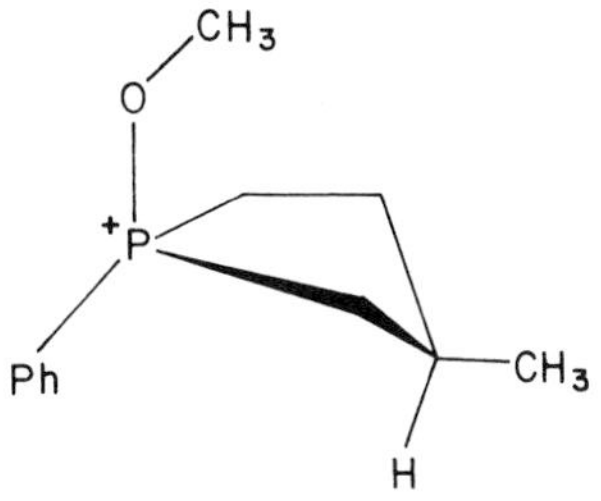

The apparent usefulness of the modeling approach suggested that possible active site interactions important in understanding the mode of action of the well-characterized enzymes, ribonuclease (16) and staphylococcal nuclease (17), may be revealed. Both have been the subject of extensive crystallographic studies (18,19) with suitable inactive substrates in place. We considered the first step of hydrolytic action of ribonuclease (RNase) on the dinucleotide substrate uridylyl-(3'-5')-adenosine(UpA). Our results (20) on the enzyme mechanism were consistent with the main features summarized by Roberts et al (21). The first step is a transphosphorylation leading to cleavage of the phosphodiester bond yielding a 2'-3' cyclic phosphate intermediate. Based on the initial x-ray coordinates (18), we show (20) a low energy pathway for this cleavage, with lysine-41 of limited importance until the cyclic intermediate is formed. An in-line attack occurs giving a pentacoordinated transition state in the low energy path. Consideration of adjacent attack and possible pseudorotations result in relatively high activation energies for hydrolytic cleavage.

In our current study, exploring the possible mechanism of action of staphylococcal nuclease, computer modeling is based on the use of x-ray coordinates (19) determined for the enzyme-inhibitor complex with deoxythymidine 3'-5' diphosphate (dTdp) and calcium ion. X-ray coordinates of the atoms of dTdp and of enzyme residues involved at the active site, namely Asp21, Asp40, Glu43, Arg35, Tyr113, Hys84, Arg87 and Tyr85 are refined and the energy of the system minimized with respect to the positions of all atoms involved. Several enzyme constraints are introduced into the calculation to mimic the possible enzyme influences on substrate conformation. By the addition of a para-nitro phenyl group to the 5'-phosphate, the inhibitor dTdp is changed into an active substrate. Factors which control nonenzymatic and enzymatic hydrolysis of deoxy thymidine-3'-phosphate-5'-para-nitrophenyl phosphate are explored by simulation of two reaction pathways (Fig. 1).

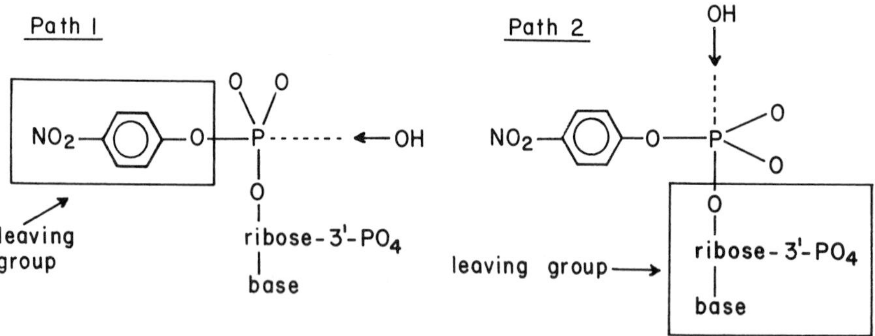

Figure 1. Reaction pathways.

The results of the computer simulation show, in agreement with experimental data, that the poorer leaving group is cleaved in enzymatic action, ie. Path 2 is the low energy path. This contrasts with nonenzymatic cleavage where the leaving group is the nitrophenoxide ion, ie. Path 1. Simulation of Path 1 for the enzyme produces a high energy route which is due to the presence of Arg35 blocking the approach of hydroxide ion to in-line attack.

Literature Cited

1. Holmes, R.R. "Pentacoordinated Phosphorus. Structure and Spectroscopy", Vol. I, ACS Monograph 175, Washington, D.C., 1980.
2. Holmes, R.R. "Pentacoordinated Phosphorus. Reaction Mechanisms", Vol. II, ACS Monograph 176, Washington, D.C., 1980.
3. Day, R.O., Holmes, R.R. Inorg. Chem. 1980, 19, 3609.
4. Day, R.O., Husebye, S., Holmes, R.R. Inorg. Chem. 1980, 19, 3616.
5. Day, R.O., Holmes, R.R., Tautz, H., Weinmaier, J.H., Schmidpeter, A. Inorg. Chem. 1981, 20, 1222.
6. Richman, J.E., Day, R.O., Holmes, R.R. J. Am. Chem. Soc. 1980, 102, 3955.
7. Deiters, J.A., Gallucci, J.C., Clark, T.E., Holmes, R.R. J. Am. Chem. Soc. 1977, 99, 5461.
8. Holmes, R.R., Deiters, J.A. J. Am. Chem. Soc. 1977, 99, 3318.
9. Holmes, R.R. Acc. Chem. Res. 1979, 12, 257.
10. Brown, R.K., Day, R.O., Husebye, S., Holmes, R.R. Inorg. Chem. 1978, 17, 3276.
11. Meunier, P.F., Day, R.O., Devillers, J.R., Holmes, R.R., Inorg. Chem. 1978, 17, 3270.
12. Day, R.O., Husebye, S., Deiters, J.A., Holmes, R.R. J. Am. Chem. Soc. 1980, 102, 4387.
13. Gallucci, J.C., Holmes, R.R. J. Am. Chem. Soc. 1980, 102, 4379.
14. Marsi, K.L. J. Org. Chem. 1975, 40, 1779.
15. Marsi, K.L., Clark, R.T. J. Am. Chem. Soc. 1970, 92, 3791.
16. Richards, F.M., Wyckoff, H.R. "The Enzymes", P.D. Boyer, Ed., 3rd ed., Vol. IV, Academic Press, New York, 1971, pp. 647-806.
17. Anfinsen, C.B., Cuatrecasas, P., Taniuchi, H. ibid., pp. 177-204.
18. Richards, F.M., Wyckoff, H.W. "Atlas of Molecular Structures in Biology, 1. Ribonuclease-S", D.C. Philips and F.M. Richards, Eds., Clarendon, Oxford, 1973.
19. Cotton, F.A., Hazen, Jr., E.E., Legg, M.J. Proc. Natl. Acad. Sci. U.S.A. 1979, 76, 2551, and references cited therein.
20. Holmes, R.R., Deiters, J.A., Gallucci, J.C. J. Am. Chem. Soc. 1978 100, 7393.
21. Roberts, G.C.K., Dennis, E.A., Meadows, D.H., Cohen, J.S., Jardetzky, O. Proc. Natl. Acad. Sci. U.S.A. 1979, 62, 1151.

RECEIVED July 7, 1981.

Ligand Effects on the Reaction of Alkoxide Ions with Organophosphorus Derivatives Containing Multiple Leaving Groups

KENNETH E. DEBRUIN, CHARLES E. EBERSOLE, MORGAN M. HUGHES, and DAVID M. JOHNSON

Department of Chemistry, Colorado State University, Fort Collins, CO 80523

Nucleophilic displacement reactions on tetracoordinate phosphorus compounds containing both alkoxy and alkylthio ligands proceed with widely varying stereochemistry. In the particular case of oxyanions as nucleophiles, hydrolysis of phosphonium salts (1-3) alkoxide displacements on phosphonothioates (2, 4-6) and alkoxide displacements on phosphorothioates (5, 6) have been observed to displace the alkylthio ligand with retention, inversion, and retention stereochemistry respectively. Presumably, these stereochemical differences reflect changes in the structure of the kinetically formed trigonalbipyramid intermediates with inversion resulting when the alkylthio ligand is co-axial with the attacking nucleophile while retention requires the alkylthio group to be in the equatorial position (equation 1). The varying stereochemistry therefore implies that the non-displaced ligands (A and B in equation 1) have a major influence on relative energies of the transition states leading to the two intermediates.

$$\text{A,B}{>}\text{P}{<}^{\text{SR}}_{\text{OR}} \xrightarrow{^{-}\text{OR}'} \underset{\mathbf{1}}{\text{A,B}{>}\text{P(SR)(OR)(OR}')} \;\; \text{vs} \;\; \underset{\mathbf{2}}{\text{B,A}{>}\text{P(OR)(SR)(OR}')} \qquad (1)$$

Suggestive that the origin of the stereochemical crossover from displacement with inversion in the phosphonothioate system (A = Ph, Me; B = O, S) to retention in the phosphorothioate system (A = OR, B = O) is a function of the electronegativity of the ligand A, is the observation that phosphoramidothioates (A = NRR') undergo displacement of the alkylthio ligand with net inversion but considerable racemization (7, 8). The NRR' group is of intermediate electronegativity between Ph and OR and appears to give intermediate stereochemistry between inversion and retention. However, no mention is made of possible OR displacement which was observed in the reaction of phosphonothioates (A = Ph) and would underestimate the amount of intermediate 2 formation or whether the racemization may have occurred after product formation.

0097-6156/81/0171-0543$05.00/0

We have carried out a detailed product, stereochemistry, and rate analysis of the reaction of O,S-dimethyl N-(1-phenylethyl)-phosphoramidothioate with sodium ethoxide in ethanol according to the Scheme below. Concentrations of all four species were followed by gas chromotography and the stereochemistry of the initial displacement products were determined and corrected for isomerization by isolation of various time intervals. [1]H-nmr analysis of

A—P(=O)(OMe)—SMe $\xrightarrow[-\ominus SMe]{\ominus OEt}$ A—P(=O)(OEt)—OMe

$\ominus$OEt ↓ $-\ominus$OMe (left); Isomeriz. ($\ominus$OEt / $-\ominus$OEt); $\ominus$OEt ↓ $-\ominus$OMe (right)

A—P(=O)(SMe)—OEt $\xrightarrow[-\ominus SMe]{\ominus OEt}$ A—P(=O)(OEt)—OEt

the diastereomeric P-O<u>Me</u> or P-S<u>Me</u> groups afforded isomer ratio and extrapolation to time zero gave the initial reaction stereochemistry (7). The results are indicated in Table I and compared to phosphonothioate and phosphorothioate systems. It appears that the competition between the two modes of attack by ethoxide ion on phosphoramidothioates is virtually identical to that for phosphonothioates and not intermediate in behavior.

Table I. Products and Competitions for Forming Intermediate in the Reaction of Alkoxide Ions with Thioate Esters of Phosphorus

A(O=)P(SMe)(OMe) $\xrightarrow{\ominus OR}$ $^{-}$O(A)P(SMe)(OMe)(OR) + $^{-}$O(A)P(OMe)(SMe)(OR)

A	$^{-}$OR	%	Products	%	Products	Ref.
Ph	OMe	85	100%-SMe	15	100%-OMe	4
PhCHMeNH	OEt	80	100%-SMe	20	95%-OMe 5%-SMe	
iPrO	OEt	0		100	100%-SMe	5
pNO_2Ph	OEt	62	100%-SMe	38	100%-OMe	
Ph	OEt	78	100%-SMe	22	100%-OMe	
pMe_2NPh	OEt	83	100%-SMe	17	100%-OMe	

As a further evaluation of the possible electronegativity control by an equatorial ligand on relative stabilities of the transition states leading to attack by alkoxide ion co-axial with an SMe group *vs* an OMe group, we investigated the products from the reaction of sodium ethoxide with para-substituted O,S-dimethyl phenylphosphonothioates. Assuming complete inversion of configuration, product ratios corrected as above according to the Scheme reflect modes of attack. The results are listed in Table I. Clearly, for large electronic variations in ligand A (rate constants span four powers of 10) a minor variation in modes of intermediate formation is observed; again suggestive that the electronegativity of ligand A is a minor contributor to determining competitive attacks by a nucleophile.

In the absence of specific ligand-ligand interactions controlling competitive modes of reaction, the kinetically preferred intermediate from attack by a nucleophile on a tetracoordinate organophosphorus ester must reflect the relative affinities of ligands to occupy an axial site versus equatorial sites. To evaluate these affinities, we have measured the rates of reaction of sodium methoxide in methanol with a variety of compounds. The results are given in Table II as logarithm of rate constants for attack by methoxide co-axial with ligand Y.

A comparison of the rate constants k(Y=SMe) and k(Y=OMe) for placing SMe *vs* OMe in the axial position of the transition state respectively indicates a *ca*. 50-100 fold rate preference by the SMe group for all compounds studied. Therefore, the origin of the stereochemical crossover between phosphonothioates and phosphorothioates lies not in the kinetic affinities of the two groups for an axial position. In the phosphono- system (A=Ph) replacing an equatorial group B=OMe by B=SMe produces a *ca*. 10 fold rate acceleration while in the phosphoro- system (A=MeO) replacing B=OMe by B=SMe produces a *ca*. 500 fold rate acceleration. Thus, phosphonothioates kinetically prefer to place the SMe group axial (100/10) while phosphorothioates prefer placing the SMe group equatorial (500/100) consistent with the stereochemical results.

Table II. Rate Constants for Reaction of Organophosphorus Esters with Sodium Methoxide (0.116M) in Methanol at 20.0°.

ABP(O)Y		log k				
A	B	Y=OMe	Y-SMe	σ(A)	σ(B)	Σσ
Ph	Me	-2.50	-0.32	-1.1	-0.6	-1.7
Ph	Ph	-3.19	-1.52	-1.1	-1.1	-2.2
Ph	SMe	-3.02	-0.84	-1.1	+0.9	-0.2
Ph	OMe	-4.19	-2.23	-1.1	-2.5	-3.6
MeO	SMe	-2.84	-	-2.5	+0.9	-1.6
MeO	OMe	-5.60	-	-2.5	-2.5	-5.0

To evaluate whether the 10 fold or the 500 fold rate acceleration observed by replacing B=OMe by B=SMe is the "normal" effect, we have adopted as a model the σ (inv) constants of Splitter and Calvin (9) defined for the effect of ligands on the pyramidal inversion barriers of amines. A parallel between amines and phosphorus barriers has been demonstrated (10). Presumably, in the inversion processes, hybridization changes in the bond between the center atom and a ligand upon undergoing pyramidal inversion would resemble the changes in placing a ligand into the equatorial position upon forming a trigonal-bipyramid intermediate from tetrahedral reactants in the absence of steric factors. Since a phenyl ligand has a resonance acceleration to pyramidal inversion which would be absent in nucleophilic attack at tetracoordinate phosphorus, we estimated the σ constant for phenyl (-1.1) to be intermediate between NH_2(-1.6) and Me(-0.6). A plot of the logarithm of the rate constants (Y=OMe) against the sum of the σ constants for ligands A and B gives a good linear correlation for all compounds except O,S-dimethyl phenylphosphonothioate (A=Ph, B=SMe) which reacts by a factor of 50 slower than predicted. Although the validity of this treatment may be in doubt, the results suggest that an alkylthio has a net relatively higher affinity for equatorial placement compared to an alkoxy group upon nucleophilic attack at tetracoordinate phosphorus. Thus, attack of the nucleophile co-axial with an alkoxy group should be preferred. The reason for preferred attack of alkoxide ion co-axial with the alkylthio group in phosphonothioates and phosphoramidothioates is at present unresolved but may suggest these compounds are not undergoing rate limiting formation of a pentacoordinate intermediate.

Literature Cited

1. De'Ath, N. J.; Ellis, K.; Smith, D. J. H.; Trippet, S. J.C.S. Chem. Commun. 1971, 714.
2. Farnham, W. B.; Mislow, K.; Mandel, N.; Donohue, J. J.C.S. Chem. Commun. 1972, 120.
3. DeBruin, K. E.; Johnson, D. M. J. Am. Chem. Soc. 1973, 95, 4675.
4. DeBruin, K. E., Johnson, D. M. J. Am. Chem. Soc. 1973, 95, 7921.
5. Cooper, D. B.; Hall, C. R.; Harrison, J. M.; Inch, T. D. J.C.S. Perkin I, 1977, 1969.
6. Inch, T. D.; Lewis, G. J. Carbohydrate Res. 1975, 45, 65.
7. Hall, C. R.; Inch, T. D. Tetrahedron Lett. 1977, 3765.
8. Hall, C. R.; Inch, T. D. J.C.S. Perkin I, 1979, 1646.
9. Splitter, J. S.; Calvin, M. Tetrahedron Lett. 1973, 4111.
10. Baechler, R. D.; Andose, J. D.; Stackhouse, J.; Mislow, K. J. Am. Chem. Soc. 1972, 94, 8060.

RECEIVED June 30, 1981.

Methanolysis of a Phosphate Ester

WILLIAM S. WADSWORTH, JR.

Department of Chemistry, South Dakota State University, Brookings, SD 57007

In a recent publication, we described the proton catalyzed methanolysis of 2-substituted-5-(chloromethyl)-5-methyl-2-oxo-1,3,2-dioxaphosphorinans (1). We found that at low acid concentrations the configuration at phosphorus was retained while at elevated proton concentrations both retention and inversion occur with inversion predominating. At low acid concentrations protonation takes place on phosphoryl oxygen, the most basic site. Displacement occurs via a pentavalent intermediate. At high concentrations additional protonation of the leaving group leads to both retention and inversion with the latter a direct displacement.

For metal ion catalysis under acidic conditions we confined our study to Zn^{2+}. While final isomer ratios are not unlike those found at high proton concentrations, there are specific

R

P=O

O

O

$ClCH_2$

CH_3

trans

CH_3OH

$ClCH_2$

CH_3

O

O

O

P-OCH_3

cis/trans

differences between the two systems. The zinc ion catalyzed reactions proceed at a reasonable rate at room temperature. Leaving groups which can coordinate with the ion, i.e., those containing trivalent nitrogen, are particularly effective. Reactions are not first order in catalyst, above one equivalent rates are only slightly effected while below one equivalent the retention-inversion ratio increases. In those cases where rates are low, retention is the only route. Rates are not first order in ester but fall off rapidly as reaction proceeds. Addition of product to reaction mixtures causes rates to fall dramatically and at the low effective catalyst concentrations, retention only is observed. Added trans para methoxyphenyl ester ($R=OC_6H_7 \cdot pOCH_3$) completely inhibits methanolysis.

0097-6156/81/0171-0547$05.00/0

Thus, methanolysis of phosphate esters can be catalyzed by zinc ion either rapidly with both inversion or retention in cases where both phosphoryl oxygen and leaving groups are complexed or slowly and by retention in cases where complex formation is through phosphoryl oxygen only. With phosphates having poor leaving groups, zinc ion is tightly bound and its catalytic effect inhibited.

Methanolysis of cis-2-chloro-5-(chloromethyl)-5-methyl-2-oxo-1,3,2-dioxaphosphorinan catalyzed by one equivalent of base (CH_3O^-) proceeds entirely by retention. Uncatalyzed methanolysis proceeds slowly by inversion. Phenyl esters do not react under neutral conditions but undergo methanolysis exclusively by retention in the presence of base. In this study care was taken to insure that concurrent isomerization of starting materials by liberated ion did not occur. Only in the case of 2,4-dinitrophenyl ester ($R{=}OC_6H_3 \cdot 2,4(NO_2)_2$) is concurrent isomerization a problem and the ester was not used. Methoxide ion, present in very small concentrations, is a more effective nucleophile than phenoxide ions.

Methanolysis follows first order kinetics, Table I, and gives using sigma values, an acceptable Hammett plot. For the methanolysis of the nitrophenyl ester ($R{=}OC_6H_4 \cdot pNO_2$) triethylamine and sodium nitrophenoxide are equally as effective. The lack of a common ion effect, slow methanolysis of a phosphate having a bulky leaving group ($R{=}OC_6H_3 \cdot 2.6(CH_3)_2$), and slow ethanolysis compared to methanolysis would point to the first step, formation of a pentavalent intermediate, as rate determining.

A number of anomolies exist. The high reactivity of the thiophenyl ester ($R{=}SC_6H_5$) may be due to low, compared to oxygen analogues, electron density about phosphorus, the result of poor

Table I. Base Catalyzed Methanolysis of Trans Esters[a]

R	$t_{1/2}$(hr)	R	$t_{1/2}$(hr)
$OC_6H_4 \cdot pNO_2$	12.0	$OC_6H_4 \cdot pCHO$	35
$OC_6H_4 \cdot mNO_2$	12.0	$OC_6H_4 \cdot pF$	118
$OC_6H_4 \cdot oNO_2$	8.0	$OC_6H_4 \cdot pCH_3$	142
$OC_6H_4 \cdot pNO_2$	36[b]	OC_6H_5	150
$OC_6H_4 \cdot pNO_2$	~350[c]	$OC_6H_4 \cdot pOCH_3$	270
$OC_6H_4 \cdot pNO_2$	96[b,d]	$OC_6H_3 \cdot 2,6(CH_3)_2$	550
$OC_6H_4 \cdot mNO_2$	38[b]	SC_6H_5	1.5
$OC_6H_4 \cdot oNO_2$	22[b]	$OC_6H_5 \cdot pNO_2$	∞[b,c]

[a]Solutions 0.1 molar in ester and triethylamine, room temperature.
[b]50% methanol-acetone.
[c]50% ethanol-acetone.
[d]cis ester.
[e]cis thiophosphoryl (P=S) ester.

backbonding between phosphorus and sulfur. Unlike its para analogue, the ester prepared from salicylaldehyde ($R{=}OC_6H_4 \cdot oCHO$)

does not undergo substitution but cleaves to form acetal and salt of the acid phosphate. Interaction of the ortho substituent with the phosphoryl group is most likely. A similar phenomena may explain the rather rapid methanolysis of the ortho nitro ester (R= $OC_6H_4 \cdot \underline{o}NO_2$). The lack of reactivity of the thiophosphate reflects the low polarization of the phosphoryl-sulfur bond compared to a phosphoryl-oxygen bond.

Methanolysis was extended to a few acyclic systems (2). The rate of methanolysis of p-nitrophenyl esters again reflects differences in the electrophilicity of the phosphorus atom. Ethyl di-p-nitrophenyl phosphate loses the first p-nitrophenoxide group upon dissolving in 50% methanol-acetone. Loss of the second is much slower ($t_{1/2}$=335 hrs, 0.1 molar solutions). Diphenyl p-nitrophenyl phosphate undergoes methanolysis in 50% methanol-acetone much more rapidly than a dialkyl analogue ($t_{1/2}$=2hrs,0.1 molar solutions) but as in the case of the six-membered ring phosphorinan system, ethanolysis is slower ($t_{1/2}$=22hrs). Finally, bis-dimethylamino p-nitrophenyl phosphate does not undergo base catalyzed methanolysis at room temperature.

As with acid catalyzed methanolysis, added cations have a dramatic effect on base catalyzed methanolysis as well, Table II.

Table II. Effect of Cations on Base Catalyzed Methanolysis of Phosphorinan Esters

R	Catalyst[a]	$t_{1/2}$(hr)	Ret.(%)[b]	Inv.(%)
$OC_6H_4 \cdot \underline{p}NO_2$	CH_3COONa	97	100	0
$OC_6H_4 \cdot \underline{p}NO_2$	$Hg(C_2H_3O_2)_2$	190	73	27
$OC_6H_4 \cdot \underline{p}NO_2$	$Mg(C_2H_3O_2)_2 \cdot 4H_2O$	31	70	30
$OC_6H_4 \cdot \underline{p}NO_2$	$Zn(C_2H_3O_2)_2 \cdot 3H_2O$	7	42	58
$OC_6H_4 \cdot \underline{p}NO_2$	$Pb(C_2H_3O_2)_2 \cdot 3H_2O$	1.25	12	88
$OC_6H_4 \cdot \underline{p}COCH_3$	$Mg(C_2H_3O_2)_2 \cdot 4H_2O$	38	52	48
$OC_6H_4 \cdot \underline{p}COCH_3$	$Pb(C_2H_3O_2)_2 \cdot 3H_2O$	3.4	13	87

[a]Reactions 0.1M in ester and catalyst.

Metal acetates, due to their solubility in methanol were used. Careful monitoring of reactions gave no indication that prior involvement of acetate ion occurs. Sodium ion has no effect and methanolysis under the basic conditions proceeds with complete retention. Lead ion is particularly effective, strong bonding with the leaving group, and methanolysis is diverted to direct substitution and inversion. Metal acetates ionize to different degrees which makes rate data unreliable. There is, however, a rough correlation between rates and the degree of inversion.

In all work reported herein, product mixtures were obtained by swamping alcoholic solutions with dil. HCl, extraction with methylene chloride and the extract washed with dilute KOH. Product ratios did not change under reaction or workup conditions.

To predict retention and/or inversion a number of factors are important (3).

a. Retention is favored by strong nucleophiles which are capable of backbonding to phosphorus.

b. Retention is promoted by ligands which enhance the electrophilicity of the phosphorus atom.

c. A catalyst which can complex with phosphoryl oxygen will promote retention.

d. With the diminished importance of a and b and the presence of a good leaving group which may complex with a positive ion, inversion becomes favorable.

Acknowledgement is made to the Donors of The Petroleum Research Fund, administered by the American Chemical Society, for the support of this research.

Literature Cited

1. Gehrke, S. H.; Wadsworth, W. S., Jr. J. Org. Chem., 1980, 45, 3921.
2. Bel'skii, V. E.; Kudryavtseva, L. A.; Derstuganova, K. A.; Fedorov, S. B.; Ivanov, B. E. Zh. Obshch. Khim., 1980, 50, 1997.
3. Hall, R. C.; Inch, T. D. Tetrahedron, 1980, 36, 2059.

RECEIVED June 30, 1981.

114

Reactivity of Tricoordinated Phosphorus Compounds

A Mechanistic Study with a Variety of Substrates

C. DENNIS HALL, ROBERT C. EDWARDS, JOHN R. LLOYD, PAUL D. BEER, PHILIP J. HAMMOND, ALBERTO O. D'AMORIM[1], and MELVIN P. MELROSE[1]

Department of Chemistry, King's College, University of London, Strand, London WC2R 2LS, England

Tricoordinated phosphorus compounds react with a wide range of substrates and a great deal of mechanistic information is now available within this area.[1-3] The classical work involved the quaternization of phosphines with alkyl halides[4] and the famous Arbusov reaction[5,6] which prompted intensive studies in the field of organophosphorus chemistry.

Both reactions involve nucleophilic attack of tricoordinated phosphorus on tetrahedral carbon and show all the characteristics of non-polar reactants combining through polar transition states although the solvent effects are sometimes quite modest.[2] Subsequent studies have demonstrated nucleophilic attack on activated alkenes,[7] activated alkynes,[7] the carbonyl group[1-3] and halogen[8] whilst in the Perkow reaction (eqn. 1) all four possible sites in

$$(RO)_3P \; + \; XCH_2COR' \longrightarrow (RO)_2\overset{\overset{\displaystyle O}{\|}}{P} - O\overset{\overset{\displaystyle R'}{|}}{C} = CH_2 \qquad (1)$$

$$X = Cl(Br)$$

the substrate (halogen, sp^3 C, carbonyl carbon and carbonyl oxygen) have at some time been proposed as the site of nucleophilic attack.[2,5,9] In more recent years the range of substrates has broadened to include the O - O bond in various peroxides,[10,11] the S - S bond,[12,13] the S - O bond in sulphenate esters,[14,15] the S - N bond in sulphenamides[16,17] and others which are too numerous to mention. It is the purpose of this paper to collate much of the available information and in combination with new kinetic data to offer an overall view and rationalization of the reactivity of tricoordinated phosphorus compounds.

A selection of the kinetic information available from a variety of substrates is shown in Table I in which the conditions of temperature and solvent have been maintained as uniform as possible.

[1]Theoretical contributions.

0097-6156/81/0171-0551$05.00/0

TABLE 1. Second-order rate coefficients for the reaction of P(III) compounds with four types of substrate

Substrate	$10^3 \times k_2$ ℓ $mole^{-1}$ s^{-1} at 25°C				Source
	Ph_3P	Ph_2POR	$PhP(OR)_2$	$P(OR)_3$	
MeI[a]	5.0	3.6	2.2	0.35	c,d
EtOOEt[b]	4.4×10^{-3}	29×10^{-3}	4.9×10^{-3}	f	c
PhSSPh[b]	3.3	1350	870	0.6	c
S_8[b]	4.6	12,520	9700	∿ 5.0	c,e

R = Pr^i throughout except for $S_8/P(OR)_3$ for which R = Et.
a) in CH_3CN b) in $C_6H_5CH_3$ c) this work d) Songstad J. Acta. Chem. Scand.(A) 1976, 30, 724 e) ref. 12. f) very slow.

It is immediately obvious that with MeI the rates of reaction decrease in the anticipated order of nucleophilic reactivity, i.e., $Ph_3P > Ph_2POR > PhP(OR)_2 > P(OR)_3$ whereas with the others the rate order follows the sequence $Ph_2POR > PhP(OR)_2 > Ph_3P \approx (RO)_3P$. The latter "anomalous" behaviour has been ascribed to the direct formation of pentacoordinated products from P(III) and the substrate as distinct from nucleophilic displacement by phosphorus so that the stability of the T.S. leading to the pentacoordinated molecule dictates the rate.[2,11,14] There is no doubt that several reactions which display the anomalous rate sequence (eg with peroxides, sulphenate esters and dithietenes) do give pentacoordinated products and the proposal is given further support by the reaction of diethyl peroxide with a series of cyclic phosphines in which the rate sequence parallels the stability of the P(V) products.[11] Furthermore, under the appropriate conditions pentacoordinated products are also formed from alkenes (eqn. 2) and alkynes[18] (eqn. 3) and although rate data is only available for

$$Ar_{\mathbf{n}}P(OR)_{3-n} + CH_2{=}CHY \xrightarrow{R'OH} Ar_{\mathbf{n}}P(OR')(OR)_{3-\mathbf{n}}CH_2CH_2Y \quad (2)$$

n = **1** or **2** Y = CO_2Et or CN $\delta\ ^{31}P\ (H_3PO_4)$ −50 (n = 2)
−35 (n = 1)

$$Ar_2POR + PhC{\equiv}CCO_2Et \xrightarrow{R'OH} Ar_2P(OR')(OR)C(Ph){=}CHCO_2Et \quad (3)$$

$\delta\ ^{31}P = -54$

phosphinites and phosphonites, the same reactions with phosphites and triarylphosphines are qualitatively very much slower. However, whereas the reaction with peroxide is unaffected by solvent changes,[11,19] the reactions with MeI, PhSSSPh and S_8 show increasing sensitivity to solvent polarity so that reaction with S_8 appears to proceed through a highly polar transition state despite the "anomalous" reactivity sequence from phosphines to phosphites.

The availability of a wide range of P(III) compounds from the work with activated alkenes, which is relevant to the dimerisation of acrylonitrile, facilitated the determination of ρ-values (Table II) for a number of substrates by systematic variation of the <u>para</u>-substituents in the aryl groups on phosphorus. The data

Table II Rho-values for the reaction of P(III) compounds with a variety of substrates

Substrate	ρ^a (°C)			Source
	Ar_3P	Ar_2POR	$ArP(OR)_2$	
$\rangle$C - X (X = Cl, Br, I)	-1.1 (30)	-1.2 (30)	-	b,c
CH_2=CHCN	-	-2.0 (30)	-1.8 (30)	b
EtOOEt	-0.4 (28)	-0.3 (27)	-0.4 (37)	b
PhSSSPh	-1.2 (25)	-1.2 (25)	-1.1 (25)	b
S_8	-2.5 (25)	-3.2 (25)	-3.3 (25)	b,d
TCNQ	-3.2 (25)	-3.3 (25)	-2.2 (25)	b

a) <u>Maximum</u> error throughout ± 10% of ρ-value b) this work
c) ref. 9 d) ref. 12.

reveal that all the reactions are accelerated by electron donation (i.e. have a nucleophilic component) and that there is a wide range of <u>sensitivity</u> of reaction rate to the <u>electronic effects</u> of substituents at phosphorus. It is also apparent that the ρ-values are virtually independent of the type of P(III) compound for a given substrate but change with the nature of the substrate.

It seems likely that the magnitude of ρ represents the "extent of electron demand" at phosphorus[20] in the T.S. or in other words is a measure of "electron transfer" in the T.S., a term (=z) which appears in the semi-empirical equations describing the nucleophilic reactivity of tricoordinated phosphorus.[21] This concept is reinforced by (i) the observation (from the data of Bokonov[22] and Goetz[23,24]) that ρ-values based on the pK_a values of protonated phosphines <u>increase</u> with increasing pK_a, i.e. increase with a shift of the equilibrium (eqn. 4) to the left and

$$ArR_2\overset{+}{P}H + H_2O \rightleftharpoons ArPR_2 + H_3O^+ \quad (4)$$

(ii) by the fact that reactions of higher ρ value (eg S_8) show a higher sensitivity to solvent.

Literature cited

1. Kosolapoff, G.M. and Maier, L.(eds) "Organic Phosphorus Compounds", Vols 1 and 4, Wiley Interscience, New York, 1972/73.
2. Emsley, J. and Hall, C.D. "The Chemistry of Phosphorus", Harper and Row, London, 1976.
3. Trippett, S. Specialist Reports, Royal Society of Chemistry, Organophosphorus Chemistry, Vols 1-11, 1969-1979.
4. Davies, W.C. and Lewis, W.P.G. J. Chem. Soc. 1934, 1599.
5. Arbusov, B.A. Pure and Applied Chem. 1964, 1, 307.
6. Aksnes, G. and Aksnes, D. Acta Chem. Scand. 1965, 19, 898.
7. Shaw, M.A. and Ward, R.S. Topics in Phosphorus Chemistry, 1972, Vol 7, p.11.
8. Jarvis, B.B. and Marien, B.A. J. Org. Chem. 1976, 41, 2182.
9. Borowitz, I.J.; Firstenberg, S.; Borowitz, G.B. and Schuessler, D. J. Amer. Chem. Soc. 1972, 94, 1623.
10. Denney, D.B. and Jones, D.H. J. Amer. Chem. Soc. 1969, 91, 5821.
11. Denney, D.B.; Denney, D.Z.; Hall, C.D. and Marsi, K.L. J. Amer. Chem. Soc. 1972, 94, 245.
12. Bartlett, P.D. and Meguerian, G. J. Amer. Chem. Soc. 1956, 78, 3710.
13. Feher, F. and Kurz, D. Z. Naturforsch, B. 1968, 23, 1030.
14. Chang, L.L.; Denney, D.B.; Denney, D.Z. and Kazior, R.J. J. Amer. Chem. Soc. 1977, 99, 2293.
15. Bowman, D.A.; Denney, D.B. and Denney, D.Z. Phosphorus and Sulphur, 1978, 4, 229.
16. Aida, T.; Furukawa, N. and Oae, S. Chem. Lett. 1973, 805.
17. Hammond, P.J.; Lloyd, J.R. and Hall, C.D. Phosphorus and Sulphur, 1981 (in press)
18. Beer, P.D.; Edwards, R.C.; Hall, C.D.; Jennings, J.R. and Cozens, R.J. Chem. Comm. 1980, 351.
19. Scott, G.; Hammond, P.J.; Hall, C.D. and Bramblett, J. J. Chem. Soc. Perkin II, 1977, 882.
20. Kosower, E.M. "An Introduction to Physical Organic Chemistry" 1968, J. Wiley, p.54.
21. Hudson, R.F. "Structure and Mechanism in Organophosphorus Chemistry", 1965, Academic Press, Chap. 4.
22. Stepanov, B.I.; Bokanov, A.I. and Kovolev, B.A. J. Gen. Chem. USSR (English Ed) 1967, 37, 2029.
23. Goetz, H. and Siegfried, D. Annalen, 1967, 704, 1.
24. Goetz, H. and Sidhu, A. Annalen, 1965, 682, 71.

RECEIVED June 30, 1981.

STEREOCHEMISTRY OF PHOSPHORUS COMPOUNDS

A New Stereospecific Synthesis of a P(III) Organophosphorus Ester

LEONARD J. SZAFRANIEC, LINDA L. SZAFRANIEC, and HERBERT S. AARON

Research Division, Chemical Systems Laboratory, Aberdeen Proving Ground, MD 21010

We have studied the reaction of methyl trifluoromethanesulfonate (methyl triflate) with isopropyl methylphosphinate (1), and have used this system to develop a new procedure for the stereospecific synthesis of trivalent (P^{III}) organophosphorus esters.

Methyl triflate is a powerful alkylating agent, which methylates tetracovalent P=O, P=S, and P=Se systems at the oxygen, sulfur, and selenium atoms, respectively (1,2,3). Its reaction with analogs containing a hydrogen substituent (e.g., phosphinate, phosphonate or secondary phosphine oxide species), however, appears not to have been reported.

Neat isopropyl methylphosphinate (1) reacts exothermically on dropwise addition to methyl triflate to form a phosphonium salt (2), ^{31}P NMR δ +73.4 (downfield from external H_3PO_4) J_{P-H} = 656 Hz, which yields isopropyl methyl methylphosphonite (3), when slowly added to a cold benzene solution containing excess triethylamine (TEA). On warming to room temperature, the product was obtained as a benzene/TEA solution, which separated from a heavier liquid layer that consisted mainly of amine salt byproducts in benzene/TEA. When (R)-(+)-1 (25% enantiomorphic excess) was used, a solution of (R)-(+)-3 (δ +176.6) was obtained in 60% yield, 90 mole-% pure with respect to its organophosphorus content. The specific rotation of this product was calculated to be $[\alpha]_D^{26}$ + 67.7° (c 2.6, benzene), if optically pure (+)-1 starting material had been used.

To isolate the neat product, the more volatile ether/trimethylamine combination was used in the reaction, because a higher recovery of product was obtained in trials conducted with the racemic material. However, the neat, distilled, optically active product proved to be stereochemically labile at ambient temperature, and was considerably racemized compared to that obtained directly in the benzene/triethylamine solution. Moreover, the latter was relatively stable when further diluted in benzene solution at ambient temperature, showing only a 14% decrease in optical rotation after 70 hours at 26°. For stereochemical studies, therefore, the product is better used directly in solution, as obtained, and not isolated as the neat material.

The benzene/TEA solution of the (+)-3 product described above was directly treated with sulfur and converted into (S)-(-)-isopropyl methyl methylphosphonothionate (4), δ +96.4. The specific rotation of the product thus obtained was calculated to equal $[\alpha]_D^{26}$ -1.55° (neat, uncorrected for 8 wt-% impurities), if optically pure (+)-1 had been used in the initial reaction. This rotation compares to the highest literature value, $[\alpha]_D$ 1.50°, reported for optically active 4, obtained from reaction of sodium methoxide with optically pure isopropyl methylphosphonochloridothionate (6) (4). These reactions appear to be highly, if not completely, stereospecific, and are in agreement with the assigned retention of configuration for the conversion of (+)-1 to (+)-3, based on the chemistry of the system, the known steric course of the sulfur addition (5), and the known (4,6,7) relationships between (R)-(+)-1, (S)-(-)-5, (R)-(-)-6 and (S)-(-)-4, as summarized in Scheme I.

Scheme I

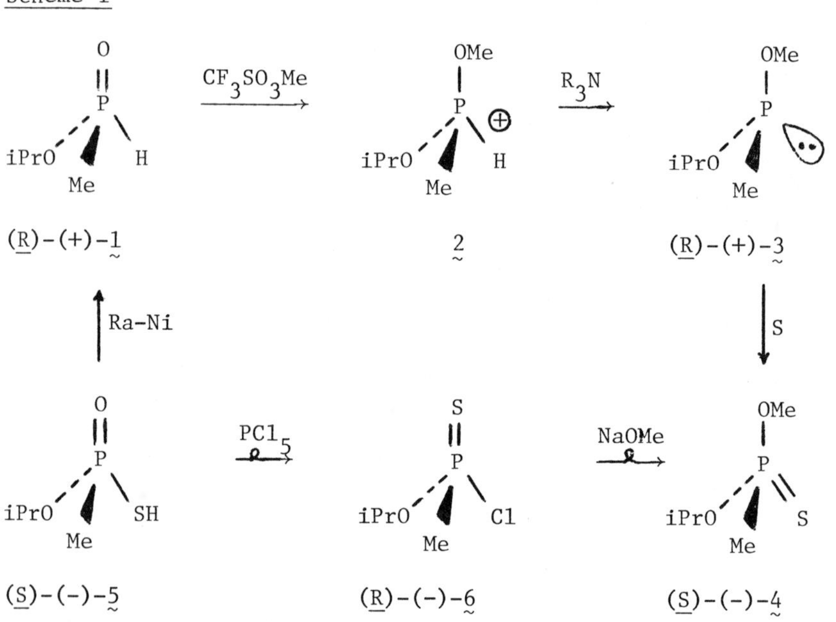

Literature Cited

1. Colle, K.S.; Lewis, E.S. J. Org. Chem. 1978, 43, 571.
2. Omelanczuk, J.; Mikolajczyk, M. J. Am. Chem. Soc. 1979, 101, 7292.
3. Omelanczuk, J.; Perlikowska, W.; Mikolajczyk, M. J. Chem. Soc., Chem. Commun. 1980, 24.
4. Mikolajczyk, M.; Omelanczuk, J.; Para, M. Tetrahedron 1972, 28, 3855.
5. Mikolajczyk, M. J. Chem. Soc., Chem. Commun. 1969, 1221.
6. Reiff, L.P.; Aaron, H.S. J. Am. Chem. Soc. 1970, 92, 5275.
7. Michalski, J.; Mikolajczyk, M. Tetrahedron 1966, 22, 3055.

RECEIVED July 7, 1981.

A Single Crystal X-Ray Diffraction Analysis of (1*R*,1′*S*)-1,1′-Ethylenebis(1,2,3,4-tetrahydro-4,4-dimethyl-1-phenylphosphinolinium) Diperchlorate

NARAYANASAMY GURUSAMY and K. DARRELL BERLIN
Department of Chemistry, Oklahoma State University, Stillwater, OK 74074

DICK VAN DER HELM and M. BILAYET HOSSAIN
Department of Chemistry, University of Oklahoma, Norman, OK 73019

Tetrahydrophosphinolinium salts **1** and **2** have confirmed carcinostatic activity[1,2] as demonstatrated by the National Cancer Institute during routine screening. Interestingly, a recent discovery

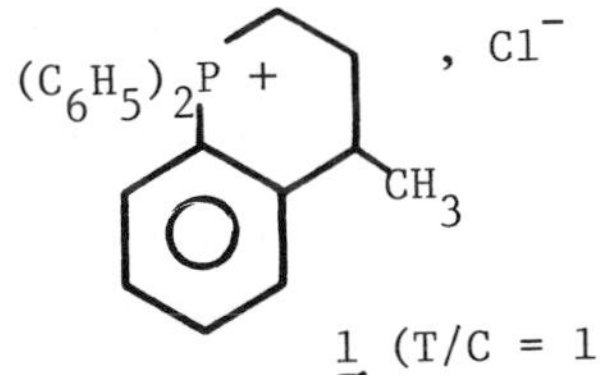

$(C_6H_5)_2P^+$, Cl^-; CH_3 — **1** (T/C = 155)

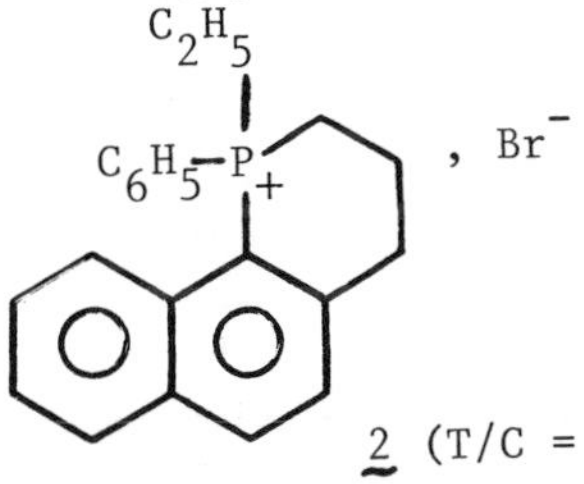

C_2H_5; C_6H_5–P^+, Br^- — **2** (T/C = 142)

that certain poly(methylene)bis(triphenylphosphonium) salts have strong anticholinergic activity[3] prompted, in part, our work on bisphosphinolinium salts which possess structural features similar to those of **1** and **2**. We report the preparation and separation of diastereoisomers of 1,1'-ethylenebis[1,2,3,4-tetrahydro-4,4-dimethyl-1-phenylphosphinolinium] diperchlorate (**3**). Although rare[4] such

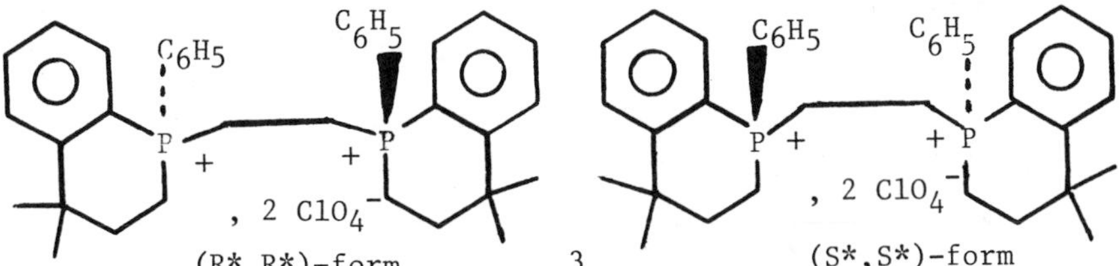

C_6H_5, C_6H_5, P +, + P, , 2 ClO_4^- — (R*,R*)-form

3

C_6H_5, C_6H_5, P +, + P, , 2 ClO_4^- — (S*,S*)-form

(meso-isomer) C_6H_5, C_6H_5, P +, + P, , 2 ClO_4^- — (R*,S*)-form

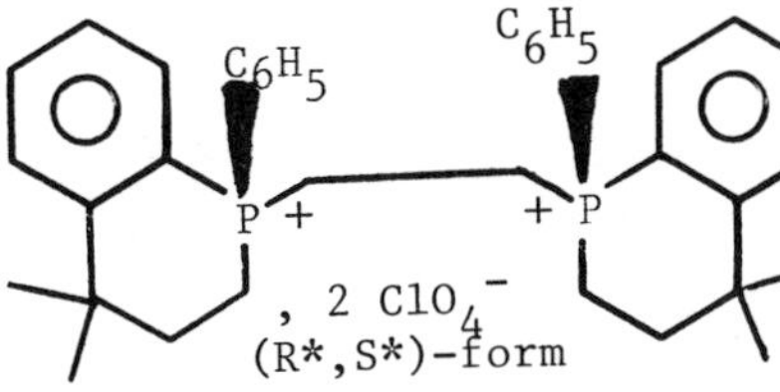

0097–6156/81/0171–0561$05.00/0

salts should be available from bisphosphines **4** which could be alkylated to produce **5**. Cyclization of the latter with 115% poly-

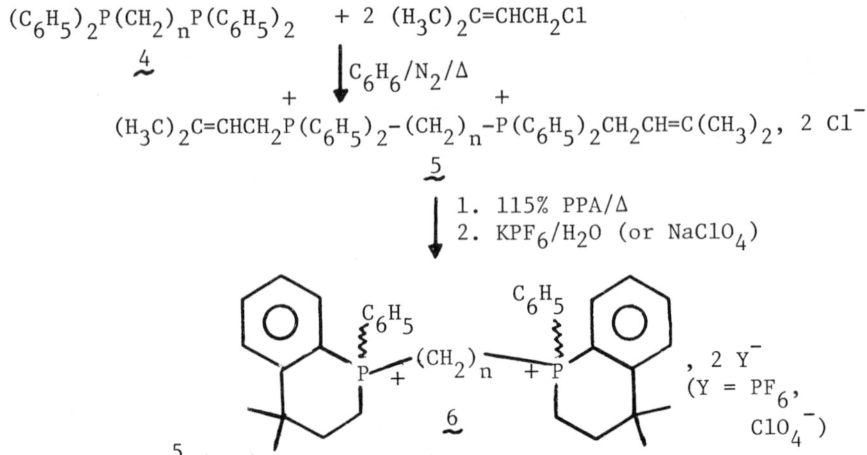

$$(C_6H_5)_2P(CH_2)_nP(C_6H_5)_2 \quad + 2\ (H_3C)_2C{=}CHCH_2Cl$$

4

$$\downarrow C_6H_6/N_2/\Delta$$

$$(H_3C)_2C{=}CHCH_2\overset{+}{P}(C_6H_5)_2{-}(CH_2)_n{-}\overset{+}{P}(C_6H_5)_2CH_2CH{=}C(CH_3)_2,\ 2\ Cl^-$$

5

1. 115% PPA/Δ
2. KPF_6/H_2O (or $NaClO_4$)

6 , 2 Y^- (Y = PF_6, ClO_4^-)

phosphoric acid[5] (PPA) gave salts **6**. Fractional crystallization of the mixture of perchlorates **3** resulted in the separation of the less soluble meso-**3** (mp 291-293°C). The residual mixture (enriched in racemic-**3**) in methanol and an aqueous solution of sodium tetraphenylborate gave the tetraphenylborates. Fractional crystallization (H_2CCl_2-ether) gave the pure racemate (mp 218-219°C). For the resolution work, meso-**3** and the racemate of the tetraphenylborate were converted to chlorides via chromatography over Dowex IX-8(Cl^-). The racemate of the chloride (hygroscopic) in methanol (anh) was treated with silver L(+)-hydrogendibenzoyltartrate (HDBT). A mixture of diastereomers precipitated, and fractional crystallization ($HCCl_3$-ether) gave pure 2 L(+)-HDBT isomer ($[\alpha]_D^{21}$ = + 60.5° (c = 1.0 g/100 mL; CH_3OH; mp 151-153°C)). Treatment of the latter with sodium perchlorate (H_2O) at room temp gave optically pure **3** ($[\alpha]_D^{21}$ = - 18.5° (c = 1.0 g/100 mL; acetone, mp 262.5-264°C)). This is the first resolution of a bisphosphinolinium salt. The meso compound (Cl^-) gave only the diastereomer 2 L(+)-HDBT [from Ag L(+)-HDBT] was obtained ($[\alpha]_D^{24}$ = + 92.5 (c = 1.0 g/100 mL; CH_3OH; mp 152-154°C)). From Ag D(-)-HDBT, the other diastereomer was found ($[\alpha]_D^{24}$ = - 91.5° (c = 1.0 g/100 mL; CH_3OH; mp 151-153°C)). Conversion to the perchlorates gave meso-**3** as expected. Optically active **3** had a ^{31}P NMR signal at + 15.40 ppm ($DCCl_3$ + drop of trifluoroacetic acid) while meso-**3** had a value of + 14.85 ppm. Thus, ^{31}P NMR analysis could be used to monitor the resolution.

Boiling meso-**6** (n =2, Y = Cl) in methanol with aqueous sodium hydroxide gave phosphine **7** and phosphine oxide **8**. All NMR spectral

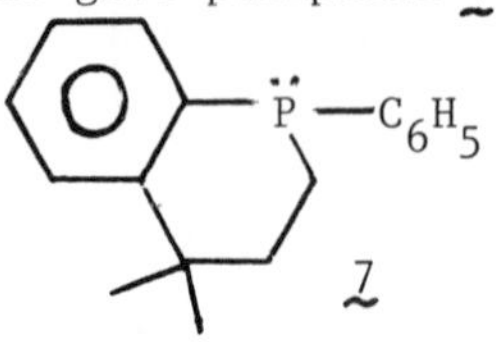

7 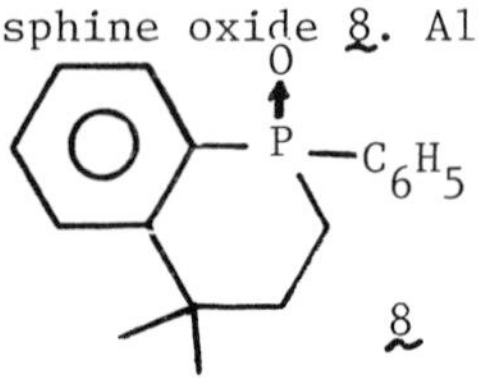 **8**

data (1H and ^{31}P) and elemental analyses supported structures 7 and 8 which in turn confirm the basic structure in 3.

A single crystal X-ray diffraction analysis of meso-3 was performed on a monoclinic crystal (space group $P2_1/c$). Bond lengths and atom numbering are given in Figure 1 and bond angles in Figure 2. Selected torsional angles are given in Table I. The two equivalent halves are designated "unprimed" and "primed" parts. In the crystalline state, the potential symmetry of the system is destroyed as can be seen from the torsional angles in the Table.

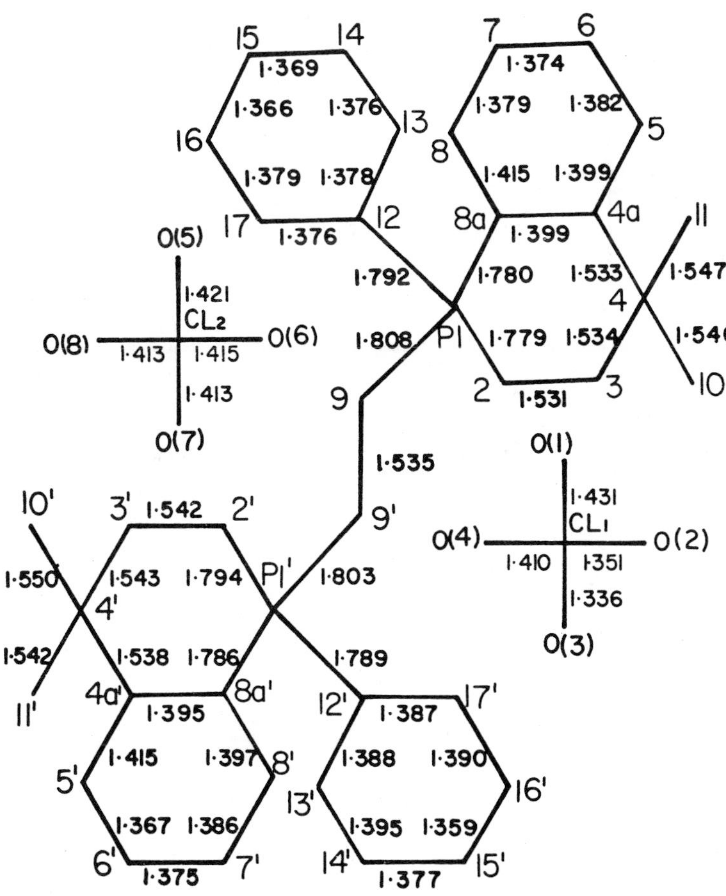

Figure 1. Bond lengths and atom number for meso-*3.*

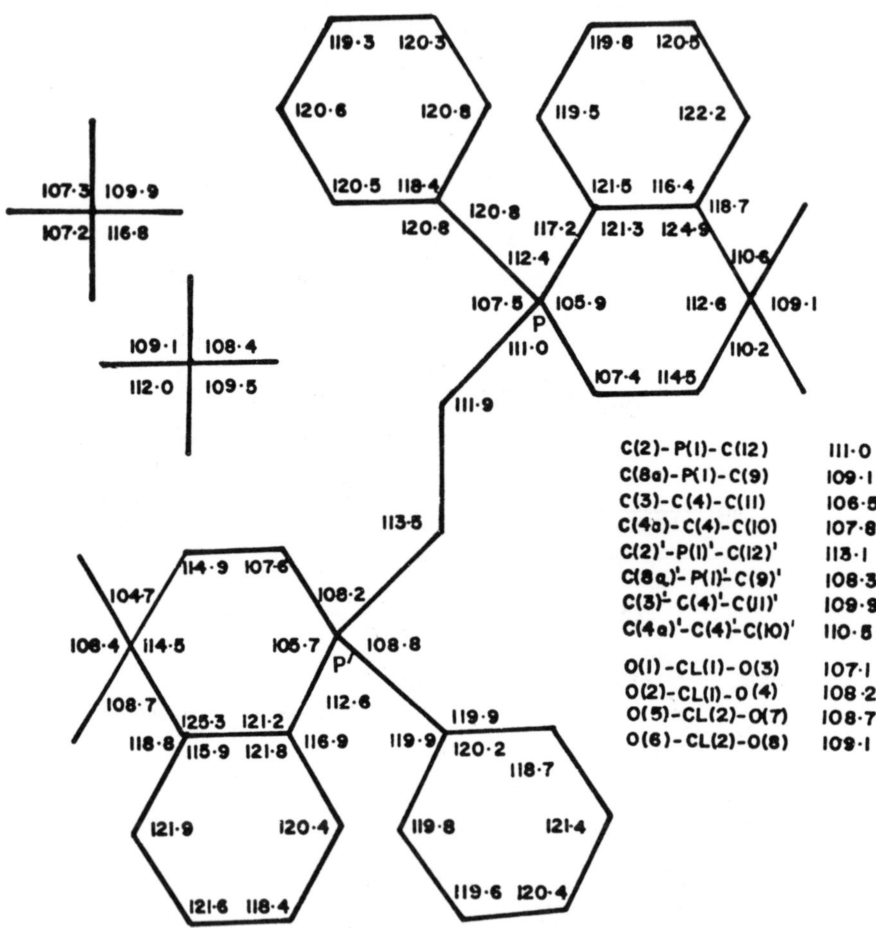

Figure 2. Bond angles for meso-*3.*

Table I. Comparison of Some Selected Torsion Angles in the Two Halves of meso-**3**

Angle	Unprimed	Primed
C(9)-P(1)-C(2)-C(3)	161.1°	-70.9°
C(9)-P(1)-C(12)-C(13)	97.3	35.8
C(9)'-C(9)-P(1)-C(2)	46.5	-73.6
C(9)'-C(9)-P(1)-C(8a)	162.8	172.3
C(9)'-C(9)-P(1)-C(12)	-75.1	49.6
P(1)-C(9)-C(9)'-P(1)'	143.6	

The heterocyclic ring assumes the familiar half-chair conformation as in somewhat related but simple systems.[1b,2,6] The relative configurations of the substituents at P(1) and P(1)' are quite different. The phenyl group assumes a pseudo axial position in the unprimed part and a pseudo equatorial position in the primed part. The considerable deviation of the dihedral angle of P(1)-C(9)-C(9)'-P(1)' from 180° to 143.6° appears unique and we have no explanation at this time. These data on the crystalline meso-**3** appear to be the first recorded on a bisphosphinolinium salt.

Acknowledgment. We gratefully acknowledge support by the U.S.P.H.S., National Cancer Institute, CA 22770.

Literature Cited

1. (a) Radhakrishna, A. S., Berlin, K. D.; van der Helm, D. Polish J. Chem. 1980, 54, 495. Holbrook, S. R.; Poling, M.; van der Helm, D.; Chesnut, R. W.; Martin, P. R.; Durham, N. N. Higgins, M. L.; Berlin, K. D.; Purdum, W. R. Phosphorus 1975, 6, 15.
2. Fink, R.; van der Helm, D.; Berlin, K. D. Phosphorus & Sulfur 1980, 8, 325.
3. McAllister, P. R.; Dotson, M. J.; Grim, S. O.; Hillman, G. R. J. Med. Chem. 1980, 23, 862.
4. Venkataramu, S. D.; Macdonell, G. D.; Purdum, W. R.; El-Deek, M; Berlin, K. D. Chem. Rev. 1977, 77, 121.
5. (a) Dilbeck, G. A.; Morris, D. L.; Berlin, K. D. J. Org. Chem. 1975, 40, 1150. (b) ibid, 1975, 40, 3763. (c) El-Deek, M.; Macdonell, G. D.; Venkataramu, S. D.; Berlin, K. D. J. Org. Chem. 1976, 41, 1403. (d) Macdonell, G. D.; Berlin, K. D.; Ealick, S. E.; van der Helm, D. Phosphorus & Sulfur 1978, 4, 187.
6. Wu, K. K.; van der Helm, D. Crystal. Struct. Comm. 1977, 6, 143.

RECEIVED July 7, 1981.

Stereochemical Investigation of Chiral Onium Hexaarylphosphates

G. P. SCHIEMENZ and J. PISTOR

Institut für Organische Chemie der Universität Kiel, FRG

In solvents of low polarity, onium salts form short-lived contact ion pairs. As a consequence, anions are NMR shift reagents for the cations (1).These may contain groups which are enantiotopic in an achiral environment but ought to become diastereotopic in ion pairs with chiral anions such as tris(2.2'-biphenylylene)phosphate (X). Indeed, in the spectrum of (4-tert.butylbenzyl)(4-tert.butylphenyl)-dimethylammonium-(-)-X (in CD_2Cl_2, 90 MHz), the N-CH_3 signals were anisochronous by 8 Hz. Much larger anisochronies were observed with sulfonium cations. The (2.5-diisopropylphenyl)-dimethylsulfonium cation contains three pairs of enantiotopic methyl groups. The (-)-X salt displayed a S-CH_3 anisochrony of 26 Hz/90 MHz, 83 Hz/270 MHz (84.5 Hz/270 MHz in the (+)-salt, zero in the (±)-salt). In addition, the low-field isopropyl signals showed a splitting of 2.2 Hz, the high-field signals one of 3 Hz.

The hexaarylphosphate proved much more efficient than (-)-bis(5.6-benzo-2.2'-biphenylylene)borate which caused only an anisochrony of 2.4 Hz/270 MHz for S-CH_3. (-)-Tris(4.4'-dimethyl-2.2'-biphenylylene)-phosphate ((-)-Y) was slightly inferior to (-)-X (S-CH_3: 69 Hz/270 MHz). The large effects induced by the two phosphorus anions allowed us to tackle a basic problem of stereochemistry. Groups must be either different or not different. While diastereotopic groups are different (non-equivalent) and homotopic groups (by definition) not different (equivalent), there is some uncertainty about enantiotopism: The Hirschmann-Hanson scheme (2) has it far away from homotopism: On equal status with diastereotopism, it belongs to heterotopism. Enantiotopic groups are then heterotopic, hence different. On the other hand, Ege (3) drew up a scheme in which enantiotopism is far

0097-6156/81/0171-0567$05.00/0

away from diastereotopism and is associated with homotopism (equitopism in Ege's notation) under the common heading of equivalence. Mislow (4) introduced still another scheme in which the notions are grouped according to isometry. Diastereotopism and constitutopism (5) fail to meet the criterion of isometry and hence are subclasses of heterotopism, but enantiotopism is segregated and forms a corresponding pair together with homotopism. This part of Mislow's scheme resembles the Ege scheme, but elsewhere Mislow stated that enantiotopic groups are non-equivalent (6). In addition, Mislow called racemic mixtures achiral (7). If this is so, then enantiotopic groups within a cation are enantiotopic not only in their salts with achiral counterions, but also in the racemic salts with chiral anions.

This view originates from the impossibility to synthesize optically active products from achiral starting material without recourse to some chiral tool. There is, however, another way of generating a racemic mixture. We took (-)-(2.5-diisopropylphenyl)-dimethylsulfonium-X in which the CH_3 groups, then, are not enantiotopic, but diastereotopic. There is a certain rotation and in the NMR spectrum an anisochrony of 83 Hz. Now increasing amounts of the (+)-enantiomer were added: The rotation as well as the anisochrony decreased, became zero and eventually reappeared with opposite sign as soon as the other enantiomer was in excess. We performed this experiment starting with the (+)-enantiomer as well, and we thus covered the whole range of rotations and anisochronies between the extreme values of the two pure salts. In the racemic mixture, the rotation as well as the anisochrony equals zero. And if in this point the environment is achiral, then the groups in this one point are enantiotopic.

These facts can be expressed with other words: in practice in a number of steps, theoretically molecule by molecule, we can set up functions of the rotation and the anisochrony vs. relative concentrations. Mathematically, these functions are steady. They pass through a common zero point, in which the groups would not be diastereotopic but enantiotopic. A point has the size zero, and hence, the probability for inactivity and enantiotopism is zero. Everywhere else those groups are diastereotopic. Enantiotopism is hence only a special case of diastereotopism.

The experimental criterion, isochrony, is not even necessarily connected with the point of optical inactivity. To a solution of (+)-(2.5-diisopropylphenyl)-

dimethylsulfonium-X, we added not its enantiomer, but (-)-(2.5-diisopropylphenyl)-dimethylsulfonium-Y. In the pure salts, the S-CH_3 groups are anisochronous by 84.5 and 69 Hz rather than 84.5 and 83 Hz, and the signal positions differ to a greater extent. Otherwise, however, everything is alike: The anisochrony decreased, as did the rotation, assumed the value zero and then increased again. But here, the point of isochrony does no longer coincide with the point of optical inactivity. Hence, the protons are diastereotopic in the point of isochrony, and the signal crossover is in no way different from cases of what may be called inherent diastereotopism. Furthermore, for different sorts of protons, the point of signal coincidence is reached at different relative concentrations.

Hence, when isochrony is observed, the solution is optically active, and when the rotation is zero, there is anisochrony. In this case, the rotation zero is what Mislow called accidental optical inactivity (7) as opposed to the 'inherent optical inactivity' of racemic mixtures, but in either case, it is a consequence of the fact that two different sorts of molecules are present in a certain ratio. The chemist's inefficiency to synthesize any other mixture of enantiomers than the 1:1 mixture does not create an alibi to tear apart what logically belongs together, and the two curves passing through a zero point are strictly analogous.

The cognition that 'enantiotopic' groups are, in fact, diastereotopic and that enantiotopism has no independent existence besides diastereotopism, is helpful for an understanding of the spectra of dibenzyl-isopropyl-sulfonium salts. The geminal protons of either benzyl group are diastereotopic; the benzyl as well as the isopropylic methyl groups are 'enantiotopic'. Salts with achiral anions and (±)-X exhibit the expected methyl doublet and a four-line signal for the CH_2 groups. Each of the enantiomeric X-salts displays triple anisochrony. The two four-line systems look different: The geminal coupling constants are about the same, but the signal separations of the inner signals are entirely different. On addition of the other enantiomer, the low-field signal group is shifted upfield, the highfield group downfield; the distance of the inner signals of the left group increases, that of the right group decreases. The racemic mixture exhibits only one four-line system in an intermediate position, and the other enantiomer has, of course, the same pattern as the first one.

All this has to be so, if 'enantiotopic' groups really are diastereotopic, hence: different. The two benzyl groups contain four different methylene protons; let us label them by A,B,C,D. They reside pairwise in two different benzyl groups which we call I and II. The coupling constant of geminal protons at sp^3-carbon with free rotation is usually ca.14 Hz and has about the same value for A/B (in I) and for C/D (in II). But since A, B, C and D are really different, there is no reason why the difference of signal positions AB and CD should be equal, and in fact, they are not. The other enantiomer gives an identical NMR spectrum, but the same signals are due to other protons: In one salt, the low-field signal group comes from A/B and the high-field group from C/D, in the enantiomeric salt vice versa.

We thus have a topomerization which is completely analogous to the valence isomerization of bullvalene. The analogy also embraces the racemic mixture. This corresponds to bullvalene at high temperature giving an averaged spectrum with only one signal belonging to protons which are neither vinylic nor cyclopropylic because of time-averaging. In our salts, the A/B low-field signals move to higher field, the C/D high-field signals to lower field, and in an intermediate position, a signal group appears which does no longer reflect the properties of benzyl group I nor those of II. The signal coincidence in the racemic mixture is not a consequence of enantiotopism, but is the result of an averaging process due to rapid topomerization of non-equivalent groups. The concept of enantiotopism is not needed for an adequate description of the process; rather is it even misleading. Actually, we can dispose of this concept entirely, because 'enantiotopic' groups are, in fact, diastereotopic, and, being so, they are not equivalent (8).

LITERATURE CITED

1. Schiemenz, G.P.; Klemm, P. Org.Magn.Reson.1974, 6, 276, and previous papers (see mainly Schiemenz, G.P. J.Magn.Reson.1972, 6, 291). - 2. Hirschmann, H.; Hanson, K.R. Tetrahedron 1977, 33, 891. - 3. Ege, G. Naturwissenschaften 1971, 58, 247. - 4. Mislow, K. Bull. Soc.Chim.Belg.1977, 86, 595. - 5. The word constitutopic (= constitutionally different) has been introduced by Ege (3). - 6. Mislow, K.; Raban, M. Topics Stereochem.1967, 1, 1. - 7. Finocchiaro, P.; Gust, D.; Mislow, K. J.Am.Chem.Soc.1974, 96, 3205. - 8. A more detailed discussion of this topomerization of non-equivalent groups will soon appear elsewhere: Schiemenz, G.P.; Pistor, J.; Wolf, M. Chem.Scr.1981, in press.

RECEIVED July 7, 1981.

SPECTROSCOPY OF PHOSPHORUS COMPOUNDS

d_π-p_π Bonding Effects on the ^{31}P NMR Chemical Shifts of *N*-Aryliminophosphoranes

PHILIP C. MURPHY and JOHN C. TEBBY

Department of Chemistry and Biology, North Staffordshire Polytechnic, Stoke-on-Trent ST4 2DE, England

Studies of the ^{31}P chemical shifts (δp) of P^{IV} (four co-ordinate phosphorus) compounds using the Letcher and Van-Wazer quantum mechanical treatment, indicates that d_π-p_π bonding has a deshielding influence in symmetrical molecules, Z_4P.[1] Egorov and Tarazevich[2] extended this approach whilst evaluating P=N π-bond orders of phosphinimines, and, with the use of constants deduced from infra-red data, concluded that an increase in the P=N π-bond order could cause δp to move to low field or to high field depending on the π-bond influence on the inbalance (non-linear shift contribution) of the phosphorus σ-bonds. N-Aryl-iminophosphoranes have been used previously to study the influence of d_π-p_π bonding on ^{31}P chemical shifts[3] but a systematic variation of the phosphorus substituents has not been undertaken. Aryliminophosphoranes are relatively stable, easy to prepare, and the electron-availability at the ylidic nitrogen can be regulated by a N-aryl substituent that is remote from the phosphorus atom.

Results and Discussion

Ten series of substituted iminophosphoranes (1-10) have been prepared, each series differing either in the steric or electronic effects of the phosphorus substituents. Within each series the N-aryl substituents, Y, (usually OMe, Me, H, F, CO_2Me, COMe, CN, NO_2) have been used to vary the electron availability at the ylidic nitrogen atom. Most substituents were para-orientated, some were meta-orientated.

The ^{31}P chemical shifts of 0.05, 0.025 and 0.0125 M solutions in deuteriochloroform were recorded, except where solubilities were limited, and the chemical shifts at infinite dilution, δP, obtained by extrapolation. The chemical shifts of the combined para- and meta- substituted compounds in each series were plotted against the various Hammett substituent constants. Using σ^- values, correlation coefficients of 0.97 to 0.99 were obtained for eight of the series and .77 and .78 for the other two.

0097-6156/81/0171-0573$05.00/0

The iminotriphenylphosphorane series (1) gave a plot with a positive slope (3.19) showing that the groups which increase electron availability at the ylidic nitrogen cause shielding of the phosphorus nucleus. The statistical parent chemical shift δP^o was 3.30.[4] This parameter relates to the general shielding of the specified series and is usually close to the chemical shift of the unsubstituted compound.

In the phosphole series (2), two phosphorus ligands are incorporated into a five membered ring. This had the effect of increasing δP^o (13.49), which was expected, and decreasing the slope (2.26), reflecting a decrease in the sensitivity of the chemical shifts to the electronic changes.

A steadier change in steric effect was investigated using the three series in which the phosphorus substituents are diphenyl (3), di-ortho-tolyl (4), and di-mesityl (5), all three series having a P-methyl group. The trend, with regard to changes in bond angles, was similar to that above and the plots had δP^o, 6.95, 6.21 and 5.27 respectively with positive slopes of 2.31, 2.83 and 3.51 respectively. Thus an increase in the bulk of the phosphorus substituents increases the sensitivity of the chemical shifts to electron availability at the ylidic nitrogen atom.

Altering the electronic effects of phosphorus substituents had a dramatic effect on the chemical shift responses. Thus the dichloro series (6) exhibited the most pronounced sensitivity (slope 10.09) and significantly was the most shielded (δp^o -15.82).

The general trends noted above, that increasing electron availability causes shielding of the phosphonium atom, is in agreement with previous observations. However, the next two series studied showed a complete reversal of this trend. Thus the diphenoxy series (7) and the bis(dimethylamino) series (8), both possessing P-methyl groups, gave plots with negative slopes (-0.93 and -2.54 respectively). Furthermore they exhibited greater deshielding of the phosphorus nucleus (δp^o 27.80 and 38.59) than any of the above series.

In the phenyldithienyl series (9), the heterocyclic rings are believed to exhibit enhanced inductive withdrawal yet maintaining good π-electron availability. The correlation graph also had a negative slope (-1.58) and compared to the iminotriphenylphosphorane series, the phosphorus nucleus was deshielded (δp^o 13.78). Thus whilst the change from a positive slope to a negative slope is favoured by a downfield shift of δp^o it is also dependent on the number of groups that can interact by d_π-p_π bonding.

The phenyldipyrrolidinophosphoranes (10) were the last series to be examined. The slope had reverted to become positive (2.69) and the statistical present shift, δp^o, was 12.09. Although there was no other aryl diheteroatom series to compare this result with, it suggests that the trends and

$\delta_P{}^o$ values are sensitive to quite subtle changes in the nature of the groups bound to the phosphorus atom.

Correlations using the infrared stretching frequencies of the P=N bond were used to ascertain the changes occurring within the ylidic bond. All the slopes were positive corresponding to a decrease in $\nu_{P=N}$ with increased electron-availability at the ylidic nitrogen. These results will be the subject of a separate report. However we feel that it is clear that, as expected, the PN bonding changes are in the same direction for all the series. Thus, if changing the electron-availability at the ylidic nitrogen induces similar changes to the P=N bond in all ten series, then the complete inversion of the slopes for the correlated ^{31}P chemical shifts in three of the series cannot be due to the sole influence of the P=N bond.

Our results clearly support the predictions of the Russian workers.[2] We have also compared the observed trends, qualitatively, with the expected changes of the contributions used in the quantum mechanical treatment, i.e. total occupancy of the d-orbitals (Σn_π), the σ-bond contribution (ζ_1), the σ-bond imbalance (σ-bond non-linear contribution), and the d_π-p_π bond imbalance. An increase in the electron availability at the ylidic nitrogen will decrease its electronegativity and thereby decrease Σn_π (causing shielding) and increase ζ_1 (causing deshielding). The change in Σn_π and ζ_1 within each series is not expected to differ very much and will also tend to cancel each other. The σ-bond imbalance could give different slopes provided the phosphorus atom has an intermediate electronegativity compared to the substituents Z. However, the predicted slopes would not be in the order observed. The order, in fact, follows the reported trends[2] in π-bonding ability i.e. Me_2N<PhO<Cl. We further note that the π-bonding ability of these substituents can be explained in terms of the combined effects of electronegativity and the number of lone-pairs of electrons on the heteroatom. Thus the ^{31}P chemical shifts appear to be strongly influenced by the π-bonding but in the reverse manner to that expected. This could be due to the presence of an independent π-bond imbalance effect. However it would be difficult to explain the trends of the carbon ylides (R_3P=CHR^1) by such an effect. Alternatively strong π-bonding might assist the transfer of sufficient electron density from the phosphorus atom to the π-donor atom through the σ-bond, to make a significant decrease in ζ_1 and cause δ_P to move upfield; when the π-bonding is weaker other factors could predominate and an increase in the σ-bond imbalance, induced by d_π-p_π bonding, could reverse the trend for the dimethylamino series.

$Z_2P(Ph)=NC_6H_4Y$

1. Z=Ph
2. Z=PhC=CPh
9. Z=2-thienyl
10. Z=N-pyrrolidino

$Z_2P(Me)=NC_6H_4Y$

3. Z=Ph
4. Z=o-tolyl
5. Z=mesityl
6. Z=Cl
7. Z=PhO
8. Z=Me_2N

Literature Cited

1. Crutchfield, M.M., Dungan, C.H., Letcher, J.H., Mark, V. and Van Wazer, J.R., "Topics in Phosphorus Chemistry", Interscience, 1972, Volume 5.
2. Egorov, Yu P., and Tarasevich, A.S., Teor.Eksp.Khim., 1973, 9, 73.
3. Kozlov, E.S., and Gaidamaka, S.N., Teor.Eksp.Khim., 1972, 8, 420; Bermann, M., "Topics in Phosphorus Chemistry", Interscience, 1972, Volume 7, p.311.
4. The parameter δp^o is defined by the linear Hammett equation $\delta p = \delta p^o - \rho\sigma^-$.

RECEIVED June 30, 1981.

^{1}H, ^{19}F, ^{31}P, and ^{13}C NMR Investigation of Diphosphanes and Triphosphanes

J. P. ALBRAND and C. TAÏEB

Laboratoire de Chimie LA CNRS N° 321, Département de Recherche Fondamentale, Centre d'Études Nucléaires de Grenoble, 85 X, F.38041 Grenoble Cédex, France

Despite recent progress in the study of the stereochemistry of tetraalkyl or aryl diphosphanes (1,2,3), there is much uncertainty in the definition of the preferred torsional angle, Ø between the lone pairs on phosphorus atoms and there is little experimental evidence to assert the presence or absence of the *trans* conformation in the liquid phase (4,5). We present here N.M.R. results obtained from two particular classes of diphosphanes, 1,2-disubstituted diphosphanes $(RPH)_2$ (R = CH_3, CH_3CH_2, c-C_6H_{11}, C_6H_5, CF_3) and CF_3 substituted diphosphanes $(CF_3PX)_2$ (X = CH_3, CN, Br, I) for which very few data are available in the literature (11).

1,2 dialkyl or diaryldiphosphanes, easily accessible through the equilibrium $nRPH_2 + (RP)_n \rightleftarrows n(RPH)_2$ (6), are interesting for stereochemical studies because they exist under both the *meso* and *d,ℓ* modifications. The variation in the steric requirement of hydrogen and the R groups can result in changes in the rotamer population, detectable by their influence on the N.M.R. parameters $^{1}J(PP)$, $^{2}J(PH)$, $^{3}J(HH)$ and $N(PC) = {}^{1}J(PC) + {}^{2}J(PC)$.

Compounds of the type $(CF_3PX)_2$ can be obtained through a coupling reaction between Hg and CF_3PXI or through the equilibrium $nCF_3PX_2 + (CF_3P)n \rightleftarrows n(CF_3PX)_2$ (8,9) formally analogous to the one used to obtain the $(RPH)_2$ derivatives. They also exist under the *meso* and *d,ℓ* modifications and can provide information on the influence of electronegative groups on the conformation, through the measurement of $^{1}J(PP)$ and $^{3}J(P,CF_3)$.

Results

With decoupling of the nuclei in the R groups, the ^{1}H and ^{31}P spectra of $(RPH)_2$ diphosphanes are simple AA'XX' patterns which give readily access to $^{1}J(PH)$, $^{2}J(PH)$ (with their relative signs), $^{1}J(PP)$ and $^{3}J(HH)$. The signs of $^{1}J(PP)$ and $^{3}J(HH)$ were related to the positive sign of $^{1}J(PH)$ by double resonance experiments in the cases of $(C_6H_5PH)_2$ and $(CF_3PH)_2$ (4). In all the other compounds of Table I, $^{1}J(PP)$ and $^{3}J(HH)$ can be reasonably assumed negative and positive respectively. In the cases of $(C_2H_5PH)_2$ and $(c\text{-}C_6H_{11}PH)_2$

0097-6156/81/0171-0577$05.00/0

where a selective decoupling of the alkyl protons is not possible, $^{1}J(PH)$, $^{2}J(PH)$ and $^{1}J(PP)$ were obtained from the spacings between the center of the well separated multiplets in the ^{31}P spectrum and $^{3}J(HH)$ was obtained from the proton spectrum.

Table I : ^{31}P Chemical shifts (ppm from H_3PO_4) and spin coupling parameters (Hz) in R(H)PP(H)R diphosphanes.

R	$\delta^{31}P$	$^{1}J(PH)$	$^{2}J(PH)$	$^{1}J(PP)$	$^{3}J(HH)$	N(PC)
C_6H_5 (a) *meso*	-67.6	+206.0	+12.6	-191.5	+4.8	13.4
d,ℓ	-71.2	+208.2	+10.0	-190.8	+11.7	15.0
CH_3	-109.5	+189.4	+11.2	167.1	5.6	6.9
	-117.	+191.	+8.5	167.1	10.4	6.5
C_2H_5	-93.8	+192.9	+11.2	185.1	3.5	4.4
	-99.6	+192.9	+6.5	179.1	11	< 1
c-C_6H_{11}	-86.3	+192.5	+12.2	181.7	2.9	
	-89.7	+191.3	+4.3	182.2	9.3	
CF_3 (b)	-90.3	+205.6	+3.2	-135.2	+9.2	
	-92.	+213.9	+15.2	-183.7	+3.0	

a) Data from references (6,7) ; b) Values from reference (4).

Except for $(CF_3PH)_2$, the values of $^{1}J(PP)$ in Table I are very close in the *meso* and *d,ℓ* isomers and this indicates that the preferred conformations have the same torsional angle Ø between the lone pairs or that if there are several conformations with different Ø angles, they should be in similar proportions for both isomers. If we consider only the contribution of *gauche* (or semi eclipsed with Ø nearer to 90°) conformations, which are probably predominant, the N(PC) values observed for $(C_6H_5PH)_2$ indicate (1) that, as in the *meso* isomer, the two possible conformers (a) and (b) of the *d,ℓ* isomer are equally populated.

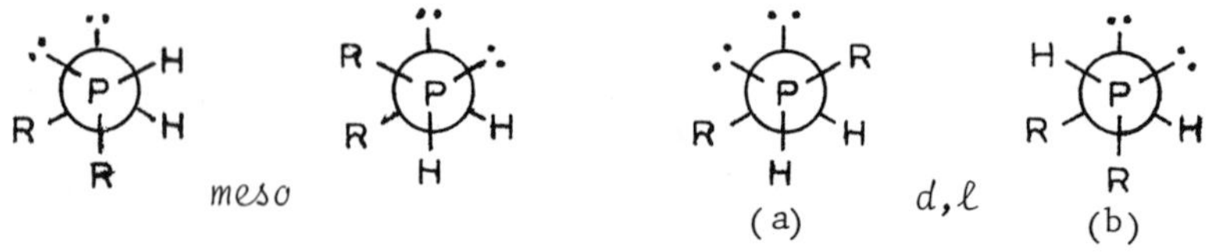

meso *d,ℓ* (a) (b)

On the other hand, in $(C_2H_5PH)_2$ the very different values of N(PC) can be explained by the absence of the (b) conformer for the *d,ℓ* isomer (1) and if it is assumed that the difference between the $^{2}J(PH)$ values has the same cause, the same conformational bias for conformer (a) could also exist in the *d,ℓ* isomer of $(c\text{-}C_6H_{11}PH)_2$.

In $(C_6H_5PH)_2$, $(CH_3PH)_2$ and $(C_2H_5PH)_2$, $^{1}J(PP)$ shows a ca.10% decrease on lowering the temperature to −120°C. This decrease could be due to the presence of a small amount of the less stable *trans* conformer at room temperature, amount which would decrease at low temperature making $^{1}J(PP)$ more negative.

The values observed for $(CF_3PH)_2$ (4) do not follow the pattern of the other $(RPH)_2$ diphosphanes of Table I and the large difference between the $^{1}J(PP)$ values has been attributed to the preference of the *meso* isomer for the *trans* conformation (4). Similar differences exist for other $(CF_3PX)_2$ diphosphanes as illustrated in Table II.

Table II : ^{19}F, ^{31}P Chemical shifts (ppm from $CFCl_3$ and H_3PO_4) and coupling constants (Hz) in $CF_3(X)PP(X)CF_3$ diphosphanes.

X	$\delta^{19}F$	$\delta^{31}P$	$^{2}J(PCF_3)$	$^{3}J(PCF_3)$	$^{1}J(PP)$	$^{5}J(FF)$
H	-45.9	-90.3	54.	17.	-135	-
	-46.3	-92.	55.5	7.3	-184	-
CH_3	-53.1	-22.	58.	13.5	173	2
	-54.7	-29.9	59.	19.5	226	2
CN	-47.	-57.1	68.	12.5	163.	2
	-48.1	-52.9	68.5	17.	192.5	2.3
Br	-53.4	-	N(PF) = 71.7		178±5	-
	-55.1	-	N(PF) = 86.		-	-
I	-47.9	-	N(PF) = 60.3		183±5	-
	-51.0	-	N(PF) = 70.3		211±10	-

The values in Table II were obtained by analysis of the $A_3A'_3XX'$ spin system observed in the ^{31}P and ^{19}F spectra, using $\{^{19}F\}$ ^{19}F homonuclear INDOR experiments. The dispersion in the $^{1}J(PP)$ values is probably due to a more important contribution of the *trans* conformation to the conformational populations in the *meso* and *d,ℓ* isomers. The $^{3}J(PCF_3)$ values show also marked differences between the *meso* and the *d,ℓ* isomer but they will be difficult to use as a stereochemical probe as long as there is no reliable way to identify these isomers.

The decomposition of pure $(CH_3PH)_2$ and $(C_2H_5PH)_2$, as that observed for $(C_6H_5PH)_2$ (10), produces triphosphanes which can be characterized by ^{31}P N.M.R.

Table III : ^{31}P Chemical shift (ppm from H_3PO_4) and coupling constants (Hz) in R(H)PP(R)P(H)R triphosphanes.

	Isomer	δP_1	δP_2	δP_3	J_{12}	J_{13}	J_{23}
CH_3	1: ABC	-83.5	-93.2	-102.5	$\mp$209.7	$\mp$202.3	±49.9
	2: A_2B	-86.6	-97.2	-97.2	$\mp$198.2	$\mp$198.2	
	3: A_2B	-79	-101.2	-101.2	$\mp$193.5	$\mp$193.5	
C_2H_5	1: ABC	-71.2	-72.8	-76.8	$\mp$219.3	$\mp$205.1	±42.0
	2: A_2B	-66.9	-81.4	-81.4	$\mp$200.2	$\mp$200.2	
	3: A_2B	-73.6	-70.4	-70.4	$\mp$201.1	$\mp$201.1	

In the threo-erythro isomer 1 of $C_2H_5(H)PP(C_2H_5)P(H)C_2H_5$, $^{2}J(PP)$ varies between +40 and +20 Hz in the temperature range -40°C/-120°C. This variation probably results from a change in the relationship of the lone pairs on the 1-3 phosphorus atoms.

Acknowledgments : We are indebted to Drs. R.C. Dobbie and P.D. Gosling (University of Newcastle upon Tyne, U.K.) who provided us with samples of the $(CF_3PX)_2$ diphosphanes.

Literature cited

1. Aime, S.; Harris, R.K.; McVicker, E.M. J.C.S. Dalton 1976, 2144.
2. McFarlane, H.C.E.; McFarlane, W.; Nash, J.A. J.C.S. Dalton 1980, 240.
3. Ali, A.A.M.; Bocelli, G.; Harris, R.K. J.C.S. Dalton 1980,638.
4. Albrand, J.P.; Robert, J.B.; Goldwhite, H. Tetrahedron Letters 1976, 949.
5. Bard, J.R.; Sandoval, A.A.; Wurrey, C.J.; Durig, J.R. Inorg. Chem. 1978, 17, 286.
6. Albrand, J.P.; Gagnaire, D. J. Am. Chem. Soc. 1972, 94, 8630.
7. Albrand, J.P.; Robert, J.B. J.C.S. Chem. Comm. 1976, 876.
8. Dobbie, R.C.; Gosling, P.D. J.C.S. Chem. Comm. 1975, 585.
9. Dobbie, R.C.; Gosling, P.D.; Straughan, B.P. J.C.S. Dalton 1975, 2369.
10. Baudler, M.; Koch, D.; Carlsohn, B. Chem. Ber. 1978, 111, 1217.
11. Kang, D.K. ; Servis, K.L. ; Burg, A.B. Org. Mag. Res. 1971, 3, 101.

RECEIVED June 30, 1981.

Principal ^{31}P Chemical Shift Tensor Components as Determined by Solid State NMR

J. P. DUTASTA[1], J. B. ROBERT, and L. WIESENFELD

Laboratoires de Chimie, Département de Recherche Fondamentale, Centre d'Études Nucléaires, 85 X, F.38041 Grenoble Cédex, France

There exists a fairly large amount of experimental data on the ^{31}P chemical shift of molecules dissolved in isotropic phases. However, one is still far from having a good rationale of the different factors which may influence the phosphorus chemical shift. The same is true for the other nuclei which may be studied by n.m.r.

The physical reason to this is that the n.m.r. screening constant σ is the trace of a tensor $\mathfrak{S}$ of which the individual components can only be obtained by recording spectra in oriented phases. This may be done by dissolving the molecules in liquid crystals (1)(2)(3), or by performing high-resolution n.m.r. experiments on powder samples or single crystals (4). The $\mathfrak{S}$ tensor components are expected to be more sensitive to minor structural or chemical changes than their isotropic counterparts (5).

We present here some results concerning the determination of the principal values of the ^{31}P $\mathfrak{S}$ tensor by means of a high resolution n.m.r. study in the solid state.

Experimental.

The solid state high resolution n.m.r. spectra were run on a CXP 200 Bruker spectrometer in which the ^{31}P nucleus resonates at 81 MHz. Samples used were finely powedered and hand-tamped in glass tubes. The spectra were recorded using the Proton Enhanced Nuclear Induction Technique (6) on the same basis of a one shot cross polarization and high power decoupling during acquisition. A capillary tube of trimethylphosphate inserted in the powder sample is used as internal reference.

[1] Current address: Department of Chemistry, University of California, Los Angeles, CA.

0097-6156/81/0171-0581$05.00/0

Results.

The results presented here concern two classes of organophosphorus molecules denoted class a) and class b). In class a), one considers a set of cyclic thioxo-phosphonates of different size, five to eight-membered rings 1, 2, 3, 4, and in which the chemical environment at the phosphorus remains the same ($(O)_2P(=S)CH_3$) (7).

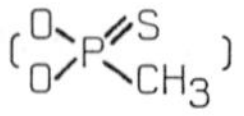

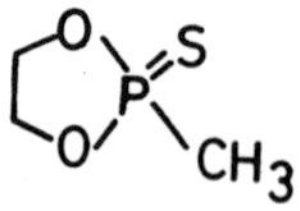

1 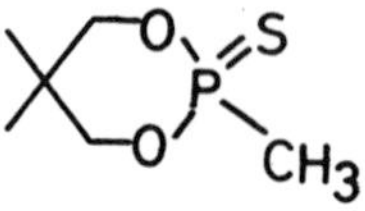 2 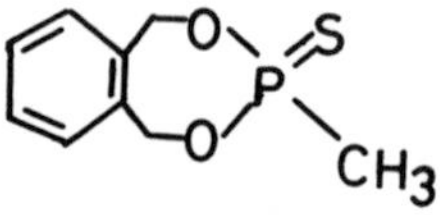 3 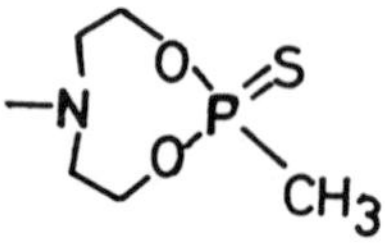 4

From the results shown in Table 1, the following conclusions can be drawn out : (i) All the spectra correspond to non axially symmetric ^{31}P tensors, (ii) the $[\sigma_{11} - \sigma_{33}]$ difference ranges from 206 to 263 ppm and is definitely larger than the chemical shift difference observed in the liquid phase between a phosphonite and the corresponding thioxo-phosphonate, (iii) a linear relationship appears when the asymmetry parameter η is plotted against the intracyclic O-P-O bond angle α.

Table 1

Principal values σ_{11}, σ_{22}, σ_{33} of the ^{31}P chemical shift tensor. σ_{ii} values are in ppm with positive values downfield from H_3PO_4 (85%). σ_{iso} is the isotropic solid state chemical shift, $\eta(\%) = (\sigma_{22}-\sigma_{11})/(\sigma_{33}-\sigma_{iso})$ is the asymmetry parameter, $\Delta\sigma = \sigma_{11}-\sigma_{33}$ characterizes the anisotropy.

Class a) : cyclic thioxo-phosphonates							
	σ_{iso}	η	σ_{11}	σ_{22}	σ_{33}	$\Delta\sigma$	α(°)
1	114	95	244	117	-19	263	98.0
2	92	55	202	124	-50	252	103.5
3	104	42	198	143	-29	227	105.1
4	92	29	173	136	-33	206	108.2

Class b) represents a set of phosphane oxides, phosphane sulfides and phosphane selenides (X = PR_3) where R is an alkyl or aryl group. The principal values of the ^{31}P chemical shift tensors are shown in Table 2. The chemical shift anisotropy values $\Delta\sigma$ are large and comparable to the ^{31}P chemical shift anisotropy variation observed in the liquid phase by change of coordination or of functionality around the phosphorus. It is noteworthy that in $P(CH_3)_3$ and $P(C_6H_5)_3$ the chemical shift anisotropies are much smaller, 8 ppm and 24 ppm respectively. For a given R group, the chemical shift anisotropies range in the order $\Delta\sigma(O\text{-}PR_3) > \Delta\sigma(S{=}PR_3) > \Delta\sigma(Se{=}PR_3)$. For aryl groups attached to the phosphorus, the $\Delta\sigma$ values are larger than or equal to those for alkyl groups.

Table 2
Principal values of the ^{31}P chemical shift.
The different terms are defined as in Table 1.

	σ_{iso}	η	σ_{11}	σ_{22}	σ_{33}	$\Delta\sigma$
$O{=}P\text{-}(CH_3)_3$	36	0	102	102	-96	198
$O{=}P\text{-}(C_6H_{11})_3$	36	17	104	85	-81	185
$O{=}P\text{-}(C_6H_5)_3$	18	9	87	75	-108	195
$S{=}P\text{-}(CH_3)_3$	59	0	87	87	4	83
$S{=}P\text{-}(C_6H_{11})_3$	60	19	117	99	-36	153
$S{=}P\text{-}(C_6H_5)_3$	43	19	107	86	-64	171
$Se{=}P\text{-}(CH_3)_3$	8	0	37	37	-49	86
$Se{=}P\text{-}(C_6H_{11})_3$	7	50	64	26	-69	133
$Se{=}P\text{-}(C_6H_5)_3$	33	48	98	56	-55	153

All the data presented here show that the ^{31}P chemical shift tensor components are extremely sensitive to small geometrical and chemical changes. The examination of their variations could provide a considerable help in the interpretation of the trends observed in the ^{31}P chemical shift values obtained in isotropic phases.

References

1. Saupe, A., Z. Naturforschung, 19a, 161 (1964) ; Saupe, A. ; Englert, G., Phys. Rev. Letters, 11, 462 (1963).

2. Emsley, J.W. ; Lindon, J.C. ; "NMR Spectroscopy Using Liquid Crystal Solvents", Pergamon Press, Oxford, 1975.

3. Albrand, J.P. ; Robert, J.B. ; Pure and Appl. Chem., 52, 1047 (1980).

4. Mehring, M. ; "High-Resolution NMR Spectroscopy in Solids", in "NMR, Basic Principles and Progress", vol. 11 ; P. Diehl, E. Fluck, R. Kosfeld, Eds., Springer-Verlag, Berlin, 1976.

5. Kohler, S.J. ; Klein, M.P. ; Biochem., 15, 967 (1976) ; Terao, T. ; Matsui, S. ; Akasaka, K. ; J. Amer. Chem. Soc., 99, 6136 (1977).

6. Pines, A. ; Gibby, M.G. ; Waugh, J.S. ; J. Chem. Phys., 59, 569 (1973).

7. Dutasta, J.P. ; Robert, J.B. ; Wiesenfeld, L. ; Chem. Phys. Lett. 77, 336 (1981).

RECEIVED June 30, 1981.

Application of the ^{18}O Shift on the ^{31}P NMR Spectrum to the Elucidation of Biochemical Phosphate Transfer Mechanisms

FRANK JORDAN, SALVATORE J. SALAMONE, and ALICE L. WANG

Department of Chemistry, Rutgers, the State University, Newark, NJ 07102

It was demonstrated in 1978 by Cohn and Hu (1) and by Lowe and Sproat (2) that substitution of ^{18}O for ^{16}O in phosphates causes an upfield shift of approximately 0.02ppm/^{18}O atom on the ^{31}P chemical shift. The magnitude of this isotopic shift varies with the environment and appears to be proportional to the P-O bond order. As part of a continuing effort to elucidate the mechanism of action of purine salvage enzymes, we employed (3) this isotopic shift to demonstrate that purine nucleoside phosphorylase (PNP) from calf spleen cleaves the C-O bond of α-D-ribose-1-phosphate (R-1-P):

$$\text{Inosine} + P_i \rightleftharpoons \text{Hypoxanthine} + \alpha\text{-D-ribose-1-}OPO_3^{2-} \qquad (1)$$

The results of that experiment allow one to synthesize α-D-ribose-1-[$^{18}O_4$]-phosphate which can be employed to determine the position of bond cleavage by other enzymes whose role is transfer of phosphate (P_i) to water or to another acceptor. We report results on a. the position of bond cleavage in R-1-P by PNP from human erythrocytes and E. coli as well as by alkaline phosphatase, acid phosphatase, formic acid and b. the position of bond making in ribose-5-phosphate by phosphoglucomutase. The earlier experiment from this laboratory employed the equilibration:

$$\text{R-1-}[^{16}O_4]\text{-P} + P^{18}O_4 \rightleftarrows \text{R-1-}[^{18}O_4]\text{-P} + P^{16}O_4 \qquad (2)$$

Assignment of the scissile bond in the sugar phosphate was made based on comparison of the chemical shifts of the phosphorus nuclei of the two isotopically distinct R-1-P's with those published for ^{18}O enriched P_i (1): 3.3-3.6 Hz (145.7 MHz) per ^{18}O atom upfield shift. Our earlier experiment has now been confirmed as follows: P_i was synthesized (4) from 50/50 atom % $^{18}O/^{16}O$ water mixture (Norsk Hydro, New York) and PCl_5 and yielded a

0097-6156/81/0171-0585$05.00/0

statistical mixture of $P^{16}O_4$, $P^{16}O_3{}^{18}O$, $P^{16}O_2{}^{18}O_2$, $P^{16}O^{18}O_3$ and $P^{18}O_4$. This statistical mixture of P_i was next equilibrated with R-1-[$^{16}O_4$]-P, hypoxanthine, inosine and PNP and yielded ^{31}P resonances for R-1-P corresponding to species containing 0,1,2,3 and 4 ^{18}O atoms (Figure 1). From such data one can unequivocally assign a 3.3-3.6 Hz (depending somewhat on pH, resolution) upfield shift per ^{18}O atom in sugar phosphates (at 145.7 MHz) just as in P_i itself. All subsequent assignments on bond cleavage or formation can be made based on the magnitude of the upfield shift of an isotopically enriched phosphate from the corresponding $^{16}O_4$ species.

Studies on Purine Nucleoside Phosphorylases

Position of Bond Cleavage: PNP (EC 2.4.2.1) from human erythrocytes (homogeneous, purified by formycin B affinity chromatography) as well as from E. coli were allowed to equilibrate a mixture of R-1-[$^{16}O_4$]-P, $P^{18}O_4$, hypoxanthine and inosine at pH 7.00 in 10 mm NMR tubes. The chemical shift differences of the ^{31}P nuclei of the two R-1-P's (13.9 Hz for the human erythrocytic and 13.4 Hz for the E. coli enzyme) as well as of the two P_i resonances (13.9 Hz for erythrocytic and 13.7 Hz for E. coli source) clearly indicated C-O bond cleavage by these enzymes as well. In addition, no evidence was found over the time course of the NMR measurements (1 hr) for purine nucleoside phosphorylase catalyzed exchange of $P^{18}O_4$ + H_2O (solvent) $\rightleftarrows$ randomized P_i. Therefore, C-O bond cleavage in R-1-P and the absence of a phosphoryl enzyme intermediate appear to be general characteristics of PNP's from a wide variety of sources: human erythrocytes, calf spleen and E. coli.

The Equilibrium Constant: K_{eq} in the direction of nucleoside synthesis had been reported by Kalckar (5) as approximately 35-50. Equilibration of known initial concentrations of R-1-P, P_i, hypoxanthine and inosine with PNP from calf spleen or E. coli at 22°C (probe temperature) allowed determination of this K_{eq} by ^{31}P NMR. Observation of the equilibrium ratios of R-1-P and P_i (after no further time-dependent changes in the integrals were noticed) allowed confirmation of the K_{eq} value in the presence of fully active enzyme.

Equilibrium $^{18}O/^{16}O$ Isotope Effect: As a part of studies to determine $^{18}O/^{16}O$ kinetic isotope effects on the reaction depicted in Eq. 1, an experiment was designed to determine the equilibrium isotope effect for the process. Initially, an approximately 50/50 mixture of $P^{18}O_4$ and $P^{16}O_4$ was equilibrated with inosine, hypoxanthine and PNP from calf spleen. Once equilibrium had been reached (22°C), as evidenced by no further changes in the integrals of the ^{31}P resonances, the integrals corresponding to R-1-[$^{16}O_4$]-P, R-1-[$^{18}O_4$]P, $P^{16}O_4$ and $P^{18}O_4$ were determined by machine integration and peak height measurements. While the results were subject to large error, the trend was clear: the

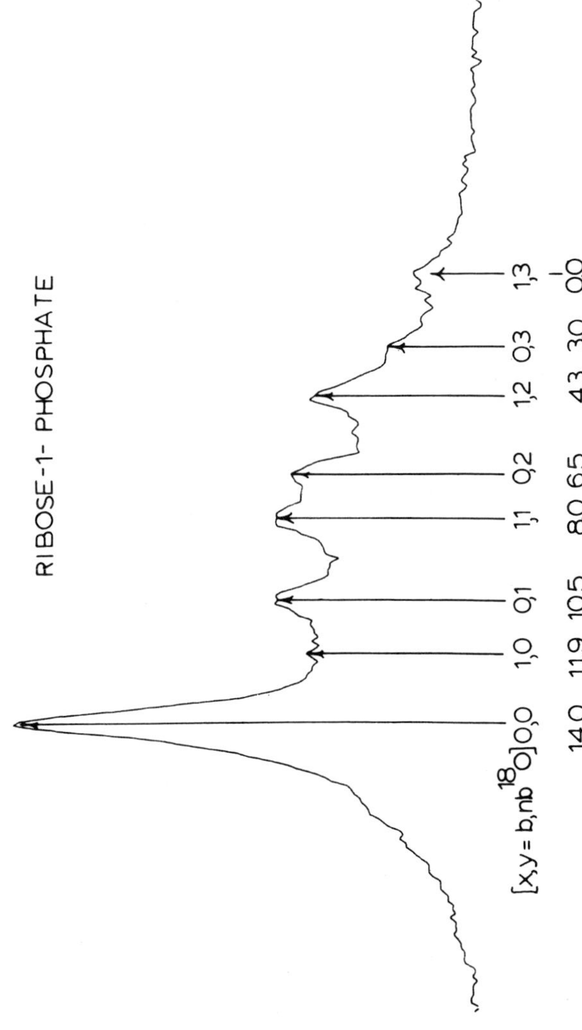

Figure 1. The ^{31}P NMR (145.7 MHz) spectrum of R-1-P produced from a PNP-catalyzed equilibration of randomized P_i (~50% ^{18}O) and R-1-P. Upper numbers denote bridging (b) and nonbridging (nb) ^{18}O's; lower numbers denote chemical shifts (in Hz) downfield from R-1-[$^{18}O_4$]P.

R-1-P was being enriched in ^{18}O, the P_i in ^{16}O. The technique shows promise, at least as an indicator of trends.

Positional Isomerization Reaction. In order to test whether the reaction depicted in Eq. 1 is stepwise or concerted the following experiment was devised. We have found that even after extensive dialysis PNP contains some tightly bound nucleoside and/or base. Therefore, detection of the exchange process R-1-P + $^{32}P_i \rightleftarrows$ R-1-^{32}P will not unequivocally imply a stepwise mechanism. Consider instead the following two reactions:

$$\underset{A}{\text{Ribose-C1-O-}\overset{\displaystyle O}{\underset{\displaystyle O}{P}}\text{-}^{18}O} \overset{k_{int}}{\rightleftarrows} \underset{B}{\text{Ribose-C1-}^{18}O\text{-}\overset{\displaystyle O}{\underset{\displaystyle O}{P}}\text{-O}} \quad (3)$$

and

$$\text{Ribose-C1-O-}\overset{\displaystyle O}{\underset{\displaystyle O}{P}}\text{-}^{18}O + P^{16}O_4 \overset{k_{overall}}{\rightleftarrows} \underset{C}{\text{Ribose-C1-O-}\overset{\displaystyle O}{\underset{\displaystyle O}{P}}\text{-O}} \quad (4)$$

If one could monitor k_{int} (for internal return) and $k_{overall}$ in the same reaction, then, aside from statistical correction, equality of the rate constants would imply concertedness. If $k_{int} >> k_{overall}$ the formation of an oxocarbocation is likely. Figure 1 demonstrates that one can distinguish ^{18}O in a bridging position (B) from ^{18}O in a nonbridging position (A) for the α-D-ribose-1-[$^{16}O_3$ ^{18}O]-phosphate species (this experiment was inspired by the observation of bridging and nonbridging ^{18}O shifted ^{31}P resonances in ATP (6)). Compound <u>A</u> was synthesized and in the presence of $P^{16}O_4$ and PNP (no added hypoxanthine or inosine), was converted to B and C at fast, identical rates and implying concertedness. While more experiments are required, either ^{31}P NMR or mass spectrometry on volatilized compounds will probably settle the question according to the above scheme.

<u>Position of Bond Cleavage or Formation in Some Other Phosphate Transfers</u>

For various mechanistic studies it would be desirable to cleave R-1-P specifically at different bonds. The following experiments were performed by first setting up PNP equilibration depicted in Eq. 2 employing appropriate amounts of hypoxanthine and inosine to maintain a nearly equal total concentration of R-1-P and P_i.

Alkaline phosphatase (EC 3.1.3.1) Sigma type VII from calf intestine at pH 10.5 was equilibrated with a mix in Eq. 2 at 25°C for 30 min. The ^{31}P NMR spectrum was recorded at pH 8.10. There was no R-1-P left and there were three P_i's found containing $^{16}O_4$, $^{16}O^{18}O_3$ and $^{18}O_4$. The integration and lack of detectable amount of $^{16}O_2{}^{18}O_2$ species (indicating no exchange with solvent of the P_i's) proved O-P cleavage.

Acid phosphatase (EC 3.1.3.2) Sigma type III from potato at pH 4.77 was equilibrated with a mix in Eq. 2 at 25°C for 3 hrs. The ^{31}P NMR spectrum was recorded at pH 7.10. All R-1-P was converted to phosphate containing $^{16}O_4$, $^{16}O^{18}O_3$ and $^{18}O_4$ indicating O-P cleavage.

Formic acid. The mix depicted in Eq. 2 was transferred into 45% (v/v) formic acid, incubated for 20 minutes and the solvents were evaporated in vacuo. The ^{31}P NMR spectrum was recorded at pH 8.10. All R-1-P had been hydrolyzed and resonances corresponding to $P^{16}O_4$ and $P^{18}O_4$ were observed, indicating C-O cleavage.

Phosphoglucomutase (EC 2.7.5.1) slowly converts R-1-P to ribose-5-phosphate. A mixture of 35 mM R-1-[$^{16}O_4$]-P, 17mM R-1-[$^{18}O_4$]-P and less than 1mg glucose-1,6-diphosphate at pH 7.33 was equilibrated with phosphoglucomutase (Sigma, P3397, rabbit muscle), at 25°C for 3 Hr. The ^{31}P NMR spectrum was recorded at pH 7.37. The enzyme converted ca. 20% to a mixture of ribose-5-phosphate $^{16}O_4$ (resonance identified by addition of authentic material) and $^{16}O^{18}O_3$ species. This demonstrated that the enzyme catalyzed formation of the O-P bond.

Acknowledgement. Supported by NIH and the Rutgers Busch Fund. We thank Dr. T. Krenitsky and R. Jordan for samples of PNP. NMR was performed at the U. of Pennsylvania supported by NIH Grant RR542.

Literature Cited

1. Cohn, M.; Hu, A. Proc.Natl.Acad.Sci. US, 1978, 75, 200.
2. Lowe, G.; Sproat, B.S. J.Chem.Soc.Perkin Trans 1, 1978, 1622.
3. Jordan, F.; Patrick J.; Salamone, S.J. J.Biol.Chem., 1979, 254, 2384.
4. a) Risley, J.M., Van Etten, R.L. J. Labelled Compd. Radiopharm., 1978, 15, 533. b) Hackney, D.D.; Stempel, K.E.; Boyer, P.D. "Methods in Enzymology", D. Purich, Ed. 1980, 64B, p.60.
5. Friedkin, M.; Kalckar, H.M. "Enzymes", 2nd Ed., Longman Green, 1961, 5, p.237.
6. Cohn, M.; Hu, A. J.Am.Chem.Soc. 1980, 102, 913.

RECEIVED June 30, 1981.

122

Phospholipase A_2 Hydrolysis of Phospholipids: Use of ^{31}P NMR to Study the Hydrolysis, Acyl Migration, Regiospecific Synthesis, and Solubilization of Phospholipids

ANDREAS PLÜCKTHUN and EDWARD A. DENNIS

Department of Chemistry, University of California at San Diego, La Jolla, CA 92093

Our laboratory has been studying the kinetics (1) and mechanism of action (2) of phospholipase A_2 which catalyzes the hydrolysis of the fatty acyl chain in the *sn*-2 position of phospholipids to give the 1-acyl lyso-phospholipid product:

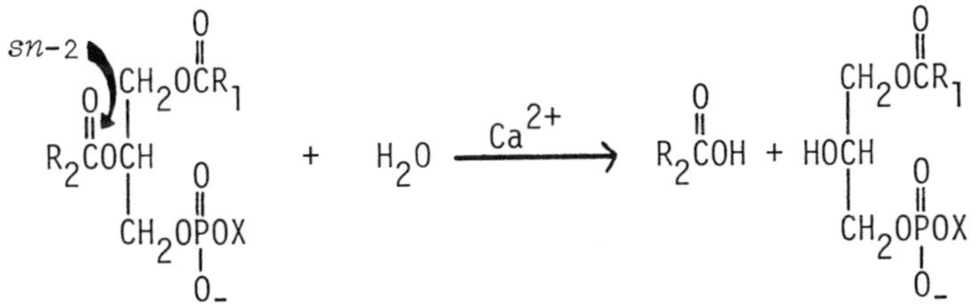

$$\begin{array}{c} CH_2OC(=O)R_1 \\ | \\ R_2C(=O)OCH \\ | \\ CH_2OP(=O)(O^-)OX \end{array} + H_2O \xrightarrow{Ca^{2+}} R_2C(=O)OH + \begin{array}{c} CH_2OC(=O)R_1 \\ | \\ HOCH \\ | \\ CH_2OP(=O)(O^-)OX \end{array}$$

(arrow marking the *sn*-2 ester bond)

Recently, we found that phosphorylcholine-containing lipids activate the enzyme from cobra venom (*Naja naja naja*) toward phosphatidylethanolamine (PE) (3,4). These studies led to the suggestion of two sites for the enzyme - an activator site with minimum specificity for phosphorylcholine and a hydrophobic chain and a catalytic site with less specificity for the polar group. We have now extended these studies to synthetic phospholipids that contain short chain fatty acyl groups and which are water soluble, such as dibutyryl and dihexanoyl phosphatidylcholine (PC). These phospholipids are monomeric below their critical micelle concentration (cmc), yet activate the enzyme. In order to carry out kinetic studies, the long chain phospholipid substrate must generally be solubilized by a detergent such as Triton X-100 which serves as an inert matrix. Further understanding of the mechanism of the activation by short-chain phospholipids requires first a quantitation of the solubilization of these compounds by detergent:

To solve this problem, the very high sensitivity of the ^{31}P-NMR chemical shift to changes in the hydrophobicity of the environ- ment has proven to be very useful. The sensitivity of the chemica shift to structural changes in the phospholipid molecule several atoms away allowed us to also examine several other mechanistic questions including the hydrolysis and specificity of phospholipas

0097–6156/81/0171–0591$05.00/0

A_2 in mixed phospholipid systems and the kinetics and migration of the acyl and phosphorus group of the lyso-phospholipid products. The new ^{31}P-NMR results also allow us to directly establish the specificity of various phospholipases by direct product observation and should also be very useful in regiospecific chemical synthesis of phospholipids.

We found the ^{31}P-NMR chemical shift of monomeric dihexanoyl PC increases upon the addition of the nonionic detergent Triton X-100. This phenomenon was used to quantitate the solubilization of this phospholipid by the detergent micelles as a function of detergent concentration using a simple phase separation model (5). Similar studies were carried out on dibutyryl PC. At a phospholipid concentration of 7 mM and 56 mM detergent, 85% of the dihexanoyl PC, but only 3% of the dibutyryl PC was incorporated into the micelles.

Having shown that dibutyryl PC is monomeric under the enzyme assay conditions, we found that the phospholipase A_2, which acts poorly on PE in mixed micelles, is activated by dibutyryl PC which is itself an even poorer substrate. ^{31}P-NMR spectroscopy was employed to show that only PE is hydrolyzed in mixtures of various compositions of these two phospholipids. The fully activated enzyme hydrolyzes PE at a similar rate to its optimal substrate, PC containing long-chain fatty acid groups. Because dibutyryl PC is not incorporated into the micelles, these results are consistent with a mechanism of direct activation of the enzyme by phosphorylcholine-containing lipids (either monomeric or micellar) rather than a change in the properties of the interface being responsible for the activation of phospholipase A_2. Therefore, two functional sites on the enzyme have to be assumed: an activator site and a catalytic site (6).

The determination of the positional specificity of phospholipases by the direct observation of product formation requires that the migration of the remaining fatty acid group in the lysophospholipid product be slow compared to the enzymatic hydrolysis reaction. Therefore a kinetic investigation of this migration was carried out. One enantiomer of each of the three possible isomeric lyso-phosphatidylcholines (1-acyl-*sn*-glycero-3-phosphorylcholine, 2-acyl-*sn*-glycero-3-phosphorylcholine and 3-acyl-*sn*-glycero-2-phosphorylcholine) were generated enzymatically. The assignment of the glycerol portion in the ^{1}H-NMR spectrum allowed an unequivocal identification of these compounds. The ^{31}P-NMR spectrum of each gave a distinct peak. From the assignments, the ^{31}P-NMR spectrum of the lyso products could also be used to show the specificity of phospholipase A_2 for the *sn*-2 position and of lipase for the *sn*-1 position directly. This eliminates the necessity of regiospecifically radio-labelled phospholipids, which are difficult to synthesize with absolute isomeric purity. ^{31}P-NMR also allowed us to demonstrate the absolute specificity of cobra venom phospholipase A_2 for the *sn*-2 position of monomeric dibutyryl PC, for which there were no specifically radio-labelled phospholipids available and they would be even more difficult to prepare because of the facile acyl migration. The monomeric phospholipids do not have the

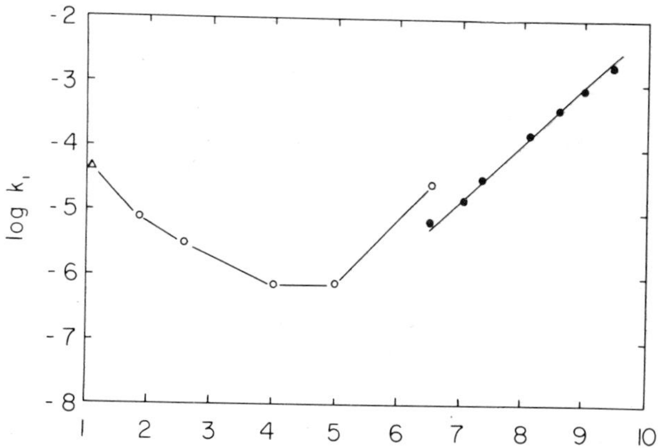

Figure 1: The pH-dependence of the acyl migration of lysophospholipids is shown. The log of the pseudo first-order rate constant k_I *for the rearrangement of 2-palmitoyl-*sn*-glycero-3-phosphorylcholine into 1-palmitoyl-*sn*-glycero-3-phosphorylcholine is plotted against the pH buffered with 50m*M *tris HCl (•), 50m*M *citrate (○), or 0.1*M *HCl and 160m*M *Triton X-100 (△) to solubilize the reaction products at acidic pH.*

sn-2 carbonyl group preferentially exposed at the interface as do micellar phospholipids (7).

Although the isomeric lyso-phospholipids can in principle be distinguished by IR, ^{1}H-NMR, ^{13}C-NMR and ^{31}P-NMR, only the latter provides the necessary sensitivity and well-resolved single resonance peaks that make it suitable as a kinetic tool to study the migration of the acyl-chain between the *sn*-1 and *sn*-2 position. The acyl migration was found to be both base catalyzed with a second order rate constant of $k_2 = 160\ M^{-1}\ s^{-1}$ and acid catalyzed, however much more slowly, with a second order rate constant of $k_2 = 4 \times 10^{-4}\ M^{-1}\ s^{-1}$ (calculated between pH 1 and 2). The pH-rate profile is shown in Figure 1. At basic pH, the hydrolysis of the fatty acyl group is approximately 6500 times slower than migration. The equilibrium mixture in base contains approximately 90% of the 1-acyl isomer and 10% of the 2-acyl isomer. The phosphoryl migration was found to be too slow to measure, except at very acidic pH. Under these conditions, however, a variety of hydrolytic reactions also take place on a similar time scale so that the quantitation of the phosphoryl migration is complicated.

Two factors make the preparation of phospholipids that are specifically labelled in one acyl chain with a high degree of purity rather difficult: Acyl-migration during the preparation of the monoacyl intermediate and acyl migration during the reacylation in organic solvents. Basic catalysts that are used in common reacylation procedures of lyso-phospholipids also catalyze the latter migration. However, the possibility of directly observing it by ^{31}P-NMR provides a simple and convenient method to optimize conditions which minimize acyl migration. ^{31}P-NMR also allows one to test the isomeric purity of α- and β-phospholipids rapidly because the chemical shift of the β-phospholipid in mixed micelles is about 0.6 ppm upfield from the α-phospholipid. These compounds have been widely used in the study of model membranes.

1. Deems, R.A.; Eaton, B.R.; Dennis, E.A. J. Biol. Chem. 1975, 250, 9013.
2. Roberts, M.F.; Deems, R.A.; Dennis, E.A. Proc. Natl. Acad. Sci. U.S.A. 1977, 74, 1950.
3. Roberts, M.F.; Adamich, M.; Robson, R.J.; Dennis, E.A. Biochemistry 1979, 18, 3301.
4. Adamich, M.; Roberts, M.F.; Dennis, E.A. Biochemistry 1979, 18, 3308.
5. Plückthun, A.; Dennis, E.A. J. Phys. Chem. 1981, 85, 678.
6. Plückthun, A.; Dennis, E.A. Fed. Proc. 1981, 40, 1805.
7. Roberts, M.F.; Bothner-By, A.A.; Dennis, E.A. Biochemistry 1978, 17, 935.

RECEIVED June 30, 1981.

PHOTOCHEMISTRY WITH PHOSPHORUS COMPOUNDS

Photolytic Rearrangement of Phosphorus, Germanium, and Silicon Azides: Evidence for New Hybridized Species

J. P. MAJORAL

Equipe de Recherche Associée au CNRS N° 926, Université Paul Sabatier, 118 Route de Narbonne, 31062 Toulouse Cédex, France

G. BERTRAND, A. BACEIREDO, and P. MAZEROLLES

Equipe de Recherche Associée au CNRS N° 829, Université Paul Sabatier, 118 Route de Narbonne, 31062 Toulouse Cédex, France

Although chemistry of unusually hybridized phosphorus compounds has been the subject of one of the most exciting areas in the last ten years, only a few papers report investigations on tricoordinated pentavalent phosphorus derivatives (1 - 3).

Photolysis of phosphorus azides involving a Curtius rearrangement seems to be a promising way for generating new metaphosphonimidates species 1

$$R-P(=O)(=N-R) \qquad \underline{1}$$

Thus photolysis of diphenylphosphine azide leads unequivocally to the transient metaphosphonimidate 1 which, by head to tail dimerization, gives rise to the phosphadiazetidine 2 and to polymers.

$$\text{Ø}_2\text{P}(=\text{O})-\text{N}_3 \xrightarrow[C_6H_6]{h\nu} \left[\text{Ø}-\text{P}(=\text{O})(=\text{N}-\text{Ø})\right]\ (\underline{1}) \longrightarrow \underline{2}$$

2: cyclic dimer (Ø(O=)P–N(Ø)–P(=O)Ø–N(Ø)–) ; $\delta^{31}P$ = +6ppm ; m.p = 290 °C

Because of its very short lifetime, no spectroscopic observations of 1 were possible even at low temperature. However, we provided evidence for the transient formation of 1 by trapping reactions with alcohols, diols, amines,methyl iodide, epoxides, dienes, ketones. Some examples are as follows.

0097-6156/81/0171-0597$05.00/0

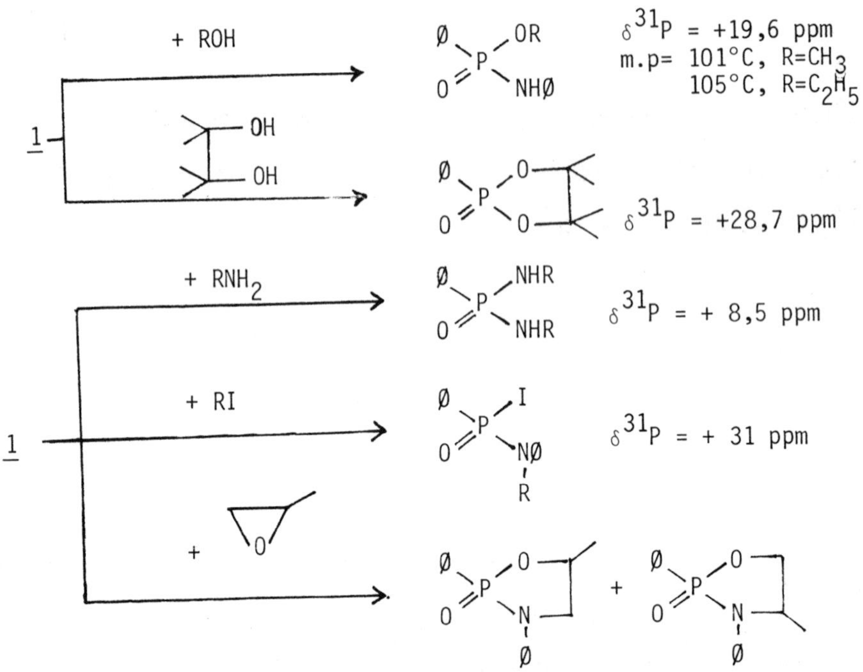

As in the carbon series, this photolytic rearrangement involved a nitrene intermediate. Indeed, in the case of thio phosphorazides, beside the formation of dimers 3 and polymers, the product 4 was characterized and arises from hydrogen abstraction from the solvent by the nitrene intermediate.

$$Ø_2P(=X)-N_3 \xrightarrow[C_6H_6]{h\nu} [Ø_2P(=X)-\bar{N}|]$$

→ (cyclic dimer: Ø(X=)P(NØ)₂P(=X)Ø) 3 + polymers

→ $Ø_2P(=X)-NH_2$ 4

$Ø_2P(=X)N_3$	yield % 3	4
X = O	>70	0
X = S	20	60
X=N-Ø	< 5	> 95

The different reactivity of phosphorus azides might be interpreted in terms of conjugation between phenyl groups and sulfur or between phenyl groups and the P=N=Ø double bond ; in these cases the rate of the phenyl migration decreases and hydrogen abstraction reactions dominate.

Efforts towards stabilization of metaphosphonimidates with bulky substituents failed. Photolysis of 5 at low temperature did not allow us to stabilize 6 or 7. Irradiation of 5 in benzene gave rise to dimers and polymers but also to the hydrogen abstraction product. Moreover we did not succeed at the present time in the preparation of the phosphorus azide 9 from 8 certainly because of steric hindrance.

5 $\delta^{31}P$ = + 51 ppm
m.p = 112 °C

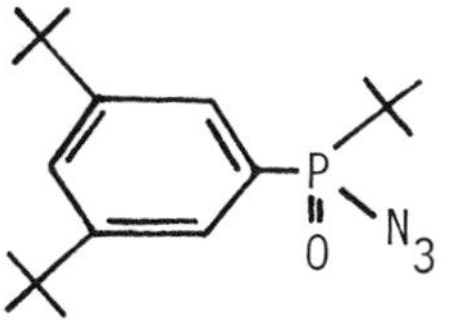

6 7

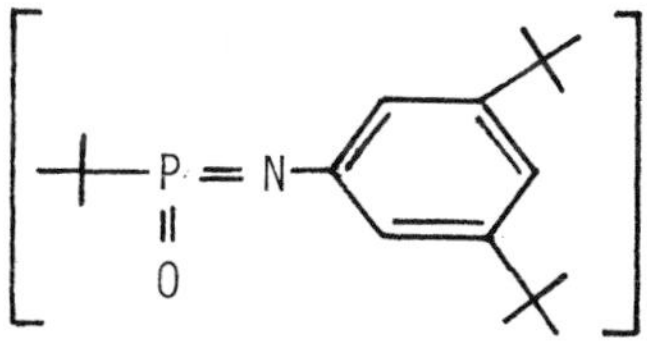

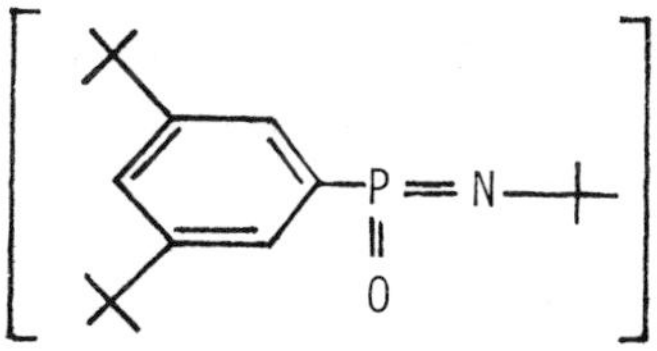

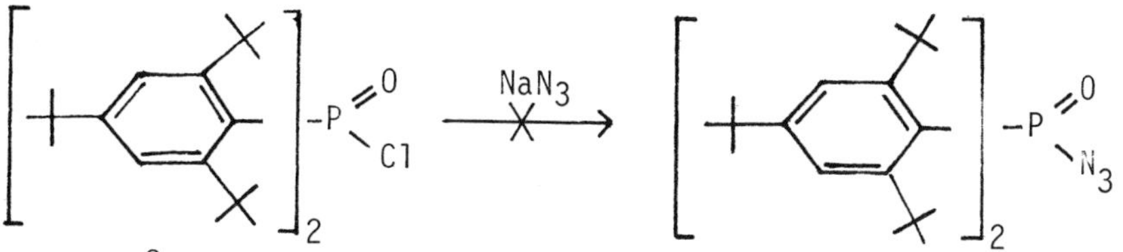

8 $\xrightarrow{NaN_3}$ (not formed) 9

Taking into account these results in phosphorus chemistry, it seemed interesting to us to study the same transposition in germanium chemistry. In fact, photolysis of the germanium azide 10 in the presence of pinacol afforded the dioxagermolanne 12. On the other hand in the absence of trapping agent, photolysis of 10 gave rise mainly to germadiazetidine 13 but also to polymer 14, biphenyle and amine. These last two products came from hydrogen abstraction from the solvent by nitrene intermediate 15. Polymer 14 was a condensed form of the corresponding germylene 16 which can be trapped with dimethyldisulfur.

Thus two possible mechanisms might explain our experimental results:

$$Ph_3GeN_3 \xrightarrow{h\nu} [Ph_2Ge{=}N{-}Ph]$$

10 → 13 (Ph₂Ge–NPh)₂ dimer; → 12; $[Ph_2Ge:]$ 16 + $[PhN]$ 15; 16 → $(Ph_2Ge)_n$ 14; 15 $\xrightarrow{PhH}$ $PhNH_2$ + Ph - Ph

Since the rearrangement of silicon azides is known, we looked for some synthetic applications in organometallic chemistry. For example, ring extension reactions involving sila.imines intermediates can be described :

$$R(N_3)Si(CH_2)_n \xrightarrow[-N_2]{h\nu} R{-}Si{=}N(CH_2)_n + R{-}N{=}Si(CH_2)_n$$

ROH → 17 (R(RO)Si–NH–(CH₂)ₙ ring) + 18 (RO(RNH)Si(CH₂)ₙ ring)

The ratio of 17/18 was dependent of the nature of the silicon substituent and thus it was possible to synthesize either a endo-exo difunctionalized silicon ring 17 or a diexofunctionalized compound 18.

Literature cited :

(1) BRESLOW R., FEIRING A., HERMAN F., J. Am. Chem. Soc., (1974) 96, 18, 5937.

(2) BERTRAND G., MAJORAL J.P. and BECEIREDO A., Tetrahedron Letters, 1980, 21, 5015 and ref. included.

(3) HARGER M.J.P., STEPHEN M.A., J. Chem. Soc.Perkin I, 1981,737.

RECEIVED June 30, 1981.

Diphenylphosphinous Acid by UV Irradiation of Aroyl Diphenyl Phosphines

K. PRAEFCKE and M. DANKOWSKI

Institute of Organic Chemistry, C 3, Technical University Berlin, D-1000 Berlin 12, FRG

Relatively little is known about the photochemistry of phosphorus containing organic substrates. Continuing our photochemical work on compounds of structure **1a** and **b** which leads to thio- and selenoxanthones **3a,b** via thiol ester-thiopyrone-, selenol ester-seleninone-transformations (1, 2, 3), or photo-Friedel-Crafts-reactions (3-6), respectively, (Scheme 1) we have extended our interest to aroyl diphenyl phosphines **1c - k** (Schemes 2 - 5) in which X in **1a** and **b** stands for P-phenyl.

hν [benzene] ; − RH ; − (C₆H₅−X)₂

1 a,b → **2** + **3 a,b**

(1)

a : X = S

b : X = Se

R = various substituents

As **1c** behaves photochemically similar to certain thiol and selenol esters (e.g. **1a** and **b**), yielding heterocycles of type **3** in the first photo-Friedel-Crafts-reaction of an organic phosphorus compound (Scheme 2), the aroyl diphenyl phosphines **1d - k** show other reactions on u.v. irradiation (7, 8, 9).

0097-6156/81/0171-0601$05.00/0

However, these latter phosphines **1d - k** also suffer photo-fragmentations. α-Cleavage occurs under formation of either aldehydes **2d - h**, via subsequent H-abstraction of aroyl radicals from solvent (Scheme 3), or hetero- and carbocycle **7** and **8**, via cyclization of the corresponding aroyl radicals under neighbouring group participation (Schemes 4 and 5).

In a competing and novel type of photoreaction, aroyl diphenyl phosphines **1d - k** yield diphenylphosphinous acid (**4**, diphenylphosphine oxide) as the photoproduct of a complex transfer of the oxygen from the carbonyl carbon onto the phosphorus atom followed by C-P-bond cleavage and hydrogen abstraction from solvent. The mass spectrometric product analysis of an u.v. irradiation experi-

1,2,5	R (X = P–Ph)
d	H
e	Br
f	OCH_3
g	SCH_3
h	SO_2CH_3

1. hv [benzene]
2. cc. [acetone]
6 (59%) + 7 (87%)
1i, hv ?, 1i', hv ?
X = P-Ph
(4)

ment with a 1:1 mixture of [^{18}O]-aroyl diphenyl and [^{16}O]-aroyl ditolyl phosphine in benzene has shown that the photoinduced oxygen transfer reaction of aroyl diaryl phosphines at least partially occurs intermolecularly.

Diphenylphosphine oxide (**4**), generated for the first time photochemically in yields up to 59%, adds in situ to various carbonyl compounds (e.g. formation of **5**, **6** and **9** by 1.2- or 1.4-additions, respectively).

1. hv [benzene]
2. cc. [acetone]
1k → [8 + 4] → 9 (36% rel. to 8) + 6 (41%)
X = P-Ph
(5)

Acknowledgements

K. Praefcke thanks the Fonds der Chemischen Industrie, Frankfurt/Main, as well as the Gesellschaft von Freunden der Technischen Universität Berlin for their financial support.

References

1. Beelitz, K; Buchholz, G; Praefcke, K. Liebigs Ann. Chem. 1979, 2043.
2. Beelitz, K; Praefcke, K; Gronowitz, S. Liebigs Ann. Chem. 1980, 1597, and J. Organometal. Chem., 1980, 194, 167.
3. Martens, J; Praefcke, K. J. Organometal. Chem. 1980, 198, 321.
4. Martens, J; Praefcke, K; Schulze, U. Synthesis, 1976, 532.
5. Praefcke, K; Schmidt, D. J. Heterocycl. Chem., 1979, 16, 47.
6. Praefcke, K; Schulze, U. J. Organometal. Chem., 1980, 184, 189.
7. Dankowski, M; Praefcke, K; Nyburg, S. C.; Wong-Ng, W. Phosphorus and Sulfur, 1979, 7, 275.
8. Dankowski, M; Praefcke, K. Phosphorus and Sulfur, 1980, 8, 105.
9. Dankowski, M; Praefcke, K; Lee, J.-S.; Nyburg, S. C. Phosphorus and Sulfur, 1980, 8, 359.

RECEIVED July 7, 1981.

BONDING AND THEORY OF PHOSPHORUS COMPOUNDS

Chemical Model Showing Three Phenomena: Phosphorane→Ylide, Ylide→ Phosphorane, and Phosphorane ⇄ Ylide

RAMON BURGADA, YVES LEROUX, and Y. O. EL KHOSHNIEH

Laboratoire des Organo-Eléments, ERA 825, Université P. et M. Curie, Tour 44-45, 4 Place Jussieu, 5230 Paris Cédex 05, France

We have shown without any doubt the formation of carbanions obtained when trivalent phosphorus compounds react with an acetylenic compound. Trapping of these carbanionic species with protic reagents, alcohol for instance, leads to an ylid A. An alternative pathway involves reaction on the phosphorus atom leading to a phosphorane B.

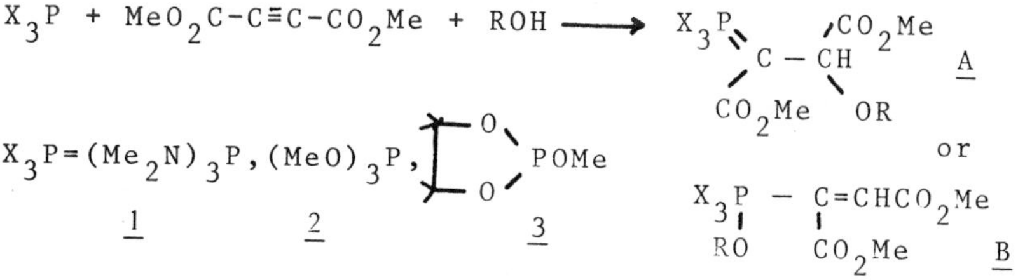

$X_3P + MeO_2C-C\equiv C-CO_2Me + ROH \longrightarrow X_3P=C(CO_2Me)-CH(CO_2Me)(OR)$ A

or

$X_3P(OR)-C(CO_2Me)=CHCO_2Me$ B

$X_3P = (Me_2N)_3P$ (1), $(MeO)_3P$ (2), cyclic pinacol phosphite POMe (3)

With cyclic phosphite 3, when the trapping reagent used in reaction is phenol, we obtain a quantitative yield of ylid. Between 0° and 20° this ylid undergoes a complete rearrangement leading to a phosphorane (1).

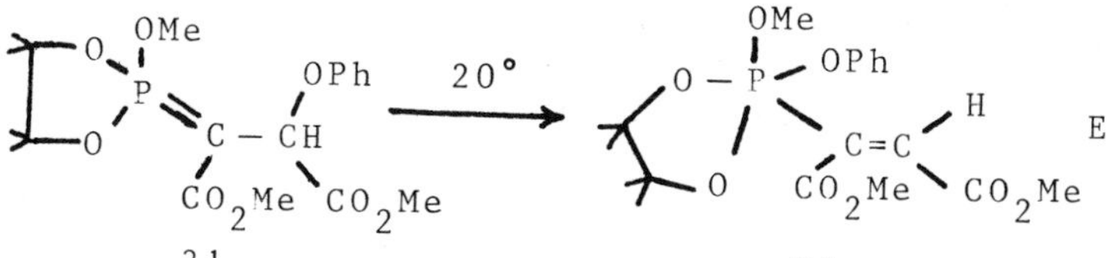

4: cyclic (pinacol) P(OMe)=C(CO_2Me)-CH(OPh)(CO_2Me) $\xrightarrow{20°}$ 5: cyclic (pinacol) P(OMe)(OPh)-C(CO_2Me)=CH(CO_2Me) E

4 ^{31}P 70.5ppm   5 ^{31}P 52ppm $J_{PC=CH}$ 23.2Hz

Conversely, a phosphorane is obtained at -20°C when the P_{III} reagent is trimethylphosphite and methanol is the trapping species. This phosphorane undergo a complete rearrangement in a few minutes at 20°C leading to an ylid.

0097-6156/81/0171-0607$05.00/0

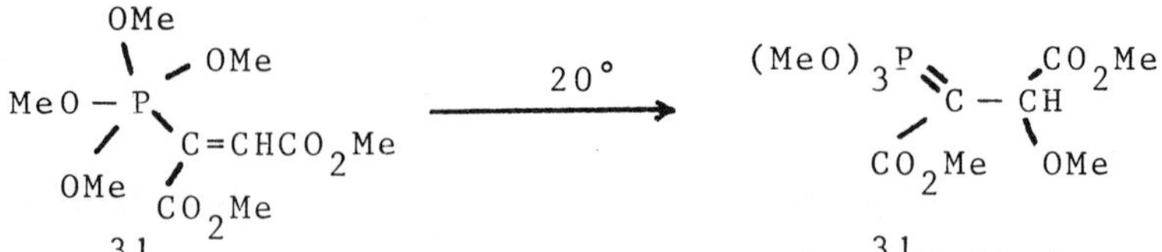

$$(MeO)_3P(CO_2Me)C{=}CHCO_2Me \xrightarrow{20^\circ} (MeO)_3P{=}C(CO_2Me)-CH(CO_2Me)(OMe)$$

6 ^{31}P -56.9ppm 7 ^{31}P 57.5ppm

When the P_{III} reagent is 3 with benzoic acid as trapping species, we observe an equilibrium between ylid and phosphorane which is strongly solvent dependent,

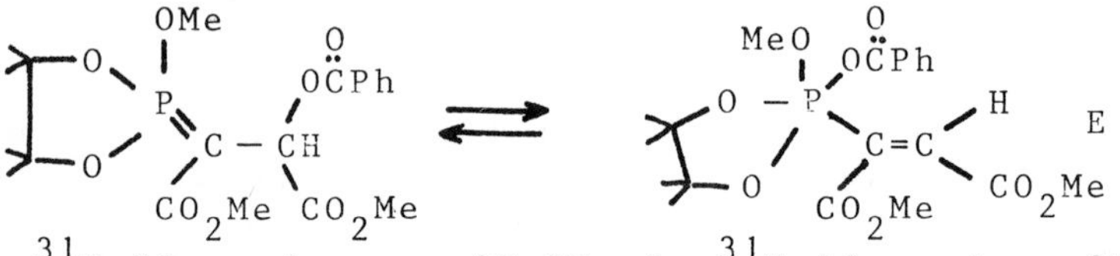

8 ^{31}P 69ppm $J_{P-C-C-H}$ 22.8Hz ⇄ 9 ^{31}P -49ppm J_{PCCH} 25Hz (E)

for instance in CCl_4 : 8 18%-9 82% and in CH_2Cl_2 : 8 37%-9 63%. If the dichloromethane solution is evaporated and a new solution made with carbon tetrachloride then the first results(8 18% and 9 82%) are found again.

In tetrachloride solution, these two species 8 and 9 remained essentially unchanged. With dichloromethane solution, the evolution takes a few hours only; NMR signals appear for 10 (Z+E) whereas the intensity of 8 and 9 signals are decreasing. When the percentage of 10 is 60% the relative ratio of ylid versus phosphorane remains unchanged (36%-64%). In the end, two isomers (Z+E) corresponding to the phosphonate 10 can be seen (^{31}P NMR).

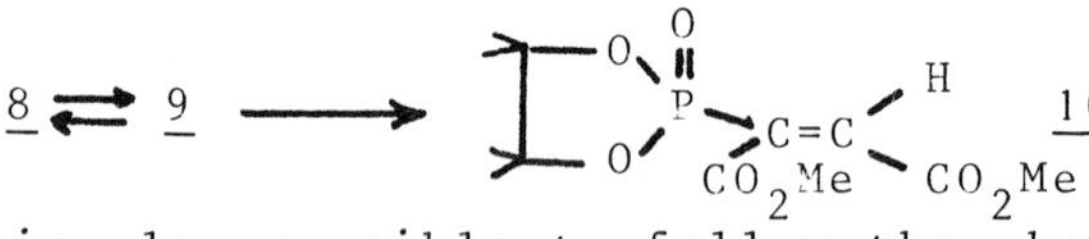

8 ⇄ 9 ⟶ 10E ^{31}P 19.8ppm $J_{P-C-C-H}$ 22.5Hz

It is also possible to follow the chemical modification of the system with ^1H NMR technique and in particular the appearance and the disappearance of doublets relative to H-C-C=P, H-C=C-P_V and H-C=C-P_{IV}. Upon heating, the equilibrium shifts from (50% Z, 50% E) of 10 to isomer E.

Another equilibrium ylid ⇄ phosphorane has been reported a few years ago. Nevertheless it was obtained by addition or loss of a reagent (2).

$$(CH_3)_3P = CH_2 \underset{-\ MeOH}{\overset{+\ MeOH}{\rightleftarrows}} (CH_3)_4POCH_3$$

Table 1 summarizes our results.

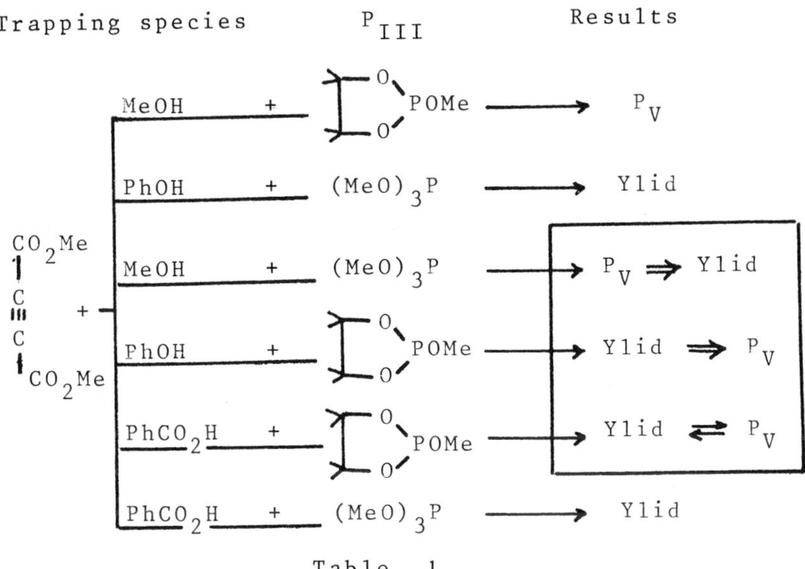

Table 1

In table 1, the three phenomena which correspond to the paper's title are encircled. We can also see that the same reagent i.e. methanol always gives a phosphorane. This one is stable when the phosphite reagent is a cyclic species. This phosphorane then undergoes a rearrangement to an ylid when the phosphite reagent is a linear species ($P(OMe)_3$).

One can see also that with the same trapping reagent, phenol for instance, the kinetic product is now an ylid. When the trapping reagent is benzoic acid we have attained thermodynamic stability between ylid and phosphorane.

At low temperature, the ylid 4 adds methanol before giving by rearrangement a phosphorane 5 which finally lead to a saturated phosphorane 11.

4 (OMe, P=C–CH, CO_2Me, CO_2Me, OPh) $\xrightarrow[-20°]{MeOH}$ 11 (MeO, OMe, O–P, O, CHCH, OPh, CO_2Me, CO_2Me)

^{31}P -49 ppm

Phosphorane 11 is stable enough in solution to be analyzed and identified. Nevertheless, in a few days 11 loses phenol and not methanol to give a vinyl phosphorane 12 (E. structure).

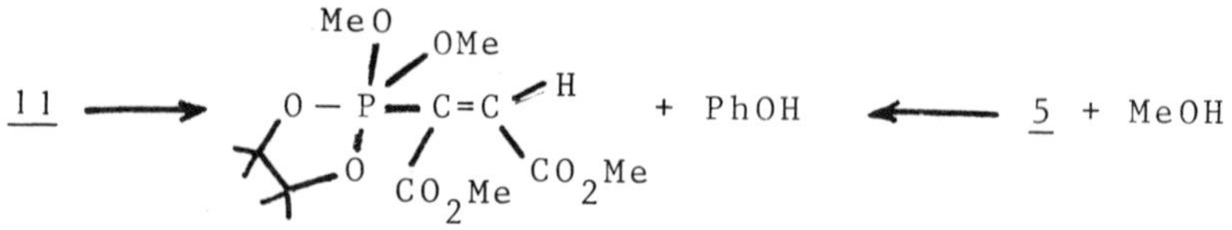

11 → [structure 12] + PhOH ← 5 + MeOH

We can obtain the same phosphorane 12 by methanolysis of phosphorane 5. Concerning the stereochemistry of the phosphorane double-bond, we observe that the E isomer is the more stable. The first reaction of Table 1 gives Z isomer, but only traces of benzoic acid are necessary to provoke instantaneous isomerisation of Z to E. The mechanism probably involves an addition-elimination.

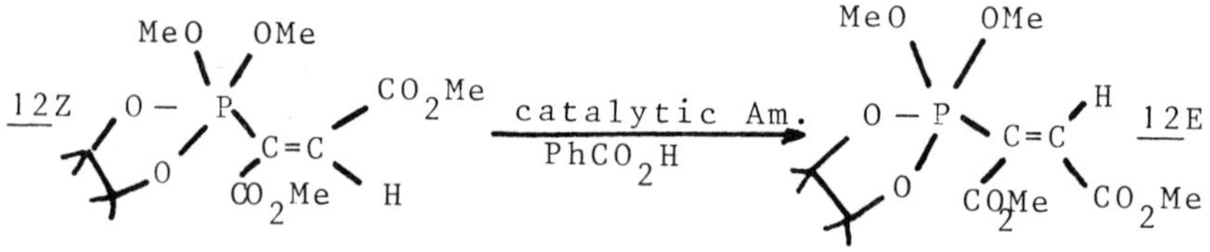

12Z → (catalytic Am. $PhCO_2H$) → 12E

Results of Table 1 and the later results seem to show that through out this work we are dealing with kinetically controlled reactions. Moreover, we demonstrate that an equilibrium between 12 and the ylid form does not exist because this should lead to the isomerisation of Z to E.

When the trapping reagent is acetylene dicarboxylate itself, the reaction yields a new phosphorane which is a pentacoordinated phosphole (3).

Literature cited

1. Burgada, R ; Leroux, Y ; El Khoshnieh, Y.O. Tetrahedron Lett. 1980, (21), 925.
2. Schmidbaur, H ; Stuhler, H. Angew. Chem. Int. Ed. 1972, 2, 145.
3. Burgada, R ; Leroux, Y ; El Khoshnieh, Y.O. Tetrahedron Lett. (in press).

RECEIVED June 30, 1981.

Base-Catalyzed Reactions of Phosphonomethylphosphinates, Bis(phosphonomethyl)phosphinates, and Bis(phosphonomethyl)phosphinic Amides with Aldehydes

W. FRANKLIN GILMORE and JOON SUP PARK

Department of Medicinal Chemistry, School of Pharmacy, University of Mississippi, University, MS 38677

One of the fascinating aspects of organophosphorus chemistry is the extent to which stereochemical and electronic factors control the rate and also the nature of the products of nucleophilic displacements at tetrahedral phosphorus. Our original goal in this work was to design a synthesis of diethylphosphonomethylalkenylphosphinates III. Unfortunately the condensation of I with aldehydes occurred primarily by path A (Scheme I) to give II

Scheme I

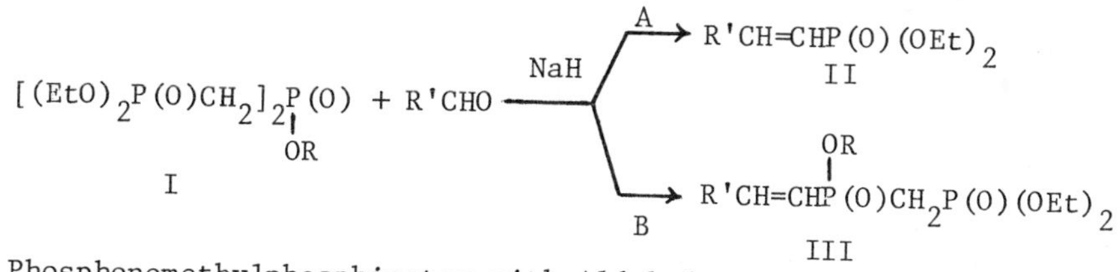

$$[(EtO)_2P(O)CH_2]_2P(O)(OR) + R'CHO \xrightarrow{NaH} \begin{cases} \xrightarrow{A} R'CH{=}CHP(O)(OEt)_2 \quad \text{II} \\ \xrightarrow{B} R'CH{=}CHP(OR)(O)CH_2P(O)(OEt)_2 \quad \text{III} \end{cases}$$

I

Phosphonomethylphosphinates with Aldehydes

Since our initial attempts to direct Scheme I towards path B by changing the group on the central phosphorus of I failed, we simplified our system to determine what factors control elimination of a phosphonate or phosphinate phosphorus during carbonyl olefination with PO ylids containing two phosphorus atoms. The results for the condensation of a series of phosphonomethylphosphinates IV with aldehydes (Scheme II) are given in Table I.

Scheme II

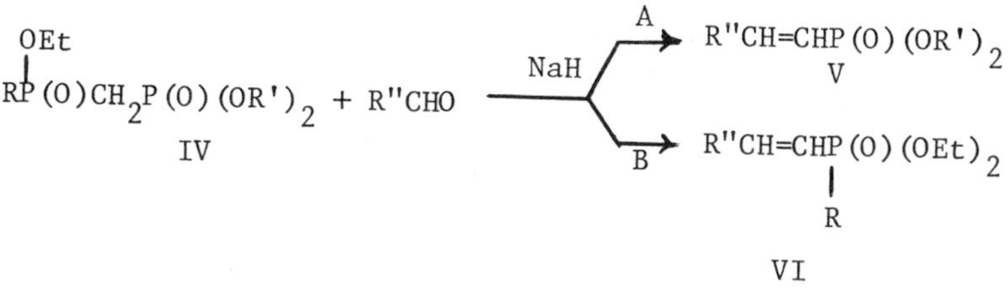

$$RP(OEt)(O)CH_2P(O)(OR')_2 + R''CHO \xrightarrow{NaH} \begin{cases} \xrightarrow{A} R''CH{=}CHP(O)(OR')_2 \quad \text{V} \\ \xrightarrow{B} R''CH{=}CHP(O)(OEt)R \quad \text{VI} \end{cases}$$

IV

0097–6156/81/0171–0611$05.00/0

Table I

Reaction of Isobutyraldehyde with Phosphonomethylphosphinates

R	R'	R"	V %	(yield) %	VI %	(yield) %
Phenyl	Ethyl	Isopropyl	11	(9)	89	(67)
2,4-Dimethoxyphenyl	Ethyl	Isopropyl	13	(13)	87	(62)
4-Methoxyphenyl	Ethyl	Isopropyl	14	(12)	86	(54)
3-Chlorophenyl	Ethyl	Isopropyl	19	(17)	81	(59)
Ethyl	Ethyl	Isopropyl	88	(68)	12	(10)
Phenyl	Phenyl	Isopropyl	0		100	(70)
Cyclohexyl	Ethyl	Isopropyl	3		97	

The data in Table I indicate that the reaction of IV with aldehydes can be directed through path A or path B by changing steric and electronic factors. The phosphorus atom having the smaller groups is eliminated (compare R=ethyl and cyclohexyl). If the difference in steric factors is not sufficient to completely control the course of the reaction, the phosphorus atom having the greater positive charge is eliminated (compare R'=ethyl and phenyl when R=phenyl).

Bis(Phosphonomethyl)phosphinates with Aldehydes

After learning that IV could be directed primarily through path B by changing steric and electronic factors, we reinvestigated the reactions of I with aldehydes. These results are in Table II.

Table II

Reaction of Aldehydes with *Bis*(Phosphonomethyl)phosphinates

R	R'	II %	(yield) %	III %	(yield) %
Phenyl	Isopropyl	95		5	
Cyclohexyl	Isopropyl	78	(48)	22	(8)
Isopropyl	Isopropyl	93	(51)	7	(1)
2,6-Dimethylphenyl	Isopropyl	86	(36)	14	(3)
(D,L)-Menthyl	Isopropyl	70	(53)	30	(11)
Ethyl	Isopropyl	100	(57)	0	

The data in Table II indicate that our goal of directing Scheme I to path B was only partially achieved. As an alternative another method of increasing both the steric hinderance and the electron density at the central phosphorus atom of I was sought.

Bis(Phosohonomethyl)phoshinic Amides with Aldehydes

To further study the effects of steric and electronic factors on the course of carbonyl olefination with these dual PO ylids, we investigated the reactions of VII with aldehydes (Scheme III). Data for these reactions are shown in Table III.

Scheme III

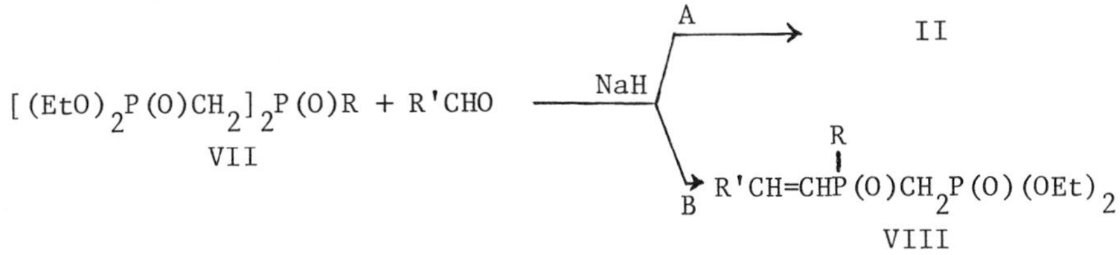

$$[(EtO)_2P(O)CH_2]_2P(O)R + R'CHO \xrightarrow{NaH} \begin{cases} \xrightarrow{A} \text{II} \\ \xrightarrow{B} R'CH{=}CHP(R)(O)CH_2P(O)(OEt)_2 \ \ \text{VIII} \end{cases}$$

VII

Table III

R	R'	II %	(yield) %	VIII %	(yield) %
Phenylamino	Isopropyl	93	(48)	7	(2)
α-Methylbenzylamino	Isopropyl	88	(57)	12	(4)
Cyclohexylamino	Isopropyl	85	(44)	14	(4)
t-Butylamino	Isopropyl	60	(54)	40	(20)
Morpholino	Isopropyl	91	(53)	9	(3)
Piperidino	Isopropyl	67	(38)	33	(11)
Di-n-propylamino	Isopropyl	12	(11)	88	(46)
N-Cyclohexyl-N-methylamino	Isopropyl	20	(20)	80	(43)

The data in Table III illustrate that condensation of VII with isobutyraldehyde is a satisfactory synthesis of VIII which can be hydrolyzed to the desired phosphonomethylphosphinic acid. This method has been tested with several other aldehydes and found to give results similar to those in Table III.

Discussion

The data presented in Tables I-III demonstrate that through proper utilization of substitutents it is possible to direct carbonyl olefination through elimination of a phosphinate phosphorus atom (PO) or a phosphonate phosphorus atom (PO_2). Condensation of the anions from I, IV and VII with carbonyl groups forms an anion similar to IX (Scheme IV). Nucleophilic attack of this anion at PO leads to two trigonal bipyramidal (TBP) oxyphosphoranes (X and XI) and attack at PO_2 leads to a single TBP oxyphosphorane (XII).

Scheme IV

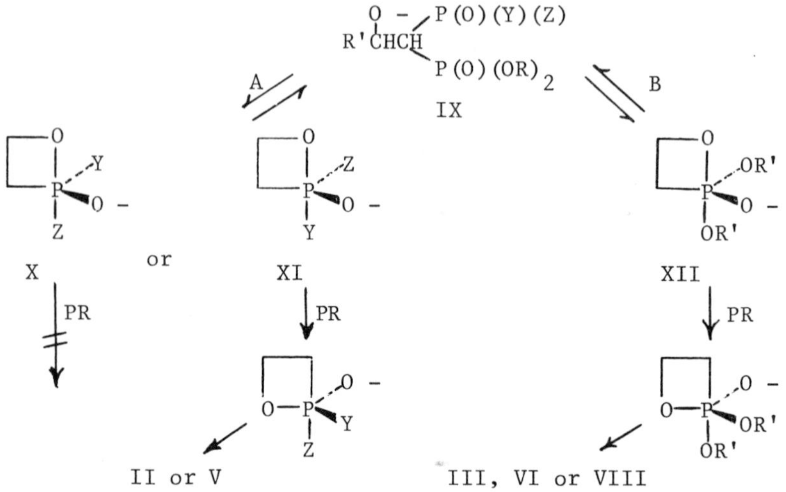

O – P(O)(Y)(Z)
R'CHCH
P(O)(OR)$_2$
IX
A
B
O Y O – Z
X
or
O Z O – Y
XI
O OR' O – OR'
XII
PR
PR
PR
O – O—P Y Z
O – O—P OR' OR'
II or V
III, VI or VIII

When Y is alkyl and Z=R=ethoxy, PO_2 is more positive than PO. Nucleophilic attack at PO_2 leads to XII which has a moderate barrier to pseudorotation (PR). Other factors being equal, path B is less favorable than path A. When Y is ethyl and Z=R=ethoxy, XI is favored due to attack from behind the CH_2 of ethyl. If Y is cyclohexyl, formation of X and XI is retarded for steric and electronic reasons; but, to the extent that either forms, XI is favored by steric factors. For Y=phenyl and Z=R=ethoxy, PO_2 is slightly more positive than PO and when R=phenoxy PO_2 is much more positive than PO. Thus, when Y=Phenyl, attack at PO_2 is not favored by either electronic or steric factors. To the extent that this occurs, attack should be from the face opposite the phenyl group with formation of XI. Formation of XI also requires a decrease in any resonance interactions which exist between the aromatic ring and PO. For I and VII, Z=alkyl or aryloxy and alkyl or arylamino, while Y is the relatively bulky diethylphosphonomethyl (DEP) group. In I and VII PO is more positive than PO_2 and XI is favored unless Z is larger than DEP. When Z is larger (methyloxy and di-*n*-propylamino), X is favored and the only reasonable path left is through XII.

Literature Cited

1. Gilmore, W. Franklin and Huber, Joseph W., III. *J. Org. Chem.* 1973, *38*, 1423-4.

RECEIVED July 7, 1981.

Structure-Reactivity Studies on Oxygen-Containing Phosphorus-Based Ligands

YUAN CHENGYE, YE WEIZHEN, ZHOU CHENGMING, and HUI YONGZHENG

Shanghai Institute of Organic Chemistry, Academia Sinica, Peoples Republic of China

Phosphorus ligands bearing phosphoryl oxygen atom are well-known extractants, which are capable of forming more stable coordination compounds with various metals in comparison with other oxygen-containing organic dentates. Structure-reactivity studies of phosphorus-based extractants can, in very practical terms, contribute both to the development of new ligands and to the progress of organophosphorus chemistry in general. As shown by our early studies, the behaviour of extractants are governed chiefly by three structural factors (1). It is proposed that, there is a significant correlation between the extraction properties and some physico-chemical constants of ligands. Therefore, it is interesting to investigate the dependence of chemical structure on the physico-chemical properties of phosphorus-based ligands. This paper describes some new aspects of this problem which have been studied in our Laboratory.

Linear Free Energy Relationships. Correlation analysis by LFER is a useful tool for structure-reactivity studies (2). The contribution of charge density of phosphoryl oxygen atom to the coordination behaviour of neutral organophosphorus compounds is well established. The Lewis basicity of such compounds, as measured by the degree of shift of OD vibration frequency ($\Delta\nu$OD) of deuterated methanol due to the association of the latter with compound under investigation correlates linearly to the nature of groups attached to phosphorus atom. In a series of phosphate-phosphine oxide, plot of νPO or $\Delta\nu$OD against Taft constant σ^* straight lines resulted in both cases. It is well illustrated by the dependence of the nature of substituents (Hammett σ value) of p-substituted phenylphosphonates on νPO and $\Delta\nu$OD (3). In the case of p-substituted phenyl dibutylphosphinates straight lines also resulted when the νPO and $\Delta\nu$OD are plotted versus the Hammett σ constants of the nuclear substituents due to the conjugation effect of ester oxygen atoms (4). The ^{31}P NMR studies of p-substituted phenyl- and benzylphosphonates show the linear free energy relationship between the nature of

0097-6156/81/0171-0615$05.00/0

substituents and chemical shift of ^{31}P nucleus in both cases though the influence of substituents is more strong in the former and the slope of straight line is evidently greater than that in the latter.

The structure-reactivity relationship of acidic organophosphorus compounds is well demonstrated by mono-esters of p-substituted phenylphosphonic acids. The acidity of these organic acids increased as the polar nature of the substituents enhanced. A linear free energy relationship exists between the pKa value and the Hammett σ constants in acidic p-substituted phenylphosphonates. When these structure parameters are plotted either against the νPOO^- asym. or against the ^{31}P chemical shift of their dicyclohexylammonium salts straight lines resulted in both cases. However, acidic p-substituted benzylphosphonates as expected fail to offer such correlation (5).

The quantitative influence of the nuclear substituents on the phosphoryl group in dialkyl p-substituted phenylphosphonates and their mono-esters may be due either to the conjugation effect of the phosphoryl group with the π-bonds of the benzene ring or to the inductive effect through the phosphorus atom. In contrast with the postulation of Kabachnik (6), we are of the opinion that the results of our experiments might also be explained by the inductive effect of the nuclear substituents toward the phosphoryl oxygen through the phosphorus atom, and the oxygen of the phosphoryl group is linked to the phosphorus by $d\pi$-$p\pi$ bond.

On the basis of correlation analysis by LFER between the chemical structure and spectral data from IR and NMR, Lewis basicity, rate constant of hydrolysis as well as pKa of various organophosphorus compounds, it is not likely that the benzene ring is conjugated with the phosphoryl group, and that the oxygen of the latter is linked to the phosphorus by a double bond, if the general concepts of conjugation in the chemistry of carbon compounds work as well in the phosphorus series.

Hükel Molecular Orbital Method. The structure-reactivity relationship of phosphorus-based ligands has been studied by Hükel molecular orbital (HMO) method. The molecule of these extractants may be considered as conjugated system containing hetero-atoms. In such case, the Coulomb and resonance integrals may be altered by known expressions in the HMO calculation. The high degree secular equation in HMO calculation was solved by the standard programing of Jacobi method, by which the characteristic root and factor of real symmetric matrix were solved. The calculation was worked out on the electronic computer CJ 719. The structure parameters, such as charge density (qx), bond order (Pc=x), π bond energy ($Ec^{\pi}_{c=x}$) as well as total energy of the molecular orbitals ($\Sigma\lambda i$) thus calculated was correlated with the extraction behaviour of such phosphorus-based ligands for uranium, thorium and rare earths (7).

The HMO calculations indicate that the charge density of oxygen (1.9453, 1.6125 and 1.5181) and bond order of PO and C=O (0.2318,

0.7344 and 0.8217) for trialkylphosphine oxide, N,N-dialkylacetamide and dialkylketone respectively. The superiority of the phosphorus-based ligands bearing oxygen over other oxygen containing organic dentates in extraction of metals is well demonstrated.

The extraction ability of neutral organophosphorus compounds closely related to the charge density of phosphoryl oxygen (qo) and phosphorus atom(qp). It was found that the distribution ratio of cerium enhanced as the qo and qp values was increased. Meanwhile a linear free energy relationship exists between the $\Sigma\lambda i$ values and the Kabachnik constant for substituents of phosphorus compounds. As estimated by least squares method, an empirical equation $\Sigma\sigma_{\phi}=-3.84+0.168\Sigma\lambda i$ was deduced

The extraction constant of neodymium, samarium, yttrium and ytterbium of various acidic organophosphorus compounds are governed chiefly by the charge density of hydroxyl oxygen atom (HOMO) in grouping >P(O)OH. The plot of Ke of rare earths versus qOH(HOMO) values of these compounds gives straight line as anticipated, owing to the direct influence of qOH value on the pKa of ligands. A quantitative relationship also exists between the $\Sigma\lambda i$ and $\Sigma\sigma_{\phi}$values of acidic organophosphorus extractants i.e.$\Sigma\sigma_{\phi}=-4.12+0.166\Sigma\lambda i$.

Pattern Recognition. In recent years, pattern recognition, an efficient computational method for the analysis of multivariant data, was developed (8). The pattern recognition processing of the influence of various structure factors of some organophosphorus compounds on their extraction behaviour of uranium is described.

Data preprocessing consists of autoscaling and weighting. The training set is classified assording to a given threshold value. Two kinds of classifying methods are used: Linear Mapping Method by using Karhunen-Loeve transformation and K-Nearest Neighbor Method (KNN). An unknown compound is classified according to the majority vote of its K-Neighbors in the feature space. Each compound is in turn taken as the unknown pattern and the rest as a training set (Leave-one-out procedure). The programs for Linear Mapping and KNN method (LTBP and NBP) were written in BASIC language and run on a DTS-131 mini-computer (9).

1. Linear Mapping Classification. The program LTBP is applied to the classification of distribution ratio of uranium in extraction of some neutral organophosphorus compounds and dialkylaminoalkyl phosphonates. The general formula of neutral organophosphorus compounds is $R_1R_2R_3PO$, where R_1, R_2 and R_3 are individually CH_3-, n-C_4H_9, iso-C_5H_{11}, n-C_6H_{13}, n-C_7H_{15}, $C_4H_9CHEtCH_2$ and $C_6H_{13}CHMe$ or corresponding alkoxyl group. The selected nine features are: number of P-O-C bonds in R_1, R_2 or R_3; total number of P-O-C bonds; degree of branch of carbon chain; number of carbon atoms in longest and shortest chain; ratio of the length of carbon chain and total number of carbon atoms. The classification threshold (Du) is 300. In the seventeen neutral organophosphorus compoounds studied, there are six with Du > 300 and eleven compounds with Du < 300. Owing to the large difference in features, the fundamentally same classified

result is obtained when the dimension of the feature space is reduced from nine to two. Therefore, by using of pattern recognition method, one can single out the principal structure feature which effect the properties of compounds under investigation, and then a classified result may be deduced according to the similarity in structure. Twenth-eight compounds with the general formula of $R_2N(CH_2)nP(O)OR'(OR'')$, where $R=C_2H_5$, $n-C_4H_9$ or $n-C_8H_{17}$; n=1 or 2; $R'=C_2H_5$, $n-C_4H_9$ or $n-C_8H_{17}$; R"=R' or H were studied analogously.

2. KNN Classification Rule. The program KNBP is applied to the classification of Du in extraction by eighty-two organophosphorus compound with eleven different structures (the number of each kind of compounds is in parathesis) $R_3PO(4)$, $R_2P(O)OR'(3)$, $RP(O)(OR')_2(7)$, $(RO)_3PO(4)$, $GC_6H_4P(O)(OR)_2(14)$, $GC_6H_4CH_2P(O)(OR)_2$ (11), $R_2N(CH_2)nP(O)(OR')_2(15)$, $R_2P(O)OH(1)$, $R_2N(CH_2)nP(O)OR(OH)(14)$, $GC_6H_4P(O)OR(OH)(5)$, $GC_6H_4(CH_2)nP(O)OR(OH)(4)$. Seventeen features are selected.

For environment descriptors, the phosphorus atom is taken as center and then the nearest atoms or bonds. For bond environment (BD): single bond=1; double bond=2; triple bond=3 aromatic bond=4; C-H bond is neglected. For atom environment (AD): atoms are assigned according to the following order C,1; O,2; N,3; S,4; F,5; Cl,6; Br,7; I,8; P, 9. For weight environment (WD), AD and BD are all taken into account. WD is the sum of BDxAD of the neaest atoms. The approach adopted in this paper is different from Jurs (10).

It is shown that the five major structure features predominating the extraction ability of uranium are: number of P-C bonds (+); AD(-); number of $-CH_2-$ (non-ester) (+); number of P-O-C bonds (-); WD (-). In the parenthese the tendency of the feature which is necessary for compounds of higher reactivity is given.

Literature Cited

1. Yuan Chengye Nucl. Sci. & Tech. 1962, 908; Kexue Tongbao 1977, 22, 465.
2. Chapman N.B. and Shorter J., Ed.; "Correlation Analysis in Chemistry", Plenum Press, New York, 1978
3. Yuan Chengye, Ye Weizhen, Ge Minjuan, Zhou Liying, Scientia Sinica 1964, 13, 1510; Acta Chim. Sinica 1964, 30, 458
4. Yuan Chengye, Yuan Qun Acta Chim. Sinica 1980, 38, 339
5. Yuan Chengye, Ye Weizhen Acta Chim. Sinica 1981, 230
6. Kabachnik M.I. Tetrahedron 1962, 17, 239.
7. Yuan Chengye, Zhou Chengming, Cheng Kongchang, Unpublished.
8. Kowaiski, B.R., Bender, C.F. J.Am. Chem. Soc. 1974, 96, 916
9. Hui Yongzhong, Wang Jizhong, Yuan Chengye, Unpublished.
10. Jurs, P.C. "Computer Assisted Studies of Chemical Structure and Biological Function" Stuper, A.J. Ed., John Wiley, New York, 1979

RECEIVED July 7, 1981.

128

Phosphoric and Carboxylic Amides

Comparison of Bonding and Reactivity

TOMASZ A. MODRO

School of Chemical Sciences, University of Cape Town, Rondebosch 7700, South Africa

Although carboxylic as well as phosphoric esters are widely utilized in nature, phosphoric amides do not parallel their carboxylic counterpartners as the widespread biological structures. This is most likely due to the low stability of the P(O)-N bond under acidic conditions; a property resulting from the different bonding at the phosphoramidate function, relative to the carboxylic system.

The geometry and protonation behavior of carboxylic amides (1) is interpreted in terms of the resonance effects, resulting in the partial double bond character of the nitrogen - acyl linkage. In amides derived from phosphoric acid, $Z_2P(O)NR_2$ (2) conjugation effects are expected to be less important because of the different atomic orbitals (2p and 3d) involved. The lesser involvement of the nitrogen lone pair in the conjugation with the phosphoryl group is taken as a factor responsible for the remarkably facile cleavage of the P-N bond under acidic conditions. (1). In this work some structural and reactivity properties of amides (1) and (2) are discussed in order to gain more insight into the comparative bonding characteristics of the $-C(O)-N<$ and $>P(O)-N\leq$ systems.

Restricted Rotation

Nitrogen - carbonyl rotational barrier in (1) is greater than 20 kcal/mole; (2). the corresponding barrier in compounds (2) has been estimated not to exceed in most cases 8 kcal/mole. (3).

Room temperature NMR spectrum of ethyl N,N-dimethylcarbamate (3) shows a single absorption for the NMe_2 group, but complexation by the Lewis acid is sufficient to cause the non-equivalence of the two nitrogen substituents. Similar complexation of the phosphoryl oxygen in diethyl N,N-dimethylphosphoramidate (4) has no apparent effect upon the free rotation of the P-N bond. The increase of the deshielding of the N-methyl protons caused by the oxygen complexation is much greater for (3) than for (4). This is an additional evidence for the poorer response of the

0097–6156/81/0171–0619$05.00/0

amide nitrogen to the electronic changes at the phosphoryl group relative to the effects in the -C(O)NR_2 function.

Polar Substituent Constants

Conjugation of the nitrogen lone pair with the adjacent phosphoryl or carbonyl function was tested by the ^{13}C NMR parameters of the N-phosphorylated and N-acetylated aniline, as well as of their complexes with Lewis acids. (4). The inductive and resonance constants for the neutral and charged amide groups were determined using the dual substituent parameter (dsp) approach. (5). Results are given in the Table.

Table. Estimated Polar Substituent Constants; Solvent $CDCl_3$

Substituent	σ_I	σ_R^o
NH_2	+ 0.07	- 0.51
NH-C(O)Me	+ 0.16	- 0.21
NH-C($OTiCl_4$)Me	+ 0.51	- 0.12
NH-P(O)Ph_2	+ 0.11	- 0.35
NH-P($OTiCl_4$)Ph_2	+ 0.86	- 0.34

Changes in the σ_R^o value show that the p_π-d_π back-donation from nitrogen to phosphorus in (2) (and its charged form) is much less important than the p_π-p_π interaction in (1).

Linear Free - Energy Relationship

Correlation between the IR stretching frequencies of the phosphoryl group and σ constants was determined for a series of ring-substituted N-aryl dimethylphosphoramidates (5). The slope of the $\Delta\nu_{PO}/\nu_o$ vs. σ plot is ca. 2/3 of that for the correspon - ding plot for ring-substituted acetanilides. (6). Dsp analysis indicates that in the phosphoramidate molecule mainly the induc - tive effects of the substituents are effectively transmitted by the nitrogen atom to the phosphoryl group.

Molecular Structures

IR studies reveal that phosphoric amides (5) have a high tendency for self-association resulting in the formation of the hydrogen - bonded dimers (6) :

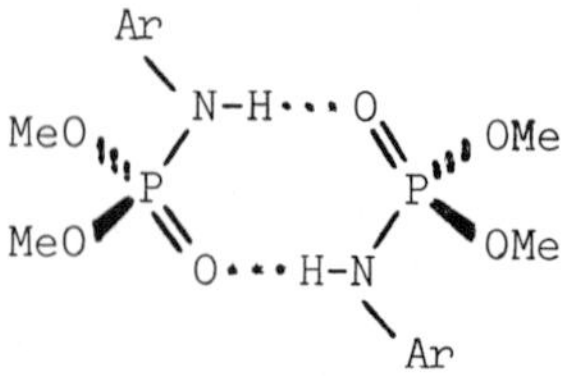

Ar, N-H···O, MeO, OMe, P, P, MeO, OMe, O···H-N, Ar (6)

For the meta- and para-substituted (5) dimerization prevails even in dilute solutions in non-polar solvents; ortho-substi -

tuents decrease the stability of (6) and make substrates (5) to exist in solutions mainly in the monomeric form.

Dimeric structure (6) has been demonstrated by analysing crystals structures of some phosphoramidates (5). The l.f.e.r. between the hydrogen bonding (measured as the intermolecular O···N distance) and σ_X constants in dimers (6) derived from amides XC_6H_4-NH-P(O)(OMe)$_2$ was established. Polar effects of substituents X modify the acidity of the NH hydrogen atom, thus vary its ability to form a hydrogen bond with the phosphoryl group of another molecule of the amide. However, the length of the phosphoryl bond in substrates (5) remains constant within the experimental error. This result demonstrates that the variation in the polarity of the N-substituent does not affect significantly the bond order of the P=O group. The crystal and molecular structures determined for the similarly substituted benzanilides ArNH-C(O)Ph demonstrate considerable variation in the carbon - oxygen distance of the carbonyl group, consistent with the model of the extensive resonance effects in the carboxyamide function.

Protonation Behavior

The comparison of the solvolysis of N-tert-butyl diphenyl - phosphinamide (7) and N-tert-butyl benzamide (8) in anhydrous trifluoroacetic acid illustrates the reactivity difference resulting from the difference in the protonation behavior. For N-tertiary-alkyl substituted amides acidic solvolysis can involve both the N-acyl and the N-alkyl bond cleavage. (7). For both types of bond fission (7) was found much more reactive than (8); $k(7)/k(8) > 10^5$ for the N-acyl cleavage, and $k(7)/k(8) = 3x10^2$ for the de-tert-butylation reaction. This result is in full agreement with the postulated structures of the respective conjugate acids. In the highly delocalised benzamidonium ion |Ph-C(OH)NH-t-Bu|$^+$ carbonyl carbon is only weakly electrophilic and the N-t-Bu bond is not significantly weakened. For the N-protonated form of (7) |Ph_2P(O)NH_2-t-Bu|$^+$ on the other hand, high electrophilicity of the phosphorus, as well as the high tendency for the unimolecular abstraction of the t-butyl carbo - nium ion would be expected.

Since there is not much of conjugation between the P=O group and the nitrogen in (2), O and N atoms can in principle behave as two independent basic centers. Medium effects upon the PMR shielding parameters of phosphoramidates demonstrated that in such a strong acid as trifluoromethanesulfonic amides (2) indeed exist, at least partly, as the O and N diprotonated species.

Solvolytic Behavior

Acid-catalysed solvolysis of both classes of amides (1) and (2) proceeds according to the A-2 mechanism. In consequence, variation in the polar effects of substituents at the nitrogen has two opposite effects. Electron donation increases the concentration of substrate's conjugate acid but at the same time slows the rate-determining step - the nucleophilic attack of a solvent molecule at the carbonyl or phosphoryl center. However, the net structure-reactivity dependence is determined by relative sensitivity of both, the preequilibrium step and the slow step, to polar effects of substituents.

For the acid-catalysed hydrolysis of benzanilides substituted in the aromatic amine moiety $\rho = 1.66$. (8). The positive sign of the reaction constant can be easily understood in terms of the hydrolysis mechanism. The protonation of the carbonyl oxygen should not be very much affected by the polarity of a remote substituent in the N-aryl group. On the other hand, the rate of approach of a nucleophile depends upon the electrophilicity of the carbonyl carbon, which in turn is modified by the electron - releasing or electron - donating properties of the N-substituent.

Acid-catalysed solvolysis of N-aryl phosphoramidates is characterised by the negative value of the reaction constant ($\rho = -1.2$). These inversed substituent effects illustrate two points discussed before. First, if the N-protonated form represents the reactive intermediate in solvolysis of (2), much stronger dependence of the protonation preequilibrium on the effect of N-substitution is expected. Secondly, if the resonance effects are poorly transmitted to the P atom through the -NH- bridge, structural variation in the N-aryl substituent should have weak effect upon the ability of phosphorus to accept a nucleophile. As a consequence of the different electronic interactions within the phosphoramidate group, P-N bond cleavage in phosphoric amides, contrary to the behavior of analogous carboxylic compounds, is accelerated by the electron - donation at the nitrogen atom.

Acknowledgements

The financial assistance of the University of Cape Town and the C.S.I.R. is gratefully acknowledged.

Literature Cited

1. Koizumi, T.; Haake, P. J. Am. Chem. Soc. 1973, 95, 8073.
2. Robin, M. B.; Bovey, F. A.; Basch, H. "The Chemistry of Amides"; Zabicky, J.; Ed., Wiley, London, 1970; p 7.
3. Burdon, J.; Hotchkiss, J. C.; Jennings, W.B. J. Chem. Soc. Perkin Trans. 1 1976, 1052.
4. Modro, T. A. Phosphorus and Sulfur 1979, 5, 331.
5. Hehre, W. J.; Taft, R. W. Progr. Phys. Org. Chem. 1976, 12, 159.
6. Peltier, D.; Pichevin, A.; Bonnin, A. Bull. Soc. Chim. Fr. 1961, 1619.
7. Modro, T. A.; Lawry, M. A.; Murphy, E. J. Org. Chem. 1978, 43, 5000.
8. De Lockerente, S. R.; Brandt, P. V.; Bruylants, A.; de Theux, T. Bull. Soc. Chim. Fr. 1970, 2207.

RECEIVED July 7, 1981.

POSTER PRESENTATIONS

Reaction of Lignin with Chlorophosphazenes

H. Struszczyk—Institute of Man-made Fibers, Technical University of Łodź, Poland
J. E. Laine—Rauma-Repola OY, Rauma, Finland

Tris(aminomethyl)phosphine Oxide and Its Derivatives

Arlen W. Frank—Southern Regional Research Center, USDA, SEA, P.O. Box 19687, New Orleans, LA 70179

Synthesis of *S*-Alkyl *S*-(Carbamoylmethyl)-ethylphosphonotrithioates

Wu Kiun-houo and Sun Yung-min—Department of Chemistry, Fudan University, Shanghai 201903, China

Cyclic Acetals of Formyl Phosphonic Acid Esters

S. Yanai, A. K. Singh, M. Halmann, and D. Vofsi—The Weizmann Institute of Science, Rehovot, Israel

Phosphorus Heterocycle Synthesis by $RPX_2 \cdot AlX_3$ Addition to $R'-C(=Z)-(CH_2)_n-C(=CH_2)R''$ ($Z = CH_2$, O, NR, . . .) Molecules

Y. Kashman and A. Rudi—Department of Chemistry, Tel-Aviv University, Ramat Aviv 69978, Israel

Synthetic Approaches to the 9-Phosphadecalin System

J. M. Cowles and S. E. Cremer—Marquette University, Milwaukee, WI 53233

Stereochemistry of Base Cleavage at Phosphorus in a 2-Phospholenium Salt

K. L. Marsi and J. A. Hagenah—Department of Chemistry, California State University, Long Beach, Long Beach, CA 90840

Synthesis and Novel Rearrangement of 2-Bromophosphetane Derivatives

S. Cremer, P. W. Kremer, and P. Kafarski—Marquette University, Department of Chemistry, Milwaukee, WI 53233

Synthesis and Stereochemistry of 3-Phenyl-3-phosphabicyclo(3.2.1)octane and Oct-6-ene Derivatives

S. Cremer, P. W. Kremer, and J. T. Most—Marquette University, Department of Chemistry, Milwaukee, WI 53233

Formation of Phosphinidenes by Thermolysis of 7-Phosphanorbornene Derivatives

L. D. Quin, K. A. Mesch, and K. C. Caster—Gross Chemical Laboratory, Duke University, Durham, NC 27706

3*H*-Benzo-2,1-oxaphospholenes: Stereochemistry of Nucleophilic Substitution at Tricovalent Phosphorus

Otto Dahl—Department of General and Organic Chemistry, The H. C. Ørsted Institute, University of Copenhagen, Universitetsparken 5, DK-2100 Copenhagen, Denmark

Phosphorus Containing Heterocycles by Reaction of Alkylsulfenyl Chlorides with Derivatives of 1,2-Alkadienephosphonic Acids

Ch. M. Angelov and K. V. Vachov—Chair of Chemistry, Higher Pedagogical Institute, 9700 Shoumen, Bulgaria

1,5-Diaza-3,7-diphosphacyclooctanes

B. A. Arbuzov, O. A. Erastov,and G. N. Nikonov —The Arbuzov Institute of Organic and Physical Chemistry, Kazan Branch Academy of Sciences, Kazan 420083, USSR

From Allenic Phosphorus Derivatives to Heterocyclic Compounds: Synthesis of a 1,2,3-Diazaphosphole Compound with a Strongly Polar NH Group Linked to Dicoordinated Phosphorus

N. Ayed, R. Mathis, B. Baccar, and F. Mathis —Université Paul Sabatier, Toulouse, France, and Faculte des Sciences, Tunis, Tunisia

Synthesis of 1-Aza-4-phosphoniabicyclo[2,2,2]-octane Salts

J. Skolimowski—Institute of Chemistry, University of Łodź, 90-136 Łodź, Narutowicz 68, Poland
M. Simalty—Laboratoire CNRS-SNPE, 2-8 Rue Henry Dunant, 94320 Thiais, France

Synthesis of New Five-, Six-, and Seven-Membered Phosphorus Heterocycles with Two Heteroelements

J. Skolimowski—Department of Organic Chemistry, Institute of Chemistry, University of Łodź, Narutowicz 68, 90-136, Łodź, Poland

Phosphatriptycene: The Influence of Geometric Constraints on Reactivity

H. J. Meeuwissen, R. Lournes, and F. Bickelhaupt—Vakgroep Organische Chemie, Vrije Universiteit, De Boelelaan 1083, 1081 HV Amsterdam, The Netherlands

New Closo Type Phosphorus Compounds

J. Navech, M. Benhammou, and J. P. Majoral—Laboratoire des Heterocycles du Phosphore et de l'Azote, Université Paul Sabatier, 118 Route de Narbonne, 31062 Toulouse Cédex, France

The Chemistry and X-Ray Structures of Compounds Containing a Phosphorus–Phosphorus Bond

H. W. Roesky, W. S. Sheldrick, and D. Amirzadeh—Asl. Anorganisch-Chemisches Institut, Universität Göttingen, Tammannstrasse 4, D-3400 Göttingen, FRG, and Gesellschaft fur Biotechnologische Forschung mbH, Mascheroder Weg 1, D-3300 Braunschweig, FRG

Δ_R-Ring Contributions to ^{31}P NMR Parameters of Transition-Metal–Phosphorus Chelate Complexes

P. E. Garrou—Dow Chemical Company, Central Research–New England Laboratory, Wayland, MA 01778

Phosphates ^{31}P Solid State NMR Experimental Data, Calculations

L. Wiesenfeld and J. B. Robert—Laboratoire de Chimie Organique Physique, Département de Recherche Fondamentale, Équipe de Recherche Associée au CNRS n° 674, Centre d'Études Nucleaires, 85 X F 38041 Grenoble Cedex, France

Quantitative Correlation of ^{31}P NMR Chemical Shift Changes on Conversion of Phosphines to P(IV) Derivatives

F. Stoneberger Pinault, A. L. Crumbliss, and L. D. Quin—Department of Chemistry, Duke University, Durham, NC 27706

^{31}P NMR of tRNAs and Duplex RNA: Stereoelectronic, Metal Ion, and Ethidium Bromide Effects

E. Goldfield and D. G. Gorenstein—University of Illinois, Chemistry Department, Box 4348, Chicago, IL 60680

Second Order ^{31}P Spectra of CAP Analogs Demonstrating Dependence on Nucleotide Bases

B. Burkes, J. Johnson, G. S. Owen, Y. Mariam, and N. K. Bose—Atlanta University, Chemistry Department, 223 Chestnut Street, Atlanta, GA 30314

^{31}P NMR Analysis of the Thermal Transition of DNA Samples of Varying Base Pair Composition

Y. Habte-Mariam—Department of Chemistry, Atlanta University, Atlanta, GA 30314
W. D. Wilson—Department of Chemistry, Georgia State University, Altanta, GA 30303

Organophosphate ^{31}P NMR of the Intact Mammalian Crystalline Lens

T. Glonek, J. V. Greiner, S. J. Kopp, and D. R. Sanders—NMR Laboratory, CCOM, 5200 South Ellis Avenue, Chicago, IL 60615, and the Department of Ophthalmology, University of Illinois Eye and Ear Infirmary, Chicago, IL 60612

Intracellular Phosphorus Pools in Algae: ^{31}P NMR and Transmission Electron Microscopy (TEM) Studies

G. A. Elgavish and A. Elgavish—Isotopes Department, The Weizmann Institute of Science, Rehovot, Israel

^{31}P and ^{13}C NMR Spectroscopic Investigations on Open Chain and Cyclic Phosphanes

J. Hahn, B. Baudler, U. M. Krause, and G. Reuschenbach—Institut für Anorganische Chemie der Universität Köln, Greinstrasse 6, 5000 Köln 41, FRG

Phosphino Radicals and Their Dimers

H. Goldwhite, R. T. Keys, and G. Millhauser—California State University, Los Angeles, CA 90032

Phosphorus Nitroxide and Iminoxy Radicals: A New Type of Organic Free Radical

A. V. Il'yasov—Institute of Organic and Physical Chemistry of the Kazan Branch Academy of Sciences of the USSR, Kazan 420083, USSR

Phosphoramidyl Radicals: An ESR Study

M. Négareche, Y. Berchadsky, and P. Tordo—S.R.E.P., Université de Provence, Rue H. Poincaré, 13397 Marseille Cédex 13, France

Vibrational Spectra and Structure of Some M^IPF_6 Fluorophosphates

B. Hájek, A. Muck, and J. Pokorný—Department of Inorganic Chemistry, Prague Institute of Chemical Technology, Suchbátarova 5, 166 28 Prague, Czechoslovakia

Electron Impact Fragmentation of Some Cyclic Ester-Amides of Phosphoric Acid

R. S. Edmundson—School of Chemistry, University of Bradford, Bradford 7, England

Mechanistic and NMR Studies on Chalcogenophosphinito Bridged Platinum(I) Dimers: Is J(PtPt) A Measure of the Pt–Pt Bond?

B. Walther, B. Messbauer, H. Meyer, A. Zschunke, and B. Thomas—Department of Chemistry, Martin Luther University, 4020 Halle (S.), Weinbergweg 16, GDR

Mechanism of Hydrolysis of Triphosphate Ion by Cobalt(III) Complexes

P. R. Norman and R. D. Cornelius—Department of Chemistry, Wichita State University, Wichita, KS 67208

Kinetic Studies of the Reactions of Chlorocyclophosphazenes with Amines

K. V. Katti, S. S. Krishnamurthy, and P. M. Sundaram—Department of Inorganic and Physical Chemistry, Indian Institute of Science, Bangalore - 560012, India

Activation Parameters for the Formation of 2,2-Dihydro-4,4,5,5-tetramethyl-2,2,2-triaryl-1,3,2-dioxaphospholanes from the Reaction of Phosphines with Tetramethyl-1,2-dioxetane

A. L. Baumstark and Kathleen M. Kral—Department of Chemistry, Laboratory for Microbial and Biochemical Sciences, Georgia State University, Atlanta, GA 30303

Interaction Mechanism of Chlorophosphines with Oxocompounds

N. A. Kardanov, N. N. Godovikov, and M. I. Kabachnik—A. N. Nesmeyanov Institute of Organo-Element Compounds of the USSR Academy of Sciences, Vavilova Street, 28, Moscow 117334, USSR

Novel Reactive Intermediates from Phosphites and Dimethyl Acetylenedicarboxylate: One-Pot Syntheses of Phospholes

J. C. Tebby and S. E. Willetts—North Staffordshire Polytechnic, Stoke-on-Trent, ST4 2DE, England
D. V. Griffiths—University of Keele, Staffs., ST5 5BG, England

Alkyl- and Phenylthiylations of P–H Phosphoranes

Wilfried Heide, Massoud Garrossian, and Wesley G. Bentrude—Department of Chemistry, University of Utah, Salt Lake City, UT 84112

Stable Chloroacyloxyphosphoranes

Itshak Granoth, Rivka Alkabets, and Yoffi Segall—Israel Institute for Biological Research, Ness-Ziona 70400, Israel

Hydroxyphosphoranes: Synthesis, Tautomerism

A. Munoz, B. Garrigues, and M. Koenig—ER CNRS N° 82 Associée à l'Université Paul Sabatier, 118 Route de Narbonne, 31062 Toulouse Cédex, France

ESR Studies of Phosphoranyl Radicals

J. H. H. Hamerlinck, P. Schipper, and H. M. Buck—Eindhoven University of Technology, Department of Organic Chemistry, The Netherlands

Synthesis of Chiral Polyphosphines from Perfluoroalkyl Phosphinidenes

Richard A. Wolcott, Larry Avens, and Jerry L. Mills—Department of Chemistry, Texas Tech University, Lubbock, TX 79409

Nucleophilic Reactions of (Silylamino)-phosphines

David W. Morton and Robert H. Neilson—Department of Chemistry, Texas Christian University, Fort Worth, TX 76129

(Silylamino)phosphines with P–H Bonds

H. Randy O'Neal and Robert H. Neilson—Department of Chemistry, Texas Christian University, Fort Worth, TX 76129

New Diphosphino- and Diphosphitoacetylenes

S. G. Kleemann and E. Fluck—Institute of Inorganic Chemistry of the University Stuttgart, Pfaffenwaldring 55, D-7000 Stuttgart 80, FRG

Reactions of Na_3P_7 and Na_3P_{11}

W. Hönle, V. Manriquez, C. Mensing, W. Bensmann, and H. G. v. Schnering—Max Planck Institut für Festkörperforschung, Stuttgart, FRG

Peroxide-Initiated Reaction of Strained-Ring Polyphosphorus Compounds with Olefins

W. E. Garwood and L. A. Hamilton—Mobil Research & Development Corporation, Research Department, Paulsboro Laboratory, Paulsboro, NJ 08066

Reactions of Phosphines and Hypophosphites with Compounds Containing Multiple Bonds

A. N. Pudovik and G. V. Romanov—The Arbuzov Institute of Organic and Physical Chemistry, Kazan Branch Academy of Sciences, Kazan 420083, USSR

Reactions of Tricoordinated Phosphorus Esters with Nitroalkanes

H. Teichmann and A. Weigt—Zentralinstitut für Organische Chemie der Akademie der Wissenschaften der DDR, 1199 Berlin-Adlershof, GDR

Phase-Transfer Catalysis in the Synthesis of Esters of Organophosphorus Acids

Yuan Chengye, Yu Weizhen, Xiang Caili, and Sun Quiyun—Shanghai Institute of Organic Chemistry, Academia Sinica, 345 Linglin Lu, Shanghai 200032, China

Oxidative Phosphinylation: A New Route to Dialkylphosphinic Acid Derivatives

Yuan Chengye, Yuan Qun, Long Haiyan, Shen Dingzhang, and Chen Wuhua—Shanghai Institute of Organic Chemistry, Academia Sinica, 345 Linglin Lu, Shanghai 200032, China

***O*-Alkylation of Phosphonic Acids by Ion-Pair Method**

Yuan Chengye, Wong Guoliang, Shen Dingzhang, Wu Fubing, and Chen Wuhua—Shanghai Institute of Organic Chemistry, Academia Sinica, 345 Linglin Lu, Shanghai 200032, China

Synthesis and Properties of Anhydrides of Phosphonous and Phosphonic Acids and Their Thioanalogs

O. N. Grishina and N. A. Andreev—The Arbuzov Institute of Organic and Physical Chemistry, Kazan Branch Academy of Sciences, Kazan 420083, USSR

Substituent Effects on the Chemistry of Phenoxaphosphine Derivatives

Jack B. Levy and Philip L. Robinson—Department of Chemistry, University of North Carolina at Wilmington, P.O. Box 3725, Wilmington, NC 28406

On Some New Inorganic and Organometallic Dihalophosphates and Dimethylphosphinates

A. F. Shihada—Department of Chemistry, College of Science, University of U.A.E., AL-Ain, United Arab Emirates

Reaction of 3,5-Dichlorophenylazophosphonic Acid, Dimethyl Ester with Diazomethane

John L. Miesel, Michael G. Chaney, and Noel D. Jones—Lilly Research Laboratories, Greenfield, IN 46140

Behavior of Phosphonic Acid and Phosphoric Acid Esters towards Halogenide Anions

I. Petneházy, Gy. Szakál, and L. Tőke—Department of Organic Chemical Technology, Technical University Budapest, Müegyetem, Budapest, Hungary 1521

Reactions of Phosphine Derivatives with Iodine in Nonaqueous Solvents

Osman Hama Amin, Habib A. Aughsteen, and Ali Mohammad—Chemistry Department, College of Science, Sulaimaniyah University, Iraq

The Reaction of Pseudohalogen Halides with Triphenylphosphine and Triphenylarsine

H. A. Aughsteen—Department of Chemistry, University of Sulaimaniyah, Iraq

Synthesis, Structure, and Properties of New Organic Compounds Containing Two Phosphorus Atoms

M. B. Gaziziv, R. A. Hierullin, and A. I. Razumov—Chemico-Technological Institute, Kazan, USSR

Mono- and Difluoro Derivatives of Methylenediphosphonic Acid: Synthesis with Perchloryl Fluoride

C. E. McKenna, P. D. Shen, and N. D. Leswara—Department of Chemistry, University of Southern California, Los Angeles, CA 90007

Phosphate-Tellurates

A. Durif—Laboratoire de Cristallographie, CNRS, 166 X, 38042 Grenoble Cédex, France

New Mixed Condensed Anions of General Formula: $Cr_nXO_{3n+4}{}^{3-}$ (X = P or As)

M. T. Averbuch-Pouchot—Laboratoire de Cristallographie, CNRS, 166 X, 38042 Grenoble Cédex, France

The Crystal Structure of $Na_5P_3O_{10} \cdot 6H_2O$

R. Hoppe, M. Jansen, and D. M. Wiench—Institut für Anorganische Chemie I der Justus Liebig Universität Giessen, Heinrich-Buff-Ring 58, 63 Giessen, FRG

The Vibrational Spectra of Hydrated $Ca_2P_2O_7$ Phases

B. C. Cornilsen, R. A. Condrate, Sr., and C. D. Gosling—NYS College of Ceramics at Alfred University, Alfred, NY 14802

Reaction Mechanism in Mixtures of Apatite and Additional Minerals with Phosphoric Acids at Heating

M. A. Veiderma, M. E. Pyldme, J. H. Pyldme, and K. O. Fynsuaadu—Tallinn Polytechnic Institute, Ehitajate Street 5, Tallinn 200026, USSR

A Study of Potassium Mono-, Di-, and Triphosphate Heterogeneous Systems in Views of Their Use as Liquid Fertilizers

M. Ebert, J. Eysseltová, I. Lukěs, and J. Nassler—Department of Inorganic Chemistry, Faculty of Sciences, Charles University, Albertov 2030, 128 40 Prague 2, Czechoslovakia

Solubility of Crystalline Phosphates as Polyenergetical Conjugation Property

M. R. Mehandjiev—State Engineering Organization "Chimcomplect", 101 Aleksei Tolstoi Boulevard, 1220 Sofia, Bulgaria

Flow Injection Analysis and High Performance Liquid Chromatography of Inorganic Phosphates

N. Yoza, Y. Hirai, and S. Ohashi—Department of Chemistry, Faculty of Science, Kyushu University 33, Hakozaki, Higashiku, Fukuoka, 812 Japan

Mercurimetry of Phosphorous Triesters and of Some Other Organic P(III) Compounds

M. C. Démarcq—Produits Chimiques Ugine Kuhlmann, Centre de Recherches de Lyon, F 69310 Pierre-Benite, France

Chromatographic Separation of Acidic Organophosphorus Compounds

Yuan Chengye, Long Haiyan, Cheng Zhichu, and Xiang Caili—Shanghai Institute of Organic Chemistry, Academia Sinica, 345 Linglin Lu, Shanghai 200032, China

New Routes to Thiopeptides Using 2,4-Bis(4-methoxyphenyl)-1,3,2,4-dithiadiphosphetane 2,4-Disulfide as Sulfuration Reagent

K. Clausen, M. Thorsen, and S.-O. Lawesson—Department of Organic Chemistry, University of Aarhus, DK-8000 Aarhus C, Denmark

A New Organophosphorus Reagent for [3 + 3] Annulation

J. Monkiewicz, K. M. Pietrusiewicz, and R. Bodalski—Polish Academy of Sciences, Center of Molecular and Macromolecular Studies, 90-362 Łodź, Poland, and Department of Chemistry, Technical University, 90-924 Łodź, Poland

Reactivity of Diazaphosphole Derivatives: Unexpected Indole Formation

G. Baccolini and P. E. Todesco—Istituto Chimica Organica, Università, Risorgimento 4, 40136 Bologna, Italy

On Regularities of Arylisocyanate Cyclization in Presence of Tervalent Organophosphorus Compounds

M. I. Bakhitov and E. V. Kuznetsov—Kazan Institute of Chemical Engineering, Karl Marx, 68, Kazan, USSR

Heterocycles from Phosphonium Salts II: Novel Methods of Synthesis of Multifunctional Imidazoles

R. Lee Webb, Clifford S. Labaw, and George R. Wellman—Smith Kline and French Laboratories, Organic Chemistry Department, 1500 Spring Garden Street, Philadelphia, PA 19101

A Model Description for Conformational Changes in DNA: A CNDO-2 Study

J. J. C. van Lier, D. van Aken, and H. M. Buck—Eindhoven University of Technology, Department of Organic Chemistry, The Netherlands

Conformational Studies of Acyclic Phosphines Using an Empirical Technique

John A. Mosbo, Mark D. McIntire, and Paul L. Bock—Department of Chemistry, Ball State University, Muncie, IN 47306

Extended Hückel Calculations on Some Phosphorus $X_2PS_2^{n-}$ Dithioanions

Ioan Silaghi-Dumitrescu and Ionel Haiduc—Chemistry Department, Babes-Bolyai University, R-3400 Cluj-Napoca, Roumania

A Pseudopotential Calculation of H_2PNH_2 and Related Molecules: A Visualization of N(*p*)–P(*d*) Conjugation: Correlation with Spectrochemical Properties

M. Barthelat, R. Mathis, and F. Mathis—Laboratoire des Hétérocycles du Phosphore et de l'Azote, Université Paul Sabatier, 118 Route de Narbonne, 31062 Toulouse Cédex, France

Steric and Electronic Structure of 2-X-1,3,2-Diheterophosphorinanes

B. A. Arbuzov and R. P. Arshinova—The A. M. Butlerov Chemical Institute, University, 18N. Lenin Street, Kazan 420008, USSR

Ylides with a P–H Bond

O. I. Kolodiazhnyi and V. P. Kukhar—Institute of Organic Chemistry, Academy of Sciences of Ukrainian SSR, Murmanskaya Street 5, Kiev, 252094, USSR

Provision of Thermodynamic Data for Gaseous PCl_4^+ and PCl_6^- Ions Arising from Studies on Ionic Isomerism in Phosphorus (V) Chloride

H. D. B. Jenkins, A. Finch, and P. N. Gates—School of Molecular Sciences, University of Warwick, Coventry, United Kingdom, and The Bourne Laboratory, Royal Holloway College, Egham, Surrey, United Kingdom

Electrical and Steric Effects of Substituents Bonded to Phosphorus: Tricoordinate Phosphorus

Marvin Charton—Pratt Institute, Brooklyn, NY 11205
Estelle Gearon—Montgomery College, Takoma Park, MD 20012
B. Charton—Pratt Institute, Brooklyn, NY 11205

Diadic, Triadic, and Tetradic Prototropic Phosphorus–Carbon Tautomerism

T. A. Mastryukova, I. M. Aladgeva, I. V. Leontyeva, and P. V. Petrovsky—A. N. Nesmeyanov Institute of Organo-Element Compounds of the USSR Academy of Sciences, Vavilova Street, 28, Moscow 117334, USSR

Configuration and Stereochemistry of Some Reactions in Optically Active *O*-Ethyl *O*-Phenylphosphorothioic Acid and Its Derivatives

Tang Chu-Chi, Wu Gui-Ping, Huang Run-Chiu, and Chai You-Xin—Institute of Elemento-Organic Chemistry, Nankai University, Tianjin, People's Republic of China

The Crystal Structure and Absolute Configuration of Quinine Salt of (+)-*O*-Ethyl *O*-Phenylphosphorothioic Acid

Dou Shi-Qi and Zheng Qi-Tai—Institute of Biophysics, Chinese Academy of Sciences, Beijing, China
Tang Chu-Chi and Wu Gui-Ping—Institute of Elemento-Organic Chemistry, Nankai University, Tianjin, People's Republic of China

X-Ray Crystal Structures of Quasiphosphonium Intermediates

Leylâ S. Shaw, K. Henrick, and H. R. Hudson—Department of Chemistry, The Polytechnic of North London, Holloway Road, London N7 8DB, England

Approach of Nucleophiles to Tetrahedral Phosphonium Centers

S. J. Archer, T. A. Modro, and L. R. Nassimbeni—School of Chemical Sciences, University of Cape Town, Rondebosch 7700, South Africa

The Molecular Structure of Tetrakis (*t*-butyl)cyclotetraphosphine

A. W. Cordes, W. Weigand, and P. N. Swepston—University of Arkansas, Fayetteville, AR 72701

The Solid and Solution Conformations of Tris-(diphenylthiophosphoryl)methane and Related Molecules

I. J. Colquhoun, W. McFarlane, S. O. Grim, and J.-M. Bassett—Departments of Chemistry, City of London Polytechnic, London EC3N 2EY, United Kingdom; University of Maryland, College Park, MD 20742; and The City University, London EC1V OHB, United Kingdom

Synthesis of Chiral Phosphite Esters with Phosphorus as the Primary Chiral Site

H. P. Abicht, J. T. Spencer, and J. G. Verkade—Gilman Hall, Iowa State University, Ames, IA 50011

New Synthesis of Polyphosphazenes

Patty Wisian-Neilson and Robert H. Neilson—Department of Chemistry, Texas Christian University, Fort Worth, TX 76129

Acetylenic and Olefinic Phosphazenes

R. A. Nissan, P. J. Harris, and H. R. Allcock—Department of Chemistry, The Pennsylvania State University, University Park, PA 16802

Reactions of $N_3P_3Cl_5(NPPh_3)$ with Primary and Secondary Amines

P. Ramabrahmam, S. S. Krishnamurthy, and A. R. Vasudeva Murthy—Department of Inorganic and Physical Chemistry, Indian Institute of Science, Bangalore 560012, India
R. A. Shaw and M. Woods—Department of Chemistry, Birkbeck College, University of London, Malet Street, London WC1 7HX, United Kingdom

Solubility Measurements of Some Halogencyclophosphazenes

V. Novobilský—Pedagogická Fakulta, Čs. mládeže 8, 400 96 Ustí nad Labem, Czechoslovakia
J. Nývlt—Ústav Anorganické Chemie ČSAV, Polská 20, 120 00 Praha 2, Czechoslovakia

Some Coordination Compounds of Novel Bis- and Trisphosphine–Phosphine Sulfide Ligands

I. J. Colquhoun, S. O. Grim, W. McFarlane, S. Nittolo, L. C. Satek, and P. H. Smith—Departments of Chemistry, University of Maryland, College Park, MD 20742, and City of London Polytechnic, London EC3N 2EY, United Kingdom

Controlling the Number of Metal Sites to which a Ditertiary Phosphine is Coordinated in Group VI Metal Carbonyls: Limitations

R. L. Keiter, N. P. Hansen, and S. L. Kaiser—Department of Chemistry, Eastern Illinois University, Charleston, IL 61920

Cobalt(II) Complexes of the Bidentate *N,N*-Bis(diphenylphosphino)phenylamine

M. C. Alexiev—Department of Chemistry, Higher Pedagogical Institute, 9700 Shoumen, Bulgaria

Synthesis and Optical Resolution of a Novel High Field Cobalt(III) Chelate Complex

James T. Spencer and J. G. Verkade—Gilman Hall, Iowa State University, Ames, IA 50011

The Dual Chemical Behavior of $RuH(\eta^2\text{-}BH_4)$-Triphosphine

Devon W. Meek and John Letts—Department of Chemistry, The Ohio State University, Columbus, OH 43210

Spectroscopic and Structural Studies of Metal Complexes of Trimesitylphosphine

Elmer C. Alyea and George Ferguson—Guelph-Waterloo Centre for Graduate Work in Chemistry, Guelph Campus, Department of Chemistry, University of Guelph, Guelph, Ontario, Canada N1G 2W1

Hybrid Multidentate Ligands: Amido-Phosphine Derivatives of Group VIII Metals

M. D. Fryzuk and P. A. MacNeil—Department of Chemistry, University of British Columbia, 2036 Main Mall, Vancouver, British Columbia V6T 1Y6, Canada

The Preparation and Structures of Large Ring Methylphosphazenes: Their Behavior as Macrocyclic Ligands

K. D. Gallicano, R. T. Oakley, N. L. Paddock, S. J. Rettig, and J. Trotter—Department of Chemistry, University of British Columbia, 2036 Main Mall, Vancouver, British Columbia, V6T 1Y6, Canada

Reactions of 2,6-Bis[di-*t*-butylphosphinomethylphenyl]rhodium Hydrochloride

W. C. Kaska, S. Nemeh, and C. Jensen—Department of Chemistry, University of California, Santa Barbara, CA 93106

Tetraphenylimidodiphosphate Complexes of Rare Earth Elements in Solution

E. Herrmann, D. Scheller, and B. Thomas—Department of Chemistry, Technical University Dresden, DDR-8027 Dresden, GDR

Synthesis and Coordination Properties of Carbamylmethylenephosphonate Ligands

S. M. Bowen and R. T. Paine—Department of Chemistry, University of New Mexico, Albuquerque, NM 87131

Tris(dialkylaminophosphine) Compounds of Silver(I) and the Molecular Structure of a Dicoordinated Complex

S. M. Socol and J. G. Verkade—Gilman Hall, Iowa State University, Ames, IA 50011

Stereoelectronic Influences of Phosphite Ester Ligands in Homogeneous and Heterogeneous Catalysis

Y. Gultneh, D. Schiff, and J. G. Verkade—Gilman Hall, Iowa State University, Ames, IA 50011

New Chiral Phosphinite Ligands Used For Asymmetric Homogeneous Hydrogenation in the Wilkinson-Type Catalyst System

S. J. Hathaway, B. S. DeHoff, M. L. Myers, and S. A. Sedivy—Department of Chemistry, Gettysburg College, Gettysburg, PA 17325

Oxygen-Donor Polydentate Phosphorus-Containing Ligands in Complex Formation with Group V Pentafluorides

E. G. Ilyin and Yu. A. Buslaev—N. S. Kurnakov Institute of General and Inorganic Chemistry, Academy of Sciences of the USSR, Leninskyi Prospekt, 31, Moscow 117071, USSR

New Dicoordinated Phosphorus Derivatives

J. P. Majoral, A. Merriem, and J. Navech—Laboratoire des Hétérocycles du Phosphore et de l'Azote, Université Paul Sabatier, 118 Route de Narbonne, 31062 Toulouse Cédex, France

Di-, Tri-, and Tetracoordinated Phosphorus Imidoderivatives with Bulky Substituents

L. N. Markovsky and V. D. Romanenko—Institute of Organic Chemistry, Academy of Sciences of Ukrainian SSR, Kiev 252660, USSR

Synthesis and Reactivity of New Low-Coordinated Phosphorus Compounds

Randall J. Thoma, Randal R. Ford, and Robert H. Neilson—Department of Chemistry, Texas Christian University, Fort Worth, TX 76129

Facile Three-Step Preparation of ^{18}O-Labeled Diastereomeric Ammonium Salts of Thymidine Cyclic 3′,5′-Monophosphate

Tadeusz M. Gajda, Alan E. Sopchik, and Wesley G. Bentrude—Department of Chemistry, University of Utah, Salt Lake City, UT 84112

Hydrolysis of P^1-(Nucleoside 5′-)P^1-Aminotriphosphates

J. Tomasz—Institute of Biophysics, Biological Research Center, Hungarian Academy of Sciences, H-6701 Szeged, Hungary

Complexes of Phosphate Esters and Nucleotides with Polyoxometalate Anions

D. E. Katsoulis and M. T. Pope—Department of Chemistry, Georgetown University, Washington, DC 20057

ATP and ADP Binding to Nitrogenase Fe Protein

C. E. McKenna, P. J. Stephens, and H. T. Nguyen—Department of Chemistry, University of Southern California, Los Angeles, CA 90007

The Application of Inorganic Ring Systems as Anticancer Drugs: A Chemical and Biochemical Study of N, P, S Heterocycles Containing a Variable Number of Aziridino Groups as Alkylating Constituents

A. A. van der Huìzen, A. P. Jekel, and J. C. van de Grampel—Department of Inorganic Chemistry, University of Groningen, Nijenborgh 16, 9747 AG Groningen, The Netherlands
A. van de Meer-Kalverkamp and H. B. Lamberts—Department of Radiopathology, University of Groningen, Bloemsingel 1, 9713 BZ Groningen, The Netherlands
B. Houwen and H. Schraffordt-Koops—University Hospital, Bloemsingel 59, 9713 EZ Groningen, The Netherlands

A Simple Method for Preparation of *N*-(Phosphonoacetyl)amino Acids

M. Soroka and P. Kafarski—Institute of Organic and Physical Chemistry, Technical University of Wroclaw, Wybrzeze Wyspianskiego 27,50-370, Wroclaw, Poland

Inhibition of *Mycobacterium leprae* by some Phosphonia Derivatives: Stereochemistry of the Drugs

P. Bricage—Centre de Recherches Biologiques sur la Lépre, Faculté des Sciences, Université de Dakar, Sénégal, Afrique de l'Ouest

Identification and Chemical Modification of Phosphorus-Containing Endgroups in Capsular Polysaccharides Isolated from *H. influenzae*

G. Zon—Department of Chemistry, The Catholic University of America, Washington, DC 20064

Dioxaphospho(III)rinanes of Carbohydrates Series

E. E. Nifant'yev, M. P. Koroteyev, and T. S. Kuhareva—V. I. Lenin State Pedagogical Institute, Malaya Pirogovskaya, 1, Moscow 119882, USSR

New Methods in the Synthesis of Phospholipids and Their Analogs on the Basis of Phosphites and Amidophosphites of Glycerine

D. A. Predvoditelev, E. E. Nifant'yev—V. I. Lenin State Pedagogical Institute, Malaya Pirogovskaya, 1, Moscow 119882, USSR

New Routes to Phosphasugars

M. J. Gallagher and H. Krawczyk—School of Chemistry, University of New South Wales, P.O. Box 1, Kensington, New South Wales, Australia 2033

INDEX

Q

R

S

T

U

V

W

X

Y

Z

Jacket design by Carol Conway.
Production by Robin Giroux and Gabriele Glang.

Elements typeset by Service Composition, Co., Baltimore, MD.
Printed and bound by The Maple Press Co., York, PA.